Radiometric Dating for Geologists

Radiometric Dating for Geologists

Edited by

E. I. Hamilton

*Radiological Protection Service,
Belmont, Sutton, Surrey,
formerly Research Fellow, Department
of Geology, Oxford*

R. M. Farquhar

*Associate Professor of Physics
University of Toronto, Canada*

1968

INTERSCIENCE PUBLISHERS
a division of John Wiley & Sons
LONDON   NEW YORK   SYDNEY

First published by John Wiley & Sons Ltd. 1968

All rights reserved, No part of this book may be reproduced by any means, nor transmitted, nor translated into a machine language without the written permission of the publisher

Library of Congress catalog card number 68–22087
SBN 470 34720 1

Printed in Great Britain by
BARNICOTTS LTD., OF TAUNTON

Preface

Radiometric dating of rocks, minerals and meteorites is becoming one of the most active fields in planetary science. The data of age determinations has accumulated mainly over the past two decades, and has had some profound effects, not only on our understanding of the *history* of our planet and the solar system, but on our ideas about the nature and extent of certain kinds of geological processes.

The rapidity with which geochronological techniques have been developed and applied has had other less desirable consequences. Research papers sometimes give confusing, apparently contradictory statements about the reliability with which various rock and mineral types may be dated; research groups cannot agree on the values of the constants to be used for computing ages; fragments of the chronologies of various areas appear from time to time scattered among a number of geophysical, geochemical or geological journals. All of these developments are typical of a young and expanding field—in fact they provide some of the challenge which attracts research workers to the subject. At the same time, these problems are particularly unsettling to the geologist or geophysicist who is not directly in touch with geochronology but who wishes to further his understanding or apply the methods of geochronology to specific problems.

Part of this difficulty can be overcome at this stage by providing texts in which the basic concepts are presented in a logical and well-defined way. Several of these are now available:

> Hurley, P. M. (1959). *How Old is the Earth*. The Science Study Series. Heinemann, London.
> Russell, R. D. and Farquhar, R. M. (1960). *Lead Isotopes in Geology*. Interscience, New York.
> Starik, I. E. (1961). Yadernaya Geokhronologiya Akademiya Nauk, SSSR, Moscow.
> Hamilton, E. I. (1965). *Applied Geochronology*. Academic Press, London.
> Schaeffer, O. A. and Zähringer, J. (1966). *Potassium–Argon Dating*. Springer Verlag, Berlin.
> Faul, H. (1966). *Ages of Rocks Planets and Stars*. Earth and Planetary Science Series (Consulting Editor, P. M. Hurley). McGraw-Hill, New York.

What still appears to be lacking is a volume in which 'case histories' are presented, so that the geologist or geophysicist has access to accounts of methods, results of age studies and related topics concerning the significance of some isotope ratios, in as full a geological and geophysical context as possible.

The present work is an attempt to fill this gap. Most of the papers are not based on unpublished data, but are attempts to interpret available data in as complete a way as is presently possible in the geological framework of the regions or geological settings concerned. It is the editors' hope that this volume will to some degree complement the basic texts of geochronology by illustrating the scope, applications and limitations of radiometric age studies in wider contexts in which the average geologist or geophysicist may feel more at ease.

E. I. HAMILTON
R. M. FARQUHAR

Contents

1 Potassium–argon dating of igneous and metamorphic rocks with applications to the Basin ranges of Arizona and Sonora
   P. E. Damon 1
2 A comparison of the isotopic mineral age variations and petrologic changes induced by contact metamorphism
   S. R. Hart, G. L. Davis, R. H. Steiger, G. R. Tilton . 73
3 Isotopic geochronology of Montana and Wyoming
   B. J. Giletti 111
4 The interpretation of lead isotopes and their geological significance
   E. R. Kanasewich 147
5 The interpretation of zircon ages
   E. J. Catanzaro 225
6 Geochronological studies in Connemara and Murrisk, western Ireland
   S. Moorbath, K. Bell, B. E. Leake, W. S. McKerrow . 259
7 Radiometric dating and the pre-Silurian geology of Africa
   T. N. Clifford 299
8 Charged particle tracks: tools for geochronology and meteorite studies
   R. L. Fleischer, P. Buford Price, R. M. Walker . . 417
9 The isotopic composition of strontium applied to problems of the origin of the alkaline rocks
   E. I. Hamilton 437
   Author index 465
   Subject index 475

Potassium–argon dating of igneous and metamorphic rocks with applications to the Basin ranges of Arizona and Sonora

PAUL E. DAMON[*]

I *Introduction*, 1. II *The potassium–argon method*, 3.
III *Application to the dating of igneous and metamorphic rock within the Basin ranges*, 35. IV *The Basin ranges in broader perspective*, 58. V *Concluding remarks*, 63.
Acknowledgments, 64. *References*, 64.

I Introduction

It is the intent of this chapter to discuss in a systematic way the merits and inherent problems associated with the potassium–argon dating method as applied to igneous and metamorphic rocks. Perhaps there is no better way to illustrate for geologists the inherent power as well as the shortcomings of the method than to apply it to actual geologic problems. For this purpose, problems in the dating of igneous and metamorphic rocks within the Basin and Range Province of Arizona and Arizona's neighbor to the south, Sonora, have been chosen.

All of Sonora and approximately half of Arizona are included within the Basin and Range Province (figure 1). The remainder of Arizona is occupied by the Colorado Plateau and a mountainous transition zone between the Colorado Plateau and the Basin ranges.

[*] University of Arizona, Tucson, Arizona, U.S.A.

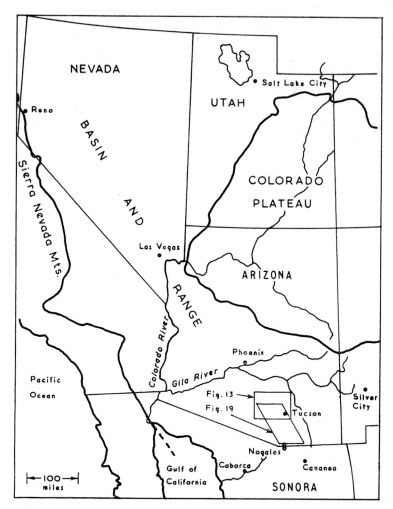

Figure 1 Location map outlining the Basin and Range Province and Colorado Plateau with pertinent reference points.

Magmatism within the Basin ranges of Arizona and Sonora has been essentially restricted to the Precambrian and Mesozoic eras. There is, as yet, no clear-cut evidence for the origin of any igneous rocks or metalliferous ore deposits within that area during the great expanse of time included within the Paleozoic era (Wilson, 1962). The Precambrian igneous rocks of the area, which have existed through a long and complex history, will serve to illustrate the limitations of the potassium–argon method, whereas the potassium–argon dating of late Mesozoic and Cenozoic rocks will serve to illustrate its virtues.

II The Potassium–Argon Method

A *Mode of decay and mathematical relationships*

There have been a number of excellent reviews of the potassium–argon method, most recently by Hamilton (1965) and Gerling (1961). The reader can refer to these works for a fuller account of the method and for references to the extensive literature. The intent here, as previously mentioned, is restricted to evaluating the method as applied to the dating of geologic events involving magmatic activity and thermal metamorphism.

There are three isotopes of potassium (^{39}K, ^{40}K, ^{41}K). The least abundant of these, potassium-40, is unstable with respect to both beta emission and electron capture. The eventual fate of each atom of potassium of atomic weight 40 is to be transmuted either by emission of a beta particle to become an atom of calcium (^{40}Ca) or by electron capture to become an atom of argon (^{40}Ar). As a result of this transmutation, the relative abundance of ^{40}K is continuously decreasing and that of ^{40}Ca and ^{40}Ar is continuously increasing with time. In 1290 m.y., the half-life ($t_{1/2}$) of ^{40}K, half of the atoms of ^{40}K in existence today will have decayed to either ^{40}Ca or ^{40}Ar. Approximately 11% of all transmutations will yield ^{40}Ar, the remainder will decay to ^{40}Ca. Thus in any mineral containing potassium, the ratios of ^{40}Ar and ^{40}Ca to ^{40}K will increase with time and be a measure of time. Naïvely, one might expect that the more frequent decay of ^{40}K to ^{40}Ca might be the most effective measure of time. Unfortunately, calcium is a major element in igneous rocks and the high abundance of nonradiogenic ^{40}Ca results in great uncertainty as to relative amounts of radiogenic to nonradiogenic ^{40}Ca. Isotopic and elemental abundances of potassium, argon and calcium are given in table 1. The abundance of calcium in igneous rocks is almost six orders of magnitude greater than the abundance of argon.

The mathematical relations governing the decay of ^{40}K are relatively simple. The probability, λ, that an atom of ^{40}K will decay in a unit of time, conveniently taken as one year, is the sum of the separate probabilities of decay to ^{40}Ca, λ_β, and ^{40}Ar, λ_K:

$$\lambda = \lambda_\beta + \lambda_K = \frac{0.693}{t_{1/2}} \qquad (1)$$

Table 1 Crustal and isotopic abundances of potassium, calcium and argon

Element	Abundance of element in igneous rocks[a] (g/ton)	Isotope	Isotopic abundance[b] (atom %)
Potassium	25,900	^{39}K	93.08
		^{40}K	0.0119[c]
		^{41}K	6.91
Calcium	36,300	^{40}Ca	96.97
		^{42}Ca	0.64
		^{43}Ca	0.145
		^{44}Ca	2.06
		^{46}Ca	0.0033
		^{48}Ca	0.185
Argon	0.04	^{36}Ar	0.337[d]
		^{38}Ar	0.063[d]
		^{40}Ar	99.600[d]

[a] Rankama (1954, Table 9.1, p. 135).
[b] Rankama (1954, Table 1.3, p. 13).
[c] More recent determinations yield 0.0118; see text for discussion.
[d] This is the isotopic abundance of atmospheric argon, which is 0.93% by volume of total atmosphere. Lithospheric argon is almost pure argon-40.

The probability, P_d, that an atom in a sample will not decay in a time, t, measured in years is

$$P_d = \frac{^{40}K}{^{40}K_t} = \exp(-\lambda t) \qquad (2)$$

where ^{40}K is the amount of that isotope measured now and $^{40}K_t$ is the amount present t years ago. From equation (2), the amount of $^{40}K_t$ at time t is

$$^{40}K_t = {}^{40}K \exp(+\lambda t) \qquad (3)$$

The fraction of ^{40}K atoms that decay to ^{40}Ar is

$$\frac{^{40}Ar}{^{40}Ar + {}^{40}Ca} = \frac{\lambda_K}{\lambda_K + \lambda_\beta} = \frac{\lambda_K}{\lambda} \qquad (4)$$

The number of ^{40}Ar atoms produced within the sample in time t is equal to the original number of potassium-40 atoms (equation 3) minus the amount measured in the sample now, ^{40}K, times the fraction of potassium-40 atoms that decay to ^{40}Ar (equation 4), which, factoring out ^{40}K, is

$$^{40}\text{Ar} = \frac{\lambda_K}{\lambda}\,^{40}\text{K}[\exp(\lambda t) - 1] \qquad (5)$$

The age equation is then derived by solving this equation for time

$$t = \frac{1}{\lambda}\ln\left[1 + \frac{\lambda}{\lambda_K}\frac{^{40}\text{Ar}}{^{40}\text{K}}\right] \qquad (6)$$

B *Radioactive clocks*

A mineral or rock containing a radioactive isotope such as ^{40}K becomes, in a very real sense, a radioactive clock or geochronometer that can be used to date geologic events. Such a geologic event may consist of the crystallization of a mineral or solidification of a rock, or more rigorously, the cooling of the rock unit to a temperature at which loss by diffusion of the daughter decay product is negligible. If the clock is thermally, mechanically or chemically perturbed, the geologic event may correspond more or less closely to the perturbation rather than solidification.

The following conditions are a prerequisite to meaningful age determinations by the K–Ar method:

(1) The decay constants, λ_K and λ_β, must be known within sufficient limits of accuracy.

(2) The potassium-40 content of the mineral must be accurately and precisely determined either directly, such as by isotope dilution or neutron activation, or by chemical determination of the potassium content and accurate knowledge of the relative abundance of the isotope potassium-40. In this case, if K is the weight of potassium, as chemically determined, and ϕ is the relative abundance of ^{40}K by weight, then the amount of potassium-40 by weight is

$$^{40}\text{K} = \phi\text{K} \qquad (7)$$

and the quantity of potassium-40 in moles is obtained by dividing equation (7) by the isotopic weight of potassium-40 which is 39.964 gram per mole.

(3) The amount of ^{40}Ar and the relative isotopic abundances of argon isotopes from the mineral must be determined precisely.

(4) An accurate estimate of atmospheric ^{40}Ar contamination must be obtainable from the measured relative abundances of argon isotopes and subtracted from the measured ^{40}Ar.

(5) There must be negligible initial ^{40}Ar (excess ^{40}Ar) present at $t = 0$ or the experimenter must be able to estimate the amount of excess ^{40}Ar.

(6) The mineral or rock must constitute an essentially closed system following the geologic event, i.e. there must be essentially no gain or loss of ^{40}Ar and ^{40}K except by radioactive decay.

(7) By knowledge of mineralogic and petrologic factors and the geologic environment, the relationship between the K–Ar date and the geologic event must be established.

C *The decay constants*

Accurate knowledge of the decay constants for electron capture and beta disintegration (prerequisite 1) depends upon accurate determinations of the β-ray emission rate which results in transmutation to ^{40}Ca and the γ-ray emission rate. One gamma quantum is believed to accompany every electron capture that results in transmutation to ^{40}Ar. The relationship between these quantities is

$$\frac{dN_\gamma}{dt} = \frac{-\lambda_K \phi N_A}{M_{40}} \quad (\gamma \text{ dis. per sec per g K}) \tag{8a}$$

$$\frac{dN_\beta}{dt} = \frac{-\lambda_\beta \phi N_A}{M_{40}} \quad (\beta \text{ dis. per sec per g K}) \tag{8b}$$

where dN_γ/dt and dN_β/dt are the specific activities for γ-ray and β-ray emission, respectively; N_A is Avogadro's number (6.0226×10^{23}); and M_{40} is the isotopic mass of ^{40}K (39.964).

Nier (1950) measured a value of 1.22×10^{-4} g ^{40}K/g K (1.19×10^{-4} atoms of ^{40}K per atom of K) for ϕ. However, Strominger and others (1958) reviewed the isotopic abundance determinations for potassium. Three separate measurements in two different laboratories agree at a value of 1.21×10^{-4} g ^{40}K/g K for ϕ (1.18×10^{-4} atoms of ^{40}K per atom of K). Therefore, in this laboratory, we have always used the latter (smaller) value. The assumption of a constant value of ϕ implies that there is negligible fractionation of potassium isotopes during geologic processes. In general, this seems to be an acceptable assumption. However, it should be noted that Schreiner and Verbeek (1965) have reported up to several percent enrichment of ^{39}K relative to ^{41}K in altered xenoliths within a granite and at the contact of the granite with sedimentary country rock.

Aldrich and Wetherill (1958) reviewed the specific activity determinations and suggested values of 3.4 γ/g sec for the specific γ-activity and 27.6 β/g sec for the specific β-activity. There has been one new determination of the specific β-activity by Glendenin (1961) who obtained a value of 28.2 ± 0.3 β/g sec (errors quoted in this paper are standard deviations unless otherwise noted). However, most laboratories, including this one, have continued to use the specific activities recommended by Aldrich and Wetherill. Two sets of decay constants are commonly found in the literature, one based on $\phi = 1.21 \times 10^{-4}$, which includes this laboratory, and the other based on $\phi = 1.22 \times 10^{-4}$. Both sets of literature values are given in table 2 along with recalculated values.

Table 2 Physical constants for potassium-40

Source	ϕ ($\times 10^{-4}$ g ^{40}K/g K)	λ_K ($\times 10^{-10}$ yr^{-1})	λ_β ($\times 10^{-10}$ yr^{-1})	$t_{1/2}$ (m.y.)
Literature	1.21	0.589	4.76	1300[a]
Recalculated	1.21	0.588	4.78	1290[b]
Literature	1.22	0.584	4.72	1310[a]
Recalculated	1.22	0.584	4.74	1300

[a] See A. G. Smith (1964) for recent review.
[b] Values recommended by this author.

The small change in the recalculated value of λ_β is caused by an error which seems to have been repeated in the literature since the review paper of Endt and Kluyver (1954). Nier (1950) who determined the value of $\phi = 1.22 \times 10^{-4}$ g ^{40}K/g K (1.19×10^{-4} atomic abundance) has since accepted the more recent lower value of ϕ (Goldich and others, 1961). Smith (1964) has suggested that a slightly higher specific activity for λ_β might be appropriate. However, until further more definitive analyses are available, the author recommends the recalculated values in table 2 based on the lower value of ϕ accepted by most physicists including Professor Nier. The age equations calculated for natural logarithms and base 10 logarithms then become:

$$t = 4.29 \times 10^9 \log_{10}[9.12\ ^{40}\text{Ar}/^{40}\text{K} + 1] \tag{9}$$

and

$$t = 1.86 \times 10^9 \ln[9.12\ ^{40}\text{Ar}/^{40}\text{K} + 1] \tag{10}$$

To convert ages calculated with the higher value of ϕ (1.22×10^{-4}) to the recommended values, multiply the $^{40}Ar/^{40}K$ ratio by 1.0085 and recalculate, using either equation (9) or (10).

Wetherill and others (1956) and Smith (1964) have calculated the effect of errors in determining the parameters λ_β, and λ_K and ϕ on the calculated K–Ar dates. The effects of errors in determining these parameters and also the $^{40}Ar/^{40}K$ ratio are given in table 3.

Table 3 Effect on calculated K–Ar dates of errors in ϕ, λ_K, λ_β and the $^{40}Ar/^{40}K$ ratio

Date (m.y.)	$\delta\phi = +5\%$ (% error)[a]	$\delta\lambda_\beta = +5\%$ (% error)	$\delta\lambda_K = +5\%$ (% error)	$\delta[^{40}Ar/^{40}K] = +5\%$ (% error)
10	+0.0	−0.0	−5.0	5.0
100	+0.1	−0.1	−4.8	4.9
500	+0.6	−0.5	−4.2	4.4
1000	+1.1	−1.0	−3.8	3.9
2000	+1.9	−1.7	−3.2	3.0
3000	+2.5	−2.2	−2.7	2.4
4000	+2.8	−2.6	−2.4	2.0
5000	+3.0	−2.9	−2.1	1.7

[a] Assuming ϕ is constant, i.e. negligible natural isotopic fractionation. Fractionation of the sample is equivalent to an error in the $^{40}Ar/^{40}K$ ratio.

It becomes apparent that, for Cenozoic potassium–argon dating problems, ϕ and λ_β are not critical parameters and only λ_K and the ratio $^{40}Ar/^{40}K$ need be precisely and accurately determined. Errors in the parameters, ϕ and λ_β, become gradually more important and vice versa for λ_K and the 'daughter–parent' ratio until for dating work involving the oldest continental nuclei, all of the pertinent parameters become equally important.

For late Tertiary and Pleistocene K–Ar dating, the age equation (6) can be simplified with negligible error to

$$t = \frac{1}{\lambda_K} \frac{^{40}Ar}{^{40}K} \qquad (11)$$

where $1/\lambda_K = 1.70 \times 10^{10}$ yr. The error introduced by use of this simplified equation is less than 0.1% for Pleistocene dating and becomes 1% at about 30 m.y.

It is important to evaluate the limits of the accuracy of K–Ar dating set by uncertainties in the determinations of the physical constants. There have been five determinations of the specific γ-emission and seven determinations of the specific β-emission of potassium-40 involving reported errors of less than 5%. The average of the γ-ray activities is 3.40 ± 0.03 γ/sec g ^{40}K and, rejecting one obviously discordant β-activity measurement, the remaining six average to 27.8 ± 0.4 β/sec g ^{40}K. The slight difference between this β-ray activity average and the currently used value of 27.6 β/sec g ^{40}K does not warrant a change that would involve laborious recalculations of many dates. The important point to be made is that, if the optimistic assumption is made that precision in determining ϕ, λ_K and λ_β is also a measure of accuracy, then the maximum uncertainty in K–Ar dating introduced by these parameters probably does not exceed $\pm 2\%$ at the 95% confidence level for rocks of any geologic age.

D Determination of the ^{40}Ar/^{40}K ratio

This brings the discussion to prerequisites (2), (3) and (4) which involve the precise and accurate determination of the ^{40}Ar/^{40}K ratio. For a full discussion of experimental details and for references the reader may refer to previously mentioned reviews (Hamilton, 1965; Gerling, 1961) and to Smales and Wager (1960).

In this laboratory we determine potassium by the flame photometric techniques described by Cooper (1963), Cooper and others (1966). The precision, as estimated by duplicate analyses of 27 samples ranging in potassium content from 0.01% to 1.83%, is $\pm 0.40\%$ of the amount present. The precision obtained for higher contents of potassium is, of course, better. The accuracy is $\pm 1.54\%$ estimated from replicate analyses of ten interlaboratory standards.

We determine argon by standard mass spectrometric isotope-dilution methods using a highly enriched ^{38}Ar (99.9% ^{38}Ar) diluent. ^{36}Ar is used as an indicator of atmospheric contamination. The precision for determining radiogenic ^{40}Ar as determined by ten duplicate analyses, with air corrections not exceeding 40%, is $\pm 1.67\%$. The accuracy as indicated by analyses of four interlaboratory standards is $\pm 0.86\%$.

For samples containing more than 0.1% potassium and with negligible air argon corrections, we report conservative errors of $\pm 3\%$ in determining the ^{40}Ar/^{40}K ratio. This corresponds to dating errors varying from 0.3 m.y. (3%) at 10 m.y. to 43 m.y. (1.44%) at 3000 m.y. The accuracy of potassium–argon dating of such samples, considering all measured parameters, is probably within $\pm 7\%$ at the 95% confidence level. This of course assumes negligible atmospheric argon correction, no excess ^{40}Ar or argon loss and, if the date is to be meaningful, that all pertinent mineralogic–geologic factors are known (prerequisites 4, 5, 6 and 7).

The graph used to estimate precision when the air argon correction becomes appreciable is given in figure 2. It is drawn on the assumption, based upon experience, that we routinely determine the atmospheric ^{40}Ar correction with $\pm 2\%$ precision. For corrections exceeding 50%, the air ^{40}Ar correction becomes the dominant error. Such large air corrections are usually restricted to very young samples or samples containing a very low content of potassium. Most of the air argon comes from the vacuum system in which the mineral or rock is fused to release its argon content, and from argon adsorbed during laboratory preparation of the sample. This argon has been derived directly from the present Earth's atmosphere and has the isotopic composition given in table 1. The amounts of ^{36}Ar and ^{40}Ar are measured relative to the known amount of isotopic diluent, ^{38}Ar. The ^{36}Ar content is multiplied by the ratio of ^{40}Ar/^{36}Ar in air (approximately 296) to obtain the amount of ^{40}Ar from the air, and this is subtracted, together with any ^{40}Ar from the isotopic diluent, to obtain the radiogenic ^{40}Ar content.

Undoubtedly, the isotopic composition of argon in the Earth's atmosphere has been gradually changing throughout geologic time as the result of degassing of radiogenic ^{40}Ar from the lithosphere. It is also possible that some of the air argon in a mineral or rock has been derived from the Earth's paleoatmosphere. However, any reasonable assumptions concerning the rate of argon degassing (Turekian, 1964; Wasserburg, 1964; Damon and Kulp, 1958a) indicate that the change in the isotopic composition of air argon during late Cenozoic and Pleistocene time would be quite negligible. Thus, for very young samples for which the air correction is highest and critically determines the precision, the constancy of the isotopic composition of air argon can be safely assumed. Air argon corrections for minerals in which potassium is a major constituent become progressively less with age, so that the assumption of air argon isotopic constancy is only in question for old minerals when the air correction is high as, for example, when attempting to date authigenic sedimentary minerals with low potassium content.

Evernden and Curtis (1965) have demonstrated that ^{40}Ar/^{40}K ratios can be measured with gratifying precision for potash feldspars less than 50,000 years old. For example, they have fused 12.6 g

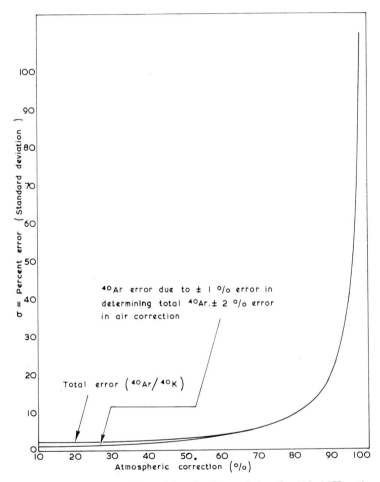

Figure 2 Estimated precision for determining the ^{40}Ar/^{40}K ratio as a function of the atmospheric correction.

feldspar (9.48% K) from the Epomeo tuff (KA 1137) obtaining only 20.6×10^{-12} mole of atmospheric ^{40}Ar. Assuming an identical feldspar fused under identical conditions but only 30,000 years old, the results shown in table 4 would be obtained.

Table 4 Analysis of hypothetical 30,000 year old feldspar

Weight (gram)	K (%)	^{40}Ar (radiogenic × 10^{-12} mole)	^{40}Ar (atmospheric × 10^{-12} mole)	^{40}Ar (atmospheric %)	K–Ar date (year)
12.6	9.48	6.36	20.6	76.3	30,000 ± 2400[a]

[a] Assuming precision as given in figure 2.

E Extraneous argon

It appears that analytical techniques *per se*, including the air ^{40}Ar correction, are not the ultimate limitation on K–Ar dating of Pleistocene minerals. This ultimate limitation will be set by three factors: (1) retention of ^{40}Ar in xenolithic minerals, (2) contamination of samples by small amounts of older minerals, (3) excess ^{40}Ar in minerals.

Dalrymple (1964) has demonstrated that the feldspars within a xenolith of granite in basalt retained 2 to 5% of their original radiogenic ^{40}Ar. Assuming 0.1% by weight contamination of feldspars from a Precambrian granite originally containing 4×10^{-8} mole/g radiogenic ^{40}Ar (a typical amount for Precambrian feldspars) and 2% retention of this argon, the 30,000 year old date for the hypothetical feldspar sample (table 4) would be increased to 77,000 years. Even 0.01% contamination of such xenolithic feldspars would make the quoted precision meaningless as an estimate of the accuracy of the K–Ar age. Furthermore, only 70 micrograms of a Lower Cambrian feldspar contaminant containing 1.4×10^{-8} mole/g ^{40}Ar would increase the apparent age by 4600 years. This is equivalent to one grain in a 28 to 35 mesh feldspar sample weighing about the same as in our example (12.6 g). Six to seven grains would double the apparent age. These problems, plus the excess ^{40}Ar problem, which will be discussed next, leave little room for optimism concerning the possibility of a meaningful comparison of K–Ar dating with carbon-14 dating. However, it does seem possible to obtain meaningful dates for most of Pleistocene time. That this is by no means an easy task is exemplified by studies of the Bishop tuff (Evernden, 1959; Evernden and others, 1957, 1964; Dalrymple and others, 1965).

When different minerals from the same deposit are dated by the potassium–argon method and by other methods or by stratigraphic dating with reference to the geologic time scale, it occasionally happens that one or more minerals may yield discordantly high apparent K–Ar ages. Beryl and cordierite, for example, invariably yield an absurdly high apparent age (Damon and Kulp, 1958b). This can best be explained by reference to the argon radioactive production equation (5) from which a term has been tacitly omitted, i.e. the term expressing the initial amount of argon-40 ($^{40}Ar_0$ present in the mineral at zero time measured from some geologic event for which independent evidence exists). Adding this term, we obtain the following modified equation:

$$^{40}Ar = {}^{40}Ar_0 + \frac{\lambda_K}{\lambda} {}^{40}K[\exp(\lambda t) - 1] \qquad (12)$$

When the event to be dated is, for example, a postcrystallization thermal metamorphic event, ^{40}Ar produced in the mineral by radioactive decay and retained from the premetamorphic history of the mineral is said to be 'inherited' ^{40}Ar. Argon-40 retained within xenoliths in igneous rocks which was accumulated by radioactive decay in the xenolith prior to solidification of the igneous rock is also referred to as 'inherited' ^{40}Ar with respect to the time of solidification. On the other hand, ^{40}Ar occluded within a mineral by processes other than radioactive decay in the mineral is referred to as 'excess' ^{40}Ar. The sum of these two components of $^{40}Ar_0$, excess and inherited ^{40}Ar, is commonly referred to as 'extraneous' ^{40}Ar. Obviously, inherited ^{40}Ar cannot always be operationally distinguished from excess ^{40}Ar.

Beryl and cordierite are clathrate-like minerals which contain tubular chains formed by six-membered silica tetrahedron rings. Argon is occluded within these tubular channels (Damon and Kulp, 1958b; Smith and Schreyer, 1962). Argon occluded in beryl and cordierite within pegmatitic or metamorphic environments is very pure radiogenic ^{40}Ar produced in the geologic environment exterior to the minerals and trapped in the minerals at the time of crystallization (Damon and Kulp, 1958b). There is also the possibility that, under confining pressure, environmental ^{40}Ar may diffuse into the minerals subsequent to crystallization.

Although the excess ^{40}Ar phenomenon is most extreme for the cyclosilicates, beryl, cordierite and also tourmaline, it has also been observed in other minerals (table 5). The error in potassium–argon dating will be most evident for minerals in which potassium is not a major component, i.e. pyroxene, plagioclase, fluorite and quartz. Excess ^{40}Ar has been looked for but not observed in the calcium mica, margarite. Fortunately, potassium micas, which are in other

Table 5 Excess ^{40}Ar in minerals

Mineral	Excess ^{40}Ar ($\times 10^{-10}$ mole/g)	Reference[a]
Chlorite	<0.009	1
Fluorite with fluid inclusions	0.0 to 0.17	2, 3
Margarite (calcium mica)	<0.45	4
	<0.031	1
Phlogopite (in xenolith)	4.0 and 11.3	5
Albite	0.19 to 0.57	6
Plagioclase	0.15 to 0.35	7
Plagioclase–quartz	0.0 to 1.21	7
Orthoclase	0.0 to 2.71	7
Sodalite	1.74 to 6.68	8
Pyroxene	5 to 50	9, 10
Pyroxene (xenolithic)	0.0 to 0.85	5
Quartz with fluid inclusions	0.016 to 5.4	3
Tourmaline	8 to 52	11
Cordierite	61 to 360	11
Beryl	20 to 14,360	11

[a] 1. Hart (1966); 2. Lippolt and Gentner (1963); 3. Rama and others (1965); 4. Damon and Kulp (1957); 5. Lovering and Richards (1964); 6. Laughlin (1966); 7. Livingston and others (1967); 8. York and MacIntyre (1965); 9. Hart and Dodd (1962); 10. McDougall and Green (1964); 11. Damon and Kulp (1958b).

respects good potassium–argon geochronometers, also usually contain negligible amounts of excess argon. Also, the degassing process upon extrusion of lava appears to be an efficient mechanism for removing most extraneous ^{40}Ar (Evernden and Curtis, 1965).

It is necessary to view the excess ^{40}Ar problem in proper perspective in order to evaluate its effect on the accuracy of K–Ar dating (figure 3). In figure 3 a range of concentrations of excess ^{40}Ar are given for different fractions of 10% K content in minerals, 10%/n K, where $n = 1, 2, 3, 4, \ldots$, i.e. any integral number from one to infinity. It can immediately be seen from table 5 and figure 3 that the cyclosilicates, beryl, cordierite and tourmaline, which in addition to occluding large amounts of excess ^{40}Ar also have low potassium contents, cannot be used for K–Ar dating. The uncertainty is also very large for low potassium minerals which contain fluid inclusions, e.g. quartz and fluorite (table 5), even for Precambrian dating work. Pyroxenes from intrusive environments also yield high ages even for Precambrian rocks, although in some intrusive igneous rocks, formed under

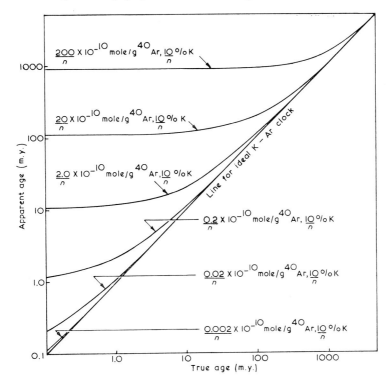

Figure 3 Departures from ideal potassium–argon clock behavior resulting from different amounts of excess ^{40}Ar in minerals of different potassium content (see also table 5). n can be any integer from 1 to ∞ (1, 2, 3, 4, 5, ..., ∞).

conditions of low partial argon pressure or in lavas, K–Ar dates for Precambrian or even younger pyroxenes may quite closely approximate the true cooling age.

It should be pointed out again that in general the degassing process upon extrusion of a magma as lava or pyroclastic ejecta is a relatively efficient process. The extraneous argon retained after extrusion is probably one to two orders of magnitude lower than the preextrusion content (Dalrymple, 1964; Evernden and Curtis, 1965; Damon, 1965). The excess ^{40}Ar data in table 5 are for intrusive igneous or hydrothermal vein environments except when otherwise indicated.

Plagioclase feldspars from intrusive igneous environments also yield maximum potassium–argon ages although for some Precambrian rocks the age may closely approximate those obtained for other radiometric geochronometers (Livingston and others, 1967). Plagioclase, with a relatively high potassium content, from volcanic rocks appears to yield reasonable dates for all but Plio–Pleistocene time (Evernden and others, 1964; Evernden and James, 1964; unpublished data from this laboratory).

Potash feldspars would be good geochronometers for Paleozoic and Precambrian dating problems involving intrusive igneous rocks, but unfortunately pre-Mesozoic feldspars rarely, if ever, quantitatively retain argon (Goldich and others, 1961). The occasionally high excess ^{40}Ar found in potash feldspars from post-Paleozoic plutonic rocks by Livingston and others (1967) together with their relative sensitivity to argon loss does not inspire confidence in K–Ar ages for these rocks when based upon potash feldspars alone. On the other hand, sanidine and other potash feldspars associated with volcanic rocks can yield reliable K–Ar dates from late Mesozoic time to late Pleistocene time (Evernden and others, 1964; Evernden and James, 1964; Evernden and Curtis, 1965; Doell and others, 1966), although, as noted above, the uncertainty increases greatly for volcanic rocks less than 100,000 years old. The lowest curve ($0.02/n \times 10^{-10}$ mole/g ^{40}Ar, $10\%/n$ K) in figure 3 appears to correspond approximately to the potash feldspars from volcanic environments.

The phyllosilicates (mica, chlorite, etc.) rarely appear to contain significant quantities of excess ^{40}Ar even when from the intrusive igneous environment. This makes the micas ideal potassium–argon geochronometers for most igneous environments for all but Pleistocene time. The tendency of the phyllosilicates to adsorb relatively large amounts of air argon frequently results in very large air corrections for very young samples.

Hornblende, an inosilicate, has very high argon retentivity and, as yet, there is no reliable evidence for excess ^{40}Ar in this mineral (Hart, 1961). However, quite certainly it will eventually be demonstrated. Excess ^{40}Ar appears to be a ubiquitous phenomenon of greater or lesser magnitude particularly in intrusive igneous environments. For example, Lovering and Richards (1964, see table 5) have observed relatively large amounts of excess ^{40}Ar in phlogopite mica from inclusions in abyssal intrusives.

In order to further increase the accuracy and precision of K–Ar dating, it will be necessary to study the excess ^{40}Ar problem in different environments. Igneous rocks from certain environments, for example the extrusive and to a lesser extent the hypabyssal environments, seem to be suitable for K–Ar dating work. On the other hand, the problem is magnified for abyssal igneous rocks. In general, the problem seems to be most severe under the following conditions: (1) the mineral has a low content of potassium and a large number of 'holes' such as tubular channels, fluid inclusions, vacancy defects or dislocations; (2) the mineral is crystallized and cooled under high pressure conditions in a dry environment, i.e. under a high partial pressure of ^{40}Ar; and (3) the time between crystallization and dating is so short that excess ^{40}Ar is not negligible in comparison to the argon produced within the mineral by radioactive decay. Or, positively stated, the accuracy of K–Ar dating is greatest for potash minerals with few defects which crystallized or cooled under conditions of low partial ^{40}Ar pressure. Added confidence in the results of K–Ar dating can be attained by the following: (1) dating different minerals with varying potassium contents from the same rock; (2) comparison of the results of K–Ar dating with results by other methods, e.g. the rubidium–strontium and uranium–isotopic lead methods; and (3) using helium-4 as an indicator of excess ^{40}Ar (Damon and Kulp, 1958b; Damon and Green, 1963; see also discussion by Damon of paper by Kulp and Engels, 1963).

There is an interesting age effect which has been observed for both helium and argon in beryl and cordierite. The magnitude of the excess ^{4}He and ^{40}Ar is greater by several orders of magnitude in the very old minerals (>2300 m.y.) than in minerals of Cenozoic age (Damon and Kulp, 1958b; Gerling, 1961). Damon and Kulp (1958b) have concluded that this is not solely the result of deeper burial or the increased rate of production of helium and argon in the past. They have suggested that this effect may have been the result of increased heat production in both the mantle and crust and consequent greater mobilization of the inert gases. Recycling of water, nitrogen and other chemically active volatiles in the crust by sedimentation and metamorphism would maintain a high proportion of these volatiles relative to the inert gases in that environment. On the other hand, the ratio of inert gases to chemically active volatiles in the mantle should constantly increase as a result of degassing because ^{4}He and ^{40}Ar will be continuously replenished by radioactive decay. Thus, mantle gas and crustal gas may have markedly different composition. This effect may possibly help to explain the high contents of excess helium and argon observed in minerals derived from deep-seated environments (Cherdyntsev, 1956, p. 86; Lovering and Richards, 1964; Naughton and others, 1966). This excess inert gas age effect may also result in occasional anomalous K–Ar ages for minerals from Archean rocks although the effect is masked in potash minerals by the large amounts of radiogenic argon produced within the minerals during such a long expanse of time.

F *Closed versus open systems*

Prerequisite (6), which is concerned with the extent to which a mineral or rock approximates a closed system, is of critical importance. An ideal geochronometer should begin its history abruptly with a very large 'parent–daughter' ratio at time equals zero corresponding to a geologic event and then accumulate the daughter product as a closed system unaffected by minor perturbations. Natural systems are never quite so simple. Four possible mechanisms by which the $^{40}Ar/^{40}K$ ratio may change other than by radioactive decay are (1) by induced nuclear reactions; (2) by mechanical abrasion; (3) by chemical alteration; and (4) by thermal diffusion. Quite possibly all mechanisms for perturbation could operate simultaneously as, for example, in a fault zone. However, it is necessary to consider them separately in order to evaluate their relative importance.

The first mechanism, i.e. induced nuclear reactions, has been investigated with regard to the excess ^{40}Ar problem (Damon and Kulp, 1958b; Damon, 1960). Terrestrial nuclear fluxes and the pertinent production cross-sections are too small to make this an effective mechanism within the Earth's lithosphere. However, iron meteorites have only minute contents of potassium and argon and are exposed to intense cosmic ray bombardment. Voshage and Hintenberger (1960) have shown that they cannot be considered closed systems with respect to nuclear mechanisms other than radioactive decay, and the argon and potassium contents are significantly changed. However, in the present context we need give this mechanism no further consideration.

Mechanical abrasion (mechanism 2) is a proven mechanism for releasing argon from minerals. Gentner and Kley (1957) observed no argon loss from pegmatitic feldspars until the minerals were crushed to 100 micron grain size. Upon crushing to a grain size of 5 to 10 μ, argon losses of 30 to 50% occurred. They ascribed the loss to release of argon from grain boundaries resulting from unmixing of the perthitic feldspars (see also Stevens and Shillibeer, 1963). Gerling and others (1961) have demonstrated that argon loss and water uptake accompany crystal damage during severe grinding of both phlogopite and muscovite mica. They also observed that severely ground micas are more susceptible to argon loss by thermal diffusion. This is an important factor because in geologic environments such as fault zones mechanical abrasion is almost invariably accompanied by some elevation in temperature. This fact has not yet been fully exploited to date fault zone activity. However, examples will be given in later sections. Another point is that care must be taken in the laboratory so as not to damage minerals in the process of separation. Occasional discordant ages may stem from laboratory separation procedures.

Chemical alteration (mechanism 3), particularly during weathering, is of constant concern to the geochronologist concerned with K–Ar dating. For example, in dating volcanic rocks he should always be alert to devitrification or hydration of glass, weathering stains or obvious deterioration of primary minerals. Petrographic study of thin sections is of great importance. Deuteric effects should, of course, be distinguished from weathering. Chemical alteration which occurs penecontemporaneous with the event to be dated is not pertinent unless it affects the retentivity for argon of the mineral to be dated.

Base exchange is the most common cause initially for the alteration of micas, in particular, the trioctahedral micas (Bassett, 1960). The effect of base exchange on K–Ar ages has been investigated under laboratory conditions by Kulp and Bassett (1961) and Kulp and Engels (1963). These authors have shown that the $^{40}Ar/^{40}K$ ratio for biotite (a trioctahedral mica) changes very slowly as the potassium content of the mineral is exchanged with calcium or magnesium. After a reduction in potassium content of 85%, the $^{40}Ar/^{40}K$ ratio was only decreased by about 15%. With further exchange at 80°C, the $^{40}Ar/^{40}K$ rapidly decreased until it was lowered by 82% after removal of 93% of the original potassium content. According to Kulp and Engels (1963, p. 226) 'the base-exchange process attacks one potassium layer (cleavage plane) at a time. When this is opened up essentially all of the potassium and argon are lost from this layer so that the $^{40}Ar/^{40}K$ ratio in the rest of the mineral remains unchanged.' This may account for the fact that micas from weathered rocks frequently give reasonable K–Ar ages for the time of petrogenesis.

Thermal diffusion of argon from minerals has been studied under laboratory conditions by Reynolds (1957), Gerling and Morozova (1957, 1958), Evernden and others (1960), Fechtig and others (1960, 1961), Hart (1960), Amirkhanoff and others (1961), Gerling and others (1961) and Baadsgaard and others (1961); and under geologic field conditions by Hurley and others (1962) and Hart (1964).

Diffusive losses of argon from minerals are dependent upon crystal chemical factors, the size and geometry of the grains which are acting as diffusing units and, of course, temperature. The rate of diffusion within the individual grains is expressed in terms of the diffusion coefficient D (cm^2/sec) which is temperature (T) dependent

$$D = D_0 \exp(-E/RT) \tag{13}$$

D_0 is an empirical rate factor, E is the activation energy usually expressed in kilocalories per mole and R is the gas constant.

The flux (J) of the diffusing component is given by Fick's first law of diffusion (see Jost, 1960, for complete mathematical treatment of diffusion)

$$\mathbf{J} = -D \operatorname{grad} c \tag{14}$$

where grad c is the concentration gradient. The divergence of $\mathbf{J}$ in turn is equal to the time (t) rate of change of concentration

$$\frac{\delta c}{\delta t} = \operatorname{div} \mathbf{J} = \operatorname{div}(D \operatorname{grad} c) \tag{15}$$

which is Fick's second law of diffusion.

When diffusion takes place in the x direction alone and D is constant throughout the mineral, Fick's second law reduces to

$$\frac{\delta c}{\delta t} = D \frac{\delta^2 c}{\delta x^2} \tag{16}$$

Liebhafsky (Jost, 1960, p. 39–42) has solved the problem (figure 4) of diffusion out of a slab of infinite length and finite width, with

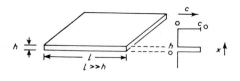

$c = c_0$ for $0 < x < h$ at $t = 0$
$c = 0$ for $x = 0$ and $x = h$ at $t > 0$

Figure 4 Slab model for phlogopite assuming D/x^2 is much greater perpendicular to the cleavage. The diffusion coefficient is of course always greater parallel to the cleavage but $l \gg h$. (After Everndcn and others, 1960)

initial homogeneous concentration c_0 when the concentrations at its faces are kept at zero

$$\bar{c} = \frac{c_0}{h} \int_0^h (c/c_0) \, dx = \frac{8c_0}{\pi^2} \sum_{n=0}^{\infty} \frac{1}{(2n+1)^2} \exp(-[2n+1]^2 y) \quad (17)$$

where $y = D\pi^2 t/h^2$ and $\bar{c}$ is the average concentration at time t.

For sufficiently large values of t, the first term in the series gives a good approximation

$$\bar{c} = \frac{8c_0}{\pi^2} \exp(-t/\tau) \quad (18)$$

$$\tau = \frac{h^2}{\pi^2 D}$$

$1/\tau$ is similar to a decay constant such as the radioactive decay constants. A directly analogous approximate relationship would involve a constant, λ_d, the probability that an atom of argon will escape from the slab by diffusion in a unit period of time

$$\bar{c} = c_0 \exp(-\lambda_d t) \quad (19)$$

Recalling that

$$\lambda_d = \frac{0.693}{(t_{1/2})_D}$$

we can solve equation (18) for λ_d when $c_0/\bar{c} = 2$, i.e. at $t = (t_{1/2})_D$, and this can be done also using approximate equations for a sphere of radius, r_s, and a cylinder of radius, r_c, with infinite length

$$\lambda_d = 14.2 \frac{D}{h^2} \text{ (slab)} \quad (20a)$$

$$\lambda_d' = 12.5 \frac{D'}{r_c^2} \text{ (cylinder)} \quad (20b)$$

$$\lambda_d'' = 35.7 \frac{D''}{r_s^2} \text{ (sphere)} \quad (20c)$$

In practice, the experimenter measures diffusive losses from a mineral and calculates D/x^2 assuming an appropriate geometry. To determine D, some assumption must be made concerning the dimensions of the diffusing grain. Because this is seldom known with certainty, D/x^2 is a more fundamental quantity than D itself.

The advantage of the previous equations, where D/x^2 is expressed as an escape probability in terms of a decay constant, λ_d, is that the standard equations for radioactive decay can now be used where argon becomes analogous to a radioactive (escaping) daughter product. The rate of change of the argon content of a mineral is then given by the following simple linear differential equation of the first order:

$$\frac{d\ ^{40}Ar}{dt} = \lambda_K\ ^{40}K - \lambda_d\ ^{40}Ar \qquad (21)$$

i.e. the rate of change of the ^{40}Ar content of the mineral is equal to the rate of radioactive decay to ^{40}Ar minus the rate of diffusive loss from the mineral. The solution of this equation including allowance for excess ^{40}Ar is

$$^{40}Ar = {}^{40}Ar_0 \exp(-\lambda_d t) + {}^{40}K_0 \frac{\lambda_K}{(\lambda_d - \lambda)} [\exp(-\lambda t) - \exp(-\lambda_d t)] \qquad (22)$$

where $^{40}K_0$ is the original potassium-40 content at time zero (time line as in figure 6). More rigorous but less convenient and general equations can be formulated for specific models (Wasserburg, 1954; Amirkhanoff and others, 1961).

The value for D/x^2 has been determined experimentally for a number of minerals. In general, it can only be considered to be constant for limited ranges of temperature. The graph of log D/x^2 usually has a number of sharp breaks, each one of which is associated with activation of argon located in different states or crystal chemical transformations such as the diffusing of alkali metal cations in the unmixing of perthites. For example, the phlogopite and microcline $\log(D/x^2)$ versus t curves can usually be resolved into three straight line segments. The typical relative retentivity of potash minerals at various temperatures can be qualitatively summarized as follows, in order of decreasing retentivity:

(a) Below 250°C:
  hornblende > muscovite > biotite ~ phlogopite > sanidine > microcline > glauconite

(b) 600–700°C:
  hornblende > muscovite > sanidine ~ orthoclase > biotite ~ phlogopite > glauconite

The retentivities vary greatly within a mineral species depending on grain size, defect structures, etc. Nevertheless, the generalized order seems to be statistically valid. The critical effect of diffusing grain size can be seen by the high diffusivity (low retentivity) of glauconite which actually has diffusion coefficients similar to phlogopite but is very fine grained (Evernden and others, 1960).

The thorough study of diffusion in phlogopite by Evernden and others (1960) provides a convenient frame of reference for further discussion of argon leakage from minerals. They analyzed their data for phlogopite by comparison of two models. One model corresponds to the slab (equation 17, figure 4) with diffusion taking place perpendicular to the cleavage ($D/h^2 \gg D'/r_c^2$) and the other to the cylinder ($D'/r_c^2 \gg D/h^2$) with diffusion taking place radially, parallel to the cleavage. They found that the slab model fits the data rather well to about 500°C. From 600 to 700°C, the cylindrical model provides the best fit, whereas above 800°C the slab model again is best. The diffusion coefficient is, as expected, always greater parallel to the cleavage. However, the very short diffusion parameter, h, along the c axis between cleavage planes seems to be the deciding factor for phlogopite retentivity.

Equations (19) and (20a) have been solved for four cases corresponding to different values of h and t for phlogopite with $D_0 = 10^{-8}$ cm²/sec and $E = 28,000$ cal/mole as measured by the above workers. The results are plotted in figure 5. At an approximation, with reference to the excellent study by Hart (1964) of a plutonic contact zone, one can say that the mineral biotite behaves like a phlogopite with h equal to 10 μ and hornblende corresponds to a phlogopite with $h = 1000$ μ. Thus, virtually complete argon loss during a period of 600,000 years requires a sustained temperature of roughly 475°C for hornblende but only 230°C for biotite. Several other factors may aid the geologist in interpreting K–Ar dates: (1) minerals at the surface under present ambient conditions (∼ 20°C) would lose an insignificant amount of argon even during the Earth's entire geologic history; (2) diagenesis at less than 100°C sustained for 50 m.y. would cause less than 10% leakage from biotite and virtually no loss from hornblende; (3) low-grade metamorphism sustained for 50 m.y. at temperatures between 100 and 150°C would cause losses of from 10 to 90% for biotite but would result in less than 10% loss from hornblende; and (4) high-grade metamorphism sustained at 500°C for 500,000 years would result in complete loss of argon for all minerals used in K–Ar dating.

The above generalizations based upon laboratory investigation appear to conform more or less to the 'observed' behavior of minerals

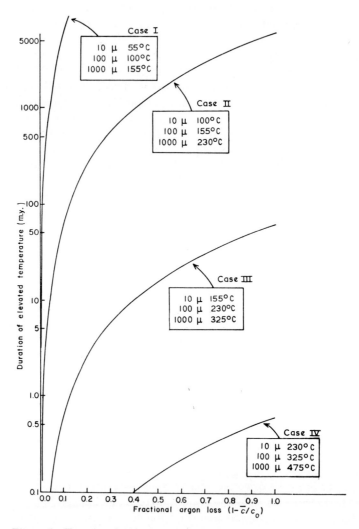

Figure 5 Fractional ^{40}Ar loss ($1-\bar{c}/c_0$) as a function of the duration of elevated temperatures. The calculations are based on $D_0 = 10^{-8}$ cm^2/sec, $E = 28{,}000$ cal/mole and slab model (figure 4; Evernden and others, 1960).

under geologic field conditions. However, a word of caution is necessary. Liebhafsky's solution for diffusion from a slab assumes that the argon concentration is zero outside of the slab. Empirically, this seems to be a fair assumption for most geologic cases. However, it will never rigorously hold because argon will, in fact, always be present in the environment of the mineral, and its partial pressure may sometimes be quite high. In fact, diffusion of argon into minerals at high temperatures is probably the origin of the excess ^{40}Ar in a number of the cases listed in table 5. Fortunately, water exerts a much higher partial pressure than argon under typical hypabyssal conditions and competes with and, in large part, excludes argon from available 'holes' in minerals. It has already been mentioned that, under abyssal conditions, the argon partial pressure may possibly be relatively very high.

Fechtig and his colleagues (Fechtig and others, 1961) and Russian workers (Amirkhanoff and others, 1961; Gerling and Morozova, 1958) have demonstrated the importance of defect structures and desorption from boundaries, particularly in low-temperature argon diffusion. For example, there is a component of argon associated with perthitic feldspars which migrates at relatively low temperatures and seems to be the cause of the characteristically low K–Ar ages relative to mica ages (Wetherill and others, 1955; Wasserburg, Hayden and Jensen, 1956; Folinsbee and others, 1956; Carr and others, 1956; Goldich and others, 1961). Amirkhanoff and his associates in some brilliant experimental research observed that the beginning of perceptible diffusion of argon in feldspars under laboratory conditions coincides with a break in the specific heat curve indicating initiation of permutations of the crystal structure. They reasoned that such transitions would take place gradually during geologic time at relatively low temperatures creating boundaries along which a component of radiogenic argon would be localized. This component of argon in the so-called 'unsteady zone' is readily desorbed by heating for two to three hours at 350 to 400°c. They also observed that a limited amount of potassium could be removed by base exchange with thallium. Presumably this potassium was also associated with argon in the 'unsteady zone.' At any rate, when the readily desorbed argon and exchangeable potassium were removed, the age discrepancy between mica and the potassium-containing feldspars no longer existed and mica–feldspar dates were concordant. From their results, it would also appear that sodic feldspar (albite) should not have characteristically low retentivities in comparison to micas. This has been borne out by measurements on Precambrian albite–mica pairs in this laboratory (Livingston and others, 1967).

The results of experimental research on potash feldspars are also borne out by other research involving behavior of minerals under geologic conditions. As previously mentioned, Hart (1964), in his investigation of a plutonic contact zone, observed the usual low retentivity of potash feldspar in the low-temperature zone at a considerable distance from the contact, but in the high-temperature zone near the contact the feldspar had consistently higher retentivity for the remaining argon presumably contained within Amirkhanoff's 'steady zone' of the feldspar. Thus, the shift in relative argon retentivities from low-temperature to high-temperature conditions is borne out both by laboratory diffusion measurements and by observed retentivities under geologic conditions. Additional work of this sort is needed to allow for more rigorous interpretation of radioactive dating data.

G Deterministic model for interpretation of potassium–argon dates

The hypothetical two-stage geologic model in figure 6 may be helpful as a frame of reference for discussion of the geologic interpretation of K–Ar dates. The model assumes the following sequence of

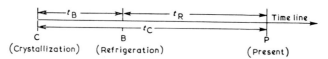

Figure 6 Time line for deterministic two-stage geologic model involving the following sequence of events: (1) mica crystals in rock are formed t_C years ago; (2) the rock remains buried at an elevated temperature $T°$c for t_B years; (3) after a time t_B, the rock is brought to the surface where it is refrigerated for t_R years.

events: (1) mica crystals in rock are formed t_C years ago; (2) the rock remains buried at an elevated temperature $T°$c for t_B years; (3) after a time t_B, the rock is brought to the surface where it is refrigerated for t_R years. Thus the radioactive clocks (mica crystals) came into existence t_C years ago but the apparent age (t_A) will be less than the 'true age' (t_C) because of diffusional loss during the time t_B. The refrigeration need only be cooling to ambient surface temperature or at least to a low enough temperature for diffusional losses of argon to become negligible.

In order to evaluate this geologic model mathematically, equation (22) must be evaluated between the limits of zero time at point C and time equals t_B at B for the period (t_B) of radiogenic accumulation and diffusional leakage of argon and the limits of t_B at B and t_C at P for the period (t_R) of argon accumulation with no diffusional loss. The resulting equation, substituting $^{40}K \exp(\lambda t_C)$ for the initial potassium at zero time ($^{40}K_0$) and t_R for ($t_C - t_B$), is

$$\frac{^{40}Ar}{^{40}K} = \frac{\lambda_K}{(\lambda_d - \lambda)} \exp(\lambda t_C)[\exp(-\lambda t_B) - \exp(-\lambda_d t_B)]$$

$$+ \frac{\lambda_K}{\lambda}[\exp(\lambda t_R - 1)] + \frac{^{40}Ar_0}{^{40}K} \exp(-\lambda_d t_B) \quad (23)$$

The first term of equation (23) is for the period of argon accumulation with diffusional leakage; the second term is for accumulation without leakage; and the third term takes account of any residual excess ^{40}Ar that may be present.

The equation has been evaluated, assuming zero Ar_0, for the four convenient cases shown in figure 7: (1) Case I: $\lambda_d = 14.2 \, D/h^2 < \lambda$, corresponding to negligible losses for short duration of elevated temperature (t_B) or $\lambda_d \ll \lambda$, and significant losses for long burial times with $\lambda_d < \lambda$. The apparent age (t_A) more or less closely corresponds to the age of crystallization (t_C or 'true' age). (2) Case II: $\lambda_d \sim \lambda$ corresponding to significant losses for burial times (t_B) exceeding 100 m.y. The apparent age (t_A) is a minimum age which corresponds more or less closely to the time of crystallization (t_C). (3) Case III: $\lambda_d > \lambda$, corresponding to the state called 'transient equilibrium' by radiochemists (Friedlander and others, 1964; Rutherford and others, 1930). When t is sufficiently long so that $\exp(-\lambda_d t)$ is negligible in comparison to $\exp(-\lambda t)$, the ^{40}Ar content during the period of elevated temperatures becomes independent of time and the daughter–parent ratio is closely approximated by

$$\left(\frac{^{40}Ar}{^{40}K}\right)_{t<t_B} = \frac{\lambda_K}{\lambda_d - \lambda} \quad (24)$$

This time-independent $^{40}Ar/^{40}K$ ratio corresponds to a 'built in' age at the time of refrigeration. For the specific parameters given in figure 7 for Case III, the 'built in' age is about 22.5 m.y. This would mean that the apparent K–Ar mica-clock age (t_A) of such a rock suddenly brought to the surface in recent time would be 22.5 m.y. or a maximum for the time since refrigeration (t_R). A K–Ar mica

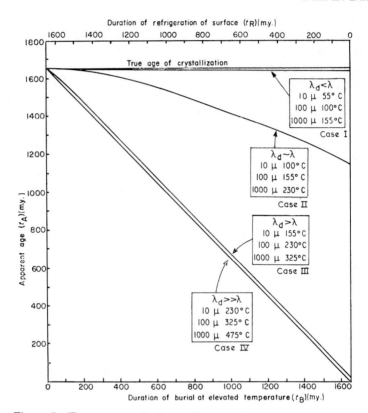

Figure 7 Two-stage geologic model with phlogopite as reference mineral ($D = 10^{-8}$ cm²/sec, $E = 28{,}000$ cal/mole) evaluated at different intensities and duration of temperature for phlogopite-like minerals with different diffusion parameters. Case I corresponds to slight argon loss. Case II corresponds to significant argon loss. Cases III and IV correspond to transient and secular equilibrium, respectively (see figure 6 for time line).

clock brought to the surface 1000 m.y. ago would 'read' 1020 m.y. or a very good approximation to t_R. (4) Case IV: $\lambda_d \gg \lambda$, corresponding to 'secular equilibrium.' Equation (24) reduces to

$$\left(\frac{^{40}\text{Ar}}{^{40}\text{K}}\right)_{t < t_B} = \frac{\lambda_K}{\lambda_d} \qquad (25)$$

For the specific conditions given in figure 7 for Case IV, the 'built in' age is about 225 thousand years. This is insignificant for pre-Cenozoic dating work, but quite significant for Pleistocene work. Thus the

K–Ar mica-clock date is a more or less good measure of the time since refrigeration which becomes increasingly accurate as the temperature of the rock before refrigeration increases and as the time since refrigeration increases.

As previously mentioned, the model assumes effectively zero ^{40}Ar pressure outside of the crystal. This would be the case in a system in which ^{40}Ar is transported away as fast as it escapes from minerals, i.e. an open system for ^{40}Ar, or a system which is open to the Earth's atmosphere. Such systems might be approximated by an open metamorphic system or when magma breeches the surface in an ash flow or extrusive lava. Excess ^{40}Ar has also not yet been considered. A number of alternative sets of conditions can be considered as follows:

(A) $C = 0$ for $x = 0$ and $x = h$ at $t > 0$, ^{40}Ar$_0 = 0$
(B) $C = 0$ for $x = 0$ and $x = h$ at $t > 0$, ^{40}Ar$_0 \neq 0$
(C) $C \neq 0$ for $x = 0$ and $x = h$ at $t > 0$, ^{40}Ar$_0 = 0$
(D) $C \neq 0$ for $x = 0$ and $x = h$ at $t > 0$, ^{40}Ar$_0 \neq 0$

We have already considered condition (A), i.e. no initial excess ^{40}Ar and negligible external effective ^{40}Ar pressures. The effect of excess ^{40}Ar is to add an exponentially decreasing term to the ^{40}Ar/^{40}K equation (23). With zero effective environmental ^{40}Ar pressure, it would appear, with reference to the excess ^{40}Ar data in table 5, that the normal effect for Precambrian potash minerals would be to counteract slightly the diffusional losses, thus making the apparent age (t_A) a slightly better approximation to the crystallization age (t_C) for Cases IB and IIB. As previously discussed, anomalously old ages resulting from excess ^{40}Ar become a severe limitation for dating very young minerals and minerals of very low potash content. Occasionally, under unusual circumstances, anomalously old ages may be observed for Precambrian potash minerals (Stockwell, 1963). Anomalously high ages resulting from excess initial ^{40}Ar will become increasingly improbable and more restricted to younger minerals as λ_d increases in Cases IIIB and IVB. This is, of course, a consequence of diffusional degassing in the presence of zero external argon pressure.

The effect of a constant external *effective* concentration of ^{40}Ar (^{40}Ar$_E$) is to add an exponential uptake term to equations (22) and (23) which results from diffusion of ^{40}Ar into the mineral. In the case of equation (23), this added term is as follows:

$$\frac{^{40}\text{Ar}}{^{40}\text{K}} = \text{equation (23)} + \frac{^{40}\text{Ar}_E}{^{40}\text{K}}[1 - \exp(-\lambda_d t)] \qquad (26)$$

Thus, ^{40}Ar will diffuse into the mineral until the *effective* concentration gradient becomes zero and

$$\frac{^{40}\text{Ar}}{^{40}\text{K}} = \frac{^{40}\text{Ar}_\text{E}}{^{40}\text{K}}$$

The equations for transient and secular equilibrium for Cases IIIC and IVC now become

$$\left(\frac{^{40}\text{Ar}}{^{40}\text{K}}\right)_{t<t_\text{B}} = \frac{^{40}\text{Ar}_\text{E}}{^{40}\text{K}} + \frac{\lambda_K}{(\lambda_\text{d}-\lambda)} \quad \text{(transient equilibrium)} \quad (27)$$

$$\left(\frac{^{40}\text{Ar}}{^{40}\text{K}}\right)_{t<t_\text{B}} = \frac{^{40}\text{Ar}_\text{E}}{^{40}\text{K}} + \frac{\lambda_K}{\lambda_\text{d}} \quad \text{(secular equilibrium)} \quad (28)$$

Thus, the 'built in' age will be increased as a result of the external argon pressure. For Cases IC and IIC, if the external *effective* ^{40}Ar pressure is equal to or less than the ^{40}Ar radiogenically generated in the mineral during time t_B, then its effect will be to minimize or nullify leakage from the mineral and the apparent age will be a better approximation to the age of crystallization. If the external *effective* ^{40}Ar pressure is greater than the radiogenically generated ^{40}Ar during time t_B, then the apparent age will be a maximum for the time of crystallization.

At this point a more specific geologic frame of reference may clarify the significance of the above generalized discussion. Let us consider a deep-seated pluton at a temperature of 700°C. The diffusion coefficient for phlogopite at this temperature will be very high, equivalent to $\lambda_\text{d} \sim 2.3 \text{ yr}^{-1}$. As a consequence, secular equilibrium will be maintained and $\lambda_K/\lambda_\text{d}$ will be very nearly equal to zero. Thus the internal ^{40}Ar content of the mineral will rapidly (within one year) equilibrate with the external ^{40}Ar. An effect of higher temperatures will be to cause even more rapid equilibration. This example is equivalent to the phlogopite in xenoliths within deep-seated intrusions which Lovering and Richards (1964) have studied. In that case, the external ^{40}Ar pressure resulted in about 1×10^{-9} mole/g of ^{40}Ar being included within the minerals. If the phlogopite in our example contains 8% potassium, the 'built in' age will be 69.5 m.y., a prohibitive figure for accurate dating of Phanerozoic rocks.

The data indicate that the external *effective* ^{40}Ar pressure must be one to two orders of magnitude less in the environment of a typical hypabyssal pluton. Such a pluton at 700°C would contain sufficient excess ^{40}Ar to result in a 'built in' age of 0.75 to 7.5 m.y. (1×10^{-10}

to 1×10^{-11} mole/g ^{40}Ar). Let us assume that such a hypabyssal pluton breeches the surface and erupts as an ash flow. The 'built in' component of ^{40}Ar will now decay by diffusional leakage according to the following relationship (atmospheric argon will also diffuse inward but this will be subtracted by the air correction)

$$^{40}\text{Ar} = {}^{40}\text{Ar}_\text{E} \exp\left(-\int_{t=0}^{t=t_R} \lambda_\text{d}\, dt\right) \sim {}^{40}\text{Ar}_\text{E} \exp(-\sum_i \lambda_{\text{d}i}\, \Delta t_i) \quad (29)$$

The relationship on the right-hand side of equation (29) is easily evaluated by an iterative graphical method. If, for example, the ash flow is assumed to cool at a uniform rate from 700 to 475°C and if the phlogopite contains 1×10^{-10} mole/g of ^{40}Ar at the instant of eruption, the resulting anomalous age will depend upon cooling time approximately as follows: $t = 0$, 7.5 m.y.; $t = 4.5$ yr, 1 m.y.; $t = 10.5$ yr, 100,000 yr; $t = 15$ yr, 15,000 yr; $t = 20$ yr, 1800 yr; $t = 25$ yr, 250 yr. Such rapid degassing of the 'built in' component probably accounts for the reasonable apparent ages obtained by various workers for Pleistocene rocks (e.g. Evernden and others, 1964; Doell and others, 1966). It should not be difficult to observe extraneous ^{40}Ar in gem-quality minerals with very low potassium contents. In fact, one such case has already been observed (Damon and others, 1966).

A terrain undergoing high-grade metamorphism would correspond to Case IV, i.e. secular equilibrium would be maintained. The 'built in' ^{40}Ar/^{40}K would be given by equation (28). For the specific parameters given for Case IV in figure 7, $\lambda_K/\lambda_\text{d}$ is approximately equal to 1.33×10^{-5}, equivalent to a 0.225 m.y. 'built in' age for minerals behaving like the 1000 μ phlogopite (475°C) which we have used as a frame of reference. To this would be added a component resulting from equilibration with ^{40}Ar from the external environment. If this component is 1×10^{-11} mole/g and the potassium content is 8%, then the total 'built in' age would be about 1 m.y. A mineral behaving like the 10 μ phlogopite would contain only a significant equilibration component ($\lambda_K/\lambda_\text{d} \sim 0$, 0.75 m.y.). If the temperature is now lowered to 400°C, the inherited component will increase by an order of magnitude and result in a 2.25 m.y. 'built in' component giving a total 'built in' age of 3 m.y.

The distinction between excess ^{40}Ar (equilibration ^{40}Ar) and inherited ^{40}Ar might seem to be idle speculation. However, these two components are, in fact, frequently operationally distinguishable. The inherited component of ^{40}Ar is dependent upon the potassium content of the mineral whereas the equilibration (excess) component is independent of the potassium content of the mineral. Thus excess argon can cause absurdly anomalous apparent ages, whereas the effect of inherited argon is more restricted. The presence of excess ^{40}Ar can be demonstrated by determining the ^{40}Ar content of minerals of greatly different potassium content (see e.g. Livingston and others, 1967). Damon and others (1963) have dated minerals from a high-grade metamorphic terrain which behaved like a Case IVA system, i.e. minerals of varying potassium content were all concordant with the exception of very large books of muscovite from pegmatites that had a small, but significant 'built in' age (see also Livingston and others, 1967). This will be discussed later.

A terrain undergoing low-grade metamorphism should behave in a quite different manner than for high-grade metamorphism. For example, at 230°C, different minerals would probably undergo a range of response varying from Case II through Case III and Case IV. For example, biotites might respond so drastically as to be in secular equilibrium, muscovite might be in transient equilibrium and hornblende might correspond to Case II with significant argon leakage. If the terrain were suddenly cooled, the biotite apparent age would be slightly high but would correspond more or less closely to the cooling time.

If the temperature of low-grade metamorphism is about 155°C, then the behavior of different minerals should shift to cover the spectrum from transient equilibrium to slight ^{40}Ar loss, i.e. Cases I to III. Hurley and others (1962) have dated different minerals from both sides of the Alpine Fault zone in New Zealand. The rocks on the inland side (southeast side) are thought to have risen from a depth of about 3 km in a short period of time and in a region of very high heat flow. This uplift has exposed biotite schists and gneisses, chlorite–muscovite schist, greywackes and argillites. Metamorphism of the terrain, probably prior to cooling in Jurassic time, reached oligoclase facies corresponding to about 550°C. The rocks on the Tasman Sea side (northwest) of the Alpine Fault zone were not deeply buried prior to the Plio–Pleistocene faulting. Thus the Paringa River granite on the northwest side corresponds to Case I with a reasonable potassium–argon apparent age of 286 m.y. However, fine-grained biotites from the high-grade metamorphic rock on the inland (uplifted) side of the fault were dated at from 4 to 8 m.y., corresponding to secular equilibrium (Case IV). Large biotite plates from a pegmatite were dated at 25 m.y., corresponding to transient equilibrium (Case III), and muscovite from the chlorite schist zone and argillite zone were dated at 76 m.y. and 133 m.y., corresponding to a case intermediate between II and III. An amazing amount of information is stored in these radioactive clocks waiting to be retrieved as our knowledge and sophistication increases. Not only is it possible to estimate in such cases the ages of crystallization, termination of high-grade metamorphism and final cooling, but eventually significant geothermetric and geobarometric data will also be obtained. We are also learning much about the role of diffusion and other processes in the transformation of rock systems.

H Probabilistic models

So far we have been considering a deterministic model for interpreting the results of potassium–argon dating. Thus, using experimental data, a model was developed as an analog to natural systems and, granting certain environmental conditions, the behavior of the model is determined and compared with actual geologic cases. An alternative probabilistic approach was originally introduced into the literature of geochronology by Wasserburg (1961). Wasserburg considered models consisting of assemblages of rocks each of which contains a radioactive clock. The clocks have a certain probability of being partially or completely reset at a later date. Granting different assumptions concerning the resetting, he rigorously developed different consequent distributions that illustrated three points of particular interest: (1) Either variable partial resetting of the clocks occurring episodically or continuously, or complete resetting occurring continuously with a given probability will result in a time band from a group of clocks all originally set at the same instant of time. (2) An event (setting of the clocks), no matter how widespread, will be subject to attrition by being rejuvenated. Thus, any event will have a certain probability of survival, depending upon the rate of resetting and the extent of the event, and will fade away with time. (3) The time band resulting from resetting may contain a maximum which does not necessarily correspond to the time of the original event. In the limiting case, a steady-state distribution of clock readings may be obtained. The maximum may then persist at a particular age, not because the clocks within the time band are intrinsically stable, but rather because as soon as a clock is reset and removed from the time band another clock is also reset and takes its place.

These ideas have been qualitatively incorporated into the distributions for the two models given in figures 8(a) and (b). The models

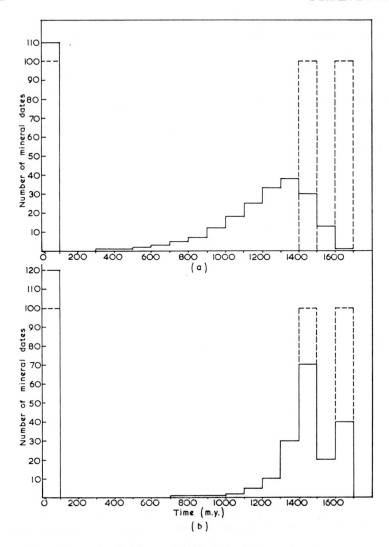

Figure 8 (a) Qualitative probabilistic model assuming three magmatic pulses involving equal amounts of newly formed igneous rock in each episode. Continuous partial argon loss is assumed to be dominant, episodic loss is of secondary importance (see Wasserburg, 1961, for rigorous treatment of probabilistic models); (b) Qualitative probabilistic model assuming three magmatic pulses as in (a). Episodic loss is assumed to be dominant, continuous partial argon loss is of secondary importance.

assume three orogenies (events) involving equal amounts of newly formed granite for each orogenic episode. In one model continuous argon leakage is assumed to be dominant and episodic loss is of secondary importance. In the second model episodic argon loss is dominant and continuous argon loss is of secondary importance. For both models there is marked attrition of the older events whereas the younger event is increased by resetting due to metamorphism.

In the case of the model with continuous argon leakage dominant (figure 8a), a time band with a single maximum results. However, the maximum does not correspond to either older event but is a minimum for both. In fact, the older events are no longer discernible on a statistical basis.

In the model for dominant episodic loss, there is a greater probability for attrition of the oldest event but all events are still distinctly discernible. It will be shown in the following section that the resetting of the K–Ar clocks in the older Precambrian of Arizona appears to conform to the model with episodic loss dominant (figure 8b).

III   Application to the Dating of Igneous and Metamorphic Rock within the Basin Ranges

A   *The Precambrian of Arizona and Sonora*

On the basis of field observations, the Precambrian rocks of Arizona have for many years been divided into the 'younger Precambrian' (Apache Group, Grand Canyon Series) and the 'older Precambrian' (Vishnu, Yavapai and Pinal Series) which include all more intensely metamorphosed and deformed Precambrian rocks whether or not they are overlain by younger Precambrian rocks (Butler and Wilson, 1938; Anderson, 1951). In some areas, e.g. the Santa Catalina Mountains of Pima County, the younger Precambrian Apache Group of rocks have been metamorphosed but may still be recognized by their characteristic lithological sequence and by the fact that they overlie the older Precambrian with a marked structural and metamorphic unconformity. In many localities the younger Precambrian rocks are intruded by vast diabase sills and dikes. Basalt flows are also present and some of the sedimentary formations are tuffaceous.

The older Precambrian consists of a great thickness of sedimentary and volcanic rocks. The oldest rocks consist predominantly of slates, thin limestones, tuffs and cherts intercalated with basaltic to rhyolitic volcanics (Yeager greenstone, Ash Creek group). These are overlain by the shales, greywackes, sandstones and conglomerates of the Alder group that are also intercalated with lavas of variable but predominantly rhyolitic composition. The youngest extensive formation of the older Precambrian is the Mazatzal quartzite which has been recognized over a wide area from northeastern through central to southeastern Arizona (Wilson, 1962).

The older Precambrian rocks were laid down in extensive geosynclinal troughs that were not confined to Arizona but extended in a northeasterly direction across the North American continent (for review see Damon and Giletti, 1961; Wilson, 1962). These troughs were folded into northeast-trending mountain systems that were probably comparable in extent and duration with the Appalachian system of the eastern United States but, until recently, evidence existed for only one orogeny, the Mazatzal Revolution (Wilson, 1936, 1939). According to Wilson (1939), this orogeny took place 'after deposition of the Mazatzal quartzite and long before Apache sedimentation' (p. 1160). Hinds (1935, 1936, 1938) had postulated a pre-Mazatzal Revolution (Arizona Revolution) but on the basis of insufficient evidence (Wilson, 1939). Stratigraphic evidence for an earlier orogeny has now been described by Philip M. Blacet of the U.S. Geological Survey (Anderson, 1963) who mapped a granodiorite gneiss basement (Brady Butte granodiorite) which unconformably underlies the Alder group of the Yavapai Series south of Prescott, Arizona. This granodiorite has been dated at approximately 1700 m.y. by E. J. Catanzaro of the U.S. Geological Survey (Anderson, 1963). Thus, stratigraphic evidence now exists for two distinct orogenic episodes during the older Precambrian. Rhyolitic volcanic rocks and granitoid intrusions also demonstrate the existence of episodic magmatism.

Sufficient isotopic dating data are now available to define distinct episodes of Precambrian magmatism and thermal metamorphism. Only data obtained by other than the potassium–argon method will now be discussed because our purpose is to evaluate the K–Ar method relative to independent age criteria. Professor L. T. Silver and his associates at California Institute of Technology have obtained U–Pb isotopic ages of between 1660 and 1760 m.y. on suites of zircons from rhyolites and granitic intrusives from the Dragoon quadrangle in Cochise County (Silver, 1963; Silver and Deutsch, 1963), from the northern Mazatzal Mountains in east central Arizona (Silver, 1964), from the Bagdad region in west central Arizona (Silver, 1966), and from the Grand Canyon of the Colorado River (Pasteels and Silver, 1965). Wasserburg and Lanphere (1965) have reported ages of about 1660 m.y. for both pegmatites and a quartz monzonite pluton from the Grand Wash cliffs region of northwestern Arizona by the mineral and whole rock mineral rubidium–strontium isochron methods, respectively. In this laboratory (see table 6), Livingston has obtained a whole rock mineral Rb–Sr isochron age of 1630 m.y. on the Madera quartz diorite which is intrusive into the Pinal schist of the Pinal Mountains in Gila County, Arizona. Concordant Rb–Sr and K–Ar mineral dates indicate that the Precambrian terrain of the Bámori district (southeast of Caborca) of Sonora was subjected to intense thermal metamorphism 1680 m.y. ago and has remained refrigerated since that time (table 6).

Rubidium–strontium whole rock or mineral isochron ages of from 1400 to 1450 m.y. have been obtained for granitoid batholiths such as the Ruin Granite of the Salt River Canyon, Gila County (Livingston, 1962a, 1962b; Livingston and others, 1967) and the Oracle granite from the northern Santa Catalina Mountains in Pinal County (Livingston and others, 1967). Erickson and Livingston (Damon and associates, 1966) have obtained a 1440 m.y. isochron on the Eaton gneissic granodiorite in the Dos Cabezas Mountains of Cochise County. A Rb–Sr whole rock mineral isochron age of 1450 m.y. was obtained by Wasserburg and Lanphere (1965) for a gneissose complex from the northern Virgin Mountains near the Arizona border in Nevada. Recently, Silver (1966) reported a uranium–isotopic lead age of 1375 m.y. for a cogenetic zircon suite from the Lawler Peak granite of the Bagdad region in Yavapai County.

Thus, there is abundant evidence for a series of magmatic–thermal and metamorphic events for the older Precambrian of Arizona and Sonora concentrated between 1630 and 1760 m.y. ago which we will tentatively refer to as the Arizonan Revolution* to distinguish it from the later Mazatzal Revolution that occurred between about 1370 and 1450 m.y. ago. There is abundant isotopic dating evidence for both of these complex events in the Precambrian of neighboring states (for review see Wasserburg and Lanphere, 1965).

The vast diabase sheet in the Sierra Ancha Mountains of east

* The designation of this geologic event as the Arizonan Revolution is not generally accepted by other geologists.

Table 6 Comparison between K–Ar dates for biotite and muscovite from the Precambrian of Arizona and Sonora

Rock unit and location	K–Ar date for igneous or metamorphic rock (m.y.)		K–Ar date for pegmatite (m.y.)	Comparison date (m.y.)	
	Biotite	Muscovite	Muscovite		
Metamorphic terrain, Bámori District, Sonora	—	1680[1]	1680[1]	1650	K–Ar on hornblende from metamorphic rock[2]
				1690	Rb–Sr on muscovite from pegmatite[3]
Pinal schist, Gila County, Arizona	—	1385[2]	1610[2]	1630	Rb–Sr isochron for Madera quartz diorite[4]
Yavapai schist, Yavapai County, Arizona	1156[2]	1440[2]	1555[2]	1740	Uranium–isotopic lead on gneissoid rhyolite and granodiorite[5]
Chino Creek granite, Yavapai County, Arizona	1330[6]	1460[2]	—		
Pinacate gneiss, Sonoyta District, Sonora	1170[2]	—	1440[2]		
Oracle granite, Pinal County, Arizona	1420[6]	—	1420[6]	1420	Rb–Sr mineral isochron[7]
Brahma Schist, Coconino County, Arizona	1240[6]	—	1410[6]	1695	Uranium–isotopic lead[8]

(1) Damon and others (1962a); (2) Unpublished data, this laboratory; (3) Livingston and Damon (1965); (4) Livingston (1966, unpublished data); (5) Silver (1966); (6) Damon and others (1962b); (7) Livingston and others (1967); (8) Pasteels and Silver (1965).

central Arizona has been dated by the uranium–isotopic lead method on cogenetic zircon suites (Silver, 1963) and by the K–Ar method on biotite (Damon and others, 1962b) at 1150 m.y. Wasserburg and Lanphere (1965) have also published Rb–Sr data for both minerals and whole rock which indicate that some of the plutonic activity in southeastern Nevada and northwestern Arizona took place about 1060 m.y. ago during the younger Precambrian.

The Grand Canyon disturbance has been associated with the intrusion of the diabase sills (e.g. see Wilson, 1962, p. 17–20). Thus the isotopic data indicate that this occurred 1150 m.y. ago. The existence of ca. 1060 m.y. plutonism suggests that later plutonism may also have been involved. As previously mentioned, there is as yet no definitive stratigraphic or isotopic evidence for Paleozoic plutonism in Arizona or Sonora.

B *Potassium–argon dating of the Precambrian rocks of Arizona and Sonora*

A histogram of K–Ar dates on mica minerals from rocks classified stratigraphically as Precambrian or for other reasons considered to be Precambrian is shown in figure 9. It is evident that all three Precambrian magmatic–thermal and metamorphic events are clearly represented by K–Ar radioactive clocks that have begun reading time or have been reset during one of these events. There is also abundant evidence of clock resetting during the intense late Cretaceous–Cenozoic magmatic–thermal and metamorphic events. However, the interpretations of the clock settings during post-Grand Canyon disturbance time are not sufficient without reference to other dating methods or field and petrographic investigations. For example, we are investigating the possibility that the peak at 300 to 400 m.y. in figure 9 may represent a real and hitherto undetected geologic event rather than merely partial clock resetting. We know that the clock reading 760 m.y. was partially reset by a basic dike that was found upon reinvestigation in the field. We have investigated a large number of cases in the field where the cause of resetting was clearly determinate. An example is given in figure 10 from the Precambrian of Sonora. A K–Ar date for the Granito Aibó of 150 m.y. was clearly the result of nearly complete resetting owing to a 105 m.y. quartz monzonite intrusion. A Rb–Sr date on feldspar yielded 700 m.y., whereas in a neighboring area the metamorphic terrain gave concordant K–Ar and Rb–Sr dates of 1680 m.y. for several different minerals (see table 6). On the other hand, in the southern Sierra Ancha Mountains of Gila County, Arizona where there is no evidence for postintrusion regional metamorphism of the Ruin

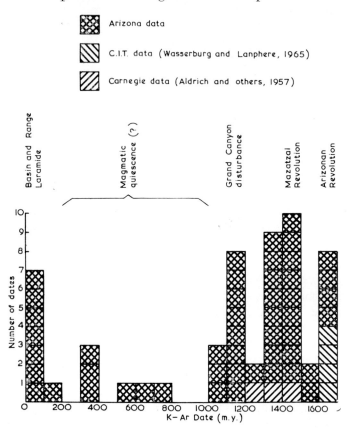

Figure 9 Potassium–argon dates for mica minerals separated from rocks assumed by geologists to have been emplaced during Precambrian time. Mineral dates within labeled peaks are believed to represent the time of emplacement or the time of episodic response to thermal events associated with known orogenies. Mineral dates falling within the quiescent period are believed to represent partial argon loss as a result of late Mesozoic and Cenozoic thermal events, although the possibility remains that some thermal events are as yet unrecognized (see text).

Granite (figure 11) or Apache Group (contact thermal effects are observable near the contacts of the diabase sill); K–Ar dates on biotite from the Ruin Granite yield the same age as obtained by whole rock or mineral Rb–Sr isochrons (Damon and others, 1962b; Livingston, 1962a, 1962b; Livingston and others, 1967). As previously mentioned, similar ages are also obtained on the diabase sill by both the uranium–isotopic lead method on a cogenetic zircon suite and by K–Ar on biotite.

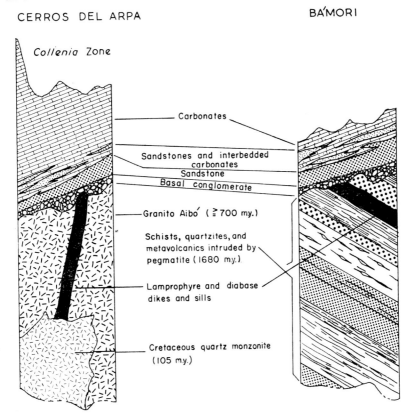

Figure 10 Correlation of Cerros del Arpa Precambrian strata with Precambrian strata of Bámori in Sonora, Mexico. Intrusion of a mid-Cretaceous quartz monzonite pluton has caused uncertainty as to the original age of the Granito Aibó (Damon and others, 1962a).

The resetting of clocks is evident by the characteristic inverted dates for coarse mica from pegmatite versus fine-grained mica from schistose country rock (see figure 12 and table 6). It is interesting that the retentivity of strontium in the micas seems to be only slightly greater than that for argon. This has also been observed by

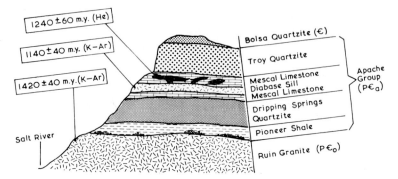

Figure 11 Younger Precambrian Apache Group overlying older Precambrian Ruin Granite in the Sierra Ancha Mountains north of the canyon of the Salt River. See text for discussion of conformable dating results by different methods on different mineral and rock samples which demonstrate that this terrain has remained unheated since emplacement of the Ruin Granite except for marginal effects due to the diabase sill. The helium date on magnetite in the Mescal limestone is less accurate than the K–Ar dates (Damon and others, 1962b; Livingston, 1962a,b; Silver, 1963).

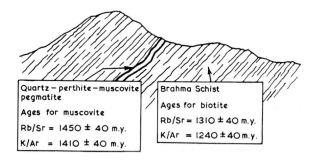

Figure 12 Pegmatite vein intruded into Brahma Schist on Kaibab Trail, Grand Canyon National Park. The dates show the typical inversion which is a result of the greater retentivity of pegmatitic muscovite relative to fine-grained biotite. Mineral genesis occurred about 1700 m.y. ago (Pasteels and Silver, 1965). The muscovite mineral has responded to elevated temperatures during the Mazatzal orogeny but has not responded significantly to the Grand Canyon disturbance as has the biotite (see table 6).

Gast and Hanson (1963) and Hart (1964). The characteristic order of argon retention seems invariably to be coarse-grained muscovite > fine-grained muscovite > fine-grained biotite (figure 6). The coarse-grained muscovites are only completely reset by an intense thermal event, whereas the fine-grained biotites respond even to the mild thermal event associated with the Grand Canyon disturbance. With reference to figure 5, fine-grained biotite seems to be responding roughly as our 10 μ reference phlogopite, whereas fine and coarse muscovite correspond roughly to the 100 and 1000 μ phlogopites, respectively. Thus, concordant K–Ar dates on minerals having widely different diffusion properties indicate clock setting followed by refrigeration, whereas each discordant K–Ar radioactive clock reading may or may not correspond to a distinct event. During periods of normal crustal temperature gradient, rocks within one mile of the surface are essentially refrigerated and negligible argon loss occurs. When the temperature rises during periods of abnormal thermal activity, there will be a more or less distinct threshold temperature for significant argon loss. For complete resetting of the K–Ar clocks during a thermal event of 100 m.y. duration, roughly 155°C is required for the fine-grained biotite, 230°C for the fine-grained muscovite, and 325°C for the coarse-grained muscovite. The temperature thresholds are far enough apart so that one may be completely reset, whereas the others are only slightly perturbed. Thus, a single K–Ar date for a mineral that has had a long and complex history yields little definite information. However, a number of K–Ar dates on different minerals provides valuable information concerning the genesis and thermal history of the mineral, and because of the episodic response of K–Ar clocks, a statistical analysis of K–Ar dates allows for a rather precise dating of thermal events.

C *Phanerozoic history of the Basin ranges*

There has been a number of excellent reviews of the geology of the western United States. For this resumé, the author has relied heavily on Eardley (1951), Gilluly (1963, 1965), Kay (1951), McKee (1951) and Osmond (1960).

During the Paleozoic era the Basin and Range Province north of about latitude 36° N was miogeosynclinal in the eastern part and eugeosynclinal to the west. The first well-documented orogeny took place in early to middle Mississippian time. As a result, a welt, known as the Manhattan geanticline or Antler orogenic belt, developed in the central part of what is now the Basin and Range Province (north of 36° N lat). Thereafter the welt extended until it included during Cretaceous time all of the northern Basin and Range Province.

During Phanerozoic time the southern part of the Basin and Range Province in Arizona, New Mexico and Sonora was essentially part of the hedreocraton, and from time to time the area was submerged by epicontinental seas. For most of this time western Arizona and northwestern Sonora were shelf and continental margin of the hedreocraton. A profound unconformity exists at the base of the lower and middle(?) Triassic Moenkopi Formation (Wilson, 1962). At the beginning of the Mesozoic era, all of the area now included in the Basin and Range Province below latitude 35° N was uplifted to form an extensive highlands area. To the north the sea regressed from the area east of the welt during the Triassic Meso–Cordilleran geanticline phase of the Manhattan geanticline. As a result of this extensive uplift of the southern province, a broad sheet of gravel, the Shinarump conglomerate, which was derived from the south, was deposited over most of northeastern Arizona and adjoining areas.

The Nevadan orogeny, which resulted in the main folding of the Sierra Nevada in late Jurassic time, was also active in Arizona. According to Wilson (1962, p. 48), 'from northeastern Arizona to the southwestward, an erosion surface at the base of the Cretaceous cuts progressively across older rocks that had been uplifted and deformed during the Nevadan Revolution.' The Nevadan uplift gave rise to the Glance conglomerate of southeastern Arizona.

As mentioned previously, during Cretaceous time the welt in the northern area spread out to include all of the Basin and Range Province. On the other hand, the Mexican (or Sonoran) geosyncline developing in southwestern Arizona invaded the southern hedreocratonic area during early Cretaceous time and spread in a west-northwesterly direction across the southern part of Arizona and Sonora. During late Cretaceous time the Sonoran geosyncline continued as a deep embayment but, in the larger perspective, it formed only a small part of the Rocky Mountain geosyncline. The great late Cretaceous sea spread until in Turonian time it not only united the Gulf of Mexico with the Arctic Ocean but its width extended from central Utah to western Iowa (Cobban, 1960). Following a maximum transgression in Turonian time, the epicontinental seas retreated until by late Maestrichtian time all of the western interior of the United States had reemerged.

The retreat of the epicontinental seas occurred during the Laramide orogeny. In the northern Basin and Range Province, the welt attained its maximum development. Throughout the entire province, deformation resulted in the uplift of mountains and the crust was invaded by synorogenic granitic intrusions.

In mid-Tertiary time the tilted fault block ranges and broad basins were formed during and following extensive outpourings of andesitic to rhyolitic volcanic rocks. It was at this time that the Basin and Range Province came into existence as a physiographic province.

Recently, Gilluly (1965) has reviewed the evidence for Phanerozoic magmatism in the western United States. In Arizona the first clear-cut evidence for Phanerozoic magmatism is the occurrence of siliceous ash beds in the Triassic Chinle Formation on the Colorado Plateau. Elsewhere in the Basin and Range Province there is considerable evidence for andesitic-rhyolitic (and basaltic) volcanism during Triassic time and, apparently, the oldest Phanerozoic plutons were emplaced during this period. Plutonism and volcanism are recurrent in late Jurassic time penecontemporaneously with the Nevadan orogeny. Thereafter there is abundant evidence for periodic magmatism during the Cretaceous and Cenozoic. The vast floods of siliceous ash flows during the mid-Tertiary orogeny are of particular importance. Basin and Range volcanism during middle to late Pliocene and Pleistocene time is primarily basaltic in composition.

Earlier workers considered the Mesozoic Nevadan, Santa Lucian, Laramide and Basin and Range orogenies to be separate and discrete geologic events. Because of a lack of definitive stratigraphic evidence, distinctions have gradually blurred until the following point of view expressed by Wilson has become quite common: 'Laramide igneous activity is marked by batholiths, stocks, dikes, plugs and volcanic rocks, especially within the Basin and Range Province. For this latter region, the orogeny is not clearly separable from the Triassic-Jurassic Nevadan Revolution; furthermore, it appears to blend with the crustal unrest and igneous activity of middle to late Cenozoic time' (Wilson, 1962, p. 58). In the following section, the chronological evidence of K-Ar dating will be used to evaluate this problem. Are these orogenies discrete events? Does magmatism within this province during Mesozoic time occur quasi-continuously or in discrete pulses?

D *Potassium–argon dating of late Mesozoic–Cenozoic events*

The University of Arizona in Tucson is located in the heart of the southern part of the Basin and Range Province (figure 1) within a broad basin surrounded by mountains which form a typical Basin and

Range topographic setting (figure 13). These nearby mountain ranges have been the subject of intensive K–Ar dating studies by members of this laboratory (Bikerman, 1965; Bikerman and Damon, 1966; Damon and Bikerman, 1964; Damon and others, 1963; Damon and

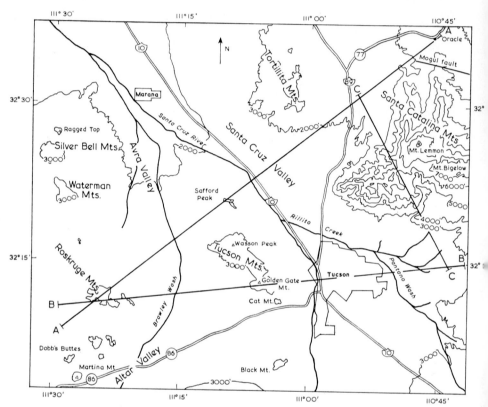

Figure 13 Location map and lines of section for the typical Basin and Range area around Tucson, Arizona (1000-foot contour interval, scale 1: 250,000).

others, 1964; Livingston and others, 1967; Mauger, 1966; Mauger and others, 1965). The approximately 2000 square mile area in figure 13 is probably the most thoroughly dated area of this size in the United States and it serves as a convenient prototype for the chronology of magmatic events within the Basin and Range Province.

A summary of dating results in the northern Tucson Mountains around Safford Peak is given in figure 14. In that area the Cretaceous(?) volcanic rocks have been too disturbed by subsequent geologic events to warrant attempts at K–Ar dating. However, fresh unaltered samples have been obtained from the Cenozoic section above the Cretaceous(?) volcanic rocks. Biotite separates from the volcanic units yield a consistent chronology beginning with the 38.5 ± 1.2 m.y. Rillito Andesite which floors the Cenozoic volcanic sequence and terminating with the intrusion of the Safford Dacite Neck at 24.5 ± 0.8 m.y. The three intervening volcanic rock units

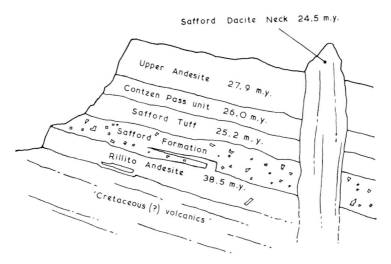

Figure 14 Idealized section of volcanic rocks from northern Tucson Mountains around Safford Peak (from Bikerman and Damon, 1966).

average 26.3 ± 1.3 m.y. and considering the precision of individual K–Ar dates are, therefore, penecontemporaneous. The entire section is tilted about 20° to the northeast which demonstrates that major Basin and Range faulting occurred during post-early-Miocene time.

An idealized section through the southern Tucson Mountains is shown in figure 15. Cretaceous sedimentary rocks have been intruded by Laramide granite and quartz monzonite plutons (71–75 m.y.) and are overlain by a chaotic breccia and ash flow sequence which appear from field and dating evidence to be younger than the plutons. The last event of the Laramide sequence is the extrusion of the Shorts Ranch andesite which has been dated at 56.8 m.y. on biotite. Unconformably overlying the Shorts Ranch andesite there is a sequence of basaltic andesites flooring and capping a sequence composed of a colluvial gravel deposit, an air-fall tuff and two ash flows representing distinct cooling units. The youngest unit, a basaltic andesite, has been dated at 19.8 m.y. on fresh whole rock. Several important generalizations appear to be warranted for the Tucson Mountains. First, magmatism appears to have occurred in two pulses,

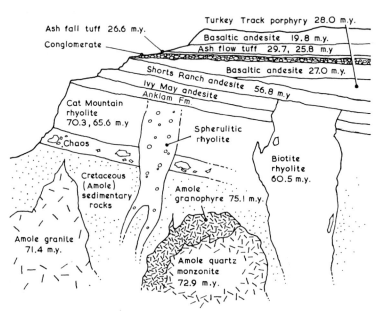

Figure 15 Idealized section through the southern Tucson Mountains. (After Bikerman and Damon, 1966)

one in late Mesozoic–early Cenozoic time and the other during mid-Tertiary time. Second, Basin and Range faulting has affected the entire volcanic sequence during post-early-Miocene time. A similar sequence has been observed in the Silver Bell mining district to the northwest, in the Roskruge Range to the west and in the Sierrita Range to the south. Idealized cross-sections across the region (figure 13) are shown in figure 16.

With reference to the line of section AA' which extends from the Roskruge Range in Pima County to Oracle in Pinal County, the section in the Roskruge Range is floored by andesitic volcanics and sediments, one andesitic unit of which has been dated at 108 m.y. on whole rock. Resting on the Cretaceous sedimentary–volcanic terrain are two felsic ash flow sequences separated by a thin sedimentary bed. The K–Ar dates indicate a significant lapse of time between deposition of the two ignimbritic sequences. The older sequence appears to be penecontemporaneous with the granitoid

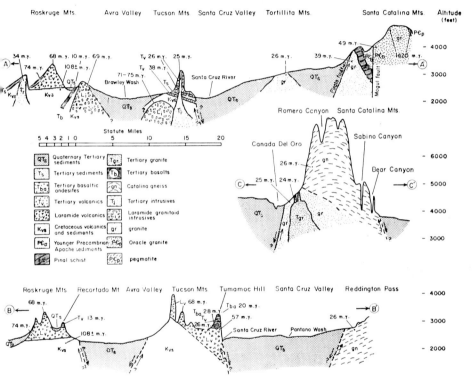

Figure 16 Idealized cross-sections with great vertical exaggeration along lines of section given in location map (figure 13).

plutons in the neighboring Tucson Mountains to the east, whereas the younger sequence is penecontemporaneous with the pluton in the Roskruge Range at Cocoraque Butte (ca. 69 m.y.) and with a similar ignimbritic sequence, the Cat Mountain rhyolite, in the Tucson Mountains. The mid-Tertiary volcanic pulse, which begins in the Roskruge Range with a granodioritic plug at La Tortuga Butte, includes a basaltic andesite sequence ranging in age from 15 to 24 m.y.

and terminates with the extrusion of a small ash flow with well-developed eutaxitic texture (the Recortado ash flow dated at 13 m.y. on sanidine: section BB'). According to Bikerman (1965), there may have been some uplift of the Recortado ash flow above the present valley floor during the past 13 m.y. i.e. during post-late-Miocene time, but it is essentially flat lying and has not been noticeably affected by Basin and Range faulting. Following the mid-Tertiary pulse of silicic magmatism, minor amounts of postorogenic basalt were extruded.

The section going from A to A' (figure 16) continues across the Tucson Mountains, which have already been discussed, to the low-lying Tortillita Mountains. In that range a granodiorite gneiss, which appears identical in lithology to the Precambrian Oracle granite in the northern Santa Catalina Mountains around Oracle, has been dated at 26 m.y. indicating intensive heating during mid-Tertiary time. The section then crosses the Pirate fault which flanks the northern Santa Catalina Mountains. All rock units between the Pirate fault and the Mogul fault, which strikes NW-SE across the Santa Catalina Mountains, have been intensely heated during Tertiary time. Thus, Oracle granite in the shear zone immediately northeast of the fault yielded a 49 m.y. date on biotite, whereas outside of the shear zone northeast of the fault K-Ar and Rb-Sr results for muscovite, biotite and plagioclase minerals yield concordant 1420 m.y. dates.

In the southern Santa Catalina Mountains and Rincon Mountains, there is further evidence of an extensive episode of Cenozoic metamorphism (figure 16, line of section CC'). Potassium-argon dates on biotite, muscovite, plagioclase and orthoclase from the gneiss and plutonic intrusions into the gneiss are all approximately concordant at about 27 m.y. even though Catanzaro and Kulp (1964) have demonstrated a Precambrian age by the uranium-isotopic lead method for zircons extracted from the gneiss. However, a detailed analysis of the data for these samples indicates real differences in the apparent age of the different minerals as shown in table 7. The samples were collected from different localities as much as 20 miles apart.

The approximately concordant age on the lowest potassium phase, plagioclase, suggests that excess environmental ^{40}Ar is very low and thus the built-in age for the pegmatitic muscovite minerals is a result of inherited ^{40}Ar in the most retentive phase (equation 25). Damon and Bikerman (1964) have shown that the cooling of the Catalina and Rincon gneiss is closely correlative with intense magmatism from 25 to 28 m.y. ago within the area shown in figure 13. Hedge (1960), on the basis of sodium–potassium ratios in muscovite minerals coexistent with orthoclase and plagioclase, has estimated a 400°C to 500°C temperature for alkali metal equilibration within the gneiss. If the cooling of the Santa Catalina–Rincon block began about 28 m.y. ago with the temperature equal to 400°C at that time, the built-in age of the pegmatitic muscovite is equivalent to that expected for a 1000 μ 'standard' phlogopite at that temperature (figures 5 and 7).

Table 7 Average apparent ages for different minerals from the Santa Catalina and Rincon gneissic complex

Mineral	Number of different samples	Average K–Ar date for samples (m.y.)
Biotite	5	25.2 ± 0.5^a
Orthoclase	1	26.8 ± 0.8^b
Muscovite (fine-grained from gneiss)	4	27.0 ± 0.9^a
Plagioclase	1	29.3 ± 0.9^b
Muscovite (large books from pegmatites)	3	31.7 ± 0.5^a

[a] Standard deviation of mean.
[b] Standard deviation for precision of single analysis.
Data from Damon and others (1963); Livingston and others (1967); Mauger and others (1966).

If we assume that the biotite minerals in the gneiss are equivalent to 10 μ 'standard' phlogopite, then a sharp limit can be set on the duration of elevated temperatures. Thus, the array of argon dates for different minerals in table 7 is quite consistent with an exponential cooling rate from 400°C at 28 m.y. ago to 100°C at 23 m.y. ago.

About 15 miles east of point C' (figure 13), a middle Tertiary unit, the Mineta Formation, crops out at Mineta Ridge on the east flank of the Rincon Mountains (Chew, 1962). A black, fetid freshwater limestone unit of these beds, subunit 12 of the upper unit, contains identifiable remains of a young rhinoceros (*Diceratherium*). According to Lance (1960, p. 156), this fossil is 'certainly no older than upper Oligocene or younger than middle Miocene.' Above subunit 12, according to Chew (1962), the arkosic sandstones, siltstones and fine-grained conglomerate commonly contain muscovite. According to Hedge (1960, p. 18), the sodium–potassium ratios of this muscovite are what 'one would expect if the muscovites were merely detrital remnants derived from exposure of gneiss in the Santa Catalina and Rincon Mountains which are only a few miles away.' The Mineta beds have been faulted and tilted during Basin and Range uplift in contrast to relatively undisturbed late Miocene and Pliocene beds. Thus the stratigraphic relationships are consistent with uplift and cooling of the Catalina–Rincon block during early Miocene time, followed by erosion of some gneiss before late Miocene time. Large areas of gneiss are exposed and being eroded at the present time.

Returning to the line of section AA' (figure 16), the Tortillita Mountains appear to have participated in the thermal metamorphic event along with the Santa Catalina–Rincon Mountains. However, the region near the Mogul fault was either not heated to as high a temperature or it cooled earlier than the areas to the south and southwest. Furthermore, the area around Oracle north of the fault has remained at low temperatures since Precambrian time. Obviously, these K–Ar results will provide important boundary conditions for the solution of the structural problems related to the Basin and Range orogeny. The Catalina–Rincon block is typical of many ranges containing metamorphic rocks throughout the Basin and Range Province (Armstrong and Hansen, 1966; Mauger and others, 1966). According to Mauger and others, 'the concordant argon results from individual mountain ranges fall into the 20–30 m.y. time interval and are essentially compatible with geologic estimates of a late Oligocene–early Miocene beginning for rift faulting in the Basin and Range Province' (p. A-I-5).

Turning now to the Basin and Range Province as a whole, including southern Oregon but excluding Idaho, southwestern Texas and the state of Chihuahua in Mexico (figure 1), we have found K–Ar histograms to provide a useful frame of reference. The histogram in figure 17 was prepared essentially according to the criteria used by Damon and Mauger (1966, p. 100). These criteria for selecting K–Ar dates are as follows:

(1) Only hypabyssal plutons and volcanic rocks are included because these rocks have cooled relatively rapidly and accordingly the mineral dates most closely represent the time of emplacement. Andesitic (including potassic basaltic-andesites) to rhyolitic volcanics are plotted with the hypabyssal plutons. Basalts (excluding potassic basaltic-andesites) are plotted separately. Metamorphic rocks are excluded although their distribution would not be dissimilar.

(2) Only one date is presented for a single mappable rock unit. When several dates for a rock unit were available, the average age was considered as 'one' date.

(3) All rock units were located within the Basin and Range Province.

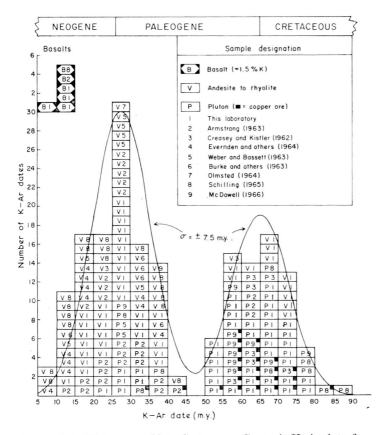

Figure 17 Histogram of late Cretaceous–Cenozoic K–Ar data for hypabyssal plutonic and volcanic rocks of the Basin and Range Province. (After Damon and Mauger, 1966)

(4) Only dates falling within the time encompassed by the period from 5 to 90 m.y. are included. This period of time encompasses early Turonian (late mid-Cretaceous) through late Hemphillian (late Pliocene) time.

(5) Plutons associated with copper ore are distinguished by a darkened square in the upper right-hand corner of the unit box to

indicate the temporal relationship between copper mineralization and plutonism.

As pointed out by Damon and Mauger (1966, p. 100), 'the resulting histogram provides a striking demonstration of the occurrence of two distinct episodes of magmatism during this time within the Basin and Range Province.' The pertinent characteristics of the histogram, with one exception, have not changed with the accumulation of more dates. They listed these characteristics as follows (p. 100):

(1) The data may be resolved, as a first approximation for reference, into two normal (Gaussian) distributions, each with a standard deviation (σ) of ± 7.5 m.y.

(2) The distributions peak at the Paleogene–Neogene boundary and in late Cretaceous (Maestrichtian)–early Cenozoic (Paleocene) time.

(3) Over 90% of the dates comprising the earlier magmatic pulse fall within the limits of the classical Laramide, i.e. the time encompassed by the Laramie, Fort Union and Wasatch formations. The close time correlation between Laramide magmatism and the Laramide as defined stratigraphically is quite impressive. At least some of the spread in ages is due to precision limits for the analyses, about ± 2.5 m.y. on the average, or one-half class interval of the histogram. All but a few of the K–Ar dates for magmatic rocks comprising the later pulse fall within the time encompassed by the Oligocene and Miocene epochs.

(4) There is magmatic quiescence during earliest Turonian through mid-Campanian time and again during middle to late Eocene time.

(5) There is a striking excess of dates over those of a simple normal distribution at the Paleogene–Neogene boundary. Furthermore, this is not the result of excessive sampling in one area. This excess seems to be indicative of prolific extrusion of magmas at this time throughout the Basin and Range Province.

(6) Basalts show up dominantly as Pliocene in age. Although the number of dates are few, field evidence points toward their extrusion as being a real and significant event which continued into Plio–Pleistocene time. Many mid-Tertiary rocks, which have been classified as basalts in the field, are actually alkali-andesites with a remarkably high potassium content (Halva, 1961; Mielke, 1965; Taylor, 1959).

(7) The Laramide is represented by more hypabyssal plutons than volcanics, whereas the opposite is true for the mid-Tertiary

magmatic pulse. This may be due to two factors: stratigraphic elevation and subsequent depth of erosion.

(8) As a second approximation, the ages comprising the pulses are not perfectly normally distributed. The normal distribution, which serves as a reference for discussion, obviously obscures real and significant complexities. The histogram for each pulse is skewed towards younger ages. There seems to be a relatively quick onset of magmatism and a relatively slow relaxation.

With the accumulation of new data, point (8) no longer seems to be valid because the Laramide magmatic pulse has tended to more closely fit a Gaussian distribution. There still remains a tendency for an excess of magmatic events relative to a simple Gaussian distribution in late Miocene and early Pliocene time. Separate histograms are presented in figure 18 for the southern and northern Basin and Range Province and that part of the entire province which is at a distance greater than 160 miles from the margin of the Colorado Plateau. These histograms suggest that the departures from Gaussian symmetry are the result of more intense magmatism in the southern province between 25–30 m.y. and more prolific magmatism in the northern province during late Miocene–early Pliocene time. There are also fewer Laramide magmatic events dated within the northern province (see also Armstrong, 1966, p. 584) and at a great distance from the Colorado Plateau. There remains the possibility that many Laramide magmatic rocks remain covered by the more recent flood of lava in the northern province.

Considering again the Basin and Range Province as a whole, in the words of Damon and Mauger (1966, p. 100–101):

> 'there is a striking time relationship between magmatism and orogeny as evidenced by the stratigraphic record. Laramide magmatism is essentially confined within the classical Laramide as stratigraphically defined. During the Eocene, erosion and basin filling is correlative with the waning of magmatism and finally with magmatic quiescence during middle to late Eocene time. The mid-Tertiary magmatism was accompanied by Basin and Range tectonism (Gilluly, 1963; Osmond, 1960). Finally, Plio–Pleistocene basaltic volcanism postdates the major uplift of the ranges (e.g. see Osmond, 1960; Taylor, 1959).'

The postorogenic extrusion of basaltic lava may be the result of the cooling and rigidifying of the crust allowing the propagation of faults into the source region of the lava beneath the Mohorovicic boundary.

The relationship between Laramide magmatism and copper mineralization has also been discussed by Damon and Mauger (1966,

p. 102–106). All of the dated late Mesozoic–Cenozoic copper porphyry deposits of the Basin and Range Province are Laramide with the possible exception of the Bingham, Utah deposit. The other mid-Tertiary pluton associated with a copper ore deposit, shown in figure 17, is the Railroad Mining District stock in the Piñon Range

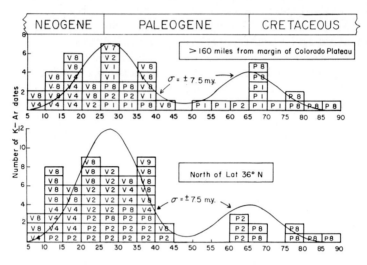

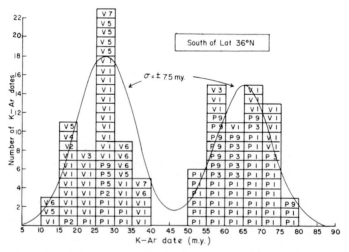

Figure 18 Histograms of late Cretaceous–Cenozoic K–Ar data for hypabyssal plutonic and volcanic rocks plotted separately for different regions within the Basin and Range Province.

of Nevada dated by Armstrong (1966) at 33 m.y. on biotite. The Railroad Mining District ore deposit is a silver–lead–copper limestone-replacement body. Thus, with the exceptions noted, the late Mesozoic–Cenozoic copper porphyry deposits of the Province are concurrent with Laramide magmatism and there appears to be a genetic relationship between ore and igneous rock. According to Damon and Mauger (1966), K–Ar dating results indicate that 'whatever processes the genetic tie may involve, the age data require

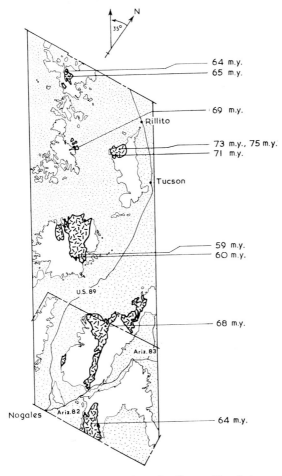

Figure 19 Potassium–argon data for Laramide plutons along a 100-mile portion of Mayo's (1958) southwestern Arizona belt (scale = 1 : 735.000).

that these processes are restricted in time to the cooling history of the host porphyry' (p. 106).

Mayo (1958) has pointed out that the ore districts of the Southwest tend to occur along or at the intersection of structural lineaments. Using the K–Ar method we have dated plutons which are located along one of Mayo's lineaments, the southwestern Arizona belt. A 100-mile portion of this belt is shown in figure 19. The belt contains many Laramide plutons and ore districts that are from southeast to northwest: Cananea (20 miles south of the Arizona–Sonora border, 59 m.y. pluton), Patagonia–Duquesne, Helvetia, Pima, Amole, Silver Bell, Bagdad (170 miles NW, 71 m.y. pluton), and Mineral Park (250 miles NW, 72 m.y. pluton). The fact that the copper porphyry plutons were not randomly distributed in space and time is of considerable interest to geologists responsible for the designing of exploration programs.

IV   The Basin Ranges in Broader Perspective

A   *Umbgrove's theory of the pulse of the earth*

Although Laramide magmatism is less intense within the northern Basin and Range Province, Laramide igneous rocks have been dated throughout the province and elsewhere within the Cordillera and Antilles of the western hemisphere. For example, with no attempt at an exhaustive compilation, plutons of Laramide age have been dated north of the province in Alberta (Baadsgaard and others, 1961), in British Columbia and the Yukon (Gabrielse and Reesor, 1964) and to the east in the Colorado mineral belt (Tweto and Sims, 1963). Laramide plutons have also been dated on the islands of Jamaica, Hispaniola, Puerto Rico and Aruba within the Greater Antilles and Lesser Antilles (Priem and others, 1966).

Documentation of evidence by earlier workers for magmatic–diastrophic pulses has been advanced by Umbgrove (1947) and, of course, by many other geologists. Damon and Mauger (1966) have evaluated the Laramide as an example of Umbgrove's concept of the 'pulse of the earth' (figure 20). Magmatism accompanies deformation and mountain building within the active tectonic belts of the earth as shown in figure 20 where the Laramide portion of the histogram in figure 17 is used as a frame of reference. Widespread regression of the epicontinental seas from the continents is accompanied by decreasing temperatures. The resulting changes in topography and climate appear to result in widespread extinction of fauna and an accelerated rate of evolution accompanied by the differentiation of flora. Following the orogeny, there is a tendency for transgression of the seas once again and for amelioration of climate. According to Umbgrove, the cause of such extensive transformations of the face of the earth must be profound transformation within the depths of the earth, below its surficial crust.

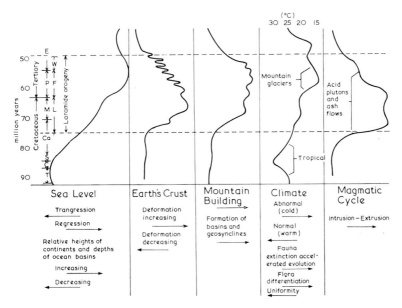

Figure 20 The Laramide as an example of Umbgrove's pulse of the earth. (After Damon and Mauger, 1966)

B Orogenic versus epeirogenic cycles

There are as yet insufficient K–Ar dating results for an adequate statistical analysis of Mesozoic magmatic events prior to the Laramide episode. However, the existing results appear to group in mid-Cretaceous time and near the boundaries of the Mesozoic periods (see Gabrielse and Reesor, 1964, and Gilluly, 1965, for reviews of North American data).

Figure 21 shows the results of an attempt to extrapolate the relationship between magmatism and the extent of the epicontinental seas throughout the Mesozoic. In order to do this, we obtained

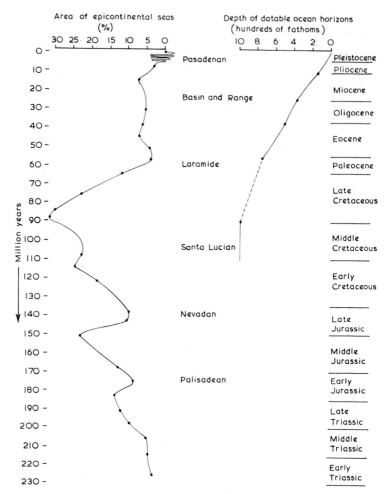

Figure 21 Depth of the Pacific Ocean and extent of the epicontinental seas. (After Damon and Mauger, 1966)

additional data by the use of paleogeographic maps (Schuchert, 1955) indicating the extent of transgression of the epicontinental seas onto the present boundaries of the North American continent including Greenland and the northern part of Venezuela, but excluding Hudson Bay.

The curve representing the area of the epicontinental seas can be resolved into two phases: (1) a transgressive phase continuing from earliest Triassic time and reaching a maximum during the Turonian stage of the Cretaceous (ca. 85–90 m.y.), and (2) a regressive phase which follows the maximum and is continuing until the present time. Taken together the transgressive and regressive phases represent an epeirogenic cycle. The epeirogenic cycle does not appear to be complete and so the duration is somewhat greater than 230 m.y., perhaps 300 m.y.

Superimposed upon the epeirogenic cycle are cyclic oscillations. Each of these oscillations appears to be correlative with classical orogenies as illustrated in most textbooks of historical geology. The time between these orogenies appears to average about 40 m.y. Orogeny is recurrent throughout the Mesozoic, but the data seem to indicate the existence of real quiescent periods between magmatic pulses (see figure 17). Thus the answer to our earlier question (section IIIC) is that the Mesozoic–Cenozoic orogenies are discrete events, but their duration is much longer than suggested by some geologists (see Gilluly, 1949, for an interesting analysis of this problem). Given the propensity of many geologists for a nonstatistical approach to geologic problems, the existence of a single magmatic event between 45 to 50 m.y. ago would be sufficient cause for some to doubt the discreteness of magmatic pulses. Certainly, the very quiescent periods seem to be short lulls between magmatic pulses which flow and ebb over a relatively long period of time. Nevertheless, the quiescent period is as statistically valid as the magmatic pulse itself and, in fact, geologists such as Lindgren (1933) and Schmitt (1933) recognized the period of erosion exposing Laramide intrusives before the mid-Tertiary flood of lava.

The results of this analysis indicate that much careful work will be required to define earlier Mesozoic magmatic pulses in detail. The results will probably conform in large part to the conclusions of classical North American geology softened somewhat by statistical considerations rather than by a rigid determinism.

C *The Basin ranges as part of the East Pacific Rise*

The second curve in figure 21 was taken from an important paper by Menard and Ladd (1963). The data represent the depths to paleontologically datable horizons on seamounts, guyots and atolls. It should be noted that this curve for the sinking of the floor of the Pacific Ocean parallels the curve for regression of the epicontinental seas. According to Menard (1964), the Darwin Rise, which was elevated as a vast bulge of the sea floor during early Cretaceous time, reversed its trend in post-mid-Cretaceous time and has been sinking ever since at the rate of a little more than 2 cm per thousand years (figure 22).

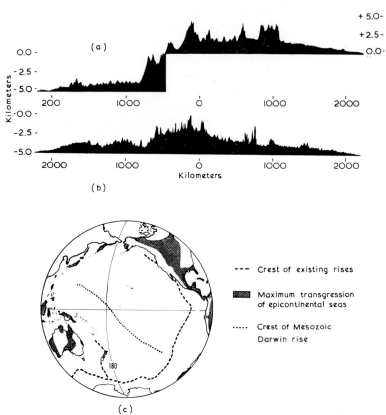

Figure 22 (a) Idealized topography of western U.S.A. (continental and offshore); (b) Idealized topography of East Pacific Rise (off Central America); (c) Oceanic and continental rises and the maximum transgression of Mesozoic–Cenozoic epicontinental seas. (After Damon and Mauger, 1966)

Concurrent with the foundering of the Darwin Rise, western North America has been rising steadily and at about the same rate (Damon and Mauger, 1966, p. 107). By Paleocene time the epicontinental seas, which once covered one-third of the total area of the present continents, had retreated from western United States to the Mississippi valley. The continuation of the upwarp throughout Cenozoic time is evidenced by the succession of marine sediments in eastern Mexico. Proceeding from the interior to the coast, a successively younger sequence of marine sediments can be observed, testifying to the continued retreat of the ocean boundary.

According to Menard (1964), as the Darwin Rise subsided, the East Pacific Rise emerged as an increasingly prominent feature of the Pacific Ocean basin. He has suggested that western United States represents the crest and eastern flank of the continental extension of the East Pacific Rise. The general shape of the rise off Central America and the warp of western North America compare favorably in bulk and slope (see figure 22). According to Damon and Mauger (1966, p. 107), 'the chronology of events affecting the floor of the Pacific Ocean shows a striking correlation with concurrent events on the continents.' Not only does crustal warping proceed concurrently and at about the same rate but volcanism appears to be concurrent. Thus, the continents and ocean basins are not completely decoupled, and gigantic warps combined with magmatism within one are coupled with similar events within the other.

When the Colorado Plateau and Basin and Range Province are viewed as part of the continental extension of the East Pacific Rise, important consequences follow naturally. The Colorado Plateau becomes an unrifted part of the flank of the East Pacific Rise which has risen steadily since Turonian time. It appears as a separate physiographic feature as a result of the rifting of the Basin and Range Province during the mid-Tertiary Basin and Range orogeny. The continued retreat of the epicontinental seas (figure 21) is an indication that uplift is proceeding at the present time.

Further discussion is not warranted here. Perhaps it is sufficient to point out that precise chronological data lead to precise correlations and these in turn may lead to a more consistent and effective framework within which to study further the evolution of our planet Earth.

V Concluding Remarks

The K–Ar method has its limitations as do all other dating methods including paleontology. For example, the excess argon problem may prohibit the precise dating of rocks derived from the abyssal environment. However, excess argon can be evaluated by analyzing minerals of low potassium content and distinguished from inherited argon. Inherited argon in turn can be used to evaluate the cooling history of a rock system.

One of the most important properties of K–Ar radioactive clocks is the remarkably different thermal diffusion properties of the different potassic minerals commonly found in rocks. For example, fine-grained biotite may be completely reset by an event which leaves a pegmatitic muscovite relatively unperturbed. Thus the dating of two or more minerals from a rock can lead to information concerning more than one geologic event.

The K–Ar method is a powerful chronological tool for the whole range of geologic time. Perhaps the greatest precision can be obtained in dating Mesozoic–Cenozoic volcanic rocks and hypabyssal plutons. However, geochronologists have applied it with success to dating Precambrian rocks and to assessing the age of meteorites; and at last, that most intractable episode of time, the Pleistocene, is yielding a precise chronology by K–Ar dating. Much interesting and rewarding work remains to be done.

Acknowledgments

This research was supported by the United States Atomic Energy Commission, Contract AT(11-1)-689, United States National Science Foundation Grant GP-3738, and the State of Arizona.

The author wishes to thank The Geological Society of America for permission to reprint Figures 14 and 15 from the *G.S.A. Bulletin*, Vol. 77, November 1966 (copyright 1966); and the Society of Mining Engineers of American Institute of Mining Engineers for permission to reprint Figures 21 and 22 from *Transactions AIME*, Vol. 235, March 1966 (copyright 1966).

References

Aldrich, L. T. and Wetherill, G. W. (1958). Geochronology by radioactive decay. *Ann. Rev. Nucl. Sci.*, **8**, 257.

Aldrich, L. T., Wetherill, G. W. and Davis, G. L. (1957). Occurrence of 1350 million-year-old granitic rocks in western United States. *Geol. Soc. Am. Bull.*, **68**, 655.

Amirkhanoff, K. I., Brandt, S. B. and Bartnitsky, E. N. (1961). Radiogenic argon in minerals and its migration. *Ann. N.Y. Acad. Sci.*, **91**, 235.

Anderson, C. A. (1951). Older Precambrian structure in Arizona. *Geol. Soc. Am. Bull.*, **62**, 1331.

Anderson, C. A. (1963). Simplicity in structural geology. In *The Fabric of Geology*. Addison-Wesley, Mass. p. 175.

Armstrong, R. L. (1963). Geochronology and geology of the Eastern Great Basin in Nevada and Utah. *Ph.D. Diss.* Yale Univ., New Haven, Conn.

Armstrong, R. L. (1966). K–Ar dating using neutron activation for Ar analysis: granitic plutons of the eastern Great Basin, Nevada and Utah. *Geochim. Cosmochim. Acta*, **30**, 565.

Armstrong, R. L. and Hansen, E. (1966). Cordilleran infrastructure in the eastern Great Basin. *Am. J. Sci.*, **264**, 112.

Baadsgaard, H., Lipson, J. and Folinsbee, R. E. (1961). The leakage of radiogenic argon from sanidine. *Geochim. Cosmochim. Acta*, **25**, 147.

Bassett, W. A. (1960). Role of hydroxyl orientation in mica alteration. *Geol. Soc. Am. Bull.*, **71**, 449.

Bikerman, M. (1965). Geological and geochemical studies of the Roskruge Range, Pima County, Arizona. *Ph.D. Diss.* Univ. Arizona, Tucson.

Bikerman, M. and Damon, P. E. (1966). K/Ar chronology of the Tucson Mountains, Pima County, Arizona. *Geol. Soc. Am. Bull.*, **77**, 1225.

Burke, W. H., Kenney, G. S., Otto, J. B. and Walker, R. D. (1963). Potassium argon dates, Socorro and Sierra Counties, New Mexico. *New Mexico Geol. Soc., Guidebook Ann. Field Conf.*, p. 224.

Butler, B. S. and Wilson, E. D. (1938). General features of some Arizona ore deposits. *Ariz. Bur. Mines Bull.*, **145**, 11.

Carr, D. R., Damon, P. E., Broecker, W. S. and Kulp, J. L. (1956). The potassium–argon age method. *Nuclear Processes*, Publ. 400. Natl. Acad. Sci.– Natl. Res. Council. p. 109.

Catanzaro, E. J. and Kulp, J. L. (1964). Discordant zircons from the Little Belt (Montana), Beartooth (Montana) and Santa Catalina (Arizona) Mountains. *Geochim. Cosmochim. Acta*, **28**, 87.

Cherdyntsev, V. V. (1956). *Abundance of Chemical Elements*. Univ. Chicago Press. Transl. from Russian 1961.

Chew, R. T. (1962). The Mineta Formation, a middle Tertiary unit in southeastern Arizona. *Ariz. Geol. Soc. Dig.*, **V**, 35.

Cobban, W. A. (1960). Upper Cretaceous. *McGraw-Hill Encyclopedia*. McGraw-Hill, New York. p. 545.

Cooper, J. A. (1963). The flame photometric potassium determination in geological materials used for potassium–argon dating. *Geochim. Cosmochim. Acta*, **27**, 525.

Cooper, J. A., Martin, I. D. and Vernon, M. J. (1966). Evaluation of rubidium and iron bias in flame photometric potassium determination for K–Ar dating. *Geochim. Cosmochim. Acta*, **30**, 197.

Creasey, S. C. and Kistler, R. W. (1962). Age of some copper-bearing porphyries and other igneous rocks in southeastern Arizona. *U.S. Geol. Surv., Profess. Papers*, **450-D**, D1–D5.

Dalrymple, G. B. (1964). Argon retention in a granite xenolith from a Pleistocene basalt, Sierra Nevada, California. *Nature*, **201**, 282.

Dalrymple, G. B., Cox, A. and Doell, R. R. (1965). Potassium–argon age and paleomagnetism of the Bishop Tuff, California. *Geol. Soc. Am. Bull.*, **76**, 665.

Damon, P. E. (1960). Terrestrial nuclear reactions. *McGraw-Hill Encyclopedia*. McGraw-Hill, New York.

Damon, P. E. (1965). Radioactive dating of Quaternary tephra. *Proc. Intern. INQUA Congr.*, 7th (in press).

Damon, P. E. and associates. (1966). Correlation and chronology of ore deposits and volcanic rocks. U.S. Atomic Energy Comm. Ann. Rept. No. COO-689-60. Univ. Arizona, Tucson.

Damon, P. E. and Bikerman, M. (1964). Potassium–argon dating of post-Laramide plutonic and volcanic rocks within the Basin and Range Province of southeastern Arizona and adjacent areas. *Ariz. Geol. Soc. Dig.*, **VII**, 63.

Damon, P. E., Erickson, R. C. and Livingston, D. E. (1963). K–Ar dating of Basin and Range uplift, Catalina Mountains, Arizona, Publ. 1075. Natl. Acad. Sci.–Natl. Res. Council. Nuclear Sci. Ser., Rept. 38. p. 113.

Damon, P. E. and Giletti, B. J. (1961). The age of the basement rocks of the Colorado Plateau and adjacent areas. *Ann. N.Y. Acad. Sci.*, **91**, 443.

Damon, P. E. and Green, W. D. (1963). Investigations of the helium age dating method by stable isotope-dilution technique. In *Radioactive Dating*. Intern. Atomic Energy Agency, Vienna. p. 55.

Damon, P. E. and Kulp, J. L. (1957). Argon in mica and the age of the Beryl Mt., New Hampshire pegmatite. *Am. J. Sci.*, **255**, 697.

Damon, P. E. and Kulp, J. L. (1958a). Insert gases and the evolution of the atmosphere. *Geochim. Cosmochim. Acta*, **13**, 280.

Damon, P. E. and Kulp, J. L. (1958b). Excess helium and argon in beryl and other minerals. *Am. Mineralogist*, **43**, 433.

Damon, P. E., Livingston, D. E., Mauger, R. L., Giletti, B. J. and Pantoja Alor, J. (1962a). Edad del Precambrico 'Anterior' y de otras rocas del zocalo de la region de Caborca-Altar de la parte noroccidental del Estado de Sonora. In *Estudios geocronologicos de rocas Mexicanas*. Instituto de Geologia, Mexico, *Boletin*, 64, p. 11.

Damon, P. E., Livingston, D. E. and Erickson, R. C. (1962b). New K–Ar dates for the Precambrian of Pinal, Gila, Yavapai and Coconino Counties, Arizona. 13th Field Conference, Mogollon Rim Region. *New Mexico Geol. Soc., Guidebook Ann. Field Conf.*

Damon, P. E. and Mauger, R. L. (1966). Epeirogeny-orogeny viewed from the Basin and Range Province. *Trans. AIME*, **235**, 99.

Damon, P. E., Mauger, R. L. and Bikerman, M. (1964). K–Ar dating of Laramide plutonic and volcanic rocks within the Basin and Range Province of Arizona and Sonora. *Proc. Intern. Geol. Congr., 22nd, New Delhi, India.*

Doell, R. R., Dalrymple, G. B. and Cox, A. (1966). Geomagnetic polarity epochs: Sierra Nevada data, 3. *J. Geophys. Res.*, **71**, 531.

Eardley, A. J. (1951). *Structural Geology of North America*. Harper, New York.

Endt, P. M. and Kluyver, J. C. (1954). Energy levels of light nuclei ($Z = 11$ to $Z = 20$). *Rev. Mod. Phys.*, **26**, 95.

Evernden, J. F. (1959). Dating of Tertiary and Pleistocene rocks by the potassium/argon method. *Geol. Soc. London Proc.*, No. **1565**, 17.

Evernden, J. F. and Curtis, G. H. (1965). The potassium–argon dating of late Cenozoic rocks in east Africa and Italy. *Current Anthropol.*, **6**, 343.

Evernden, J. F., Curtis, G. H. and Kistler, R. (1957). Potassium–argon dating of Pleistocene volcanics. *Quaternaria*, **IV**, 13.

Evernden, J. F., Curtis, G. H., Kistler, R. W. and Obradovich, J. (1960). Argon diffusion in glauconite, microcline, sanidine, leucite and phlogopite. *Am. J. Sci.*, **258**, 583.

Evernden, J. F. and James, G. T. (1964). Potassium–argon dates and the Tertiary floras of North America. *Am. J. Sci.*, **262**, 945.

Evernden, J. F., Savage, D. E., Curtis, G. H. and James, G. T. (1964). Potassium–argon dates and the Cenozoic mammalian chronology of North America. *Am. J. Sci.*, **262**, 145.

Fechtig, H., Gentner, W. and Kalbitzer, S. (1961). Argonbestimmungen an Kaliummineralien—IX: Messungen zu den verschiedenen Arten der Argondiffusion. *Geochim. Cosmochim. Acta*, **25**, 297.

Fechtig, H., Gentner, W. and Zähringer, J. (1960). Argonbestimmungen an Kaliummineralien—VII: Diffusionsverluste von Argon in Mineralien und ihre Auswirkung auf die Kalium–Argon-Altersbestimmung. *Geochim. Cosmochim. Acta*, **19**, 70.

Folinsbee, R. E., Lipson, J. and Reynolds, J. H. (1956). Potassium–argon dating. *Geochim. Cosmochim. Acta*, **10**, 60.

Friedlander, G., Kennedy, J. W. and Miller, J. M. (1964). *Nuclear and Radiochemistry*. Wiley, New York.

Gabrielse, H. and Reesor, J. E. (1964). Geochronology of plutonic rocks in two areas of the Canadian Cordillera. In Fitz Osborne, F. (Ed.), *Geochronology in Canada*. Univ. Toronto Press, Canada. p. 96.

Gast, P. W. and Hanson, G. N. (1963). Effects of contact metamorphism on Rb–Sr systems. In *Nuclear Geophysics*. Natl. Acad. Sci.–Natl. Res. Council. p. 25.

Gentner, W. and Kley, W. (1957). Argonbestimmungen an Kaliummineralien —IV. Die Frage der Argonverluste in Kalifeldspaten und Glimmermineralien. *Geochim. Cosmochim. Acta*, **12**, 323.

Gerling, E. K. (1961). The contemporary state of the argon method for the determination of age and its application to geology. Acad. Sci. USSR, Moscow (in Russian).

Gerling, E. K. and Morozova, I. M. (1957). Determination of activation energy of argon liberation from micas. *Geokhimiya*, No. **4**, 359.

Gerling, E. K. and Morozova, I. M. (1958). The kinetics of argon liberation from microcline-perthite. *Geokhimiya*, No. **7**, 775.

Gerling, E. K., Morozova, I. M. and Kurbatov, V. V. (1961). The retentivity of radiogenic argon in ground micas. *Ann. N.Y. Acad. Sci.*, **91**, Art. 2, 227.

Gilluly, J. (1949). Distribution of mountain building in geologic time. *Geol. Soc. Am. Bull.*, **60**, 561.

Gilluly, J. (1963). The tectonic evolution of the western United States. *Quart. J. Geol. Soc. London*, **119**, 133.

Gilluly, J. (1965). Volcanism, tectonism, and plutonism in the western United States. *Geol. Soc. Am., Spec. Papers*, **80**.

Glendenin, L. E. (1961). Present status of the decay constants. *Ann. N.Y. Acad. Sci.*, **91**, Art. 2, 166.

Goldich, S. S., Nier, A. O., Baadsgaard, H., Hoffman, J. H. and Krueger, H. W. (1961). The Precambrian geology and geochronology of Minnesota. *Minn. Geol. Surv., Bull.*, **41**.

Halva, C. J. (1961). A geochemical investigation of basalts in southern Arizona. *M.S. Thesis*. Univ. Arizona, Tucson.

Hamilton, E. I. (1965). *Applied Geochronology*. Academic Press, London and New York.

Hart, S. R. (1960). Some diffusion measurements relating to the K–A dating method. Eighth Ann. Prog. Rept. U.S. Atomic Energy Comm., Massachusetts Institute of Technology, Cambridge. p. 87.

Hart, S. R. (1961). The use of hornblendes and pyroxenes for K–Ar dating. *J. Geophys. Res.*, **66**, 2995.

Hart, S. R. (1964). The petrology and isotopic-mineral age relations of a contact zone in the Front Range, Colorado. *J. Geol.*, **72**, 493.

Hart, S. R. (1966). A test for excess radiogenic argon in micas. *J. Geophys. Res.*, **71**, 1769.

Hart, S. R. and Dodd, R. T., Jr. (1962). Excess radiogenic argon in pyroxenes. *J. Geophys. Res.*, **67**, 2998.

Hedge, C. E. (1960). Sodium–potassium ratios in muscovites as a geothermometer. *M.S. Thesis*. Univ. Arizona, Tucson.

Hinds, N. E. A. (1935). Ep-Archean and Ep-Algonkian intervals in western North America. Carnegie Inst. Washington, Publ. 463, 1.

Hinds, N. E. A. (1936). Uncomphagran and Beltian deposits in western North America. Carnegie Inst. Washington, Publ. 463, 53.

Hinds, N. E. A. (1938). Pre-Cambrian Arizonan revolution in western North America. *Am. J. Sci.*, **35**, 445.

Hurley, P. M., Hughes, H., Pinson, W. H., Jr. and Fairbairn, H. W. (1962). Radiogenic argon and strontium diffusion parameters in biotite at low temperatures obtained from Alpine fault uplift in New Zealand. *Geochim. Cosmochim. Acta*, **26**, 67.

Jost, W. (1960). *Diffusion in Solids, Liquids, Gases*. Academic Press, New York.

Kay, M. (1951). North American geosynclines. *Geol. Soc. Am., Mem.*, **48**.

Kulp, J. L. and Bassett, W. H. (1961). The base-exchange effects on potassium–argon and rubidium–strontium isotopic ages. *Ann. N.Y. Acad. Sci.*, **91**, Art. 2, 225.

Kulp, J. L. and Engels, J. (1963). Discordances in K–Ar and Rb–Sr isotopic ages: Discussion. In *Radioactive Dating*. Intern. Atomic Energy Agency, Vienna. p. 235.

Lance, J. F. (1960). Stratigraphic and structural position of Cenozoic fossil localities in Arizona. *Ariz. Geol. Soc. Dig.*, **III**, 155.

Laughlin, A. W. (1966). Excess radiogenic argon in minerals from the Amelia, Virginia pegmatites. *Trans. Am. Geophys. Union*, **47**, 197.

Lindgren, W. (1933). Differentiation and ore deposition, Cordilleran region of the United States. In *Ore Deposits of the Western States*. Am. Inst. Mining and Met. Engineers, New York.

Lippolt, H. J. and Gentner, W. (1963). K–Ar dating of some limestones and fluorites. In *Radioactive Dating*. Intern. Atomic Energy Agency, Vienna. p. 239.

Livingston, D. E. (1962a). Older Precambrian rocks near the Salt River Canyon, central Gila County, Arizona. *New Mexico Geol. Soc., Guidebook Ann. Field Conf.*, p. 55.

Livingston, D. E. (1962b). Strontium isotopes and rubidium–strontium ratios of igneous rocks and minerals from the Precambrian of Gila County. Abstract. *Trans. Am. Geophys. Union*, **43**, 447.

Livingston, D. E. and Damon, P. E. (1965). Isotopic ages from northern Sonora, Mexico. Abstract. *Geol. Soc. Am. Program, Ann. Meeting*, p. 96.

Livingston, D. E., Damon, P. E., Mauger, R. L., Bennett, R. and Laughlin, A. W. (1967). ^{40}Ar in cogenetic feldspar–mica mineral assemblages. *J. Geophys. Res.*, **72**, 1361.

Lovering, J. F. and Richards, J. R. (1964). Potassium–argon age study of possible lower-crust and upper-mantle inclusions in deep-seated intrusions. *J. Geophys. Res.*, **69**, 4895.

McDougall, I. and Green, D. H. (1964). Excess radiogenic argon in pyroxenes and isotopic ages on minerals from Norwegian eclogites. *Norsk Geol. Tidsskr.*, **44**, 183.

McDowell, F. W. (1966). Potassium argon dating of Cordilleran intrusives. *Ph.D. Diss.* Columbia Univ., New York.

McKee, E. D. (1951). Sedimentary basins of Arizona and adjoining areas. *Geol. Soc. Am. Bull.*, **82**, 481.

Mauger, R. L. (1966). A petrographic and geochemical study of Silver Bell and Pima Mining Districts, Pima County, Arizona. *Ph.D. Diss.* Univ. Arizona, Tucson.

Mauger, R. L., Damon, P. E. and Giletti, B. J. (1965). Isotopic dating of Arizona ore deposits. *Trans. AIME*, **232**, 81.

Mauger, R. L., Damon, P. E. and Livingston, D. E. (1966). Cenozoic argon ages on metamorphic rocks from the Basin and Range Province. U.S. Atomic Energy Comm. Ann. Rept. No. COO-689-60. Univ. Arizona, Tucson. p. A-I-1.

Mayo, E. B. (1958). Lineament tectonics and some ore districts of the Southwest. *Trans. AIME*, **211**, 1169.

Menard, H. W. (1964). *Marine Geology of the Pacific.* McGraw-Hill, New York.

Menard, H. W. and Ladd, H. S. (1963). Oceanic islands, seamounts, guyots and atolls. In *The Sea*, Vol. 3. Interscience, New York. p. 365.

Mielke, J. E. (1965). Trace element investigation of the 'Turkey Track' porphyry, southeastern Arizona. *M.S. Thesis.* Univ. Arizona, Tucson.

Naughton, J. J., Funkhouser, J. G. and Barnes, I. L. (1966). Fluid inclusions in potassium–argon age anomalies and related inert gas studies. *Trans. Am. Geophys. Union*, **47**, 197.

Nier, A. O. (1950). A redetermination of the relative abundances of the isotopes of carbon, nitrogen, oxygen, argon, and potassium. *Phys. Rev.*, **77**, 789.

Olmsted, F. H. (1964). Personal Communication.

Osmond, J. C. (1960). Tectonic history of the Basin and Range Province in Utah and Nevada. *Mining Eng.*, **12**, 251.

Pasteels, P. and Silver, L. T. (1965). Geochronologic investigations in the crystalline rocks of the Grand Canyon, Arizona. Abstract. *Geol. Soc. Am. Program, Ann. Meeting*, p. 122.

Priem, H. N. A., Boelrijk, A. I. M., Verschure, R. H., Hebeda, E. H. and Lagaay, R. A. (1966). Isotopic age of the quartz-diorite batholith on the Island of Aruba, Netherlands Antilles. *Geol. Mijnbouw*, **45e**.

Rama, S. N. I., Hart, S. R. and Roedder, E. (1965). Excess radiogenic argon in fluid inclusions. *J. Geophys. Res.*, **70**, 509.

Rankama, K. (1954). *Isotope Geology.* McGraw-Hill, New York.

Reynolds, J. H. (1957). Comparative study of argon content and argon diffusion in mica and feldspar. *Geochim. Cosmochim. Acta*, **12**, 177.

Rutherford, E., Chadwick, J. and Ellis, C. D. (1930). *Radiations from Radioactive Substances.* Cambridge Univ. Press, Cambridge.

Schilling, J. H. (1965). Isotopic age determinations of Nevada rocks. *Nevada Bur. Mines Rept.*, No. **10**.

Schmitt, H. (1933). Summary of the geological and metallogenetic history of Arizona and New Mexico. In *Ore Deposits of the Western States*. Am. Inst. Mining and Met. Engineers, New York.

Schreiner, G. D. L. and Verbeek, A. A. (1965). Variations in ^{39}K/^{41}K ratio and movement of potassium in a granite–shale contact region. *Proc. Roy. Soc. (London), Ser. A*, **285**, 423.

Schuchert, C. (1955). *Atlas of Paleogeographic Maps of North America*. Wiley, New York.

Silver, L. T. (1963). The use of cogenetic uranium–lead isotope systems in zircons in geochronology. In *Radioactive Dating*. Intern. Atomic Energy Agency, Vienna. p. 279.

Silver, L. T. (1964). Mazatzal orogeny and tectonic episodicity. Abstract. *Geol. Soc. Am. Program, Ann. Meeting*, p. 185.

Silver, L. T. (1966). U–Pb isotope relations and their historical implications in Precambrian zircons from Bagdad, Arizona. *Geol. Soc. Am. Program, Ann. Meeting*, p. 52.

Silver, L. T. and Deutsch, S. (1963). Uranium–lead isotopic variations in zircons: A case study. *J. Geol.*, **71**, 721.

Smales, A. A. and Wager, L. R. [Eds.] (1960). *Methods in Geochemistry*. Interscience, New York.

Smith, A. G. (1964). Potassium–argon decay constants and age tables. The Phanerozoic Time-scale. Symp. Geol. Soc. (London). Bartholomew Press, England. p. 129.

Smith, J. V. and Schreyer, W. (1962). Location of argon and water in cordierite. *Mineral. Mag.*, **33**, 226.

Stevens, J. R. and Shillibeer, H. A. (1963). Loss of argon from minerals and rocks due to crushing. *Proc. Geol. Assoc. Can.*, **8**, 71.

Stockwell, C. H. (1963). Third report on structural provinces, orogenies, and time-classification of rocks of the Canadian Precambrian shield. *Geol. Surv. Can. Paper*, **63-17**, 125.

Strominger, D., Hollander, J. M. and Seaborg, G. T. (1958). Table of isotopes. *Rev. Mod. Phys.*, **30**, 585.

Taylor, O. J. (1959). Correlation of volcanic rocks in Santa Cruz County, Arizona. *M.S. Thesis*. Univ. Arizona, Tucson.

Turekian, K. K. (1964). Degassing of argon and helium from the earth. In Brancazio, P. J. and Cameron, A. G. W. (Eds.), *The Origin and Evolution of Atmospheres and Oceans*. Wiley, New York.

Tweto, O. and Sims, P. K. (1963). Precambrian ancestry of the Colorado mineral belt. *Geol. Soc. Am. Bull.*, **74**, 991.

Umbgrove, J. H. F. (1947). *The Pulse of the Earth*. Martinus Nijhoff, The Hague.

Voshage, H. and Hintenberger, H. (1960). Calculation of cosmic-ray ages in the iron meteorites 'Carbo' and 'Treysa'. *Nature*, **185**, 88.

Wasserburg, G. J. (1954). Argon-40: Potassium-40 dating. In Faul, H. (Ed.), *Nuclear Geology*. Wiley, New York. p. 341.

Wasserburg, G. J. (1961). Crustal history and the Pre-Cambrian time scale. *Ann. N.Y. Acad. Sci.*, **91**, Art. 2, 583.

Wasserburg, G. J. (1964). Comments on the outgassing of the earth. In Brancazio, P. J. and Cameron, A. G. W. (Eds.), *The Origin and Evolution of Atmospheres and Oceans*. Wiley, New York.

Wasserburg, G. J., Hayden, R. J. and Jensen, K. J. (1956). ^{40}A–^{40}K dating of igneous rocks and sediments. *Geochim. Cosmochim. Acta*, **10**, 153.

Wasserburg, G. J. and Lanphere, M. A. (1965). Age determinations in the Precambrian of Arizona and Nevada. *Geol. Soc. Am. Bull.*, **76**, 735.

Wetherill, G. W., Aldrich, L. T. and Davis, G. L. (1955). ^{40}Ar/^{40}K ratios of feldspars and micas from the same rock. *Geochim. Cosmochim. Acta*, **8**, 171.

Wetherill, G. W., Wasserburg, G. J., Aldrich, L. T. and Tilton, G. R. (1956). Decay constants of ^{40}K as determined by the radiogenic argon content of potassium minerals. *Phys. Rev.*, **103**, 987.

Wilson, E. D. (1936). Pre-Cambrian Mazatzal Revolution in central Arizona. Abstract. *Geol. Soc. Am. Proc.*, p. 112.

Wilson, E. D. (1939). Pre-Cambrian Mazatzal Revolution in central Arizona. *Geol. Soc. Am. Bull.*, **50**, 1113.

Wilson, E. D. (1962). A résumé of the geology of Arizona. *Ariz. Bur. Mines Bull.*, **171**.

York, D. and MacIntyre, R. M. (1965). Excess ^{40}Ar in sodalite. Abstract. *Trans. Am. Geophys. Union*, **46**, 177.

A comparison of the isotopic mineral age variations and petrologic changes induced by contact metamorphism

S. R. HART,* G. L. DAVIS,† R. H. STEIGER‡ and G. R. TILTON§

I *Introduction*, 73. II *Geologic setting*, 74. III *Description of the Eldora Stock and contact zone*, 76. IV *Contact temperatures from heat flow theory*, 77. V *Sample collection and analytical procedures*, 79. VI *Effect of the contact metamorphism on the mineralogy*, 80. VII *Mineral age variations*, 84. VIII *Age variations as diffusion phenomena*, 97. IX *Comparison of mineral age stabilities and their relationship to metamorphic facies*, 106. *Acknowledgments*, 108. *References*, 109.

I Introduction

It is uncommon to find complete agreement between two or more mineral ages determined for a given geologic unit. Discordant mineral ages are almost a certainty in areas which have had metamorphic or polymetamorphic histories. 'True' ages can frequently be obtained in these areas by proper interpretation of zircon U–Pb and whole rock Rb–Sr results. A real understanding of the complete geologic history, however, depends on an adequate knowledge of the ways in which the various mineral ages are affected by geologic events.

* Dept. of Terrestrial Magnetism, Carnegie Institution of Washington, U.S.A.
† Geophysical Laboratory, Carnegie Institution of Washington, U.S.A.
‡ Division of Geological Sciences, California Institute of Technology, Pasadena, California, U.S.A.
§ Dept. of Geology, University of California, Santa Barbara, California, U.S.A.

Many studies have been made of discordant mineral age patterns in regionally metamorphosed areas (Tilton and others, 1958; Long and Kulp, 1962; Giletti, Moorbath and Lambert, 1961; Jager, 1962; Wetherill and others, 1962; Lanphere and others, 1964; Aldrich, Davis and James, 1965). Generally it is not possible in these areas to relate the age patterns to any specific environmental parameters. Temperatures can only be inferred from the mineral assemblages. Many mineral ages are affected at temperatures below those at which reliable mineralogic indicators of metamorphic grade develop. We decided to study mineral age relationships in a contact metamorphic zone, where relative temperatures are clearly established and absolute temperatures can be estimated from heat flow theory. Then mineral ages can be compared with each other, as a function of temperature, as well as being compared with the mineralogic changes which occur in response to the contact metamorphism. The contact chosen for study was near Eldora in the Front Range of Colorado, where the Eldora Stock of Tertiary age cuts Precambrian gneisses and schists of 1200 m.y.–1600 m.y. age. The ages of biotite, hornblende, potassium feldspar and zircon were measured by K–Ar, Rb–Sr and U–Pb isotopic techniques. Various aspects of this study have been reported previously (Hart, 1961; Doe and Hart, 1963; Hart, 1964; Steiger and Hart, 1966).

II Geologic Setting

The crystalline core of the Front Range is dominantly granite, schist and gneiss of Precambrian age. The oldest of these rocks is the Idaho Springs Formation, a thick series of metasediments and metavolcanics ranging in type from pelitic schist to amphibolite. The original sediments and volcanics were regionally metamorphosed and intruded by a series of Precambrian granitic and pegmatitic rocks. The grade of metamorphism is rather uniform throughout, and is characterized by the sillimanite–almandine subfacies of the almandine–amphibolite facies.

Cutting the Precambrian rocks are about ten major Tertiary stocks (figure 1). These Laramide stocks, the largest of which is 15 square miles in exposure, occupy a narrow belt some five miles wide extending southwestward across the Front Range. Many smaller irregular bodies and dikes and most of the early Tertiary ore deposits of the Front Range occur in a zone just southeast of the line of stocks. The linear structures within the stocks and the topographic exposure indicate that they have been intruded upward at a steep angle. Some of the stocks, such as the one west of Ward, Colorado, probably represent old volcanic throats.

Orogenic activity during the Laramide consisted mainly of broad arching, faulting and intrusion of the stocks. There is no evidence for any regional thermal metamorphism since Precambrian time. Further details concerning Front Range geology may be found in Lovering and Goddard (1950).

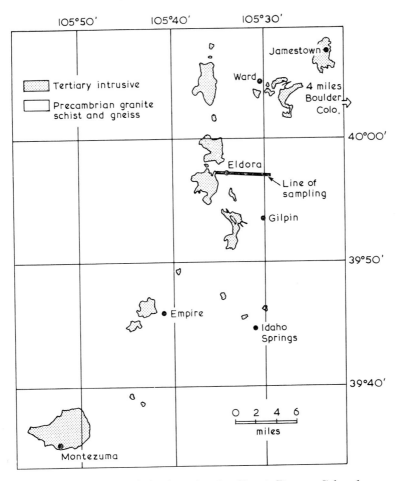

Figure 1 Tertiary intrusives in the Front Range, Colorado. Sampling traverse shown running east–west through Eldora, Colorado. (After Lovering and Goddard, 1950)

III Description of the Eldora Stock and Contact Zone

This contact was chosen because it is well exposed; the country rocks are unaltered and show wide compositional variation so that many different minerals can be investigated. A detailed geologic and

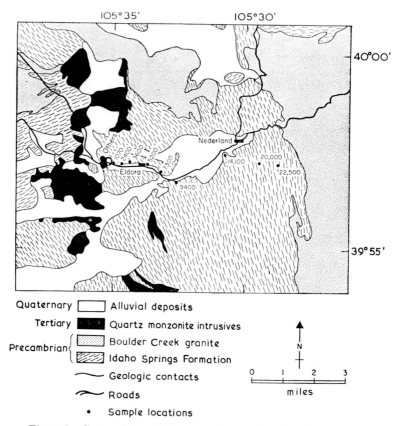

Figure 2 Composite geologic map showing Tertiary intrusives in the Eldora area, and locations of mineral age samples. (After Cree, 1948; Lovering and Goddard, 1950)

petrographic study of the Eldora Stock (Bryan Mountain Stock) has been made by Cree (1948). Figure 2 is a map showing the Eldora Stock and the other nearby Laramide intrusives.

The composition of the Eldora intrusive as a whole is not uniform. Quartz monzonite, granodiorite and syenodiorite predominate and grade into one another. According to Cree (1948), these do not represent separate intrusions but are formed from a common magma by differentiation and variable assimilation of country rock. The presence of this assimilation process is well demonstrated by the abundant inclusions and by the striking contact breccia developed along the western contact. Cree (1948) describes windows of stock within the Precambrian rocks, and inclusions and roof pendants of schist and gneiss within the intrusive. This suggests that the present erosion surface is not very far below the original roof of the stock.

The configuration of the Eldora Stock at depth and its relationship to the other nearby stocks is less certain. While the topographic relief reveals near-vertical contacts on all sides, it is clear that a heat source of some other configuration at depth cannot be ruled out. The heat flow results presented in a later section could prove oversimplified in this respect.

The sampling traverse starts near the northeast corner, where the intrusive is a quartz monzonite containing abundant xenoliths and ghost structures. Here the contact is steeply dipping and varies from very sharp to transitional over a foot or so.

The country rock in this vicinity is the Idaho Springs Formation. Its foliation is generally conformable with the contact, though vague and poorly defined in the first several feet from the contact. Compositionally, the dominant rock type is a quartz–feldspar–biotite gneiss. This alternates with biotite–quartz–feldspar schists representing segregation banding and primary layering. Amphibolites and rocks intermediate between these and the quartzo-feldspathic gneisses are also common. In addition to these metasediments, small bodies of pegmatite are ubiquitous. They are seldom more than a few feet thick, of small lateral extent, and are dominantly concordant. They consist of feldspar or quartz or both, with biotite but without muscovite and are interpreted as 'sweat' or 'secretion' pegmatites formed during the regional metamorphism of the Idaho Springs Formation. They are distinct from the large discordant pegmatites found further to the east which contain muscovite and are related to the Precambrian batholith on the eastern edge of the Front Range.

IV Contact Temperatures from Heat Flow Theory

Before discussing the variations in mineralogy and age, it will be helpful to develop heat flow models that can be applied to the Eldora contact zone. Of the variables involved in heat flow calculations,

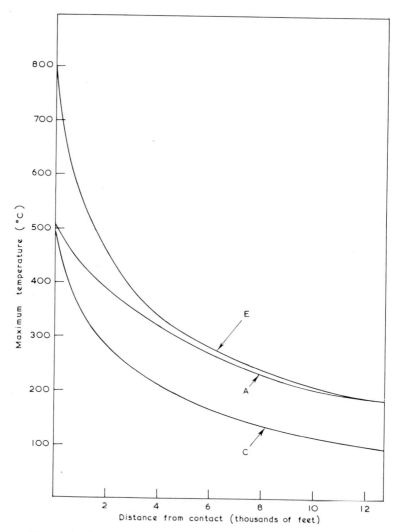

Figure 3 Curves of maximum temperature versus distance for three intrusive models. (A) infinite dike, 10,000 ft thick; (C) brick, 10,000 ft thick, 26,000 ft long, 3000 ft below roof; (E) Model A with convective cooling.

good estimates can be made of the magma temperature, the conductivities of intrusive and host rock and the latent heat of fusion. The largest uncertainty arises in the specification of the actual shape of the intrusive. Calculations can only be made for simple geometric forms. In the present case, calculations were made for two shapes which were chosen to bracket the actual Eldora case. In addition a third model was evaluated in which convective cooling was used.

Model A. Dike with infinite vertical extent, infinite length in the N–S direction and with a thickness equal to the average E–W dimension of the Eldora Stock. This model clearly represents the upper limit of temperature which can be attained at a given distance, assuming nonconvective cooling.

Model C. Brick with same E–W thickness as Model A, N–S length equal to size of Eldora and Caribou intrusives, and with large vertical extent downward only. Upward vertical extent limited to 3000 ft above present surface.

Model E. Same geometry as Model A, but with convective (isothermal) cooling for the first 200°c.

For all models a magma temperature of 780°c was used. This corresponds to the melting temperature of a granitic melt under a water-load pressure of about 6000 ft. The latent heat was taken as 80 cal/g, the ambient wall rock temperature as 35°c and the diffusivities of intrusive and wall rock as 0.009 cm²/sec. Details of the calculations are given in Hart (1964).

The curves for maximum temperature attained at a given distance versus distance from the contact for these models are shown in figure 3. At most distances, the temperature difference between the models is less than 100°c. Near the contact, Model E predicts very high temperatures, temperatures which are in fact incompatible with the petrographic observations, especially those regarding the homogenization of the perthitic feldspars. While there is evidence in the mineral age patterns for some convective cooling, the Model E temperatures are excessive in this respect. It is possible that temperatures several feet from the contact were in excess of 600–700°c, but otherwise we believe that a more realistic upper limit to contact temperatures is given by Model A.

V Sample Collection and Analytical Procedures

The age measurements were carried out using stable isotope-dilution techniques, described fully in Hart (1964) and Tilton and others (1957). In most cases the analytical uncertainty in a given age is approximately ±3%.

A suite of samples for age determination was collected along an east–west line, starting from the northeast contact (figure 2). These samples are referred to by numbers which are distances from the contact in feet. Pegmatitic feldspars were also collected along five other traverses for mineralogical study. The results of the petrographic and mineralogical study of the contact effects will be presented first, followed by a discussion of the various mineral age variations.

VI Effect of the Contact Metamorphism on the Mineralogy

The mineralogical changes brought about by the contact heating are very slight, consisting mainly of alteration effects within a few feet of the contact, a slight color change in the biotites at moderate distances, and a transition in the potassium feldspars from microcline to orthoclase.

A *Amphibolites*

Table 1 lists the mineral assemblages observed in the amphibolites. For the most part this varied mineralogy is believed to be a

Table 1 Summary of amphibolite mineral assemblages

Major minerals given in order of abundance, plagioclase given by composition, quartz and magnetite common in most samples	
2	Hornblende, oligoclase–andesine, biotite
11	Hornblende, An 48
134	Hornblende, An 45, diopside
248	Hornblende, An 48, diopside
230	Hornblende, An 44
950	Hornblende, An 60, diopside
1070	Hornblende, diopside, An 52
1130	Hornblende, An 48
2400	An 48, biotite, hornblende
3600	Hornblende, plagioclase
5200	Diopside, An 62
7300	Hornblende, biotite, An 28
9400	Hornblende, diopside, An 48
20,000	Hornblende, plagioclase

Isotopic mineral age variations and petrologic changes 81

function of rock composition and not a contact effect. The assemblages are consistent with the almandine–amphibolite facies Precambrian regional metamorphism of the area. Sample 2 (2 ft), however, shows marked contact effects. The hornblende has been partially altered to biotite, the twinned oligoclase has untwinned sodic overgrowths and the sample contains abundant sphene.

B Biotites

A faint color change was observed in biotites from both the pegmatites and the gneisses, the color varying from orange-brown beyond 3000 ft to brown nearer the contact. This color change is correlated with the presence of abundant rutile clusters in the near-contact biotites, whereas biotites from beyond 3000 ft are clear and free of rutile. We believe this change to be related to the contact metamorphism, with the orange color being due to titanium in solution in the biotite which has exsolved on heating to form rutile.

C Feldspars

The most striking mineralogical change brought about by the contact heating is the transition of microcline to orthoclase. This change will be described in some detail because of its bearing on the geometry and the temperature distribution around the intrusive stock.

A number of field studies have attempted to correlate geologic temperatures with the microcline–orthoclase transition. In 1952 MacKenzie and Laves each demonstrated the variable obliquity of microcline. Since then a general correlation between obliquity and temperature of formation has been suggested by several investigators. In the only published study of these relationships in a contact aureole, Heier (1957) showed that the obliquity of microcline had been considerably lowered in the thermal aureole around the Permian igneous rocks of the Oslo region. He also correlated the boundary between regional amphibolite and granulite facies rocks with the transition from microcline to orthoclase and suggested a temperature of 500°C for this. There are very few definite experimental data concerning the temperature or nature of this transition. A temperature of 500°C has been more or less accepted since Goldsmith and Laves (1954) found that microcline could be hydrothermally altered to orthoclase at 525°C but not at 500°C. More recently, Tomisaka (1962) found that the obliquity of microcline heated at 460°C under water pressure slowly decreased with time, but this still represents an upper limit since the reaction has never been reversed.

At Eldora, this transition has been observed along seven different traverses, at distances ranging from 800 to 3000 ft from the contact. The transition was studied in the feldspar samples, usually from pegmatites, by both optical and x-ray methods. (A complete discussion of techniques and results may be found in Steiger and Hart,

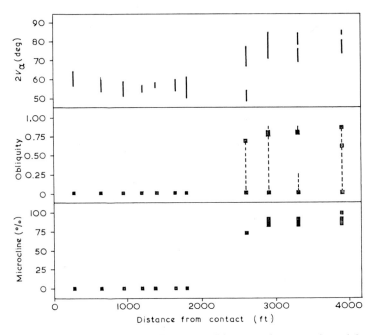

Figure 4 Ute Mountain Traverse: Diagram shows optic axial angles, obliquity determined by x-ray from $\Delta(1\bar{3}1$–$131)$, and microcline–orthoclase proportions as a function of distance from the contact.

1966.) The variation of optical and x-ray properties along one of the traverses is illustrated in figure 4. The other traverses show similar though often less sharp transitions in the properties shown in figure 4. The areal distribution of feldspar types on all the traverses is shown in figure 5. There is a rather well-defined zone near the contact where orthoclase is predominant. A line marking the first appearance of orthoclase could be drawn which would represent an orthoclase 'isograd.' This line would also represent an isothermal contour, since the orthoclase–microcline transition is sensitive to temperature and relatively insensitive to changes in bulk composition and water pressure. The isograd occurs at various distances from the contact, but can be seen to be related to the shape of the intrusive. Near convex contacts (Bryan Mountain) the 'isograd' is closest to the contact. Near concave contacts (Ute Mountain) the 'isograd' can be at considerably greater distances because of the higher heat flux in these

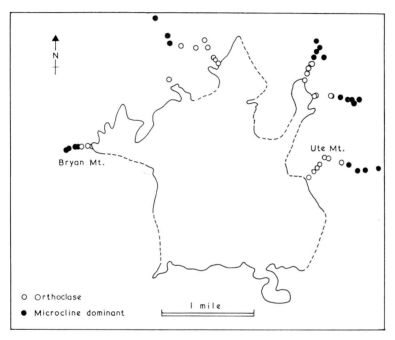

Figure 5 Map of Eldora intrusive showing types of feldspar along five traverses.

areas. Within the limits of the observations the 'isograd' appears to be symmetric on both sides of the intrusive, showing that the subsurface shape of the intrusive is not grossly unsymmetric. This supports the use of simplified shapes in the heat flow calculations.

The temperature of the transition may be estimated from the heat flow curves of figure 3. The eastern contact, which most closely approaches the planar dike case, has the transition about 3000 ft from the contact. The temperature for the transition then would be 400°c or less, even using the extreme Model E convection case. This is

50–100°C lower than the laboratory values and suggests that the transition is being kinetically limited at low temperatures in the laboratory experiments. In any event, the feldspar transition in this contact zone provides a good marker against which the mineral age can be compared.

VII Mineral Age Variations

To properly evaluate the age results in the contact zone, the background level of ages in the unaffected Precambrian must be established first. Previous investigators (Aldrich and others, 1958; Giffin and Kulp, 1960) have reported Rb–Sr and K–Ar mica ages from the Front Range ranging from 1000–1500 m.y. Our own results on the Idaho Springs Formation and its included pegmatites are summarized in table 2. The spread of ages clearly indicates a complex Precambrian

Table 2 Ages of minerals from the Idaho Springs Formation

Unit	Age (m.y.)[a]	Mineral and method
Idaho Springs Fm.	1585	Zircon, ^{207}Pb–^{206}Pb age
	1610	Zircon, 'concordia' intercept
	1375	Hornblende, K–Ar
Secretion pegmatites in Idaho Springs Fm.	1400	Potassium feldspar, Rb–Sr isochron
	1170	Biotite, K–Ar
	1150	Muscovite, K–Ar
	1280	Biotite, Rb–Sr
	1350	Muscovite, Rb–Sr

[a] Ages have analytical uncertainties of ± 30–40 m.y.

history for these rocks. The Pb–Pb zircon ages of 1600 m.y. may represent the time of the high-grade regional metamorphism or they may reflect the age of the source rock, since these zircons were originally detrital. The agreement of the hornblende K–Ar, muscovite and feldspar Rb–Sr ages at about 1350–1400 is probably meaningful, since these minerals are normally quite resistant to metamorphic effects. We believe this age of 1350–1400 m.y. is the time when the secretion pegmatites were formed, and is therefore the age of the regional metamorphism. The younger biotite K–Ar and Rb–Sr ages in the region then represent either a later heating event or the time when regional temperatures had decreased to a point where the biotite could retain its argon and strontium.

It is significant that no ages have been found in the Front Range which are less than 1000 m.y., except in the immediate vicinity of the Tertiary intrusives. We can therefore conclude that the dispersion of Precambrian ages mentioned above is associated with events occurring within Precambrian time and is not a result of any Laramide regional metamorphism.

The time of the contact metamorphism at Eldora is obtained by measuring the intrusive itself. A coarse biotite fraction gave a K–Ar age of 55 m.y., and a finer biotite–hornblende fraction gave an age of 54 m.y. At Eldora, therefore, we have a Precambrian terrain showing ages of 1200–1600 m.y., intruded by a stock 55 m.y. old. The ages of all the minerals from the contact zone analyzed in this study clearly show the effects of this 55 m.y. heating event. The different ways in which the minerals have responded to the contact metamorphism are illustrated in figures 6, 7, 8 and 9. It is convenient to discuss these curves in terms of the individual minerals first, then draw comparisons later.

A *Biotite ages*

The age results are given in table 3 and shown in figure 6. The coarse biotite age curves are very regular, with the effects of the heating noticeable out to more than 7000 ft. The unaffected Rb–Sr ages are greater than the highest K–Ar ages (1280 vs. 1170 m.y.), perhaps due to uncertainties in the rubidium decay constant or to variable response during the Precambrian metamorphic events. Even accounting for this difference, the Rb–Sr ages in the contact zone have been much less affected than the K–Ar ages, showing significantly higher ages at all distances. Even 20 ft from the contact, where the K–Ar age represents almost complete loss of argon, the Rb–Sr age is still in excess of 200 m.y.

The fine biotites are not plotted in figure 6, but the data are given in table 3. The coarse biotites are dominantly from the pegmatitic phases of the Idaho Springs Formation. The fine biotites are from various phases of the quartzo-feldspathic schists and gneisses and represent smaller original grain sizes, not simply a finer sieve fraction. The effect of the grain size is clearly shown by comparing the fine biotite ages and the coarse biotite ages at all points. The actual difference in grain size is more than sufficient to account for the age differences if one assumes that the loss of strontium and argon follows a regular diffusion law with its inverse radius-squared dependence.

A simple experiment was performed to demonstrate this grain-size effect in biotite and to show that the low Rb–Sr ages are indeed

Table 3 Mineral ages in the contact metamorphic zone. (Reproduced, by permission, from Hart, 1964)

Sample number	Coarse biotite (m.y.)		Fine biotite (m.y.)		Hornblende (m.y.)	Feldspar (m.y.)
	K–Ar	Rb–Sr	K–Ar	Rb–Sr	K–Ar	K–Ar
BCOA (intrusive)	54.8	63 ± 9	53.7			
2					120	78.5
11					956	
20		201				298
58	68.4	304	76.0	149		
85						270
134					1172	
248	80.6	435			1180	516
630				102		
950			59.0	97.3	1080	
1070			203	277		272
1130	397	788			1220	
2400	540	831	407	716		477
Repurified		853				
3600	771	1089				
5200	1023	1204				
7300	1058	1247				
9400					1375	
14,100	1172	1275				
			Muscovite			
22,500	1168	1291	1150	1365		916

Rb–Sr ages were calculated using a decay constant of 1.39×10^{-11} yr^{-1}. Other constants used can be found in Hart (1964).

caused by a diffusion loss of strontium and not by addition of rubidium instead. A number of biotite flakes, about 8 mm diameter × 0.25 mm thickness, were chosen from sample 248. These were punched so as to provide a central disc and several concentric rings in sufficient amount to allow analysis for Rb and Sr. Figure 7 shows the variation in the concentration of K, Rb and Sr and in the Rb–Sr age between the centers of these flakes and the outside edges. The

Rb–Sr age of the centers is more than twice that of the edges, the difference being clearly caused by a drop in the radiogenic strontium content toward the edge. The rubidium (and potassium) is essentially constant throughout. This is exactly what one expects if diffusion in biotite takes place parallel to the cleavage and is controlled by the physical grain size and not by smaller mosaic units. In theory one

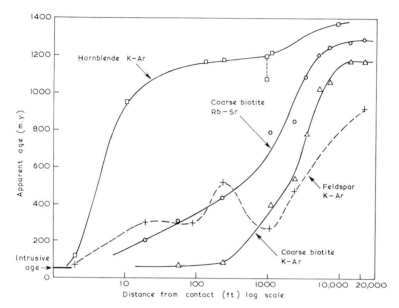

Figure 6 Variation of mineral ages as a function of distance from intrusive contact. (Reproduced, by permission, from Hart, 1964)

could use a test such as this to differentiate between true or metamorphic ages and mixed 'in-between' ages. Only the biotite in which partial loss of strontium or argon had taken place would show a difference in age between center and edge.

B *Hornblende ages*

The K–Ar ages for hornblende are presented in table 3 and shown in figure 6. As in the case for biotite the variation in age is quite regular, but with the largest effects limited to the zone within 10 ft of the contact. The ages between 100 ft and 1000 ft are lower than the age of 1375 m.y. on the most distant sample, but it is not possible to

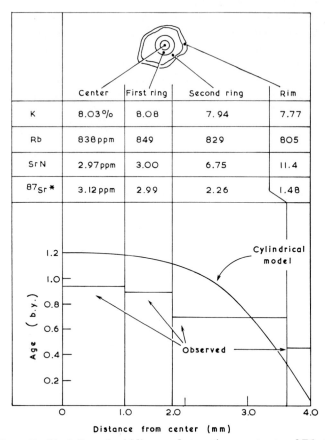

Figure 7 Variation of rubidium and strontium content and Rb–Sr age within biotite plates as a function of distance from the center of the plates. The theoretical curve for diffusion parallel to cleavage (cylindrical) is indicated for comparison. (Reproduced, by permission, from Hart, 1964)

tell whether this is a contact effect or a difference relating back to the Precambrian metamorphic events. The fact that three of the samples between 100 and 1000 ft are the same age (1200 m.y.) within analytical error suggests that these samples have not been affected by the contact metamorphism. The low age of 1080 m.y. on sample 950 is caused by the presence of biotite impurity, which if corrected for would raise that hornblende age to about 1200 m.y. also.

Since grain size was found to be important in the biotite, some simple heating experiments were conducted on hornblende sample

134 to see if grain size is also important in the case of hornblende. Two fractions were prepared, one of grains ranging from 140 to 330 μ in size and another of grains finely ground to sizes from 44 to 53 μ. This is an average difference in radius of about 5, or an expected difference in diffusion loss during heating of about a factor of 25 if the diffusion follows a radius-squared dependence. Both samples were heated at 860°c for about 10 hours, and both lost between 6 and 7% of their argon. In other words, the diffusion loss was essentially independent of grain size. This can only mean that diffusion is occurring from crystalline subunits of hornblende not related to the physical grain size. This is similar to the more extensive results obtained on microcline and sanidine by Evernden and others (1960), who showed that diffusion in these feldspars was also controlled by a subunit mosaic unrelated to the gross physical size. As these subunit sizes probably depend strongly on the thermal and stress history of the mineral, the retention of argon by hornblende may vary considerably. The very high argon retention of the Eldora hornblendes is not representative of all hornblendes, as some do not retain argon any better than biotite. We may infer that these nonretentive hornblendes are structurally very imperfect, with subunits of very small size. The study of mineral imperfections by argon diffusion experiments may some day yield very interesting data on thermal-stress histories of rocks.

C Feldspar ages

For the potassium feldspars, in addition to K–Ar and Rb–Sr ages, model or common lead ages were also determined, thus enabling comparisons to be made between the diffusion behavior of the three elements, argon, strontium and lead.

The K–Ar ages are given in table 3 and shown in figure 6. Unlike the biotite and hornblende curves, the K–Ar feldspar curve is rather irregular. (The curve serves only to join the experimental points and does not imply loci for ages at other points.) It should first be pointed out that potassium feldspar argon ages are usually 20–40% lower than those of other minerals, even in areas where no metamorphic history is indicated. This effect has been studied and commented on by several earlier investigators (Wetherill, Aldrich and Davis, 1955; Wasserburg, Hayden and Jensen, 1956; Sardarov, 1957). The age of 916 m.y. at 22,500 ft is probably related to this mica–feldspar discrepancy and not to the contact heating. The other feldspar K–Ar ages, though, are undoubtedly related to the contact metamorphism. It is interesting to note that three of the feldspars in the contact zone show argon ages that are higher than the biotite K–Ar ages at the same distance. This is the reverse of the usual feldspar–mica discrepancy and has been noted before in metamorphic areas by Goldich, Baadsgaard and Nier (1957). The age of 78 m.y. for the 2 ft feldspar sample indicates some slight retention of argon in spite of the high temperatures at that point. In addition, the feldspars in the zone from 20 to 1000 ft show partial retention of argon even though they have undergone the microcline to orthoclase transition, and have been at least partially homogenized from their original perthitic state. These facts suggest that at least some part of the argon in feldspar is bound very tightly.

Experimental proof of the tightly bound nature of part of the argon in feldspar is given by Amirkhanov, Brandt and Bartnitskii (1959). They found that all of the argon in feldspar does not diffuse at a single value of the diffusion coefficient but could be separated into three or more 'phases'. Typically, most of the argon was contained in the first two phases, with a small amount (5%) in the third phase. The difference in value for the diffusion coefficients of the different phases of argon was frequently four orders of magnitude! We believe the irregular behavior of the feldspar argon ages in the contact is caused in part by the existence of these various 'phases' of argon. Variations in the subunit diffusion size of the feldspar from sample to sample may also contribute to the irregular nature of the age curve.

The Rb–Sr data for the feldspars is presented graphically in figure 8. Ages were not calculated individually because the feldspars are all less than 6% radiogenic. Taken together, however, the data provide some information regarding the effects of contact metamorphism on feldspar Rb–Sr ages. All but one of the samples lie on a 1400 m.y. isochron, within the limits of analytical error. Because four of the points are colinear and do define a reasonable age, we conclude that they have been unaffected by the metamorphism. The 20 ft sample, however, lies distinctly above the line and on that basis may be said to have been affected by the contact heating. The departure from the line is not a result of loss of rubidium, as all the contact feldspars have a rubidium content of 300 ppm, with less than $\pm 3\%$ variation. The fact that sample 20 lies above the line (i.e. has gained radiogenic strontium instead of losing it) can be explained by the simple mineralogy of the pegmatites. Any radiogenic strontium which diffuses out of the biotite should end up in the other pegmatite minerals, especially those that accept strontium readily into their structure. The feldspars contain a large quantity of original or 'common' strontium so it is clear that they will accept radiogenic strontium as well. Epidote and apatite are minerals which also should readily accept radiogenic strontium into their structure. Small amounts of epidote and apatite were separated from the 2400 ft pegmatite sample and as shown in

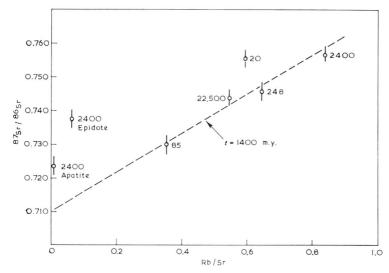

Figure 8 Isochron plot of feldspar Rb–Sr data. Sample numbers are distances from contact in feet. Estimated ordinate error indicated by vertical line. (Reproduced, by permission, from Hart, 1964)

figure 8 they do in fact contain excess amounts of radiogenic strontium (^{87}Sr). This shows that the isotopic composition of strontium in epidote and apatite is more easily affected by contact metamorphism than that in feldspar since the feldspar from 2400 ft lies on the isochron and has presumably been unaffected.

Because most pegmatitic feldspars contain easily measurable amounts of lead, it was possible to determine the isotopic composition of lead in these contact feldspars. Because feldspar contains very little uranium, the lead isotope ratios will not change appreciably with age. In undisturbed systems these initial ratios can be used to calculate model lead ages and to study various aspects of the petrogenesis of igneous rocks. It is therefore important to know to what extent a metamorphic episode will disturb the feldspar lead system. The data are plotted in figure 9 on a ^{206}Pb/^{204}Pb–^{208}Pb/^{204}Pb diagram. A similar diagram has been plotted for ^{207}Pb/^{204}Pb–^{206}Pb/^{204}Pb and may be found in Doe and Hart (1963), along with the original data and discussion.

In figure 9, the points for samples 2400, 7300 and 22,500 lie rather close together and probably represent the composition of the original

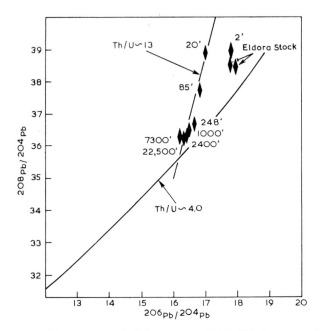

Figure 9 Plot of ^{208}Pb/^{204}Pb versus ^{206}Pb/^{204}Pb for potassium feldspars from Precambrian pegmatites and Eldora Stock. The line labeled Th/U ~ 4.0 is the normal growth curve for conformable galenas. The line labeled Th/U ~ 13 is a best-fit mixing line for samples 20 to 22,500. Size of diamond represents experimental uncertainty.

lead in these feldspars. The other points are more widely separated, reflecting the effect of the contact heating. If the samples had been altered by the activity of solutions carrying intrusive-type lead the points would lie on a line joining the intrusive lead and the unaffected leads. Except for sample 2, the points lie on a line which does not pass through the intrusive lead point, showing that the lead which was added to the feldspars came from some other source. Since the slope of the mixing line indicates a thorium–uranium ratio of about

Isotopic mineral age variations and petrologic changes 93

13, the source of this lead was probably local minerals or interstitial sites rich in thorium. The feldspar at 2 ft has a lead content similar to the intrusive lead and probably represents metasomatic activity at that distance, supporting the petrographic conclusions cited earlier.

Another way to illustrate the contact effects on lead in feldspar is to calculate the model lead ages for each sample and plot these

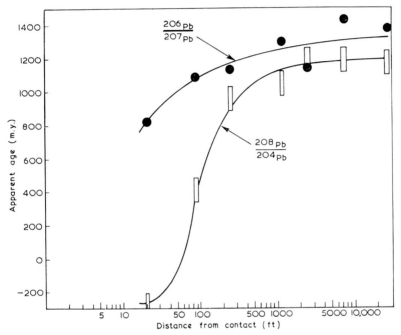

Figure 10 Variation of feldspar model lead ages as a function of distance from the contact. Analytical uncertainties in the model ages indicated by the vertical extent of the circles and boxes.

against distance from the contact. Figure 10 shows the $^{206}Pb/^{207}Pb$ isochron model ages and the $^{208}Pb/^{204}Pb$ model ages, taken from Doe and Hart (1963). The small scatter in the ages beyond 1000 ft is due to original variations of lead isotopic composition, reflecting the complex Precambrian history. Sample 248 shows a rather small contact effect, but samples 85 and 20 show very distinct effects, especially in the $^{208}Pb/^{204}Pb$ model ages.

It appears that lead in feldspar is more easily disturbed than strontium in feldspar, because the 20 ft sample was the only one in which the strontium was altered. A similar case has been described from a regional metamorphic area by Doe, Tilton and Hopson (1965), where the lead isotopic composition in feldspar has been altered without changing the Rb/Sr age of the feldspar. Lead and strontium have the same charge, about the same ionic radius and both are easily accepted into the feldspar structure. (The contact feldspars averaged about 60 ppm common lead and 500 ppm common strontium.) On the basis of the feldspar itself there seems to be no reason why the two elements should behave differently. The flux of a diffusing species depends on the chemical potential gradient as well as the diffusion coefficient. If the diffusion coefficients for lead and strontium in feldspar are similar, then we must resort to a high external chemical potential for radiogenic lead to explain its high mobility.

D *Zircon ages*

Zircon ages present something of an enigma when their resistance to metamorphism is considered. Concordant zircon ages have been reported (Tilton and others, 1958) in an area of kyanite grade regional metamorphism, showing that zircons can be very resistant to thermal metamorphism. However, highly discordant zircon ages have also been found in many areas where the absence of even very low-grade metamorphism is demonstrable (Tilton, 1960). Tilton suggested that, in the latter case, continuous diffusion of lead from zircon had taken place over long periods of time at low temperatures, and that the diffusion was characterized by a low activation energy. This means that the diffusion is not strongly temperature dependent, so that zircons could then effectively withstand short periods at high temperature. A contact metamorphic zone is an ideal place to study this problem.

Figure 11 shows the zircon ages ($^{238}U-^{206}Pb$ and $^{207}Pb/^{206}Pb$) in the Eldora contact zone. (Data and further discussion may be found in Tilton and others, 1964, and Davis, Hart and Tilton, 1967.) It is clear that the ages have been strongly affected in the zone within 25 ft of the contact. Another way of looking at the ages is by use of the concordia diagram, figure 12, developed by Wetherill (1956). Concordant zircons, that is, zircons in which the $^{238}U-^{206}Pb$, $^{235}U-^{207}Pb$ and $^{207}Pb/^{206}Pb$ ages are all equal, would lie on the curve labeled 'concordia.' The contact zircons are discordant, as shown above, so they do not lie on 'concordia.' However, they do lie on a single straight line which intersects concordia at two points—1600 m.y. and 60 m.y.

The 1600 m.y. intercept represents the original 'unaffected' age for the Eldora zircons. The 60 m.y. intercept represents the time at which the zircon U–Pb systems were disturbed, in this case by the contact

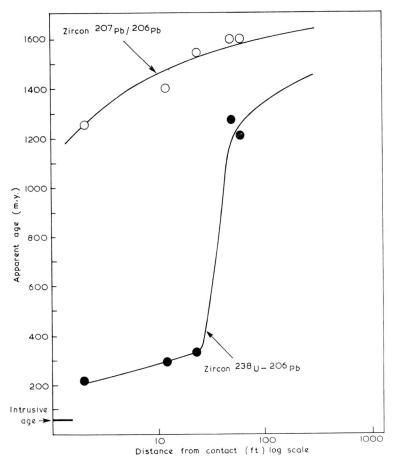

Figure 11 Variation of zircon ages with distance from the contact. The ^{235}U–^{207}Pb points are not shown, but would lie somewhat above the ^{238}U–^{206}Pb curve.

metamorphism. More complex interpretations are possible but do not seem justified in view of the very close fit of the zircon points to the single line. The curve for continuous diffusion is marked 'discordia' and it is clear that the Eldora zircon data cannot be interpreted in terms of continuous diffusion alone.

While it is straightforward to relate the Eldora zircon ages to the contact metamorphism, it is more difficult to pinpoint the exact mechanism by which the zircon ages were altered. The lowering of age is caused both by gain of parent (U, Th) and loss of the daughter (Pb) elements, as shown in figure 13. The loss of lead is about a factor of three, the gain of uranium a little less than a factor of two, and the gain of thorium about a factor of seven. Does this mean that the

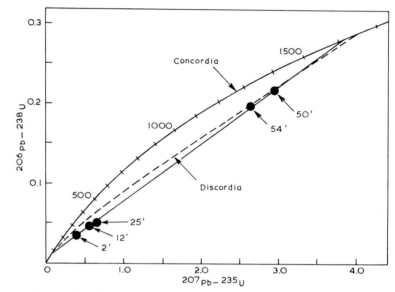

Figure 12 Concordia diagram showing uranium–lead ratios for zircons from the Eldora contact zone. The chord fitted to the points illustrates the pattern expected for 1600 m.y. zircons that have lost varying proportions of lead episodically about 70 m.y. ago.

original zircons incorporated uranium (and thorium) into their structure or does it mean that new zircons were actually formed during the metamorphism? The zircons change morphology markedly on approaching the contact. In sample 2 the zircons are all euhedral, in sample 25 about 50% of them are euhedral and at 50 ft they are essentially all rounded. This change clearly indicates at least a recrystallization of the original detrital zircons if not actual crystallization of new zircon. The fact that this recrystallization effect did not also occur previously during the Precambrian regional metamorphism suggests either that the temperatures within 30–40 ft of the contact reached higher levels than those which prevailed during the regional Precambrian heating, or that the intrusive supplied new material from which zircon could be crystallized.

Figure 13 also shows that the material added to the zircons during the contact heating was derived from a source with a high Th/U ratio, since the thorium increases much more than the uranium on nearing the contact. Because the zircons were detrital, their original Th/U ratio might be expected to be out of equilibrium with the Th/U

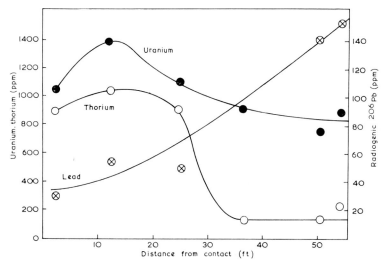

Figure 13 Change in concentration of uranium, thorium and radiogenic lead in zircon as a function of distance from the contact.

ratio of the rock in which the zircons now reside. Unusual amounts of monazite are common in the Idaho Springs Formation, and this would certainly tend to raise the Th/U ratio of the environment of the zircon. Reequilibration of the zircons and host rock might then have taken place during the contact metamorphism. Evidence for a high Th/U ratio in the country rocks was also remarked upon earlier, in the discussion of the common lead variation in the feldspars.

VIII   Age Variations as Diffusion Phenomena

For most of the minerals studied in the Eldora contact zone the age variations can be described as a simple diffusion loss of daughter product (argon, strontium and lead). For the feldspars this statement is oversimplified as an addition or mixing of the lead and strontium daughter products is indicated. However, the transfer can still be described as a simple diffusion process. Zircon is the only mineral for which the evidence demonstrates a movement of both parent and daughter elements. It is also the only mineral which shows morphological or petrographic evidence of recrystallization or new growth. Therefore, in constructing models for the observed age variations we feel a diffusion mechanism is sufficient for description of these variations, without resorting to the more complex reaction-rate mechanisms.

A *Diffusion in the contact biotites*

The process of diffusion is quantitatively described by the diffusion coefficient of a given element. This diffusion coefficient is a function of temperature, and relates the amount of diffusive loss to the duration of heating. We can calculate diffusion coefficients for each of the contact biotites by using the ages to determine the amount of loss and by using the heat flow calculations to derive the length of time of heating. This has been done, assuming spherical diffusion geometry. (For biotite it would perhaps be more accurate to use a cylindrical geometry, but the differences are not important as long as a consistent geometry is used throughout.) The diffusion coefficient for each sample is then plotted against a characteristic temperature for the sample, which in this case is taken to be the maximum temperature attained during the heating as given by the models of figure 3. These results are shown in figure 14 for argon and strontium in biotite.

On a log diffusion coefficient versus reciprocal absolute temperature plot such as figure 14, simple diffusion mechanisms appear linear, with an absolute slope proportional to 'activation energy.' The points for argon and strontium both fit straight line relationships rather well, proving that the data are consistent with a simple diffusion mechanism. They are also consistent in this respect with first- and second-order reaction-rate mechanisms, as can be shown by similar calculations. The choice of heat flow geometry is not critical as both Model A and Model C result in linear arrays. The slopes differ for these two models, however, with Model A showing activation energies of about 30 kcal/mole and Model C showing activation energies of 16 kcal/mole. It should be noted that for either case, the slopes of the argon and strontium curves are essentially identical. This means that the 'activation energy' for argon and strontium in these biotites is

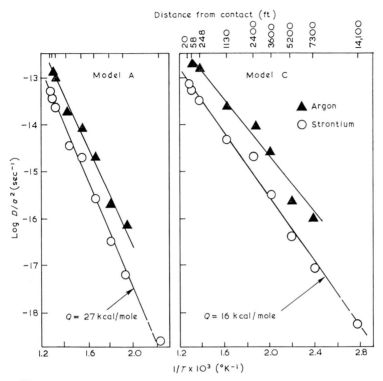

Figure 14 Diffusion coefficients (spherical) of argon and strontium in biotite plotted versus reciprocal absolute temperature. The temperatures are derived from the heat flow models A and C.

the same. The diffusion coefficients, however, differ by about a factor of five and it is this factor which has caused the large differences in the K–Ar and Rb–Sr ages of these biotites.

B Biotite comparison with laboratory diffusion results

It is possible to measure the diffusion coefficient of argon in the laboratory directly. A complete curve for one phlogopite has been reported by Evernden and others (1960), for temperatures above 300°C. We have determined single points for four different biotites at a temperature of 240°C and several points at 660°C, for which the data are given in table 4. The low-temperature samples were heated in evacuated glass tubes and the escaped argon measured after heating. The high-temperature samples were heated in air, and the

Table 4 Diffusion data for heated biotites

Sample	Locality	Average grain size (mm)	Temp. of heating (°C)	Time of heating (hour)	Fraction ^{40}Ar[a] released (%)	D/a^2 (sec^{-1})[b]	D/a^2 (sec^{-1})[c]
C6C	Eldora, 7300 ft	0.78	660 / 240	39 / 2710	14.5 / 0.016	1.40×10^{-8} / 2.1×10^{-16}	4.4×10^{-8} / 4.2×10^{-13}
CC26	Rainy Lake, Ontario	0.22	660 / 240	39 / 2710	8.5 / 0.021	4.68×10^{-9} / 3.9×10^{-16}	8.6×10^{-10} / 1.9×10^{-14}
LS78	Kenora, Ontario	0.17	240	2710	0.0090	7.2×10^{-17}	4.6×10^{-15}
4088	Idaho Springs, Colorado	0.65	300	1560	0.016	4.0×10^{-16}	8.4×10^{-13}

[a] Radiogenic component.
[b] Calculated for spherical geometry, with initial concentration uniform.
[c] Corrected for previous natural diffusion history and normalized to grain size of 0.78 mm (C6C).

argon remaining in the samples was determined after heating. There are two major sources of uncertainty in laboratory measurements of this kind. First, for the low-temperature samples, a very small fraction of the total argon is released. It is possible that this released argon is not representative of the main portion of the argon, but may diffuse more readily. This would give diffusion coefficients which are too large. On the other hand, if argon concentration gradients already exist in the biotite because of previous natural diffusion events, the calculated diffusion coefficients will be too small.

The first case is difficult to evaluate except to note that the contact biotites show a constant diffusion coefficient for argon over the range from 10% loss to 99% loss. Whether or not this constancy can be extrapolated to the 0.02% loss range of the heating experiments is uncertain at present.

The second problem can be crudely evaluated through estimates of the sample's geologic history. C6C, for example, is a biotite from 7300 ft from the contact zone, and its age suggests a previous natural diffusion loss of argon of 13%. CC26 and LS78 are both from the Superior Province of the Canadian Shield and show ages of 2.5–2.6 b.y. This is very near the true metamorphic age for this region, but losses of the order of 5% cannot be excluded. First-order corrections for this effect have been made on this basis.

The values in table 4 are listed in terms of D/a^2 (diffusion coefficient over grain radius squared), but since the four samples have different grain sizes, direct comparison of their D/a^2 values is not meaningful. By using the average grain sizes listed in table 4, CC26, LS78 and 4088 were normalized to the grain size of C6C, which represents the grain size of the contact biotites. These values, given also in table 4, can then be plotted on the same figure as the contact biotite diffusion curves of figure 14. This is done in figure 15, with the phlogopite curve of Evernden and others (1960) included for comparison.

It is clear, first, that the laboratory data cover a rather large range of values. For a given temperature, the spread in D/a^2 values is more than two orders of magnitude. Some of this spread may be a result of the imprecise nature of the corrections for previous history discussed above. There may also be some uncertainty involved in the identification of physical grain size as the size of the diffusion unit. If biotites have any tendency toward mosaic structure, the grain-size normalization made above would tend to spread the apparent values of D.

Despite the spread in D/a^2 values, there is a remarkable similarity

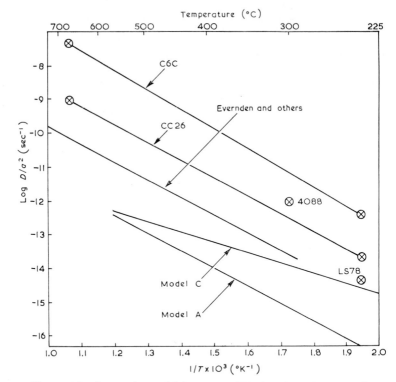

Figure 15 Comparison of laboratory diffusion data for argon in biotite with the contact diffusion models.

in slope for the laboratory curves, including the phlogopite curve of Evernden and others (1960). This suggests that the activation energy for argon diffusion in biotite may be rather uniform from sample to sample. It is interesting to note that the contact biotite curve for Model A has a slope similar to the laboratory curves, whereas the slope for Model C is quite different. It is certainly premature to use the laboratory data to establish absolute heating temperatures for natural environments. However, the comparison of activation energies does suggest that Model A provides a reasonable approximation to the thermal conditions of the Eldora intrusive.

C *Diffusion in the contact hornblendes*

The hornblende age curve can be treated the same way as the biotite age curves. In this case, however, the result is not a linear array of points. The difficulty probably lies with the breakdown of the thermal models very near the contact.

To illustrate this point, the hornblende ages at 2 ft and 11 ft may be compared. The thermal models predict a temperature difference of less than 10°c between these points, yet the ages are grossly different (120 m.y. versus 950 m.y.), implying an impossibly high activation energy. By altering the thermal models to allow for some convective cooling of the magma, a much steeper temperature gradient can be produced between 2 ft and 11 ft without significantly affecting the temperatures at greater distances. For example, after two weeks of convection (noncrystallizing contact), the temperature at 2 ft will be 200°c higher than predicted by Model A. The hornblende ages further out will not be much affected by this preheating. The diffusion parameters for these samples are plotted in figure 16. These points suggest high but not unreasonable activation energies. Here the data are in agreement with the laboratory measurements of Gerling and others (1965) who determined activation energies ranging from 100–200 kcal/mole for different samples of hornblende.

Activation energies of this magnitude are very unusual for simple diffusion processes, which normally have activation energies in the range 20–50 kcal/mole. High activation energies are more characteristic of reaction rates, and this suggests that the rate controlling step in the loss of argon from hornblende is the liberation of the argon from its initial structural site. Regardless of the exact explanation, this high activation energy makes hornblende particularly useful for K–Ar dating in regions with a metamorphic history.

D *Diffusion in the contact feldspars*

Thus far we have discussed only argon and strontium diffusion phenomena. It is clear that the irregular K–Ar age curve for feldspars does not allow the development of a diffusion model for argon in feldspar, and very little can be done with the strontium data since only one sample was affected. The model lead ages in feldspar, however, show a rather large variation, and a diffusion model can be developed for them. This was done using the $^{208}Pb/^{204}Pb$ ratio as the index of the variation. The $^{208}Pb/^{204}Pb$ ratio of samples 2400, 7300 and 22,500 was taken to be the initial unaffected value, and the ratio of sample 20 was assumed to represent 90% of a final equilibrium value. The calculations are only meaningful if the distribution of lead isotopes in the environment surrounding the feldspars is initially the same for all samples. The data are plotted in figure 16, using the Model A temperature curve. Considering the uncertainties in the data and calculations, we believe a simple linear dependence of the data

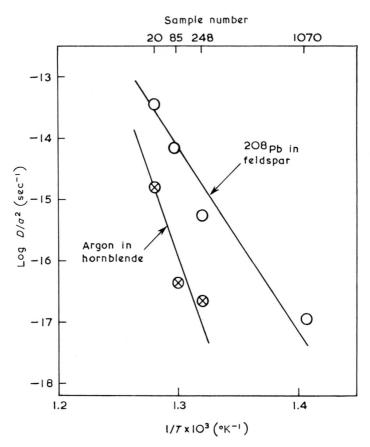

Figure 16 Diffusion coefficients of argon in hornblende, and lead in feldspar calculated from the contact results, plotted against Model A heat flow temperatures.

is the best representation that can be justified. The activation energy for the best-fit line is about 150 kcal/mole.

For comparison, the most reliable part of the hornblende curve is indicated also. While the slopes of the feldspar and hornblende curves depend in detail on the thermal model, the relative slope for the two minerals is independent of thermal models to first order.

Therefore, the data show that the activation energy of lead in feldspar is somewhat less than that of argon in hornblende. The feldspar case is complicated by the fact that it is not a simple loss of a rare gas being considered. The lead has been added to the feldspar from some other phase and the high activation energy may refer to this outside phase and not to the feldspar. That is, the rate controlling step may be the release of radiogenic lead from its initial position in the external source. This is a general point which can be applied to strontium also. In the case of mineral dating where the movement of lead or strontium depends on other phases for its take up or release, the apparent age stability of the dated mineral may depend on the particular mineral assemblage it is in. To give a simple example, the Rb–Sr age of a muscovite from a pure quartzite could appear more resistant to metamorphism than a muscovite from a marble.

E *Diffusion in the contact zircons*

This represents a more complicated case than the others because of the recrystallization or growth of the zircon and because of the apparent movement of both parent and daughter elements. It may be noted, however, that the zircon ages show the sharpest drop of any of the mineral ages, going from 1250 m.y. at 50 ft to 300 m.y. at 25 ft. This rapid change indicates a very high activation energy for whatever process is controlling the uranium–lead age pattern. The activation energy must be higher even than that for argon in hornblende, so it is probably in excess of 200 kcal/mole.

This very high activation energy may be compared with the low energy (~ 10 kcal/mole) required by Tilton (1960) for the continuous diffusion of lead from zircon over long periods of time at low temperature. This apparent anomaly between high-temperature and low-temperature activation energies was discussed by Tilton (1960), who concluded that low-temperature lead diffusion was controlled by a temperature-independent process such as the formation of vacancies during crystallization of the zircon, or in part due to the radiation damage from the uranium parent element. Only at high temperatures, then, would normal temperature-dependent volume diffusion take over. Wasserburg (1963) showed that low-temperature diffusion controlled by radiation damage was consistent with many observed zircon age patterns. The contact results clearly show that zircon ages can be affected by short periods of time at high temperatures. While the data are consistent with a simple diffusion process running with very high activation energy, it is not possible at present to distinguish this mechanism from a recrystallization or new growth mechanism.

IX Comparison of Mineral Age Stabilities and their Relationship to Metamorphic Facies

Here we will summarize the relative stabilities of mineral ages as determined from this contact study and attempt to relate them to the general framework of metamorphic facies. The age response curves for each of the minerals discussed earlier are compared in figure 17. A list could be made of the mineral ages as a function of relative stability to metamorphism, but this would be strictly applicable only to the present situation. The order of stability of mineral ages in a contact zone adjacent to the Duluth gabbro was found by Hanson, Signer and Gast (1964) to be the same as that at Eldora. Under different environmental conditions, however, the position on such a list might change considerably. For example, some hornblendes show retention of argon which is not significantly better than that of biotites. Feldspars lose some of their argon very easily, but retain small amounts even under very severe conditions. Some zircons have been shown to retain their lead even when contained in magma (Stern, 1964), whereas many appear to lose lead continuously at low temperatures. In some areas, biotite argon ages may be significantly older than biotite strontium ages. While all these variances have reasonable explanations, it is clear that no single age pattern will prevail in all areas under all conditions. It is important that this fact be understood, because each inconsistency contains extra information which may be of particular geologic importance.

In addition to the relationships of the mineral ages to each other, we would like to know their relationship to metamorphic facies. On figure 17 we could draw a band at 1000–2000 ft representing the microcline–orthoclase transition and in this way split the mineral ages into two categories—those affected below this transition temperature (biotite K–Ar and Rb–Sr, and feldspar K–Ar) and those affected above this transition temperature. In regional metamorphic terrains, the split between microcline-bearing rocks and orthoclase-bearing rocks comes at about the boundary between almandine–amphibolite facies rocks and granulite facies rocks. In contact aureoles, orthoclase is generally restricted to the pyroxene hornfels facies whereas microcline is common in the hornblende hornfels

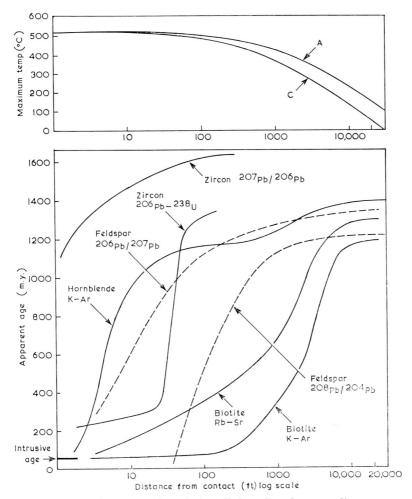

Figure 17 Summary plot relating all the mineral ages to distance from the contact. Temperatures derived for Models A and C are shown at the top.

facies. At Eldora, however, orthoclase has formed in rocks which clearly do not belong to the pyroxene hornfels facies. Furthermore, the heat flow calculations at Eldora suggest a maximum temperature of 400°c for the feldspar transition whereas much higher temperatures are generally assigned to the granulite facies (for example, Fyfe, Turner and Verhoogen, 1958, suggest 700°c for granulite facies).

We believe this anomaly to be related to the ease with which orthoclase converts back to microcline in certain natural environments. At Eldora, for example, the orthoclase in rocks near reentrant parts of the intrusive has apparently reverted back to microcline during cooling of the stock. Whether this depends on the slower cooling rates or the presence of higher water pressures in these zones is uncertain. Since granulites are notably anhydrous assemblages, we would correlate the persistence of orthoclase in these rocks with the lack of water. In amphibolite facies rocks, then, which might actually represent similar temperature regimes to granulite rocks, cooling in the presence of water would allow the conversion of orthoclase back to microcline. This means that the feldspar transition does not really provide a direct link between the contact assemblages and assemblages in terrain that has been subjected to regional metamorphism.

Alternatively, we may use the temperatures from the thermal models (figure 3) for comparison with regional facies. This is less satisfactory because of the large uncertainties in the assignment of temperatures to regional facies. In the zone within 500 ft of the contact, model temperatures run about 400–450°c, which in regional terrains would be greenschist facies (biotite zone) according to Fyfe, Turner and Verhoogen (1958). It is in this zone that feldspar model lead ages begin to be affected. Within 20–50 ft of the contact, temperatures are probably in the range 500–550°c, which would correspond to low almandine–amphibolite facies (garnet–staurolite zone). It is at this level that hornblende, zircon and feldspar Rb–Sr begin to show significant alteration of ages. Biotite argon and strontium ages show effects out to 7000 ft, where temperatures are in the range 200–250°c. This is very low-grade metamorphism indeed, corresponding to the zeolite facies of Fyfe, Turner and Verhoogen (1958). Effects of regional metamorphism at this level are petrographically discernible only in favorable cases and will frequently go unnoticed. Thus biotite ages may prove to be one of the most sensitive indicators available for detection of this 'incipient' metamorphism.

Whether these correlations with metamorphic facies will prove to be general or atypical remains to be shown. They do serve to illustrate the common ground that exists between the fields of metamorphic petrology and geochronology.

Acknowledgments

This study was initiated by one of us (S. R. Hart) at the Massachusetts Institute of Technology with the encouragement and supervision of P. M. Hurley. The feldspar lead work was carried out by

B. R. Doe while a fellow at the Geophysical Laboratory. We are indebted most of all to L. T. Aldrich for his ready assistance in a great many roles.

References

Aldrich, L. T., Davis, G. L. and James, H. L. (1965). Ages of minerals from metamorphic and intrusive rocks near Iron Mountain, Michigan. *J. Petrol.*, **6**, 445.

Aldrich, L. T., Wetherill, G. W., Davis, G. L. and Tilton, G. R. (1958). Radioactive ages of micas from granitic rocks by Rb–Sr and K–Ar methods. *Trans. Am. Geophys. Union*, **39**, 1124.

Amirkhanov, I. I., Brandt, S. B. and Bartnitskii, E. N. (1959). Diffusion of radiogenic argon in feldspars, *Dokl. Akad. Nauk SSSR, Geochem. Ser.*, No. **6**, 125.

Cree, A. (1948). Tertiary intrusives in the Hessie-Tolland area, Boulder and Gilpin Counties, Colorado. *Ph.D. Thesis*. Univ. Colorado. Chronic catalogue No. 2046.

Davis, G. L., Hart, S. R. and Tilton, G. R. (in preparation). The effect of contact metamorphism on zircon ages.

Doe, B. R. and Hart, S. R. (1963). The effect of contact metamorphism on lead in potassium feldspars near the Eldora Stock, Colorado. *J. Geophys. Res.*, **68**, 3521.

Doe, B. R., Tilton, G. R. and Hopson, C. A. (1965). Lead isotopes in feldspars from selected granitic rocks associated with regional metamorphism. *J. Geophys. Res.*, **70**, 1947.

Evernden, J. F., Curtis, G. H., Kistler, R. and Obradovich, J. (1960). Argon diffusion in glauconite, microcline, sanidine, leucite, and phlogopite. *Am. J. Sci.*, **258**, 583.

Fyfe, W. S., Turner, F. J. and Verhoogen, J. (1958). Metamorphic reactions and metamorphic facies. *Geol. Soc. Am., Mem.*, **73**.

Gerling, E. K., Koltsova, T. V., Petrov, B. V. and Zulfikarova, Z. K. (1965). Investigation of the suitability of amphiboles for the determination of absolute age of rocks by the K–Ar method (in Russian). *Geokhimiya*, No. **2**, 219.

Giffin, C. E. and Kulp, J. L. (1960). Potassium–argon ages in the Precambrian basement of Colorado. *Bull. Geol. Soc. Am.*, **71**, 219.

Giletti, B. J., Moorbath, S. and Lambert, R. St. J. (1961). A geochronological study of the metamorphic complexes of the Scottish Highlands. *Quart. J. Geol. Soc. London*, **117**, 233.

Goldich, S. S., Baadsgaard, H. and Nier, A. O. (1957). Investigations in $^{40}A/^{40}K$ dating. *Trans. Am. Geophys. Union*, **38**, 547.

Goldsmith, J. R. and Laves, F. (1954). The microcline–sanidine stability relations. *Geochim. Cosmochim. Acta*, **5**, 1.

Hanson, G. N., Signer, P. and Gast, P. W. (1964). The effect of thermal metamorphism on the K–Ar and Rb–Sr ages of various minerals in the Snowbank stock. *Trans. Am. Geophys. Union*, **45**, 115.

Hart, S. R. (1961). Mineral ages and metamorphism. In Geochronology of Rock Systems. *Ann. N.Y. Acad. Sci.*, **91**, 159.

Hart, S. R. (1964). The petrology and isotopic-mineral age relations of a contact zone in the Front Range, Colorado. *J. Geol.*, **72**, 493.
Heier, K. S. (1957). Phase relations of potash feldspar in metamorphism. *J. Geol.*, **65**, 468.
Jager, E. (1962). Rb–Sr age determinations on micas and total rocks from the Alps. *J. Geophys. Res.*, **67**, 5293.
Lanphere, M. A., Wasserburg, G. J. F., Albee, A. L. and Tilton, G. R. (1964). Redistribution of strontium and rubidium isotopes during metamorphism, World Beater complex, Panamint Range, California. In Craig, H., Miller, S. L. and Wasserburg, G. J. (Eds.), *Isotopic and Cosmic Chemistry*. North-Holland, Amsterdam. p. 269.
Laves, F. (1952). Phase relations of the alkali feldspars. *J. Geol.*, **58**, 548.
Long, L. E. and Kulp, J. L. (1962). Isotopic age study of the metamorphic history of the Manhattan and Reading Prongs. *Bull. Geol. Soc. Am.*, **73**, 969.
Lovering, T. S. and Goddard, E. N. (1950). Geology and ore deposits of the Front Range, Colorado. *U.S. Geol. Surv., Profess. Papers*, **223**.
MacKenzie, W. S. (1952). Optical and X-ray studies of alkali feldspars. *Carnegie Inst. Wash. Yearbook*, **51**, 49. (Ann. Rept. Geophys. Lab.)
Sardarov, S. S. (1957). The preservation state of radiogenic argon in microclines. *Geokhimiya*, No. **3**, 193.
Steiger, R. and Hart, S. R. (in press). The microcline–orthoclase transition within a contact aureole. *Am. Mineralogist*.
Stern, T. W. (1964). Isotopic ages of zircon and allanite from the Minnesota River Valley and La Sal Mountains, Utah. Abstract. *Trans. Am. Geophys. Union*, **45**, 116.
Tilton, G. R. (1960). Volume diffusion as a mechanism for discordant lead ages. *J. Geophys. Res.*, **65**, 2933.
Tilton, G. R., Davis, G. L., Hart, S. R., Aldrich, L. T., Steiger, R. H. and Gast, P. W. (1964). Geochronology and isotope geochemistry. *Carnegie Inst. Wash. Yearbook*, **63**, 250. (Ann. Rept. Geophys. Lab.)
Tilton, G. R., Davis, G. L., Wetherill, G. W. and Aldrich, L. T. (1957). Isotopic ages of zircon from granites and pegmatites. *Trans. Am. Geophys. Union*, **38**, 360.
Tilton, G. R., Wetherill, G. W., Davis, G. L. and Hopson, C. A. (1958). Ages of minerals from the Baltimore Gneiss near Baltimore, Md. *Bull. Geol. Soc. Am.*, **69**, 1469.
Tomisaka, T. (1962). On order–disorder transformation and stability range of microcline under high water vapor pressure. *Mineral. J. (Tokyo)*, **3**, 261.
Wasserburg, G. J. (1963). Diffusion processes in lead–uranium systems. *J. Geophys. Res.*, **68**, 4823.
Wasserburg, G. J., Hayden, R. J. and Jensen, K. J. (1956). ^{40}A–^{40}K dating of igneous rocks and sediments. *Geochim. Cosmochim. Acta*, **10**, 153.
Wetherill, G. W. (1956). Discordant uranium–lead ages, I. *Trans. Am. Geophys. Union*, **37**, 320.
Wetherill, G. W., Aldrich, L. T. and Davis, G. L. (1955). ^{40}A–^{40}K ratios of feldspars and micas from the same rock. *Geochim. Cosmochim. Acta*, **8**, 171.
Wetherill, G. W., Kuovo, O., Tilton, G. R. and Gast, P. W. (1962). Age measurements on rocks from the Finnish Precambrian. *J. Geol.*, **70**, 74.

Isotopic geochronology of Montana and Wyoming 113

the time of deposition of the Belt Series; and what were the actual ages of the Cretaceous–Tertiary 'Laramide' intrusives and their relative times of emplacement?

There were four basic reasons for the study of the questions listed above in this particular region. The first was to develop a chronology for a major area in the United States. Secondly, this region could then be correlated with others in North America to try to develop a more complete picture of the evolution of the North American continent. Thirdly, the orogenic aspects of the events being dated in the northern Rocky Mountain states could be related not only to other areas in North America, but to orogenies elsewhere in the world in order to test hypotheses on the world-wide synchroneity of orogeny. The fourth reason has to do primarily with the questions a geochemist is most concerned with. These are the study of the chemical processes which occur during orogeny and the evolution of a continent. They can be studied by chemical and isotopic analysis of some rock systems, but such studies must have a firm basis in an established chronology of events for the region. These questions will be discussed further following the specific discussion of the chronology.

Various authors have published data on ages from the northern Rockies and various radioactive decay constants have been employed. However, all the ages reported here are based on one set of decay constants. Where different decay constants had been used, the ages have been recalculated. The following are the decay constants used herein.

^{40}K $\lambda_\beta = 4.72 \times 10^{-10}(yr^{-1})$, $\lambda_e = 0.585 \times 10^{-10}(yr^{-1})$

^{87}Rb $\lambda = 1.47 \times 10^{-11}(yr^{-1})$; ^{238}U $\lambda = 1.54 \times 10^{-10}(yr^{-1})$

^{235}U $\lambda = 9.72 \times 10^{-10}(yr^{-1})$; ^{232}Th $\lambda = 4.88 \times 10^{-11}(yr^{-1})$

This will permit comparison of all the data with the knowledge that relative ages will always be in the correct order even if some changes in decay constants are made in the future.

In general, it is assumed that the precision of the age measurements is about 3 to 5%, unless stated otherwise.

Although something is known of the geology of the basement rocks of all the mountain ranges shown in figure 1, detailed field mapping has been carried out in only a few areas. The first Precambrian areas mapped in detail were in the Beartooth Range, where interest centered on the Stillwater igneous complex (Jones, Peoples and Howland, 1960; Hess, 1960) and on the igneous and metamorphic complex of rocks at the southeastern end of the range (Eckelmann and Poldervaart, 1957; Spencer, 1959; Harris, 1959). Both groups of rocks underlie the Middle Cambrian unconformably.

Since 1960, detailed geological studies have appeared for a number of the areas of interest here, notably: the Ruby and Gravelly Ranges (Heinrich, 1960; Heinrich and Rabbit, 1960), the Teton Range (Reed, 1963), the Medicine Bow Range (Houston and Parker, 1963) and the Tobacco Root Range (Reid, 1963).

Concurrent with the increased geological control on the region, isotopic dating has developed from simple large-scale reconnaissance to increasingly sophisticated studies. This is in contrast with some of the earliest work in this area which was carried out not only to measure the ages of the rocks, but to test the radiometric methods being employed.

It was suggested by Aldrich and his associates (1957) that an episode of regional metamorphism and igneous intrusion occurred to the south of the region discussed here about 1350 m.y. ago. The data were for rocks from the southern Rocky Mountains Precambrian in Arizona, New Mexico, Colorado and the Front Range in southeastern Wyoming. These authors found older ages in central Wyoming and the Black Hills of South Dakota.

In 1958, Aldrich and others published data to show that rocks in central Wyoming (near the Wind River Canyon) might be as old as 2400 m.y. and that the rocks at the Bob Ingersoll Mine, in the Black Hills, were probably 1600 m.y. old (see tables 1 and 2). The data were for K–Ar and Rb–Sr mineral analyses in both cases, and included a concordant U–Pb analysis on uraninite in the Black Hills.

II The Terrane Yielding 2600 m.y. Ages

Gast, Kulp and Long (1958) measured K–Ar and Rb–Sr mineral ages from the southeastern end of the Beartooth Range in the area described by Eckelmann and Poldervaart (1957) and Harris (1959). The data are shown in table 1. These authors concluded that the time of metamorphism, granitization and formation of migmatites occurred 2750 ± 150 m.y. ago. The decay constant for ^{87}Rb used in this chapter would make the data somewhat more concordant and would result in a more probable age of 2600 ± 150 m.y. This was in agreement with a nearly concordant uranium–lead uraninite age of 2700 m.y. from a nearby pegmatite.

Isotopic geochronology of Montana and Wyoming 115

These results served to show that there were rocks to be found in the Rocky Mountains as ancient as any in the Canadian Shield. This would have to be taken into account by any hypothesis of continental growth.

These data also demonstrated what geologists had long suspected, but could not show. The term Archean, when used to mean the oldest basement rocks in an area, did not have a single time connotation. Rather, some Archean rocks gave ages that were twice those of other rocks also called Archean.

The different ages obtained in different parts of the northern Rockies immediately raised the question of the areal extent of the various age terranes. The result was the initiation of the reconnaissance type of study.

The 2600 m.y. age terrane has been shown to include a large portion of the state of Wyoming. The data of Giletti and Gast (1961), Bassett and Giletti (1963) and Hills, Gast and Houston (1965), in addition to those of Aldrich and others (1958) and Gast, Kulp and Long (1958), are shown in table 1. A value of 2600 ± 200 m.y. for an episode of regional metamorphism and igneous intrusion would satisfy all these results, with the possible exception of some in the northern Wind River Range. The latter will shortly be discussed in connection with data from the Teton Range to the west. Catanzaro and Kulp (1964) refer to additional K–Ar work at the Lamont laboratories which 'suggest an age of about 2700 m.y. for the last event' in the Beartooth Range. Although the specific data were not published, the conclusion is in agreement with the above.

In addition, Catanzaro and Gast (1960) extracted lead from four potassium feldspars derived from some of the pegmatites reported in table 1. The isotopic compositions of these leads, when plotted on a Holmes–Houtermans common lead development curve yield ages ranging from 2500 to 2850 m.y. These data are also in agreement with the 2600 m.y. age, especially when the lead model uncertainties are recalled.

Hills, Gast and Houston (1965) in discussing the Medicine Bow Range state: 'Whole rock Rb–Sr isotope compositions of three samples of granitic gneiss from the older terrain fit a 2500 ± 100 m.y. isochron ($\lambda = 1.47 \times 10^{-11}$ year^{-1}). Five whole-rock samples of the foliated granite, intrusive into the older terrain, fit a 2450 ± 100 m.y. isochron'. These measurements in southern Wyoming serve to extend the 2600 m.y. terrane practically to Colorado.

Table 1 Ages from the Wyoming and Montana 2600 m.y. age terrane

Location	Rock	Mineral	Age (m.y.)		Reference[a]
			K–Ar	Rb–Sr	
SE Beartooth Range	Pegmatite	Muscovite	2470 ± 50	2640 ± 50	1
		Microcline	—	2540 ± 100	1
SE Beartooth Range	Pegmatite	Muscovite	2520 ± 50	2540 ± 70	1
SE Beartooth Range	Pegmatite in norite	Biotite	2290 ± 60	2400 ± 50	1
		Microcline	—	2570 ± 70	1
SE Beartooth Range	Amphibolite	Biotite	2340 ± 50	2590 ± 60	1
SE Beartooth Range	Amphibolite	Hornblende	2450	—	5
Shoshone River Gorge, Cody, Wyo.	Pegmatite	Biotite	—	2560 ± 70	1
Northern Bighorn Range	Gneiss	Biotite	2540	2600 ± 70	1
Bonneville, Wyo.	Pegmatite	Muscovite	2250	2280	2
Wind River Canyon, Wyo.	Granite	Biotite	1350	1830	2
Wind River Canyon, Wyo.	Pegmatite	Muscovite	—	2560	3
N of Bonneville, Wyo.	Pegmatite	Microcline	—	2490	3
Powder River Pass, Bighorn Range	Banded gneiss	Biotite	—	2410	3

Location	Rock type	Mineral			Ref.
Casper Mt., Casper, Wyo.	Biotite gneiss	Biotite	—	2310	3
Southern Wind River Range	Quartz monzonite	Biotite	—	2230	3
Northern Wind River Range	Pegmatite	Muscovite	—	2400	3
Northern Wind River Range	Pegmatite	Muscovite	2410	2700	4
		K Feldspar	1570	2250	4
Northern Wind River Range	Pegmatite	Biotite	1810	1930	4
Northern Wind River Range	Pegmatite	Muscovite	2410	2520	4
		Muscovite	—	2320	4
Northern Wind River Range	Pegmatite	Muscovite	2470	—	4
		K Feldspar	1850	2390	4
Northern Wind River Range	Pegmatite	Muscovite	2460	2470	4
		Biotite	2250	2350	4
Northern Wind River Range	Metamorphic rock	Biotite	2100	—	4
Northern Wind River Range	Metamorphic rock	Biotite	2050	—	4
Northern Wind River Range	Metamorphic rock	Biotite	1840	—	4
Northern Wind River Range	Metamorphic rock	Biotite	1770	—	4
Northern Wind River Range	Metamorphic rock	Biotite	2200	—	4

[a] 1. Gast and others (1958); 2. Aldrich and others (1958); 3. Giletti and Gast (1961); 4. Bassett and Giletti (1963); 5. Hayden and Wehrenberg (1960).

Table 2 Ages from the Wyoming, Montana, South Dakota and Utah 1600 m.y. terranes

Location	Rock	Mineral	Age (m.y.) K–Ar	Age (m.y.) Rb–Sr	Reference[a]
Bob Ingersoll Mine, Keystone, S. Dak.	Pegmatite	Lepidolite	1380	1560	1
		Muscovite	1550	1630	1
Bob Ingersoll Mine, Keystone, S. Dak.	Pegmatite	Lepidolite	—	1530	2
Medicine Bow Range	Biotite gneiss	Biotite	—	1620	2
Medicine Bow Range	Pegmatite	Biotite	—	1560	2
Medicine Bow Range	Banded gneiss	Biotite	—	1430	2
Bountiful Pk., Salt Lake City, Utah	Pegmatite	Muscovite	—	1490	2
Norris, Mont.	Biotite–plagioclase–quartz gneiss (Pony(?) gneiss)	Biotite	1630	—	3
Tendoy Range	Dillon granite gneiss	Biotite	1410	1480	4
		Whole rock	—	1460	4
Ruby Range	Dillon granite gneiss	Biotite	1660	—	4
Ruby Range	Dillon granite gneiss	Biotite	1550	—	4
Tobacco Root Range	Pony gneiss	Biotite	1720	—	4
Ruby Range	Cherry Creek gneiss	Biotite	1610	1530[b]	4
		Muscovite	1680	—	4
Ruby Range	Cherry Creek gneiss	Biotite	1520	1560	4
Tobacco Root Range	Cherry Creek gneiss	Biotite	—	1530	4
Tobacco Root Range	Cherry Creek gneiss	Muscovite	1660	1590	4
	Pegmatite	K Feldspar	—	1620	4
		Biotite	1520	—	4

Location	Rock type	Mineral			Ref
Blacktail Range	Pre-Cherry Creek gneiss	Biotite	1600	1540	4
Blacktail Range		Whole rock	—	3080	4
Blacktail Range	Pre-Cherry Creek gneiss	Biotite	1330 ± 150	1480	4
Gravelly Range	Granite gneiss	Muscovite	1610	—	4
Anceney, Mont.	Biotite schist	Biotite	1550	—	4
Bridger Range	Amphibolitic gneiss	Biotite	1730	—	4
Gallatin Canyon	Granitic gneiss	Biotite	1690	—	4
Little Belt Range	Pinto meta-diorite	Hornblende	1470	—	5
Little Belt Range	Pinto meta-diorite	Hornblende	1420	—	5
Little Belt Range	Pinto meta-diorite	Hornblende	1480	—	5
		Biotite	1380	—	5
Little Belt Range	Pinto meta-diorite	Hornblende	1820	—	5
		Biotite	1720	—	5
Little Belt Range	Pinto meta-diorite	Hornblende	1780	—	5
		Biotite	1700	—	5
Little Belt Range	Migmatite	Hornblende	1740	—	5
Little Belt Range	Migmatite	Hornblende	1610	—	5
Little Belt Range	Migmatite	Hornblende	1780	—	5
		Biotite	1500	—	5
Little Belt Range	Augen gneiss	Hornblende	1620	—	5
Little Rocky Mountains	—	Hornblende	1710	—	6
		Biotite	1750	—	6
Eastern Mont. drill core	47° 38′N 105° 34′W	Biotite	1720	—	7
		Hornblende	1770	—	7
		K Feldspar	—	1790	7

[a] 1. Aldrich and others (1958); 2. Giletti and Gast (1961); 3. Hayden and Wehrenberg (1960); 4. Giletti (1966a); 5. Catanzaro and Kulp (1964); 6. Burwash, Baadsgaard and Peterman (1962); 7. Peterman and Hedge (1964).
[b] Mineral isochron.

III The Terrane Yielding 1600 m.y. Ages

Several areas peripheral to the 2600 m.y. age terrane just discussed have given mineral ages of about 1600 m.y. This younger age was first observed by Aldrich and others (1958) in the Black Hills, as noted earlier. Table 2 shows their results together with all other published data for the areas to be discussed in this section.

The first ages to be measured in the Medicine Bow Range yielded 1600 m.y. values (Giletti and Gast, 1961). Hills and his colleagues (1965) have reported that the Medicine Bow Range presents a rather complex geology, including the old rocks mentioned above, the 1600 m.y. age metamorphism and a possibly younger event at 1470 m.y.

One muscovite Rb–Sr age of 1490 m.y. published by Giletti and Gast (1961) suggests that there may be a similar age terrane in the Wasatch Range just east of Salt Lake City. This is still a quite tentative suggestion.

Most of the Precambrian basement rocks west of the Beartooth Range in southwestern Montana give mineral ages of about 1600 m.y. Table 2 shows all of these data except for some to be discussed later which appear in table 5. Several geological studies of portions of this area have been published and the data will be discussed below in the context of these results. However, it can be stated here that the 1600 m.y. age for southwestern Montana represents a time of regional metamorphism and igneous intrusion.

In addition to the data in table 2, Catanzaro and Gast (1960) measured the isotopic composition of lead in pegmatitic feldspars from the Black Hills. Three specimens gave model lead ages of 1300 ± 150, 1500 ± 150 and 1550 ± 125 m.y. Again, these may be considered to support the 1600 m.y. age. Note that the quoted uncertainties are analytical and do not include errors in the model.

The Little Belt Range in the northern part of the region being discussed has yielded K–Ar ages on biotite of 1380 to 1720 m.y. (four samples) and on hornblende of 1420 to 1820 m.y. (nine samples) as shown in table 2. This might be indicative of a regional metamorphism which is somewhat older than the 1600 m.y. event to the south. These authors actually suggest a 1920 m.y. metamorphism when these data are combined with their uranium–lead ages on zircons. This area will be discussed below, but there is no unequivocal evidence that there was a 1600 m.y. event here despite the fact that some of the K–Ar ages have that value.

The only other data from the regions being treated in this section are either related to the Laramide intrusives or are considerably older. In either case, they will be treated below.

Finally, some data from adjacent regions has some direct bearing on this region. Burwash, Baadsgaard and Peterman (1962) and Peterman and Hedge (1964) report K–Ar and Rb–Sr data from the Little Belt Range, Little Rocky Mountains and from drill cores in eastern Montana and North Dakota, as well as the contiguous Canadian provinces. These all cluster around 1700 m.y. except for samples from Manitoba and the eastern part of North Dakota.

It can be seen that the 1600 m.y. terrane appears to wrap around the 2600 m.y. terrane in a zone from southwestern Montana to western North and South Dakota and the Medicine Bow Range.

No major Precambrian event that was younger than 1600 m.y. has been reported from this region.

IV The Belt Series

A second group of Precambrian rocks is exposed in Montana. This is the Belt Series which is, in the main, a sequence of unmetamorphosed sedimentary rocks. This series was considered part of the 'Algonkian' at the turn of the century, and there has been considerable debate since that time as to the correct assignment. The problems are similar to those of igneous and metamorphic terranes in that diagnostic fossils are lacking. It is not proposed to treat the stratigraphic problems of the Belt here. Rather, the application of radiometric ages has set some specific limits on the time of deposition which should be of some value to the student of the Belt. For a comprehensive treatment of the geology of the Belt Series see Ross (1963).

Four methods of approach are possible in the dating of ancient sedimentary rocks by radioactive methods. The best, in principle, is the dating of a syngenetic mineral that formed in the sediment or was formed as part of the sequence of deposition such as a mineral in an interstratified volcanic rock. A second approach is to place maximum age limits on the sedimentary unit by dating the underlying rocks. The third approach is similar to the latter in that the minimum age can be obtained by dating overlying or intrusive rocks. Finally, it might be possible to date some of the detritus in the sedimentary unit and, if this represents the date of the formation of the older rock from which the detritus was derived, this too would set a maximum age limit on the time of deposition.

The earliest ages obtained for the Belt Series were from the Coeur d'Alene district, in Idaho. Uranium–lead ages on uraninite gave the following ages in m.y.

$^{206}Pb/^{238}U$	$^{207}Pb/^{235}U$	$^{207}Pb/^{206}Pb$	
712 ± 10	785 ± 15	885 ± 50	Kerr and Kulp (1952)
700 ± 10	830 ± 15	1190 ± 30	Kerr and Kulp (1952) recalculated by Eckelmann and Kulp (1957)
805 ± 10	860 ± 20	1035 ± 35	Eckelmann and Kulp (1957)

The uranium-bearing veins cut folded and metamorphosed Belt sedimentary rocks, which means that the time of mineralization is a minimum age for the Belt. The results of Kerr and Kulp as they appeared in 1952 differ from those in the table above because different decay constants were in use at that time. The second set of data are really the same analyses as the first. However, Eckelmann and Kulp (1957) had obtained and analyzed galena from the same deposit, which permitted them to make a more accurate common lead correction on the raw data. The third set of data are for a different specimen from the same mine. Eckelmann and Kulp conclude that the mineralization occurred '1100–1200 m.y. ago.' Deposition of the Belt sediments would, therefore, have been earlier than this in this area.

The Belt rocks rest unconformably on older Precambrian in exposures in the Little Belt Range and the northern areas of the Tobacco Root Range. There has been some question as to the distinction between the two types of basement rocks (see Ross, 1963, p. 90) exposed in these two areas. In terms of the ages of these rocks, however, the difference can be stated in quantitative terms. The La Hood Formation of the Belt Series must be younger than 1600 m.y. old because it rests unconformably on rocks which were metamorphosed at that time in the area of the Tobacco Root Range. Catanzaro and Kulp (1964) suggest that the latest Precambrian event for which they have evidence in the Little Belt Range occurred 1920 m.y. ago. This event was followed by the deposition of the Belt Series in that area. It can be seen from this that at least some of the Belt deposition occurred since 1600 m.y. ago, perhaps all of it.

Gulbrandsen, Goldich and Thomas (1963) have reported data on glauconite from the Belt Series. If the glauconite was syngenetic, its true age is the time of deposition. They obtained ages of 1090 m.y. and 1050 m.y. by the K–Ar and Rb–Sr methods respectively. These data are for a part of the Missoula Group of the Belt. Gulbrandsen, Goldich and Thomas state that they consider 1070 m.y. a minimum age for the unit, which is in agreement with the observation that glauconites generally yield either the age expected on the basis of other dating methods or a lower age.

Long, Silverman and Kulp (1960) reported additional data on the question of Belt deposition. They studied the isotopic composition of lead in galena samples from the Coeur d'Alene mining district. These deposits occur in veins in the Belt. They found that the Holmes–Houtermans model ages they measured averaged 1400 m.y. The error in the individual age calculations is ± 100 m.y. However, there is a greater uncertainty in the model as a whole, which means that this age should be taken as confirming the general pattern of ages for the Belt. The authors' statement that the Belt Series is older than 1400 m.y. is too strong because of the model uncertainty. Further, there is evidence, which they cite, that some of the lead occurs in veinlets which cut Laramide age rocks. Long, Silverman and Kulp argue that this represents minor remobilization of a small portion of the lead at that time.

It can be seen from the sum of these data that the Belt Series was probably deposited between 1600 m.y. and 1200 m.y. ago. Certainly, portions of the series were deposited between those times.

V The Laramide Intrusions

Intrusive into the rocks of the region being discussed are several large igneous bodies of late Cretaceous or later age. The most notable of these is the Boulder Batholith, but others, including the Tobacco Root Batholith, have also been studied.

There were several objectives in studying these rocks from an isotope geochronological point of view. These included: the establishment of the absolute chronology of the events in the area; the relation of the times of intrusion of the various intrusives to each other; the duration of an episode of intrusion which results in the formation of a batholith; the relative timing of the various phases of the intrusion; and the effect of the intrusion on the surrounding rocks. Owing to the known low ages of these intrusives, the primary tool has been the K–Ar method.

A summary of the data for the Boulder Batholith of western Montana has recently been published by Knopf (1964). Table 3 lists those data together with all the other published data for the young intrusives in this area.

Table 3 Ages of Laramide intrusives

Description of rock	Mineral	Age (m.y.) K–Ar	Age (m.y.) Rb–Sr	Reference[a]
Boulder Batholith				
Basic granodiorite	Biotite	77.8 ± 1.6[b]	—	1
		76.4 ± 2.3[b]	—	1
Clancy granodiorite	Biotite	73.2 ± 2.2[b]	65 ± 7	1
	Biotite	82 ± 4	—	2
	Biotite	81	—	1
	Hornblende	113[b]	—	1
	Hornblende	76	—	1
Adamellite	Biotite	73.5 ± 1.4[b]	74 ± 10	1
Granite	Biotite	70.2 ± 1.5[b]	71 ± 7	1
Marysville Stock				
Granodiorite	Biotite	78 ± 4	—	2
Tobacco Root Batholith				
Quartz monzonite	Biotite	75	—	3
Quartz monzonite	Biotite	52	—	3

[a] 1. Knopf (1964); 2. Baadsgaard and others (1961); 3. Giletti (1966a).
[b] See text for discussion.

Clearly, the age of the Boulder Batholith is of the order of 75 to 80 m.y. However, different laboratories do not agree on the age of one phase of intrusion which they all dated. The dates marked with a 'b' were all obtained by Geochron Laboratories, a commercial laboratory. They represent the most complete set of data for the four phases of intrusion listed.

Comparison of the Clancy granodiorite figures is possible as all samples come from the same quarry. Three laboratories contributed data, Baadsgaard and his coworkers at the University of Alberta (1961), J. Evernden, of the University of California at Berkeley and

Geochron. Although there is a difference of eight or nine million years between Geochron and the other two laboratories, the source of the difference is difficult to assess. The analytical data have not been completely supplied. In particular, only one potassium analysis has been published. Baadsgaard and coworkers report 4.86% potassium. No data on the atmospheric argon correction subtracted from any of the samples are available. It would appear that two samples are involved. The first was dated both by Baadsgaard and coworkers and by Evernden. A second sample was then apparently obtained which Geochron measured.

Contrary to what appears, at first sight, as a basis for evaluating the fine structure in the data, all that can be concluded from the Clancy granodiorite results is that Berkeley and Edmonton agreed when supplied with the same sample.

Comparison of all the Geochron data led Knopf to suggest that the time required to emplace the Boulder Batholith was about 7 to 8 m.y. However, as he pointed out, there are enough different data to show that the resolving power of the method is just not good enough as yet. The uncertainties in the ages obtained by the Rb–Sr method are so large that they merely confirm the dates obtained by the argon method.

The Marysville Stock and the Tobacco Root Batholith can be seen (table 3) to have intruded at about the same time as the Boulder Batholith. Little more can be added at this point concerning these intrusives.

The effect of these intrusives on the rocks surrounding them is significant. Three measurements on the Precambrian rocks near the intrusives yielded quite low ages (Giletti, 1966a).

The Rochester mining district in southwestern Montana occurs in Precambrian rocks including a muscovite-bearing granitic gneiss. The muscovite from this gneiss gave a K–Ar age of only 175 m.y. The mineralization in the area is thought to derive from the Boulder Batholith intrusion. The surface contact of the Precambrian with the Boulder is five miles northeast of the sample locality. A nearby biotite-bearing schist, also of Precambrian age, yielded a biotite K–Ar age of 75 m.y. This specimen is also from about five miles from the surface contact. It is quite possible that the Boulder Batholith extends to some point below the sample localities and was, therefore, much closer to the samples than five miles.

A similar phenomenon is observed northeast of the Tobacco Root Batholith where a Rb–Sr age of less than 80 m.y. was measured on biotite from the Precambrian below the Cambrian Flathead quartzite. In this case, the surface location of the contact is eleven miles away from the sample. However, a small intrusive is observed about five miles east of the sample, that is, still further away from the batholith contact. It is possible that this is related to the Tobacco Root and may be connected at some shallow depth to the batholith.

It should be noted that the exact distance at which an intrusive will modify isotopic ages in minerals is not known. In the area under discussion, the data from near the Tobacco Root Batholith set some limits. Several dates from the Precambrian southeast of the batholith give 1600 m.y. ages consistent with the regional value. These samples are from about seven miles from the surface contact.

In a study of the contact effects of the Eldora Stock in Colorado, Hart (1964) showed that at a distance of about two miles from the contact no appreciable effects in the argon or strontium systems could be observed. The stock is small, measuring about two miles across. This suggests that the reduction of mineral ages might extend about one intrusive width away from the intrusive. However, the effect must be a function of the geometry of the body as well as of its temperature. If a batholith has minimum outcrop dimensions of about 20 by 60 miles, as does the Boulder, this is not a vertical intrusive as might be supposed in the case of the Eldora Stock. Rather it is a slab where the vertical dimension is the thin one. The thickness here would be of the order of 5 to 10 miles, at most. The temperature history of the surroundings of a cooling igneous slab has not been worked out for cases where the point of interest lies beyond the end of the slab. However, as most of the heat would be lost vertically, the thermal effects beyond the edge would clearly not extend in direct proportion to the lateral extent of the slab. For a discussion of this sort of thermal event see Jaeger (1964).

As a fairly safe working rule in the collection of specimens for isotopic dating, rocks several intrusive widths away from the intrusive contact are probably unaffected by the intrusive heating. If the intrusive is very large, the distance is probably no longer a function of width. A few miles may suffice. However, the real problem in these cases is probably the true distance to the intrusive. That is, does the intrusive extend to some point directly below the sample locality? Needless to say, the contact metamorphic effects in terms of changed mineralogy all occur much closer to the contact. There is no apparent effect visible in hand specimen or thin section at the larger distances being considered here.

VI Geological Applications

In addition to the establishment of chronologies for different parts of a region, additional geological information can be obtained by the application of the above data and other types of isotopic data. This discussion will be limited to that information derived from radioactive isotope systems, but this is only one aspect of the total potential of chemical and isotopic studies. In order to treat these geological questions, more attention must be devoted to the geological data. Specific areas will be discussed, each of which illustrates particular problems.

A Southwestern Montana

Most of the geological literature for the Precambrian of this area can be found in the references cited in Giletti (1966a) and only selected papers will be noted here. In the area of the Ruby and Gravelly Ranges (see figure 1) Heinrich (1960) and Heinrich and Rabbit (1960) provide detailed maps showing the relations of the various Precambrian rock types. They suggest that the sequence consists of: Pre-Cherry Creek gneiss, Cherry Creek gneiss, Pony gneiss and Dillon granite gneiss.

The basis for the stratigraphical assignment is that the Cherry Creek apparently overlies the Pre-Cherry Creek; the Pony is physically above the Cherry Creek; and the Dillon is an intrusive granite which cross-cuts the Cherry Creek. The Pony–Cherry Creek contact is somewhat arbitrary and the two units are considered conformable. All of these units give 1600 m.y. mineral ages as discussed above.

The Pre-Cherry Creek gneiss is a foliated granitic gneiss of variable texture, usually pink with occasional layering. Characteristically it is coarse grained. It appears in contact with Cherry Creek rocks in the Ruby Range and in the Gravelly Range. It is structurally below and probably stratigraphically below the Cherry Creek.

The Cherry Creek gneiss is really a layered sequence consisting mostly of metamorphosed sedimentary rocks including marbles and quartzites. It is characterized by quartzo-feldspathic gneisses and schists, sometimes alternating with amphibolitic layers. In some outcrops it is not possible to distinguish this unit from the one above, the Pony, which has similar layered gneiss sequences.

The Pony gneiss is distinguished in the field by the absence of marble and the scarcity of quartzite horizons. In most other respects it closely resembles the Cherry Creek gneiss. The distinction has been made, therefore, only where sufficient thickness of the section is visible to permit the determination of the presence or absence of these marker horizons.

The Dillon granite gneiss is a medium- to fine-grained biotite-bearing granite. The Dillon is often foliated, but is not normally layered. Heinrich considers this to have intruded the terrane in the late phases of the regional metamorphism and deformation.

It should be noted that the sequence of units described here has not been carried through the whole of the area of southwestern Montana. This is partly owing to lack of detailed studies in some areas. At least as important is the difficulty of correlation of units based on the lithologic character of the units. The usual difficulty attendant on this sort of correlation applies: lack of continuous outcrop in going from one mountain range to the next raises the questions of facies transitions within a unit, pinching out of entire units, or the appearance of stratigraphically higher or lower units which have the same composition.

In a detailed study of the northern end of the Tobacco Root Range (see figure 1), Reid (1963) postulated a sequence of deformational and metamorphic events which should bear some relation to the ages obtained in the area. Table 4 lists these events as described by Reid.

Table 4 Postulated deformational and metamorphic events in the Tobacco Root Range, Montana. (After Reid, 1963)

Deformation	Metamorphism
1. Isoclinal folding, axial-plane schistosity	a. Unknown, but probably high grade
Intrusion of basalt dikes and sills	
2. Shearing deformation along preexisting axial-plane schistosity	b. Granulite facies
3. Shearing deformation still parallel to axial planes	c. Amphibolite facies
4. Formation of boudins, shearing off of fold crests	
5. Open upright folds at 45° from previous trends	d. Minor epidote–amphibolite retrogressive metamorphism
6. Laramide folding	e. Greenschist facies

No serious effort has been made to measure ages in the area he mapped. The main reason for this is the set of low ages obtained in the vicinity of the Tobacco Root and Boulder Batholiths described above. It appears likely that rocks in the northern Tobacco Root Range would have been disturbed by the intrusion of the batholith and any data obtained could not be interpreted uniquely.

Because the area of detailed structural and mineral paragenetic studies does not coincide with the areas of isotopic study, some question must arise as to the correlation of the two types of data. Some additional comment on the geology is, therefore, in order.

The basis for the separation of the deformations and metamorphisms into the sequence of table 4 was twofold. Reid studied the sequence of deformations in the field by such techniques as observation of refolded folds, direction of fold axes, etc. The intrusion of the basalt dikes provided a key to the identification of differences in the timing of deformations. The second aspect of his study dealt with the presence of nonequilibrium mineral assemblages. For example, he described pyroxenes rimmed by amphiboles as part of the evidence which indicated a retrograding of a granulite facies terrane to an amphibolite facies terrane.

Reid does not suggest that each deformational or metamorphic episode he describes is separate from the others and represents a separate orogeny. He considers it likely that some of these were parts of a single orogenic episode. Further, the amount of deformation during deformations 2 and 3 is not known, as the evidence for the deformation is from the shearing noted in the thin sections.

A similarity was noted by Reid between the events known as metamorphisms b and c in his area and the sequence of Scourian and Laxfordian metamorphisms in Scotland described by Sutton and Watson (1950). The similarity lies in the metamorphic grades and the order in which they occurred. In that case, the two events were separated by at least 800 m.y. (Giletti, Moorbath and Lambert, 1961).

In order to define the sort of time differences that may exist between orogenies, it is useful to consider the sequence of events for a much better understood terrane. It is not essential that the exact ages of the following be known for the purposes of this discussion. Various authors might disagree on the details, but the overall picture is correct.

Ages have been measured in the Appalachian Mountains of the eastern United States which suggest orogenies at approximately 450,

360, 250 and 180 m.y. ago. Although each one had a finite duration, they are distinct events in the geological record. It has been argued that they may all represent parts of one major orogenic cycle.

In order to appreciate the analogy with the rocks discussed here, add 1300 m.y. to each of those ages arbitrarily. The resulting sequence would be 1750, 1660, 1550 and 1480 m.y. All of these ages have been measured in southwestern Montana (see table 2). Some of this variation is due to analytical error, which is about $\pm$ 40 m.y. in this age range. However, the data spread too much for the range to be strictly caused by analytical error. Clearly, other factors enter.

It is not the purpose of this paper to argue that the behavior of the Appalachian geosyncline and its related orogenies can be translated directly to the Precambrian. In fact, there is evidence to suggest that orogeny in the Precambrian may not have had the same geometry. It is suggested, however, that since a closely spaced sequence of orogenies occurred in the Paleozoic, the same may have occurred in Montana about 1600 m.y. ago. In that case, the time interval between the 'peaks' of orogeny could have been comparable to the analytical uncertainty. Further, in the Piedmont of the southeastern United States, many ages exist in the range between 360 and 250 m.y., two of the key ages for that area. Noncritical tabulation of the dates for this area would reveal an apparent continuum. The greater number of measurements plus the greater geological control avoid this problem. This type of terrane may have its counterpart in Montana, but it still awaits clarification.

At present there is no way to analyze the existing data to determine whether they spread over the range they do because of multiple orogenies or because of other causes such as minor weathering or groundwater ion exchange which cannot be detected by thin section study. Clearly, minor effects of this sort amounting to only a few percent could account for the spread in ages.

If it can be assumed that the pains taken to collect fresh materials were sufficient, and that there was no recent alteration of the rocks, then it might be useful to try to assign different times to the different episodes described by Reid. The data show an irregular trend from younger to older ages in going northeast from the Tendoy Range to the Gallatin Canyon and beyond. However, the general structural trend of the lithologic units is also to the northeast in the Ruby, Gravelly, Madison and Tobacco Root Ranges.

It was noted by Reid that the amphibolite metamorphism, c, increased in intensity to the southwest. That is, the fraction of

Isotopic geochronology of Montana and Wyoming

completely amphibolitized granulite facies rocks increased from northeast to southwest from 10 to 70% of the mafic rocks at the expense of mafic rocks whose pyroxenes were only rimmed by amphibole.

In a long, narrow north–south trending strip of Precambrian exposure in the Gravelly Range there is a notable increase in metamorphic grade from south to north in the Cherry Creek gneiss. This area, incidentally, includes the type locality of the Cherry Creek. The area has not been studied from the point of view or the detail that was employed in the Tobacco Root Range study, hence it is difficult to correlate the events in the two areas. The fact that the isotopic ages are about 1600 m.y. in both cases might argue for a correspondence of the amphibolite facies metamorphism in the Tobacco Roots with the metamorphism in the Gravellies. The facies, of course, need not be the same.

Only K–Ar ages have been calculated in the eastern part of southwestern Montana, so that it is possible that some of the 1700 m.y. ages to the northeast reflect an older time of metamorphism. Rb–Sr measurements might pick out this older event, which might have been the granulite facies metamorphism b of Reid.

There is good isotopic evidence that the 1600 m.y. age metamorphism dies out to the southeast of the area being discussed. Both in the Gravelly Range and the Gallatin Canyon (see figure 1) older ages, up to 3250 m.y., are found (see figure 1, table 5, and Giletti, 1966a,b).

Table 5 Mineral ages for the Teton Range, Wyoming

Description	Mineral	Age (m.y.) K–Ar	Age (m.y.) Rb–Sr	Reference[a]
Older gneiss (Bright-eyed gneiss, Death Canyon)	Biotite	1410	1290	1, 2
Pegmatite cutting older gneiss (above) (thought to be younger generation)	Microcline	—	1870	1
	Muscovite	1970	2500	1, 2
Pegmatite cutting older gneiss (thought to be younger generation)	Muscovite	2240	2350	2
Quartz monzonite	Muscovite	1680	—	2

[a] 1. Giletti and Gast (1961); 2. Menzie (1966).

Work in these transitional areas continues, and some complications in the pattern of ages have arisen, but the argument that these are at the margins of the 1600 m.y. age terrane still appears to hold. These results are not yet published.

Table 6 K–Ar mica ages from the southwestern Montana transition zone

Description	Mica	Age (m.y.)	Reference[a]
Gravelly Range N–S traverse			
Kyanite schist	Biotite	1520	1
Granite gneiss	Muscovite	1610	1
Granitic gneiss	Biotite	2620	1
Gallatin Canyon NW–SE traverse			
Granitic gneiss	Biotite	1690	1
Psammitic schist	Biotite	1790	1
Psammitic schist	Biotite	2070	2
Granitic layer in gneiss	Biotite	2430	2
Granitic layer in gneiss	Biotite	3250	1

[a] 1. Giletti (1966a); 2. Giletti (1966b).

All of the published isotopic data have been listed in tables 1, 2 and 6. If the results are now combined with those of Reid, any of the following three interpretations is possible:

(1) Three separate deformational–metamorphic episodes occurred. Referring to table 4, metamorphism c produced the isotopic mineral ages observed. In that case, the Pony and Cherry Creek may be much older than 1600 m.y. as there were two earlier metamorphisms. No clear-cut isotopic data bear on the latter point.

One whole rock analysis of Cherry Creek gneiss (see table 2) gave a mineral isochron of 1530 m.y., consistent with the regional pattern. The initial $^{87}Sr/^{86}Sr$ ratio was 0.744. Although this is quite high, the age obtained by assuming a normal value for the initial strontium is not much higher than the mineral age. Taking initial values of $^{87}Sr/^{86}Sr$ of 0.700 or 0.710 yields ages of 1700 or 1660 m.y. respectively. Thus a 1700 m.y. age would appear to be maximal for the Cherry Creek. However, the rock used violated one of the basic assumptions of the whole rock method, the premise that it was a homogeneous rock. This sample was of a layer in a layered gneiss. It is possible that the layer did not behave as a closed system to rubidium and strontium. If this happened, the 1530 m.y. mineral age is still correct, but the whole rock age could be equal to 1530 m.y. or older, and the 1700 m.y. date obtained would have little significance. Until evidence is available to demonstrate the validity of the whole rock age in this case, no reliance will be placed on it.

The rocks in the age transition area have not been correlated with any of the named units to the west. If they could be so correlated, then the older ages might indicate that the Pony and Cherry Creek are as old as the 2600 m.y. age rocks to the east in the Beartooth Range. There would be no reason to limit the extent of the earlier metamorphisms to the same area as that of the 1600 m.y. event. The argument that the metamorphic events need not coincide is also applicable, of course, to the Gravelly Range. The Cherry Creek may have existed prior to the earlier metamorphisms, as it must according to Reid, and still not have been seriously metamorphosed at the southern end of the Gravelly Range.

Dates could also be assigned to the earlier events of Reid, but with less assurance. The Pre-Cherry Creek gneiss gives a single whole rock age of 3100 m.y. (table 2). This, coupled with the 3250 m.y. K–Ar age in the eastern part of the area (table 5) and data to be discussed below in relation to the Beartooth Range, suggests that some event occurred in this general area about 3200 to 3000 m.y. ago. This could be Reid's metamorphism a.

The regional event, discussed earlier in this chapter, at about 2600 m.y. might then be assigned to the granulite facies metamorphism b of Reid. Again, the granulite facies need form only a part of the whole metamorphic terrane. In fact, the 2600 m.y. regional metamorphism may have extended into most of the terrane now giving 1600 m.y. ages.

Finally, the metamorphism d of Reid would simply be a part of the final stages of the 1600 m.y. event.

(2) A second possible interpretation would be similar to the first except that metamorphisms a and b occurred at times about 1600 m.y. ago as well as the c event. This would be the idea if this area behaved as did the Appalachian area during the Paleozoic. Again, the metamorphism d would simply be a late stage of c.

This interpretation would have the Pre-Cherry Creek in existence since about 3100 m.y. ago and the contact with the Cherry Creek would then be unconformable. This would imply that no new continent was formed in this area 1600 m.y. ago as the Pre-Cherry Creek was presumably already part of the continental crust. The same argument applies to interpretation (1), of course.

(3) The third possibility is that the critical metamorphism which reset all the ages in the area to 1600 m.y. was the 'Minor epidote–amphibolite facies retrogression d' of Reid (1963). In this case, the metamorphisms a, b and c are all displaced backward in time by one step. The reader can easily formulate the various possibilities.

Several positive statements can be made on the basis of the present data which are not part of the above interpretations. The amphibolite facies metamorphism c must have been intense enough to affect mineral ages drastically. The age of 1600 m.y. which is found in the area must represent either this or a later event. The only recorded later event is the epidote–amphibolite facies retrogression. This may have been minor as suggested by Reid. If it were both minor and local, then it is of little importance. In that case, however, it may not yet have been dated. If it was important enough to reset the ages, then this is the only one that has been dated. Clearly there is scope here for structural, stratigraphical studies in areas removed from the Laramide batholiths and the age transition area.

Recently, a similar situation has been reported from the Medicine Bow Range in Wyoming (see figure 1) by Hills, Gast and Houston (1965). Whole rock rubidium–strontium measurements have shown that an older terrane, 2500 ± 100 and 2450 ± 100 m.y. ages for two units in it, was metamorphosed at about 1500 m.y. ago and also intruded by a granite at about 1470 ± 50 m.y. ago. The data from the younger rocks appear to behave almost identically in the Montana and Wyoming areas. Details on this interesting area in Wyoming have not yet appeared in print.

B The Beartooth Range

Some of the early work in the region was that of Gast, Kulp and Long (1958) in the Beartooth Range. Their data were discussed above in connection with the 2600 m.y. event(s) in this region. This episode of regional metamorphism in the southeastern Beartooth at about 2600 m.y. is now established.

In 1964, Catanzaro and Kulp published uranium–lead data for zircons from the Beartooth Range. Their concordia plot is reproduced in figure 2. In referring to the 2600 m.y. event, which they took as 2700, they stated 'More recent measurements by the K–Ar method at this laboratory confirm a date of about 2700 for the latest metamorphic event, but also suggest an earlier metamorphism in the area as early as 3100 m.y.' These argon dates are not yet published.

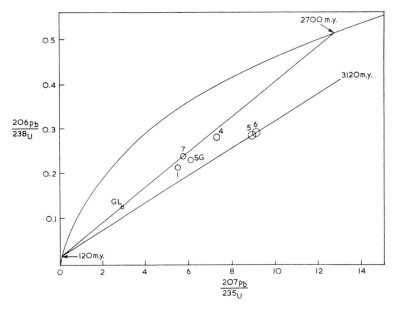

Figure 2 Beartooth Mountains—episodic lead loss interpretation. (After Catanzaro and Kulp, 1964)

Based on these results, Catanzaro and Kulp then listed a sequence of events taken from Eckelmann and Poldervaart (1957) with some modification, and assigned dates to these. Their table follows (see p. 119, Catanzaro and Kulp, 1964).
(1) Deposition of Archean sediments (source rocks > 3100 m.y. old).
(2) Metamorphism (>3100 m.y.).
(3) Intrusion of ultramafic and Stillwater complex.
(4) Last metamorphism and pegmatite formation (2700 m.y.).
(5) Uplift and peneplanation.
(6) Deposition of Paleozoic and younger sediments.
(7) Uplift, thrusting and emplacement of felsic porphyry dikes (Cretaceous).

These authors apparently suggest the >3100 m.y. dates on the basis of the 120 to 3120 m.y. chord in figure 2, plus the K–Ar data they refer to. It should be noted that Eckelmann and Poldervaart (1957) proposed a single metamorphic episode which included the granitization they described. This would correspond to the 2600 (or 2700) m.y. episode.

The zircon data are interpreted on the basis of an episodic lead loss (or uranium gain) model. There are Cretaceous intrusives in the area to which these authors assign an age of 120 m.y. They then suggest that significant lead was lost at that time. The loss mechanism is not described in detail, although they state that it is not a simple case of loss of lead increasing with proximity to the intrusives.

If significant lead was lost 120 m.y. ago, then a lead loss chord can be passed through the 120 m.y. point on the concordia curve. This was done and the line was also passed through the sample 7 analytical data point. From this, the youngest age zircon which could give the data values of sample 7 and which suffered lead loss 120 m.y. ago would have to have been 2700 m.y. old.

A chord was also passed through the same 120 m.y. point and then drawn through sample points 5 and 6. This yields an intercept on concordia at 3120 m.y. The authors' argument is that there were really two times of episodic lead loss. The earlier was at 2700 m.y. and the later at 120 m.y. ago. The reader is referred to Catanzaro's discussion in this volume for more details.

The result of these episodic losses would be as follows. Depending on the fractions of lead lost at the 2700 m.y. and the 120 m.y. episodes, the sample points could lie anywhere in the triangle determined by the 3120, 2700 and 120 m.y. points. The sample points do, of course, lie within this triangle by construction. Further, samples 7 and GL would have lost respectively all and practically all their lead during the 2700 m.y. event. Samples 5 and 6 would have lost none of their lead 2700 m.y. ago.

In order to allow for all possibilities, it must be recognized that even samples 5 and 6 might have lost some lead 2700 m.y. ago. If they had, then their original age would have been greater than 3120 m.y. For this reason, Catanzaro and Kulp list ages greater than 3120 m.y. for their events 1 and 2 in the table. Similarly, if samples 7 and GL had not lost all their lead 2700 m.y. ago, the 2700 m.y. event might be younger.

A second possibility must be considered in connection with the interpretation of these data, however. The zircon separates used in this study came from a variety of different rocks, some of which were metasedimentary. Further, these authors agree with Eckelmann and Poldervaart (1957) that the zircons from this area are often rounded and overgrown. The rounding is suggestive of a detrital origin. If the overgrowths were added at the 2600 (or 2700) m.y. event, their addition might well appear as simple uranium gain by the zircon (mathematically equivalent to lead loss).

The critical point here, is that the detrital zircons need not, in fact probably did not, come from one source with one age. The populations analyzed might have been combinations of different age zircons, where the different samples had different proportions of the different source age zircons. In this case, it would not be surprising if the points scattered as they do. There would still be a likelihood of an igneous or metamorphic event 3120 m.y. ago or earlier, but it would have occurred in the rocks of the source area of the sediments and not necessarily where the samples were collected.

From the arguments given above, it is clear that the potassium–argon ages mentioned by Catanzaro and Kulp are critical. Although these have not been published, Butler (1966) reports them in summary form: '(1) Minerals from pegmatites just west of the Cathedral Peak area give ages of about 2700 m.y. or slightly younger. (2) Biotite from pegmatite cutting the ultramafic zone of the Stillwater complex has an apparent age of 2660 m.y. (3) Biotite from hornfels in the aureole of the Stillwater complex has an apparent age of 3065 m.y.'

The argument for a metamorphism in the rocks adjacent to the Stillwater at some time previous to 3100 m.y. ago probably is that the biotite in the hornfels gives the intrusive age or a date younger than the intrusion. The preintrusion metamorphism must, therefore, have been older. Some of these data will be discussed further in connection with the Stillwater complex.

It might also be noted that the 120 m.y. lead loss event is assigned that age because of the Cretaceous intrusives in the area. As no geometric relation between lead loss and proximity to the intrusives exists, this might be another case of discordant lead–uranium ages where, if the zircons had had a simple earlier history, they would all fall on a chord which passed approximately through the origin. As numerous such cases exist and are not well understood, there would be about as much justification in drawing the chords in figure 2 through the origin as through the 120 m.y. point. This would have the effect of reducing the 2700 and 3120 m.y. dates by a few tens of millions of years.

One of the major units in the Beartooth Range is the Stillwater complex, a mafic, layered intrusive. This rock intruded a sequence of metamorphosed sediments which can be seen at the base of the Stillwater at its western end. Howland (1954) suggested that the Stillwater intruded the sedimentary rocks before they were metamorphosed. The hornfelsed zone below the Stillwater was produced in unmetamorphosed or low-grade metamorphic rocks. However, Butler (1966) suggests that a period of folding, regional metamorphism, and granitization occurred prior to the intrusion. Another regional metamorphism and associated pegmatite formation occurred after the intrusion of the Stillwater.

All of the data from the Stillwater were obtained by the K–Ar method. The quotation from Butler's paper includes a reference to a 3065 m.y. biotite age in the hornfels zone of the Stillwater contact. This can be taken as the minimum age for the Stillwater.

Butler also mentions other measurements by Kulp. One is for a biotite from a pegmatite in the Stillwater, which gives a 2660 m.y. age. Schwartzman (1966) reports K–Ar ages from the eastern end of the Stillwater complex (Mouat Mine area). Two biotites from a mafic pegmatite and a poikilitic harzburgite in the ultramafic zone of the Stillwater gave ages of 2470 and 2600 m.y. respectively.

It may be that the 2600 m.y. regional metamorphism affected the southeastern end of the Stillwater to give the ages just discussed, while the effect of this event to the west was smaller or absent. This would explain the difference in the measured ages in the east and west. The heating of the Stillwater at about 2600 m.y. ago could have driven argon from already existing biotites. Hence, no new minerals or rocks need have formed in the Stillwater 2600 m.y. ago.

It can thus be stated that the Stillwater complex is at least as old as the 3065 m.y. biotite in the hornfels. The 2600 m.y. metamorphism was probably the agent responsible for the ages of about 2600 m.y. observed in the eastern Stillwater.

C The Little Belt Range

This area was discussed earlier in connection with the 2600 m.y. age terrane of the Montana–Wyoming region and is discussed by Catanzaro in another part of this volume. It is also of interest because Catanzaro and Kulp (1964) reported an event whose age is different from those discussed thus far in the region. This age of 1920 m.y. derives from a study of lead–uranium isotopic ages from this area (see figure 3).

There is a clear demarcation, geographically, between the samples which fall on the 120–2470 m.y. chord and those to the south which fall on the 120–1930 m.y. chord. These authors suggest that two metamorphic events occurred in the area at 2470 or more million years ago and at some time prior to 1800 m.y. ago, probably 1920 m.y.

ago. Additional evidence for the 1920 m.y. event can be found in table 2 where several K–Ar ages were measured and several gave values of approximately 1800 m.y.

It appears likely that some metamorphic event occurred at about 1900 m.y. ago. The spread in the K–Ar dates from 1380 to 1820 m.y. makes it difficult to establish whether one or more events took place at about that time.

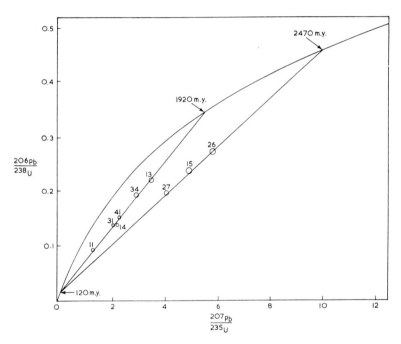

Figure 3 Little Belt Mountains—episodic lead loss interpretation. (After Catanzaro and Kulp, 1964)

There is always some question about excess, or inherited, radiogenic argon when amphiboles are used. The effect of this inherited argon would be to yield an age which is greater than the true age. This is especially pronounced for low potassium content minerals. In this case, a plot of percent potassium vs. age reveals no systematic behavior. The only curious fact about these data is that the potassium contents are all rather low for the biotites although the potassium contents for the amphiboles are not. The four biotites have potassium contents of 2.99, 5.74, 3.20 and 4.27% in the order listed in table 2.

The nine amphiboles range from 0.98 to 2.27% potassium. The significance of this is not clear.

This is the first instance of data suggesting a metamorphic or other event at about 1900 m.y. in this area. If such an event occurred to the south, it has been obscured by the 1600 m.y. metamorphism. Owing to the isolated nature of the Little Belt Range relative to the other Precambrian basement in Montana, it is possible that the 1900 m.y. event was areally quite significant, but that the evidence for it is now buried. The data all derive from a north–south traverse about ten miles long, so that no areal inferences can be drawn with regard to that part of Montana.

D *The Teton Range*

The Teton Range of western Wyoming has proved to be one of the more complex areas in this region from a geochronological viewpoint. The geological history was recently described by Reed (1963) in a preliminary report. Reference to other workers may be found in that paper.

The sequence of events as given by Reed is:
(1) Deposition of a thick sedimentary sequence.
(2) Regional metamorphism, intense shearing, isoclinal folding and some granitization.
(3) One or more folding episodes where older folds were refolded.
(4) Intrusion of quartz monzonite and pegmatite.
(5) Long interval of cooling and fracturing.
(6) Intrusion of diabase dikes.
(7) Erosion, subsidence, Cambrian deposition.

The earliest measurements made on these rocks were by the rubidium–strontium method (Giletti and Gast, 1961). They were highly discordant as can be seen in table 5.

Additional data have been obtained, principally by the K–Ar method, by Menzie (1966) and the present author. These have not been published as yet, but they are consistent with the following interpretation of the data.

No clear evidence has been obtained on the age of the older metamorphism and granitization. The pegmatites related to the quartz monzonite intrusion (the younger generation pegmatites) give muscovite Rb–Sr ages of 2350 and 2500 m.y. The highest K–Ar ages measured in the Tetons are for muscovites. The suggestion made in Giletti and Gast (1961) that the discordances resulted from the different ways minerals respond to a given metamorphic event or reaction with groundwater still appears valid. This would place the muscovite as the most resistant to change, the potassium feldspar next and the biotite least resistant.

This interpretation would make all the ages minimal, and make the pegmatites at least 2500 m.y. old. If so, then all the events described by Reed as having occurred prior to the intrusion of the quartz monzonite would have to be older than 2500 m.y.

It is tentatively suggested that the quartz monzonite intruded this terrane at about 2600 m.y. ago. The older events may have been earlier episodes in the same time span or they may be related to older events such as the possible 3100 m.y. events to the north and west of the Tetons.

The cause for the lowering of all, or nearly all, of the ages in the Tetons remains unclear. Events subsequent to the pegmatite formation include intrusion of a limited number of dolerite dikes, formation of epidote veins and the deep burial of the rocks, attested to by their present vertical displacement along the Teton fault on their eastern margin.

The diabase dikes form an insignificant fraction of the total rock in the Tetons. Several of the samples are remote from the dikes and could not have been affected by them. This, then, does not appear to explain the discordances.

The epidotization occurs as veins which are quite narrow, up to a few millimeters thick, and have a nonuniform areal distribution. These may be the only recorded evidence of a significant event at some time in the past, but this has not been demonstrated.

There is little doubt that the Precambrian in the Tetons was rather deeply buried. The Teton fault along the eastern margin of the range has a vertical displacement of at least 14,000 feet and occurred since the Cretaceous (Edmund, 1956). If a normal geothermal gradient of about 30°c/km depth is assumed, some of these rocks may have been at temperatures above 150°c for some time. It is likely that the extensive volcanic activity in the nearby areas of Idaho and Wyoming is an indication of an abnormally high thermal gradient. If this is true, burial may well be the critical factor. However, this remains to be demonstrated.

If the age data are used to set time limits on when the event which caused loss of radiogenic nuclides took place, the event must have occurred at some time since about 1300 m.y. ago. This is about the youngest age obtained. If more than one event is responsible for the discordances, the latest must have occurred since 1300 m.y. ago.

At present this limit is not particularly useful, but it does suggest that if the 1600 m.y. event affected the Tetons something happened since then to remove significant evidence of its existence.

There are a few ages in the Wind River Range which are lower than the 2400 to 2700 m.y. values. It is possible that the events which lowered the Teton ages also affected those in the Wind Rivers.

VII Regional Summary

It is sometimes useful to have some idea as to the extent of continental crust at different times in the past. In order to obtain such a map, it becomes necessary to define the continental crust. It will be assumed for purposes of the following discussion that plutonic, igneous and/or metamorphic rocks of granitic composition or compositions approaching granite will constitute portions of a continent. The question of the areal extent which such rocks must possess in order to justify a continental assignment can be difficult, but will largely be avoided in this discussion as really small areas are not considered.

Owing to the nature of radiometric dating, it is sometimes not possible to determine the maximum age of a rock or terrane. In those cases, all that can be said is that continental crust has existed in that locality since a certain time, but may have existed for some time before that. As new developments and refinements of technique occur, some of these vague times predating the present limits of detection may be defined.

The chronology of this region begins with the least well defined ages. Scattered Rb–Sr, K–Ar and uranium–lead zircon ages from the southern part of Montana all suggest that some event occurred at about 3000 to 3200 m.y. ago. Going from west to east, the Pre-Cherry Creek gneiss in the Blacktail Range, which lies between the Tendoy and Ruby Ranges (figure 1) has given a single whole rock age of 3080 m.y. (see table 2). Biotite from the Gallatin Canyon has given a K–Ar age of 3250 m.y. (table 6) and a hornblende from this same area has given a K–Ar age of 3470 m.y. As both of the latter are in the age transition zone mentioned earlier, the exact significance of these values is not yet clear.

Brown and Brookins (Abstracts. *1965 Geol. Soc. Am.*) reported that rocks as old as 2900 m.y. may exist in the Jardine–Crevasse area. This area is immediately southwest of the main Beartooth Range block (figure 1) and is just north of the Montana–Wyoming boundary.

Isotopic geochronology of Montana and Wyoming 143

In the central and eastern Beartooth Range, zircon analyses suggest that the area may have undergone a metamorphism at about 3100 m.y. or earlier. This is subject to the qualifications discussed above. Catanzaro and Kulp allude to K–Ar data supporting this suggestion.

If all of the above areas were continental at about 3100 m.y. ago, they would have formed a strip of land about 300 km long running about east–west near the southern boundary of Montana. The tenuous nature of this suggestion must be stressed.

If 2400 m.y. be taken as a time when most of the events which cluster around 2600 m.y. were essentially completed, then the continental area can be defined with much greater assurance than for 3100 m.y. time. The area includes nearly all of the Precambrian discussed in this chapter. It includes all of the areas in, and bounded by the Blacktail Range on the west, the Little Belt Range on the north, the Bighorn and Medicine Bow Ranges on the east and south and the Wind River and Teton Ranges on the southwest and west.

At 1400 m.y. ago, the area had undergone the possible regional event at 1900 m.y. in the Little Belt Range and the 1600 m.y. regional metamorphism and igneous intrusion in the Tendoy, Blacktail, Ruby and Madison Ranges in the west, the Black Hills on the east and the Medicine Bow Range on the south. Extensive orogeny was still going on in the southern Rocky Mountain states.

The continent now extended, probably continuously, from the area just outlined to the Canadian Shield. Deep well samples from North Dakota have shown this (Burwash and others, 1962; Peterman and Hedge, 1964). These workers also place the maximum extent of the 1600 (or 1700) m.y. terrane at a line running about north–south approximately at the 102° W meridian. The Superior Province of the Canadian Shield, of about 2600 m.y. age, lies to the east of that line.

Deposition of the Belt Series of sediments began after 1600 m.y. ago and was probably complete by about 1200 m.y. ago.

Numerous dolerite dikes intruded the Precambrian of the region. Some have since been metamorphosed. Little information is available on the isotopic ages of these dikes. In some instances, such data would greatly enhance the knowledge of the chronology of the region or of an area.

The next major igneous event in the area was the emplacement of the batholiths at about 75 m.y. ago. These include the Boulder and the Tobacco Root Batholiths.

The relation to continental growth of these dated regions is worth discussing briefly. The idea of lateral accretion by continents to produce ever larger continental areas as suggested by J. T. Wilson might be applied here. A continental nucleus would have grown until it joined up with another which we call the Canadian Shield. The period of growth would have been from a small start at about 3100 m.y. ago to about 1600 m.y. ago. There is no reason to believe that the Montana–Wyoming region was not all continent after the 1600 m.y. events.

One curious fact emerges from the data of Catanzaro and Kulp (1964), however. Although the area studied is quite small, the position of the different age rocks is the reverse of what a continental growth hypothesis would imply. The rocks which give a 2470 m.y. zircon age lie to the north of those which yield a 1920 m.y. age (see Catanzaro, figure 7, this volume). This may be the first clue in Montana that the younger events (1900 and 1600 m.y.) occupied a belt within the 2600 m.y. age terrane. A similar occurrence in southwestern Montana where the 1600 m.y. event was superimposed on what are probably 3100 m.y. rocks (the Pre-Cherry Creek gneiss) raises the same question.

There is little doubt that new rocks formed at each of the times discussed in this chapter. It is not clear if these rocks formed new continent or merely displaced or buried preexisting continental rocks. The latter process seems to have been the significant one at 1600 m.y. ago in the Little Belt, Blacktail, Gallatin and Medicine Bow Ranges and possibly in most of southwestern Montana.

References

Aldrich, L. T., Wetherill, G. W. and Davis, G. L. (1957). Occurrence of 1350 million-year-old granitic rocks in western United States. *Bull. Geol. Soc. Am.*, **68**, 655.

Aldrich, L. T., Wetherill, G. W., Davis, G. L. and Tilton, G. R. (1958). Radioactive ages of micas from granitic rocks by Rb–Sr and K–A methods. *Trans. Am. Geophys. Union*, **39**, 1124.

Baadsgaard, H., Folinsbee, R. E. and Lipson, J. (1961). Potassium–argon dates of biotites from Cordilleran granites. *Bull. Geol. Soc. Am.*, **72**, 689.

Bassett, W. A. and Giletti, B. J. (1963). Precambrian ages in the Wind River Mountains, Wyoming. *Bull. Geol. Soc. Am.*, **74**, 209.

Burwash, R. A., Baadsgaard, H. and Peterman, Z. E. (1962). Precambrian K–Ar dates from the western Canada sedimentary basin. *J. Geophys. Res.*, **67**, 1617.

Butler, J. R. (1966). Geologic evolution of the Beartooth Mountains, Montana and Wyoming. Part 6. *Bull. Geol. Soc. Am.*, **77**, 45.

Catanzaro, E. J. and Gast, P. W. (1960). Isotopic composition of lead in pegmatitic feldspars. *Geochim. Cosmochim. Acta*, **19**, 113.

Catanzaro, E. J. and Kulp, J. L. (1964). Discordant zircons from the Little Belt (Montana), Beartooth (Montana) and Santa Catalina (Arizona) Mountains. *Geochim. Cosmochim. Acta*, **28**, 87.

Eckelmann, F. D. and Poldervaart, A. (1957). Geologic evolution of the Beartooth Mountains, Montana and Wyoming. Part 1. *Bull. Geol. Soc. Am.*, **68**, 1225.

Eckelmann, W. R. and Kulp, J. L. (1957). Uranium–lead method of age determination. Part II. *Bull. Geol. Soc. Am.*, **68**, 1117.

Edmund, R. W. (1956). Resumé of structures and physiography in the northern Teton Mountains, Wyoming. *Wyoming Geol. Assoc., Guidebook Ann. Field Conf.*, 151.

Gast, P. W., Kulp, J. L. and Long, L. E. (1958). Absolute age of early Precambrian rocks in the Bighorn Basin of Wyoming and Montana, and southeastern Manitoba. *Trans. Am. Geophys. Union*, **39**, 322.

Giletti, B. J. (1966a). Isotopic ages from southwestern Montana. *J. Geophys. Res.*, **71**, 4029.

Giletti, B. J. (1966b). Isotopic effects at the margin of a regional metamorphism. Abstract. *Trans. Am. Geophys. Union*, **47**, 195.

Giletti, B. J. and Gast, P. W. (1961). Absolute age of Precambrian rocks in Montana and Wyoming. *Ann. N. Y. Acad. Sci.*, **91**, 454.

Giletti, B. J., Moorbath, S. and Lambert, R. St. J. (1961). A geochronological study of the metamorphic complexes of the Scottish Highlands. *Quart. J. Geol. Soc. London*, **117**, 233.

Gulbrandsen, R. A., Goldich, S. S. and Thomas, H. H. (1963). Glauconite from the Precambrian Belt Series, Montana. *Science*, **140**, 390.

Harris, R. L. (1959). Geologic evolution of the Beartooth Mountains, Montana and Wyoming. Part 3. *Bull. Geol. Soc. Am.*, **70**, 1185.

Hart, S. R. (1964). The petrology and isotopic-mineral age relations of a contact zone in the Front Range, Colorado. *J. Geol.*, **72**, 493.

Hayden, R. J. and Wehrenberg, J. P. (1960). ^{40}A–^{40}K dating of igneous and metamorphic rocks in western Montana. *J. Geol.*, **68**, 94.

Heinrich, E. W. (1960). Pre-Beltian geology of the Cherry Creek and Ruby Mountains areas, southwestern Montana. Part 2. *Montana Bur. Mines Geol., Mem.*, **38**, 15.

Heinrich, E. W. and Rabbit, J. C. (1960). Pre-Beltian geology of the Cherry Creek and Ruby Mountains areas, southwestern Montana. Part 1. *Montana Bur. Mines Geol., Mem.*, **38**, 1.

Hess, H. H. (1960). Stillwater igneous complex, Montana. *Geol. Soc. Am., Mem.*, **80**.

Hills, A., Gast, P. W. and Houston, R. S. (1965). Chronology of some Precambrian igneous and metamorphic events of the Medicine Bow Mountains, Wyoming. Abstract. *Geol. Soc. Am., Spec. Papers*, **82**, 92.

Houston, R. S. and Parker, R. B. (1963). Structural analysis of a folded quartzite, Medicine Bow Mountains, Wyoming. *Bull. Geol. Soc. Am.*, **74**, 197.

Howland, A. (1954). Relations of regional and thermal metamorphism near the base of the Stillwater Complex, Montana. *Bull. Geol. Soc. Am.*, **65**, 1264.

Jaeger, J. C. (1964). Thermal effects of intrusions. *Rev. Geophys.*, **2**, 443.

Jones, W. R., Peoples, J. W. and Howland, A. L. (1960). Igneous and tectonic structures of the Stillwater complex, Montana. *U.S. Geol. Surv. Bull.*, **1071-H**, 281.

Kerr, P. F. and Kulp, J. L. (1952). Pre-Cambrian uraninite, Sunshine Mine, Idaho. *Science*, **115**, 86.

Knopf, A. (1964). Time required to emplace the Boulder Bathylith, Montana: a first approximation. *Am J. Sci.*, **262**, 1207.

Long, A., Silverman, A. J. and Kulp, J. L. (1960). Isotopic composition of lead and Precambrian mineralization of the Coeur D'Alene district, Idaho. *Econ. Geol.*, **55**, 645.

Menzie, J. C. (1966). Potassium–argon ages of the Teton Range, Wyoming. *M.A.T. Thesis*. Brown Univ., Providence, R. I.

Peterman, Z. E. and Hedge, C. E. (1964). Age of basement rocks from the Williston Basin of North Dakota and adjacent areas. *U.S. Geol. Surv., Profess. Papers*, **475-D**, D100.

Reed, J. C., Jr. (1963). Structure of Precambrian crystalline rocks in the northern part of Grand Teton National Park, Wyoming. *U.S. Geol. Surv., Profess. Papers*, **475-C**, C1.

Reid, R. R. (1963). Metamorphic rocks of the northern Tobacco Root Mountains, Madison County, Montana. *Bull. Geol. Soc. Am.*, **74**, 293.

Ross, C. P. (1963). The Belt Series in Montana. *U.S. Geol. Surv., Profess. Papers*, **346**.

Schwartzman, D. W. (1966). Excess argon in minerals from the Stillwater complex. *M.Sc. Thesis*. Brown Univ., Providence, R. I.

Spencer, E. W. (1959). Geologic evolution of the Beartooth Mountains, Montana and Wyoming. Part 2. *Bull. Geol. Soc. Am.*, **70**, 467.

Sutton, J. and Watson, J. (1950). The pre-Torridonian metamorphic history of the Loch Torridon and Scourie areas in the northwest highlands, and its bearing on the chronological classification of the Lewisian. *Quart. J. Geol. Soc. London*, **106**, 241.

The interpretation of lead isotopes and their geological significance

E. R. KANASEWICH*

I *Interpretation theory*, 147. II *The search for ordinary leads*, 171.
III *The interpretation of anomalous lead results*, 185.
Appendix 1: Theory of multi-stage models, 208.
Appendix 2: A table for plotting single-stage growth curves, 214.
Appendix 3: Lead isotope data on deposits used for defining a single-stage growth curve, 215. *References*, 216.

I Interpretation Theory

A Introduction

Significant advances have been made in the last few years in the field of geochronology and, indeed, in all branches of the earth sciences. The instrumentation has been greatly improved and chemical procedures have been developed which allow the preparation of much purer samples. In the study of lead isotope abundances, precision greater by nearly an order of magnitude is being obtained by both gas- and solid-source techniques as compared to results several years ago. In the interpretation of old data, researchers necessarily relied on much averaging and an overall statistical treatment before reaching conclusions of any validity. The new data are falling into striking patterns which are amenable to mathematical analysis, and the number of models which fit this pattern is quite restricted. The net result of these developments is that the physical and chemical properties of the crust–mantle system may be outlined more accurately with reference to the elements of uranium, thorium and lead.

* Dept. of Physics, University of Alberta, Edmonton, Alberta, Canada.

Geophysical application of radioactivity began with Boltwood's (1907) use of lead, uranium and thorium analyses to estimate the age of minerals. In 1911 Soddy obtained experimental proof that there existed atoms of different atomic weight but identical chemical properties and in 1913 he proposed the word isotopes for such pairs of elements as ^{232}Th and ^{230}Th. The end products of the ^{238}U, ^{235}U and ^{232}Th radioactive series are the stable isotopes ^{206}Pb, ^{207}Pb and ^{208}Pb respectively. A fourth stable isotope of lead, ^{204}Pb, occurs naturally and has not been generated by any radioactive decay. Mass spectrometers can only measure relative abundances so it is useful to express all the other lead isotope abundances in terms of the number of ^{204}Pb atoms present.

Only *common leads* will be considered here. These are leads which now have an insignificant amount of uranium and thorium associated with them. The isotopes of ^{206}Pb, ^{207}Pb and ^{208}Pb in any lead may be thought of as consisting of two parts:

Common lead = Primeval lead + Radiogenic lead

The isotopic composition of lead existing at the time when the Earth formed is termed primeval lead and within present-day experimental error it appears to be the same for the Earth and meterorites. Radiogenic lead consists of the isotopes generated by uranium and thorium decay since the formation of the Earth. These radiogenic atoms, which are indistinguishable from primeval atoms, become mixed with the original lead.

The parent radioactive nuclei of ^{238}U, ^{235}U and ^{232}Th each produce another unstable nucleus so that a decay series results with the end product being a stable lead isotope. The decay systems may be indicated symbolically as

$$^{238}\text{U} \xrightarrow{\lambda_{238} = 1.54 \times 10^{-10} \text{ yr}^{-1}} {}^{206}\text{Pb} + 8 \ {}^{4}\text{He} + \text{Energy}$$

$$^{235}\text{U} \xrightarrow{\lambda_{235} = 9.72 \times 10^{-10} \text{ yr}^{-1}} {}^{207}\text{Pb} + 7 \ {}^{4}\text{He} + \text{Energy}$$

$$^{232}\text{Th} \xrightarrow{\lambda_{232} = 0.499 \times 10^{-10} \text{ yr}^{-1}} {}^{208}\text{Pb} + 6 \ {}^{4}\text{He} + \text{Energy}$$

Each nucleus is characterized by its own probability of decay, λ. The decay probabilities of the intermediate members of the thorium and uranium series are large in comparison to those of the parent isotopes, the smallest being the 2.8×10^{-6} yr^{-1} value for ^{234}U in the ^{238}U decay series. Therefore after about one million years the rates of decay of the intermediate members are controlled by the rate of decay of the parent uranium and thorium nuclei.

For the purposes of geochronology it is convenient to let the present be $t = 0$ and count time positively into the past. Let N_0 be the number of atoms at the present time, then the number of radioactive atoms at any other time is

$$N = N_0\, e^{\lambda t} \tag{1}$$

The variation due to radioactive decay is easily calculated and is illustrated in figure 1, for some typical abundances.

For the moment let us assume that the uranium, thorium and lead have remained in a locally closed system from time t_0, when the Earth formed, to a time t_1, when mineralization isolated the lead from the uranium and thorium. Under these conditions the number of ^{206}Pb

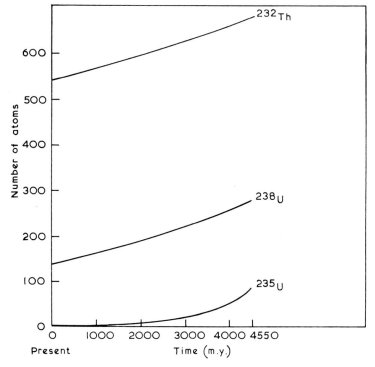

Figure 1 Radioactive decay of uranium and thorium isotopes as given by equation (2). The amount of ^{235}U is taken to be unity at the present time. The present-day ratio of ^{238}U/^{235}U is 137.8 in all natural substances. The average ratio of Th/U for the Earth is 3.9.

atoms present at time, t_1, in any such closed system is

$$N(^{206}\text{Pb})_{t_1} = N(^{206}\text{Pb})_{t_0} + N(^{238}\text{U})_{t_0} - N(^{238}\text{U})_{t_1}$$

Both sides of the equation may be divided by the number of ^{204}Pb atoms, while the uranium abundances at times t_0 and t_1 may be expressed in terms of equation (1).

$$\frac{N(^{206}\text{Pb})_{t_1}}{N(^{204}\text{Pb})} = \frac{N(^{206}\text{Pb})_{t_0}}{N(^{204}\text{Pb})} + \frac{N_0(^{238}\text{U})}{N(^{204}\text{Pb})} e^{\lambda t_0} - \frac{N_0(^{238}\text{U})}{N(^{204}\text{Pb})} e^{\lambda t_1}$$

In terms of the symbols and constants shown in table 1 this equation for the ^{206}Pb/^{204}Pb ratio becomes

$$x = a_0 + \mu(e^{\lambda t_0} - e^{\lambda t_1}) \tag{2}$$

Similar equations may be written for y and z, the ^{207}Pb/^{204}Pb and ^{208}Pb/^{204}Pb ratios.

$$y = b_0 + \frac{\mu}{137.8}(e^{\lambda' t_0} - e^{\lambda' t_1}) \tag{3}$$

$$z = c_0 + W(e^{\lambda'' t_0} - e^{\lambda'' t_1}) \tag{4}$$

The method by which one obtains the primeval ratios, a_0, b_0 and c_0 and the age of the Earth, t_0, will be discussed in a later section. These constants are nearly universally agreed to at the present time and are even used by the Berne group (see Houtermans, Eberhardt and Ferrara, 1964). Therefore we will not repeat the arguments over which 'age of the Earth' to use, particularly as these have been adequately discussed many times in the past (see Richards, 1962). If the lead is separated from the uranium and thorium at time t_1 by the formation of a crystal of galena, the isotope ratios are frozen at the values x, y, z and serve as fossil indicators of this event. A plot of y against x or z against x will define a series of *growth curves* showing how the isotope ratios changed with time. In the present-day ratio of ^{238}U/^{204}Pb, μ may be eliminated between equations (2) and (3) as noted by Houtermans (1946).

$$\frac{y - b_0}{x - a_0} = \phi = \frac{(e^{\lambda' t_0} - e^{\lambda' t_1})}{137.8(e^{\lambda t_0} - e^{\lambda t_1})} \tag{5}$$

On a graph of y against x this equation defines a straight line known as Houtermans' *isochron*.

$$y = b_0 + \phi(x - a_0) \tag{6}$$

The isochron passes through the primeval abundances, a_0 and b_0 and has a slope determined only by the times t_0 and t_1. If mineralization

Table 1 Symbols and constants used in age determination

Isotope ratio	4.55 billion years ago	At present time
$\dfrac{^{206}\text{Pb}}{^{204}\text{Pb}}$	$a_0 = 9.56^a$	x
$\dfrac{^{207}\text{Pb}}{^{206}\text{Pb}}$	$b_0 = 10.42^a$	y
$\dfrac{^{208}\text{Pb}}{^{204}\text{Pb}}$	$c_0 = 30.0^b$	z
$\dfrac{^{238}\text{U}}{^{204}\text{Pb}}$	$2.0124\,\mu$	μ
$\dfrac{^{238}\text{U}}{^{235}\text{U}}$	3.33	137.8^c
$\dfrac{^{232}\text{Th}}{^{204}\text{Pb}}$	$1.2548\,W$	W

Parent atom	Decay constant in 10^{-9} yr^{-1}	Half-life in units of 10^9 yr
^{235}U	$\lambda' = 0.9722^d$	0.713
^{238}U	$\lambda = 0.1537^e$	4.49
^{232}Th	$\lambda'' = 0.0499^f$	13.9

[a] Murthy and Patterson (1962).
[b] Kanasewich and Farquhar (1965).
[c] Inghram (1947).
[d] Fleming, Ghiorso and Cunningham (1952).
[e] Kovarik and Adams (1955).
[f] Picciotto and Wilgain (1956).

occurred contemporaneously for a group of locally closed systems with differing initial amounts of uranium, then the measured ^{206}Pb/^{204}Pb and ^{207}Pb/^{204}Pb ratios will lie on a straight line as indicated in figure 2.

The model outlined above is much too simple to describe the geochemical history of many leads which are known to exist and a somewhat more realistic model will be described in the next section.

B Multi-stage lead model

If the history of uranium and thorium concentration for any system is known, it is possible to calculate the lead isotope composition with an integral equation derived from the theory of radioactive decay

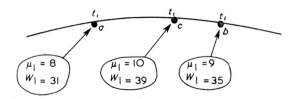

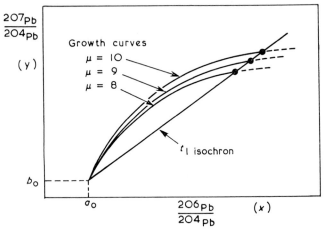

Figure 2 The upper diagram shows three locally closed systems within the Earth's crust with mineralization occurring at time t_1. The lower graph illustrates the growth curves along which the lead isotopes develop. After mineralization the isotope ratios will lie along the t_1 isochron.

(see appendix 1, equations A6, A7 and A8). In practice, it is the isotopic composition of a group of common leads which is known and the inverse problem must be solved to determine the history of lead, uranium and thorium concentration in the area. The ^{238}U/^{204}Pb and ^{232}Th/^{204}Pb ratios change because of radioactive decay and also by the addition or subtraction of uranium and thorium to the system. The ratios, μ and W, specifying the amount of radiogenic parent in the system, are given in terms of their present-day abundances. They may vary from place to place but, because metamorphic and orogenic events are episodic *locally* and brief in comparison to the half-lives of uranium and thorium, it seems reasonable to treat them as constants for discrete intervals of time. The intervals defining each stage may be as short as desired so this form of a multi-stage lead model is not only very simple but also quite general.

Each stage of lead isotope development is, in a closed system, characterized by the parameters μ_1, W_1 at time t_0 when the Earth formed; μ_2, W_2 at a subsequent time, t_1; μ_3, W_3 at time t_2; ..., ..., ...; μ_m, W_m at time t_{m-1}. The time of mineralization, t_m, is when the lead is separated from the uranium and thorium, usually by the formation of a crystal of galena. Since the lead isotope ratios do not change from time t_m to the present, they serve as fossil indicators preserving an integrated history of the radiogenic development from the time the Earth formed to the period of final mineralization. As such, they yield only indirect information about the processes occurring during the formation of a mineral deposit.

Model lead age determinations are fundamentally different from potassium–argon dates on biotites and other minerals and they are also quite distinct from lead–uranium ages on zircons. A flow diagram illustrating this difference is shown in figure 3.

Zircons containing uranium but negligible common lead may be formed during an orogenic phase of the Earth's developmental history. No information is contained about the period prior to mineralization but the crystal may survive subsequent metamorphic episodes to yield a true lead–uranium age. Even if lead or uranium is lost by chemical leaching or diffusion, the $^{207}Pb/^{235}U$ and $^{206}Pb/^{238}U$ ratios are modified and with analyses on many zircons, it is possible to enrich the historical record with these details. Biotites may lose most of their argon during a period of metamorphism and, since there is only one radioactive decay system, it is difficult to verify if diffusion has occurred. The last phases of tectonic activity in any area are usually determined most precisely by potassium–argon dating. The lead isotope ratios from common leads supply information on the very earliest stages of development of the Earth's crust and also on the subsequent episodes of orogenesis. However, the analysis of a single sample does not supply enough data to make such a comprehensive calculation and many very precise isotope determinations are necessary. It must also be remembered that periods of sulfide mineralization may not be related to processes of biotite and zircon generation.

C *Ordinary leads*

An ordinary lead is one which has been produced in a single-lead–uranium–thorium system between the time of formation of the Earth and the time of mineralization. Gerling (1942), Holmes (1946)

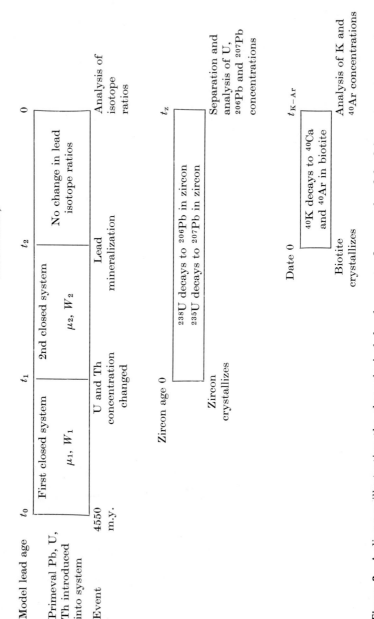

Figure 3 A diagram illustrating the chronological development of a sample of lead in a two-stage system and its possible relationship to lead–uranium and potassium–argon dates.

and Houtermans (1946) assumed that the ratios of ^{238}U, ^{235}U and ^{232}Th to ^{204}Pb remained constant in locally closed systems and the lead isotope ratios developed along a series of growth curves as already shown in figure 2. If mineralization occurred contemporaneously for all the locally closed systems then the measured ratios of ^{206}Pb/^{204}Pb and ^{207}Pb/^{204}Pb will lie on an isochron having a slope related to the age of the Earth and the time of mineralization, t_1.

The interpretation presented above was known to be an oversimplification of geological history but it seemed to be borne out by a large number of analyses which were made in the decade prior to 1960. This is illustrated by figures 4 and 5 in which all the measured

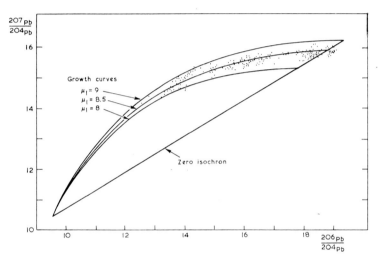

Figure 4 Lead isotope ratios from all over the world which appear to fit a simple single-stage model.

ratios which could be dated on the basis of this model have been plotted. The data were obtained from the tables compiled by Russell and Farquhar (1960a) and indicate an apparent spread of about ± 7 percent about the mean value of the ^{238}U/^{204}Pb ratio. However, several suites of samples have been reanalyzed by the author and several coworkers and it has been found that most of the variation along any particular isochron may be attributed to experimental error.

The isotope abundance of ^{204}Pb is always less than 2 percent of the total lead content and constitutes a very difficult measurement.

Errors are easily made, not only because of chemical impurities, but because of inadequate instrumental resolution in the mass spectrometer, long-term drift because of changes in resistances, temperature variations and even personnel differences in measuring peak heights. In solid-source mass spectrometry, Doe, Tilton and Hopson (1965)

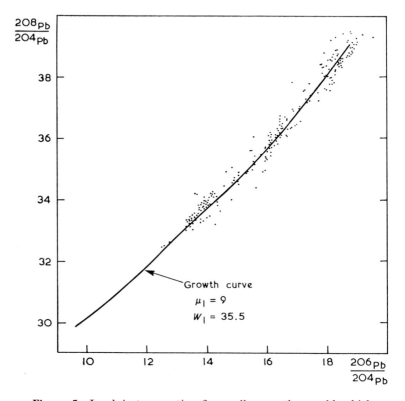

Figure 5 Lead isotope ratios from all over the world which appear to fit a simple single-stage model.

have found that the isotopic composition is a function of filament material and sample impurity, particularly due to alkalis. Mass-dependent variation due to analytic uncertainties may reach 1 percent in $^{208}Pb/^{204}Pb$ for samples run on different filament materials.

When plotted on a graph of $^{206}Pb/^{204}Pb$ against $^{207}Pb/^{204}Pb$, samples of lead with errors in the measured ^{204}Pb abundances fall on a line passing through the origin. Such a line may be called a ^{204}Pb

error line and its relationship to isochrons is illustrated in figure 6. When old data from the Bluebell Mine, British Columbia, and Ivigtut, Greenland, which seemed to have ^{204}Pb errors were analyzed (Kanasewich, 1962; Kanasewich and Slawson, 1964), nearly an order better precision was obtained by preparing samples with high chemical purity and eliminating short-term variations by an intercomparison of sample and standard as recommended by Kollar, Russell and

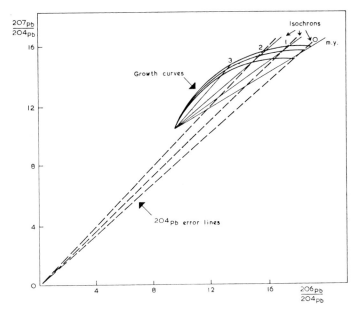

Figure 6 Relationship between single-stage isochrons and ^{204}Pb error lines.

Ulrych (1960). The linear relationship has disappeared in these new data (figures 7 and 8) and the samples from different locations have nearly identical isotopic composition. Similar results have been obtained on other suites from St. Magloire, Quebec (Kanasewich, 1962) and from the Svecofennian deposits in the Baltic Shield (Whittles, 1962). The suite from the Baltic Shield has sometimes been referred to as an example of an isochron (Schütze, 1962) or a modified isochron (Vinogradov and others, 1959). As seen in figures 7 and 9, the results of the older analyses fit an error line better than an isochron and no geological significance should be attached to this variation. The experimental results indicate that single-stage isochrons as proposed by Houtermans are either very short or do not exist.

Ages obtained with Houtermans' isochron equation will be valid if the leads have developed in a single-stage system. This appears to be the case with the group of 1800 m.y. deposits from the Baltic

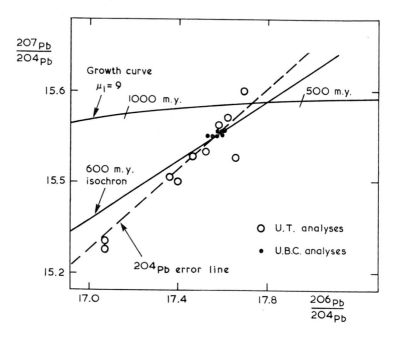

Figure 7 Comparison of lead isotope ratios from the Bluebell Mine, British Columbia, as measured at the University of Toronto (Russell and Farquhar, 1960a) and at the University of British Columbia (Kanasewich, 1962b). The modern analyses show that there is very little variation in the isotope ratios throughout the mine. The older analyses on the same samples fit a ^{204}Pb error line better than an isochron.

Shield (figure 9). The ages obtained for the deposits at the Bluebell Mine and at Ivigtut have no relationship with known geological events, and this is to be expected since lead isotope ratios from the surrounding area indicate that the ages must be calculated with a more complicated multi-stage equation. In general then, the data in figures 4 and 5 must be interpreted with caution since some of the

The interpretation of lead isotopes and their geological significance 159

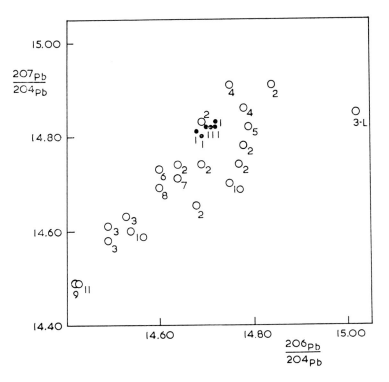

Figure 8 Comparison of lead isotope ratios from Ivigtut, Greenland, as measured by different laboratories. The data fit an error line indicating that the scatter is due to uncertainties in the determination of ^{204}Pb only.
1. University of British Columbia (Kanasewich and Slawson, 1964).
2. Columbia University (Mohler, 1960; Russell and Farquhar, 1960a). 3. Oxford University (Moorbath and Pauly, 1962). 4. University of Toronto (Russell and Farquhar, 1960a). 5. U.S. National Bureau of Standards (Mohler, 1960). 6. Oak Ridge Laboratories (Mohler, 1960). 7. U.S. Geological Survey (Mohler, 1960). 8. California Institute of Technology (Mohler, 1960). 9. Harwell Laboratories (Moorbath, 1962). 10. University of Minnesota (Russell and Farquhar, 1960a). 11. AB Atomenergi, Stockholm (Wickman and others, 1963).

scatter is due to ^{204}Pb error and some of it is due to physical and geological processes. The subject of ordinary leads will be discussed at greater length in section II.

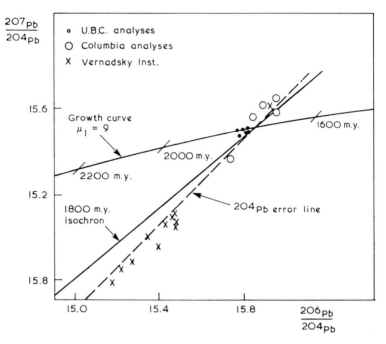

Figure 9 Lead isotope ratios having an age of 1800 m.y. in the Baltic Shield. Samples analyzed at Columbia University (Kouvo and Kulp, 1961) and at the University of British Columbia (Whittles, 1962; Kanasewich, 1962b) are from the same mines. Samples analyzed by Vinogradov and others (1959) at the Vernadsky Institute are from a wider area and are not directly comparable to those analyzed at Columbia University. They are included to show their relationship to the ^{204}Pb error line.
Analyses used are 1, 2, 116, 4, 6, 7, 8, 9, 10, 12, 15, 16, 34, 35, 41, 46, 87, 88, 89 as listed by Vinogradov and others (1959) and marked by crosses.

D *Meteorites*

The interpretation of common leads was greatly advanced by the determinations of the isotopic composition of microgram quantities of lead in meteorites with the use of a solid-source mass spectrometer. Patterson (1953) measured the isotope ratios from two troilites which had such small quantities of uranium and thorium that their lead isotope composition had not changed since they were formed. Thus, the troilite (FeS) phase of Canyon Diablo meteorite had 18 ppm of lead but only 0.009 ppm of uranium. In a bold extrapolation Houtermans (1953) and Patterson (1953, 1955) assumed that these meteoritic lead isotope ratios could be used for the primordial lead abundances, a_0, b_0, c_0, of the Earth. Combining terrestrial lead ores and primordial meteoritic abundances, they obtained a value of 4500 m.y. for the age of the Earth.

Considerations about the origin of the solar system are beyond the scope of this work. However, the evidence from lead isotopes definitely indicates that it is possible to find a unique time when meteorites cooled from a quasi-fluid state to a solid state before final disruption. Patterson (1955, 1956) found that the isotopic composition of stone and iron meteorites could be linearly related, showing that they represented isolated closed systems with widely varying uranium–lead ratios since a time t_0 when they became differentiated. The line formed by the measured abundances of ^{206}Pb/^{204}Pb and ^{207}Pb/^{204}Pb which developed in a single closed system was called an isochron by Houtermans (1946). The time t_0 is given by a single-stage formula which involves only the slope of the meteoritic or zero isochron and the decay constants of uranium. This turned out to be 4550 ± 70 m.y.

Since Patterson's pioneer work there have been 56 analyses on 33 meteorites. The linear relationship (figure 10) is still good although some of the analyses show definite evidence of terrestrial contamination either during weathering or in the laboratory. Many of the determinations were made on samples with less than 1 ppm lead and errors of several percent may easily be made in the chemistry or the mass spectrometry. In addition, there are interlaboratory differences as indicated on the analyses of Ivigtut lead in figure 8. Only sporadic attempts have been made to resolve these differences by measurements on standard samples.

Regression analyses were made on the meteorite data to determine the best-fitting straight line. Since both ratios have errors, the square of the deviations was mimimized perpendicular to the line itself. The results of these calculations together with the determination by Murthy and Patterson (1962) on a selected suite of measurements is presented in table 2. Murthy and Patterson selected their data on the basis of low apparent laboratory contamination and evidence from U–He and K–Ar dating that the meteorites have remained closed

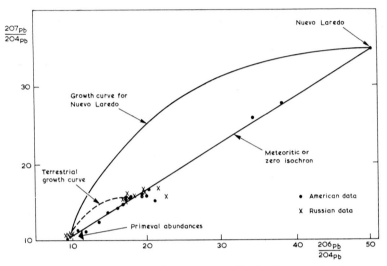

Figure 10 Lead isotope data on trace leads from meteorites. Zero or meteorite isochron is related to the age of meteorites. Patterson (1955, 1956), Chow and Patterson (1961), Murthy (1961) (see Murthy and Patterson, 1962), Staryk and others (1958, 1960), Murthy (1964), Marshall and Hess (1961), Marshall (1962), Edwards and Hess (1956), Hess and Marshall (1960), Staryk, Shatz and Sobotovich (1958).

isolated systems. The result of all calculations indicates that the age of meteorites is 4550 ± 30 m.y. This compares with a rubidium–strontium age of 4370 m.y. by Gast (1962) on four achondrites and four chondrites. These model ages involve the calculation of an isochron for a series of closed systems with a decay constant of 1.47×10^{-11} yr^{-1} (Flynn and Glendenin, 1959). It assumes that all variations in $^{87}Sr/^{86}Sr$ are related to a single event and that this time corresponds with a primordial $^{87}Sr/^{86}Sr$ ratio which is the same for all meteorites. A recent rubidium–strontium determination by Shields (1964) gives an age of 4210 m.y. The discrepancy between strontium and lead determinations may be related to uncertainties in the decay constants or because meteorites have had a more complicated history than is assumed in the model equations.

The $^{208}Pb/^{204}Pb$ and $^{206}Pb/^{204}Pb$ ratios are also linearly related, indicating that there is a constant Th/U ratio for all meteorites. The data are plotted in figure 11 and the various statistical methods yield average Th/U ratios of 3.6 ± 0.1 and 3.8 ± 0.1. A direct measurement of the Th/U ratio is of considerable interest and progress has

Table 2 The age of meteorites from lead isotope data

Statistical procedure	Number of analyses	Equation of line	Standard deviation on slope	Th/U ratio	Age (m.y.)
Least squares with no constraints	53	$y = 4.73 + 0.595x$ $z = 20.79 + 0.925x$	0.011 0.024	3.7 ± 0.1	4560 ± 30
Least squares with line constrained to pass through $a_0, b_0, c_0,$ from Murthy and Patterson	53	$y = 4.73 + 0.595x$ $z = 20.91 + 0.920x$	0.008 0.016	3.6 ± 0.1	4560 ± 20
Least squares using same data as Murthy and Patterson and constrained to pass through a_0, b_0, c_0	11	$y = 4.69 + 0.599x$ $z = 20.78 + 0.934x$	0.010 0.026	3.7 ± 0.1	4570 ± 30
Calculation by Murthy and Patterson based on weighted length of line through a_0, b_0, c_0	11	$y = 4.78 + 0.59x$	0.01	3.8 ± 0.1	4550 ± 30

The symbols and constants are defined in table 1.

been made by means of neutron-activation techniques (Lovering and Morgan, 1964). A determination on the achondrite Nuevo Laredo by Morgan and Lovering (1964, 1965) gave a value of 3.6. More precise measurements will have to be made by both isotopic and neutron-activation techniques before any conclusions may be drawn about the individual variation of the results.

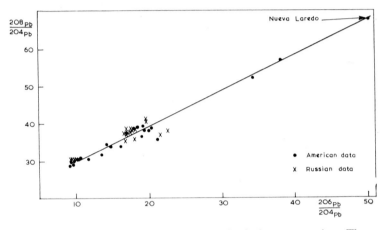

Figure 11 Lead isotope data on trace leads from meteorites. The slope of the line is related to the average Th/U ratio for meteorites.

E *Anomalous lead isotopes*

Leads which have developed in more than one closed uranium–thorium–lead system are called *anomalous* leads. The existence of anomalous leads was anticipated from the beginning by Holmes (1947) who cautioned that a 'lead ore might be due to concentrations of rock-lead plus lead from older ore deposits' or that 'a lead ore might come from a source to which additions of the radioactive elements had been made at some time after the origin of the source but before the date of ore formation'.

Anomalous leads are most frequently recognized by the linear relationship of the ^{206}Pb/^{204}Pb and ^{207}Pb/^{204}Pb ratios on a compositional diagram. When this occurs it is possible to use multi-stage equations for calculations of the geochemical properties of the source regions and the time of final mineralization. The age calculations may also be done graphically and the techniques will be illustrated in section III with actual case histories.

Because of the variable uranium/lead and thorium/uranium ratios in a crustal source it appears that a linear relationship is always present. However, it may take very precise measurements and widespread sampling to discover this, since local homogenization may occur. Two examples of local homogenization have already been illustrated in figures 7 and 8. Ulrych (1964) has analyzed the lead from the area surrounding Ivigtut and found a linear relationship between the lead isotope ratios, as predicted by Kanasewich (1962b). The author has analyzed seven samples of galena from the Comfort

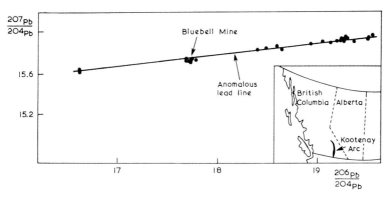

Figure 12 Lead isotope ratios from the Kootenay Arc, British Columbia (Sinclair, 1964). The data fit an anomalous lead line indicating their development in a multi-stage system.

and Kootenay Chief zones of the Bluebell Mine, British Columbia (figure 7), and found that the isotope ratios varied by less than $\frac{1}{2}$ percent. Geologic evidence indicates that this local homogenization of mineralizing solutions may have occurred in one of the associated batholiths in the area during Mesozoic times. The Bluebell Mine occurs on a geologic feature called the Kootenay Arc. Sinclair (1964) has obtained the isotope ratios from 17 lead–zinc mines along this feature and found that they are all linearly related. The results are shown in figure 12 together with the anomalous lead line passing through the data from Bluebell and the other mining districts.

Frequent geological mixing of continental materials has often been taken as a justification for the application of the assumption of a single-stage interpretation. Thus Alpher and Herman (1951) assumed that multiple orogenies in the histories of lead samples involved small changes in the lead–uranium ratios and that increases or decreases were equally probable. According to this assumption they were able to consider values of μ and W for the entire surface regions of the Earth. Since the calculation of Alpher and Herman did not involve interpretation of individual samples, there is some validity to their assumptions when applied to ^{206}Pb/^{204}Pb and ^{207}Pb/^{204}Pb ratios only. Several years later, Shaw (1957) argued that 'The crust is admittedly heterogeneous, but crustal processes, especially gradation, diastrophism and vulcanism tend to restore homogeneity'. Moorbath (1962) also justifies his application of the single-stage formula to individual British galenas on the basis that 'regional homogenization processes' tend to average out crustal heterogeneities over a period of time.

There is a basic flaw in all of the arguments presented above which invalidates their application to all isotopic data. This is that the chemical properties of parent and daughter elements are invariably different and this leads to differentiation, particularly under the influence of weathering and oxidation. The use of rubidium and strontium, and in particular the initial ^{87}Sr/^{86}Sr ratio, to trace the history of differentiation of sialic rocks is well known (Hurley and others, 1962). With reference to common leads it is possible to take advantage of differences between the two parent elements, uranium and thorium. Under oxidizing conditions the uranium does not exist in the 4+ valence state but can assume the hexavalent state, usually forming the very stable divalent uranyl ion (UO_2^{2+}). Minerals with hexavalent uranium are considerably more soluble than those with Th^{4+} so that the thorium migrates chiefly with colloids and fragments of rock. Leonova (1963) has noted that uranium accumulates in granites while thorium accumulates in alkalic rocks such as syenite. In general, crustal processes do not lead to the restoration of homogeneity but to an irreversible differentiation of both the elements and their isotopes.

F *The frequently mixed lead model*

Since ^{235}U and ^{238}U have identical chemical properties it was decided that the process of repeated mixing of source rocks should be subjected to detailed quantitative analyses. Isotopic compositions have been calculated by Russell and others (1966), Kanasewich (1962b) and Russell (1963) for sixty hypothetical leads which were assumed to have developed in two or more successive systems, each closed with respect to transfer of uranium, thorium and lead. It is generally known that studies of the ages of terrestrial rocks and minerals fail to show evidence for many events prior to about 3000 m.y. ago, and it may be supposed that surface conditions of the Earth prior to about this time were such that rocks were not preserved in any abundance. Therefore in these calculations, it has been assumed that mixing was so intense prior to that time that heterogeneities in lead isotope abundances were not produced and that lead grew isotopically in a regular way between the common starting time for the Earth–meteorite systems and the time 3000 m.y. ago.

For the first example, isotopic compositions were calculated for sixty hypothetical leads each of which grew in a closed system between time 3000 m.y. ago and the present. For each of the hypothetical lead samples the value of μ_2 was chosen at random using tables of random numbers from the distribution shown in figure 13. The

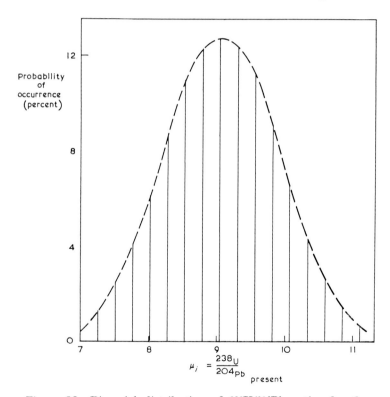

Figure 13 Binomial distribution of $^{238}U/^{204}Pb$ ratios for the source regions of the frequently mixed model. The relative amount of uranium in each stage and each of the sixty leads was chosen at random from this distribution.

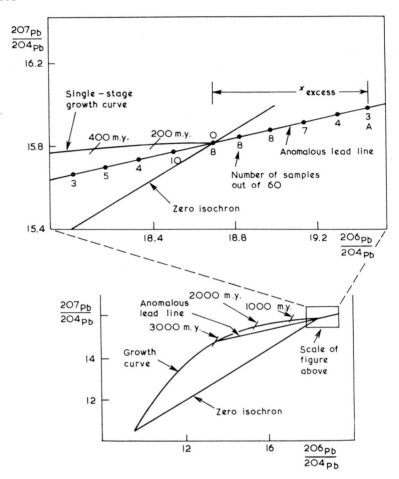

Figure 14 Graph of lead isotope ratios for a theoretical two-stage model in which $t_1 = 3000$ m.y. and t_2, the time of mineralization, is 0 m.y. The upper diagram is an enlargement of a portion of the lower one.

population has a binomial distribution with an arithmetic mean of 9.09 and a standard deviation of 0.77. The distribution approximates reasonably well to a Gaussian of the same parameters. This example is a slightly special case of the two-stage model on which Kanasewich (1962a) based his interpretations of anomalous leads. It is well known that such anomalous leads have isotopic compositions that fall on a chord, or its extension, to the single-stage growth curve between two points representing the times enclosing the duration of the second stage (equation A18). Figure 14 shows the linear relationship of the calculated points. The points are not distributed continuously along the line because the distribution chosen permitted only discrete values of μ_2. For these points, the average ratios lie close to the zero point time on the growth curve for $\mu_2 = 9.09$. This would be precisely true if an infinite number of examples had been calculated. This

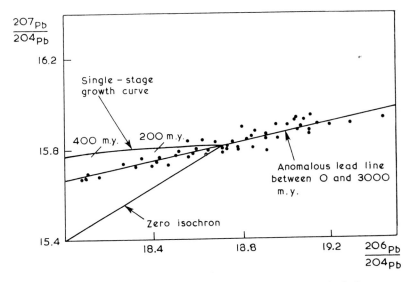

Figure 15 Graph of lead isotope ratios for a theoretical three-stage model. Diastrophism has caused a random mixing of uranium and lead at 3000 and 1500 m.y. ago. Mineralization occurs at 0 m.y.

result is a consequence of the fact that the μ_1 values for the first growth stage coincided with the mean value, μ_2, for the distribution assumed for the second stage. This is true for all the examples to follow.

For the second example, the second time interval has been divided into equal parts, forming a three-stage model. The μ values for the second and third stages were each drawn independently from the distribution shown in figure 13. This corresponds to the physical assumption of complete mixing between the second and third stages. The sixty points calculated for the three-stage model are shown in figure 15 where it is seen that the points tend to follow the same trend as before, but that now there is a small lateral spread.

As the total number of stages increases to eleven (figure 16), the distribution of points becomes shorter and broader, but the cluster is always about the same straight line.

From the calculations made, it must be concluded that the 'frequently mixed model' for lead isotope abundances approximates much more closely to the two-stage models applied to anomalous leads, than to the single-stage model with variable μ values. That is, the concept of isochrons through the primeval lead abundances has

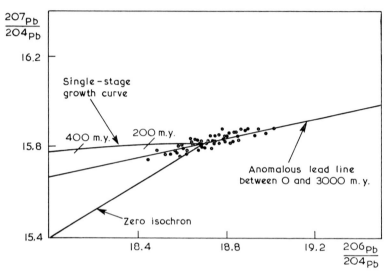

Figure 16 Graph of lead isotope ratios for a theoretical eleven-stage model. Diastrophism is postulated to occur every 300 m.y. between 3000 m.y. ago and the present time when mineralization takes place.

little significance for such leads. Under certain circumstances the use of a two-stage model may provide a good first approximation to multi-stage leads. The calculations give additional reasons for the desirability of determining as precisely as possible the quality of fit of the isotope ratios to the best straight line through them.

Eberhardt, Geiss and Houtermans (1955) have defined two classes of anomalous leads: J-type which give apparent ages which are younger than the rocks in which they are found, and B-type which give apparent ages which are older than the enclosing rocks. The theoretical examples given in figures 14 to 16 illustrate the shortcomings of such definitions of anomalous leads. The time of mineralization for all the examples was 0 m.y. ago but the apparent ages of individual samples as calculated with a single-stage (isochron) equation vary from $+400$ to -400 m.y. That is, both B- and J-type leads appear to occur together. The apparent negative or positive age is a meaningless parameter and only tells us that the interpreter has used the wrong equation.

If one requires a parameter to express the amount of excess radiogenic lead produced in a multi-stage system without assumptions about the details of the model, one can give the apparent excess $^{206}Pb/^{204}Pb$ over a meteoritic system:

Excess apparent radiogenic lead in percent
$$= \left\{ \frac{x_{\text{sample}} - x_{\text{meteorite}}}{x_{\text{meteorite}}} \right\} \times 100$$

where x is the $^{206}Pb/^{204}Pb$ ratio. The $^{207}Pb/^{204}Pb$ ratio of the zero or meteoritic isochron is assumed to be the same as that of the measured sample. Since the $^{207}Pb/^{204}Pb$ ratio has changed very little over the last billion years, this procedure eliminates most of the ^{204}Pb error. For sample A in figure 14 the 'apparent excess radiogenic lead' is approximately 2%.

The most important conclusion to be derived from this theoretical example is that the assumption of frequent mixing of the rocks from which leads are derived is not sufficient for the *point-by-point* interpretation of such abundances on the basis of a single-stage (isochron) model.

II The Search for Ordinary Leads

A *Lead ores*

Collins, Russell and Farquhar (1953) assumed that there is a single average growth curve to which all lead isotope ratios of ordinary or single-stage leads fit to within a few percent. Under this model the $^{238}U/^{204}Pb$ and $^{232}Th/^{204}Pb$ ratios, reduced to their present-day values, have remained essentially constant over the entire world. Evidence for a unique growth curve has increased from the analysis of many types of lead sulfide deposits, particularly when the isotope ratios were analyzed with great precision. As will be noted later, present-day results from leads in oceanic basins indicate that a different model may have to be used for oceanic and continental leads.

As early as 1956, Russell suggested that the first stage of development occurred in the mantle with a nearly constant ^{238}U/^{204}Pb and Th/U ratio. Russell and Farquhar (1960b) used Patterson's meteoritic data together with the isotopic composition of galenas from Bathurst, New Brunswick to derive a single-stage growth curve. Stanton and Russell (1959) have argued that only leads which have not traversed crustal rocks are most likely to satisfy the requirements for this model. Mineral deposits which satisfy this criterion by virtue of having only limited contact with crustal rocks which might contaminate them with extra radiogenic lead are:

1. 'Orthomagmatic deposits in mafic rocks derived from beneath the continental crust.'

2. 'Sedimentary deposits derived from depth by basalt–andesite vulcanism along, or at some distance from continental margins, and quickly isolated in volcanic sediments.'

Stanton and Russell (1959) and Stanton (1960) thought that many conformable deposits might have characteristics of ordinary or single-stage leads. The group which Stanton recommended for study included Manitouwadge, Lake Geneva, Yukon Treadwell, Sullivan, Buchans and Bathurst, Canada; Bleiberg, Austria; Broken Hill, Mount Isa, Captain's Flat, Cobar, Read Roseberry and Hall's Peak, Australia. All but Lake Geneva, Buchans and Bleiberg have now been analyzed with a precise intercomparison technique at the University of British Columbia. From this group Manitouwadge and Read Roseberry have been definitely established as anomalous and must be excluded.

Stanton's hypothesis that all of the deposits mentioned above are syngenetic in origin has been criticized by geologists more familiar with the individual mines. Furthermore, a syngenetic origin has been found to be completely inconsistent with the new lead isotope data no matter how the constants in the model lead formulas are adjusted. All of the evidence has been examined very carefully by Ostic (1963a) and he found that the geological criteria were not sufficient to identify ordinary leads.

Although the model lead ages are inconsistent with the interpretation that Stanton's group of ores are syngenetic, most of them do lie very close to a unique growth curve. Furthermore, if one interprets some of these deposits as being epigenetic in origin, the model lead ages are consistent with age determinations by other methods. It is then possible to include selected deposits from areas such as Sudbury and Cobalt, Canada; Korsnäs, Pakila, Metsamouttu, Attu and

Orijärvi, Finland since these also fit the same growth curve and have single-stage model lead ages consistent with other evidence. The isotope ratios from all these deposits are given in appendix 3 and are plotted in figures 17 and 18 to show how well they fit a unique growth band. The analyses were all measured relative to Broken

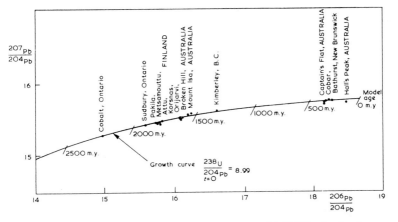

Figure 17 Graph of $^{207}\text{Pb}/^{204}\text{Pb}$ against $^{206}\text{Pb}/^{204}\text{Pb}$ for leads which are believed to have had single-stage histories. The parameters on which this curve is based are given in the text.

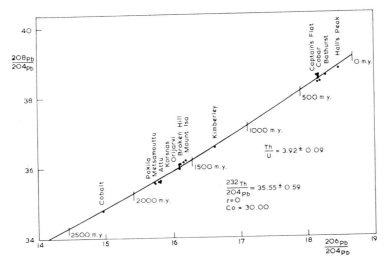

Figure 18 Graph of $^{208}\text{Pb}/^{204}\text{Pb}$ against $^{206}\text{Pb}/^{204}\text{Pb}$ for leads which are believed to have had single-stage histories. The parameters on which this curve is based are given in the text.

Table 3 Chronological data on lead sulfide deposits used for defining a single-stage growth curve

Mine area	Age of host rock (m.y.) (1)	Period of major orogenic or intrusive activity (m.y.)	Model lead age in m.y. $t_0 = 4550$ (m.y.)	Model lead age in m.y. $t_0 = 4530$ (m.y.)	Nearby K–Ar (or other) dates (m.y.)	References[a]
Hall's Peak, N.S.W., Australia	U. Permian (?) (230?) (1)	U. Cret? (63–90) U. Perm? (230)	80	170		2
Bathurst, N.B., Canada	M. Ordov. (450)	Acadian Rev. (370) (Dev.)	270	360	366, 388 395, 396	3, 4
Cobar, N.S.W., Australia G.S.A. Mine, Cobar Tharsis Mine	Silurian? (420)	Kanimblan orog. (L. Carb.—340) Tabberabberan orog.	320 310	410 390	379, 392 410	2, 5, 6
Great Cobar Mine Silver Peak Mine Queen Bee Mine		(M. Dev.—360) Sil. intrusives (405–425)	320 330 340	410 420 420		
Captain's Flat, N.S.W., Australia	Silurian (420)	Kanimblan orog. (340) Tabberabberan orog. (360) Sil. intrus. (405–425)	340	420	420, 396 390, 378 348	2, 5
Sullivan Mine, B.C., Canada	Precambrian Aldridge Fm. (Belt System)	Precambrian Cret. (63–135)	1340	1400	1580, 883 765, 580 100	7, 8 9, 10 11
Mt. Isa, Queensland, Australia Black Star Mine, Mt. Isa	Precambrian Mt. Isa Series	Precambrian	1600 1600	1600 1660	1770 1400–1450	2, 12

Location	Series		Values		Notes	Refs
Broken Hill, N.S.W., Australia	Precambrian				(Pb–U ages of 1510 and 1702–(206/207) 1360 on one muscovite Many biotites give 460–550	2, 13
Main Lode						14, 15
Pinnacles Mine	Willyama Series		1620	1670		16, 17
Little Broken Hill		Tyennan orogeny	1610	1670		
Globe Mine		(U. Camb.—500)	1610	1670		
			1610	1670		
Metsamouttu, Aijala Mine, Kisko, Finland	Precambrian	Svecofennian	1810	1860	1790	18, 19
Korsnäs, Finland	Precambrian	Svecofennian	1780	1830	1730, 1720 at Maalahti	20
Attu, Parainen, Finland	Precambrian	Svecofennian	1790	1850		
Orijärvi Mine, Kisko, Finland	Precambrian	Svecofennian	1800	1850	(Pb–U = 1850 (206/207)Lavia	21
Pakila, Finland	Precambrian	Svecofennian	1820	1870	(Rb–Sr = 1830 Kälviä K–Ar = 1820 (Perniö)	22
Sudbury, Ontario, Canada	Precambrian sediments within Sudbury basin (Vermillion Fm.)				(Rb–Sr whole rock dates): rhyolite: 2215; granite: 2065–2020;	
Errington Mine #359			1920	1970	micropegmatite 1740; norite: 1650	23
Errington Mine #358			1700	1750	K–Ar: 1770	24
						25
						26
Cobalt, Ontario, Canada	Precambrian Cobalt Series	Precambrian	2250	2290	K–Ar = 1990 Rb–Sr = 2230	27
						28

[a] 1. Kulp (1961); 2. David and Browne (1950); 3. Smith and Skinner (1958); 4. Tupper (1959); 5. Thompson (1953); 6. Evernden and Richards (1962); 7. Rice (1941); 8. Hunt (1961); 9. Swanson and Gunning (1948); 10. Lowdon (1960); 11. Lowdon and others (1963); 12. Richards, Webb and Cooper (1963); 13. King and Thompson (1953); 14. Carruthers and Pratten (1961); 15. Richards and Pidgeon (1963); 16. Collins and others (1953); 17. Greenhalgh and Jeffrey (1959); 18. Kouvo (1958); 19. Vaasjoki and Kouvo (1959); 20. Polkanov and Gerling (1961); 21. Wetherill and others (1962); 22. Gerling and Polkanov (1958); 23. Martin (1957); 24. Fairbairn and others (1960); 25. Faure and others (1962); 26. Aldrich and others (1959); 27. Thomson (1957); 28. Aldrich and Wetherill (1960).

Hill No. 1 (Toronto sample 1003) at the University of British Columbia. Therefore they contain no errors due to interlaboratory differences or time varying differences due to changes in resistances, etc. It has been suggested that samples from Balmat, New York (Doe, 1962) and some trace leads from Llano, Texas (Zartman, 1965) may also be single-stage leads. This interpretation seems very plausible although it would be useful to have their isotope ratios measured relative to the Broken Hill standard to see how well they fit this particular growth curve. Table 3 summarizes the available chronological information on the samples from 23 mines which are presented in figures 17 and 18.

As noted above, no infallible geological criteria have been found for independently distinguishing ordinary or single-stage leads from those which were produced in two or more distinct lead–uranium–thorium systems. As an extension to the model proposed by Russell and Farquhar (1960b), Kanasewich (1962a) and Kanasewich and Farquhar (1965) suggested a series of isotopic criteria for judging the appropriateness of a single-stage model lead age. Again these criteria are not infallible, but if enough analyses are made with sufficient precision it is possible to make an interpretation which is consistent with the available geologic information. In applying the criteria given below, the lead isotope ratios should be measured relative to the Broken Hill (U.B.C. 1) standard. Each sample should be plotted on graphs similar to those illustrated in figures 17 and 18, and only those samples which fall within the limits of the single-stage growth band should be considered for model age calculations using Houtermans' formula.

Criteria for recognizing single-stage leads are as follows:

1. The isotopic composition of ordinary leads from a geologically related area should be constant within about one-third of one percent. Apparently lead is brought up from a subsialic source region contemporaneously over a widespread area.

2. The $^{206}Pb/^{204}Pb$ ratio of an ordinary lead should have a value of μ_1 equal to 8.99 ± 0.07. The limits given throughout this section are 95 percent confidence limits on the growth curve. This means that the mean $^{238}U/^{204}Pb$ ratio of the subsialic or upper mantle source region appears to be constant within three-fourths of one percent under all continental areas.

3. The $^{208}Pb/^{204}Pb$ ratio of any ordinary lead should have a W_1 value equal to 35.55 ± 0.59. From μ_1 and W_1 the Th/U ratio for the source region is calculated to be 3.92 ± 0.09. This compares with a

The interpretation of lead isotopes and their geological significance 177

Th/U ratio of 3.8 ± 0.1 for meteorites (Murthy and Patterson, 1962). The geochemical properties of uranium as compared to those of thorium are quite different in the sialic source region. As a consequence the ^{232}Th/^{204}Pb ratio is a very sensitive indicator of the degree to which the single-stage model breaks down in practice because of contamination with lead which developed in an upper crustal environment.

4. The age of an ordinary lead should agree reasonably well with determinations by other age dating techniques. Since it is not always possible to classify a sulfide deposit as syngenetic or epigenetic or to determine its exact chronological relationship with the formation of biotites, zircons, etc., this agreement will not be exact. In addition, there are uncertainties in the model lead parameters a_0, b_0, t_0 and λ'. For the analyses of the twenty-three mines used in figures 10 and 11 the agreement is within 150 m.y.

B *Lead from oceanic basins*

The close relationship of the Earth and the meteoritic system was established by noting that the isotopic composition of recently mineralized oceanic leads falls in a cluster slightly to the right of the meteoritic or zero isochron (Patterson, 1955). This relationship was critically examined in a very significant study by Chow and Patterson (1959, 1962) involving the determination of isotope ratios of microgram quantities of lead from manganese nodules and pelagic sediments in the Pacific basin. The results are summarized in figure 19. On the other hand, isotope ratios from Japanese sulfide deposits which appear to have originated from a subsialic source region during the Paleozoic, Mesozoic or Tertiary eras all fall to the left of the meteoritic isochron (figure 20).

Chow and Patterson (1959) state that most of the manganese nodules which were studied are probably less than one million years old since the nodules cease to grow after being buried by sediments. The lead in manganese nodules and pelagic sediments (Chow and Patterson, 1962) represents an average sampling of a large amount of sea water. They emphasize that the leads are largely derived from continental land masses but the data are not entirely consistent with this interpretation since the most radiogenic samples are found near the American continents (points marked with a cross in figure 19). Chow and Johnstone (1963) have obtained values of ^{206}Pb/^{204}Pb between 21.63 and 25.00 in Hudson Bay. The radiogenic lead brought down by rivers must be diluted with a less radiogenic source to explain the isotope ratios found in deep sea basins. For instance, Tatsumoto (1964) has found that the lead isotopes from Hawaiian

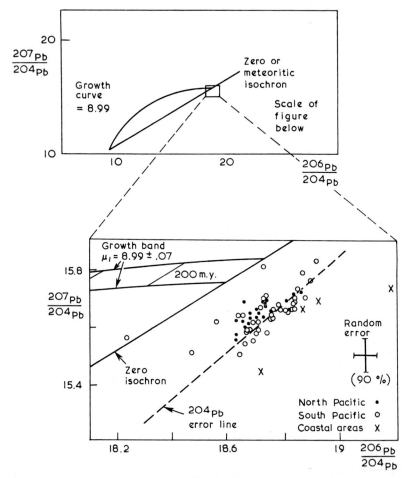

Figure 19 Isotopic composition of leads from manganese nodules and pelagic sediments in the Pacific basin (Chow and Patterson, 1959, 1962). The location of the samples is shown in figure 22.

basalts lie about the zero isochron (figure 21). This pattern is similar to that shown in the frequently mixed model (figure 16) but the slope of the line through the data is very small, indicating initial differentiation less than a billion years ago. The isotopic evolution of the oceanic

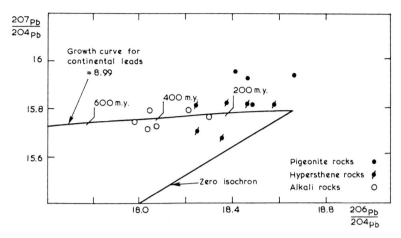

Figure 20 Isotopic composition of leads from sulfide deposits in Japan and their relationship to the meteoritic isochron. (After Masuda, 1964)

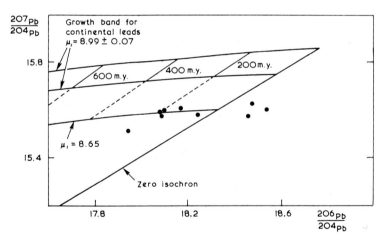

Figure 21 Lead isotope ratios of lead in volcanic rocks from Hawaii. (After Tatsumoto, 1964)

leads was interpreted by Chow and Patterson (1962) on the basis of a two-stage model with different periods of aging in the second stage.

In deep waters the average concentration of lead is only 0.03 micrograms per liter and is rather uniform (Tatsumoto and Patterson, 1963). The concentration of lead in phytoplankton is poorly known but is estimated to be 8 ppm. The uranium concentration of oceanic waters is about 3 micrograms per liter (Thurber, 1964).

While organisms have probably played a key role in concentrating the lead in the pelagic sediments on the ocean floors, it is not known how fast they remove the lead or how far the continentally derived lead is transported before final deposition.

Although a part of the lead in sea water is derived from continental areas by means of rivers and ocean currents, it seems that a significant part is also derived from volcanic exhalations within the oceanic basin and from the island arcs which partially surround the Pacific basin. The lead from a deep mantle source should plot on the zero isochron if there is a direct relationship between terrestrial and meteoritic systems. Because the rivers carry more radiogenic lead into the oceans, it is difficult to predict how closely the isotopic ratios should plot on the zero isochron. Leads close to basic volcanic sources are closer to the zero isochron as shown by the samples marked with solid circles in figure 19. The location of the samples is indicated in figure 22.

Leads in the south Atlantic (figure 23) are similar in isotopic composition to those in the Pacific but those in the north and east Atlantic basins are much more radiogenic. This may result from the closer proximity of granitic land masses with abundant anomalous leads but it may partly reflect local irregularities from within the Mid-Atlantic Ridge. Gast, Tilton and Hedge (1964) have found local variations in lead and strontium isotope ratios from rocks on Gough and Ascension Islands. The lead concentration in samples from these two islands on the Mid-Atlantic Ridge varies from 2 to 20 ppm. Most of the lead ratios from Gough Island are close to the zero isochron (figure 23) and, within experimental error, are on the growth curve used for ordinary continental leads. The lead from Ascension Island indicates that it has developed in a more radiogenic and heterogeneous environment. Similar results have also been reported by Tatsumoto (1964) on leads from volcanic rocks of Iwo Jima. These variations could arise from contamination by oceanic crustal lead but it is more likely that they represent real heterogeneities in the upper mantle. Chemical fractionation has probably gone on for 200 to 300 million years in the vicinity of the Mid-Atlantic Ridge so that local variations in the uranium/lead and thorium/lead ratios should not be surprising, particularly when one recalls the significant variations in seismic velocities and heat flow values in such areas.

The studies reviewed above indicate that lead and strontium

isotope data from oceanic areas will provide a wealth of information from which it may be possible to develop the evolutionary history of the underlying mantle. It is clear that the lead in basalts and basic volcanic rocks has not developed in a single-stage environment and

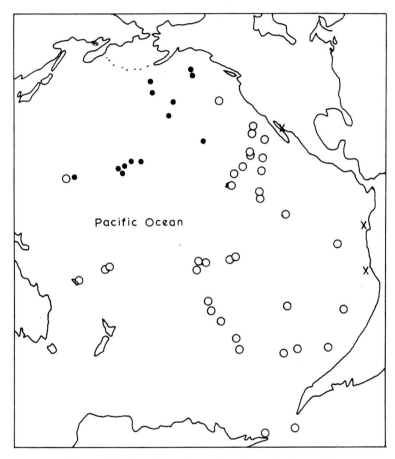

Figure 22 Location of samples shown in figure 19.

cannot be classified as ordinary. Eaton (1962) has shown from a study of periodic swarms of deep earthquakes around Kilauea, Hawaii, that the probable source of the lava is a zone 60 kilometers deep in the mantle. Since the lead isotope ratios from Hawaii show variations it seems reasonable to assume that local heterogeneities in the source region must exist.

The isotope ratios from Hawaii and also those from manganese nodules and pelagic sediments generally lie below the growth band used for continental leads. At the present time it is impossible to decide if this is a systematic error due to interlaboratory differences in measuring the ^{204}Pb peak or if the differences are real. If they are real this would represent additional evidence that the continental and oceanic upper mantle have had a different geochemical evolution as is indicated by heat flow and seismic Love and Rayleigh wave studies.

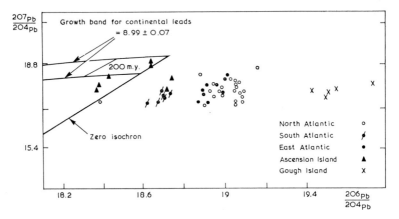

Figure 23 Isotopic composition of leads from manganese nodules and pelagic sediments in the Atlantic Ocean (after Chow and Patterson, 1959, 1962) and from two islands in the Mid-Atlantic Ridge. (After Gast, Tilton and Hedge, 1964)

C *Trace lead studies*

A large number of studies are being carried out on trace leads in rocks and minerals. Early work includes lead in olivine by Tilton, Patterson and Davis (1954); in granite by Tilton and others (1955); in ultramafic rocks by Tilton and Patterson (1956); in rocks associated with various ores by Vinogradov (1955), Golubchina and Rabinovich (1957), Vinogradov, Zykov and Tarasov (1958, 1959, 1960); in eclogites by Marshall (1960); in feldspars by Catanzaro and Gast (1960); ores and rocks from Butte, Montana by Murthy and Patterson (1961). One of the objectives of these studies was to determine the origin of lead ores and, in particular, to see if homogenization of crustal material could produce isotope ratios characteristic of ordinary leads. The issue was not resolved because the isotope ratios of microgram quantities of lead could not be measured with sufficient precision or with reference to recognized standards. Tilton and Reed (1963) used the neutron-activation and isotope-dilution techniques to remeasure lead concentrations in ultramafic rocks measured in 1956. They found the early results given by Tilton and Patterson (1956) were too high by one order of magnitude and so this early isotope data had to be discarded.

Wampler and Kulp (1964) have analyzed the lead in a number of samples of sedimentary pyrite. Their analyses from Ordovician slates

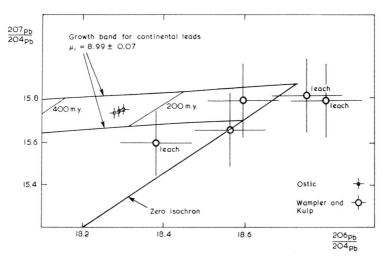

Figure 24 Lead isotope ratios of galenas (Ostic, 1963a) and trace leads in pyrites (Wampler and Kulp, 1964) from Bathurst, New Brunswick. The crosses indicate the random errors in the individual measurements.

at Bathurst, New Brunswick, are shown in figure 24 together with results from lead ores as obtained by Ostic (1963a). The pyrite samples are thought to be of syngenetic origin but may have been recrystallized during subsequent metamorphism. From these and other samples Wampler and Kulp show that the lead in pyrites is probably similar to dissolved lead in the sea at the time of formation of the pyrites provided there has been no subsequent metamorphism. The samples may be interpreted on the basis of a two-stage model as used by Chow and Patterson (1962) on pelagic leads. Figure 24 shows that ore leads (galenas) and trace leads in pyrites are not identical in isotopic composition and there is no reason to suppose that homogenization and concentration of trace leads in sediments will produce an ordinary lead or lead ore as seen in the mines at Bathurst.

Feldspars have been examined for their lead isotope composition particularly since they contain negligible uranium (0.02 to 0.38 ppm) and thorium (0.07 to 1.02 ppm) but considerable quantities of lead (13 to 521 ppm, according to Doe, Tilton and Hopson, 1965). Unfortunately Doe and Hart (1963) have shown that feldspars do not remain closed systems under even slight metamorphism. They found that lead entered crystals of potassium feldspar 300 meters from the contact of the Eldora Stock and the Precambrian Idaho Springs Formation in the Front Ranges in Colorado. However, the Rb–Sr age of the feldspar was not affected substantially. All the facts indicate that the isotope ratios were altered without solution, redeposition or melting of the feldspars. Thus the lead in feldspars should be very useful in tracing regional metamorphism but is not much help in defining ordinary leads.

D *Conclusions*

The search for ordinary or single-stage leads must continue since no conclusive examples have been found. Although it may be that no such leads exist, the lead ores presented in section II.A are certainly very close approximations to the mathematical idealization. Individual minerals such as feldspars have been shown to behave as open systems but no such problem has been discovered with most lead ores. The reservoir or system in which these leads developed is so large (several cubic kilometers) that transfer of uranium, thorium and lead in and out of the system is either insignificant or can be accounted for by a multi-stage model.

Trace element studies have not uncovered any evidence that ordinary leads might be produced by homogenizing a large volume of sialic crustal material. Indeed, homogenization of sediments now being deposited in Hudson Bay can only produce a very anomalous lead. Thus all the available evidence favors a deep subsialic or upper mantle source for ordinary lead deposits.

Seismic and heat flow studies show that the upper mantle does not have radial homogeneity. The new research on trace leads from volcanic and basaltic rocks from oceanic islands indicates that the mantle may have quite local heterogeneities of the ^{238}U/^{204}Pb ratio, μ, and the ^{232}Th/^{204}Pb ratio, W. It should be expected that local irregularities of μ and W also exist in the continental mantle. The characteristics of ordinary lead ores were that the isotopic composition from a local area was very uniform and that the source regions under all continents have a ^{238}U/^{204}Pb ratio of 8.99 ± 0.07 and a Th/U ratio of 3.92 ± 0.09. Such uniformity can only arise if homogenization has occurred in the deep source prior to mineralization. Furthermore, the average value of μ and W may not be the same in the oceanic upper mantle as in the subsialic or upper mantle regions under continents. This means that different lead isotope growth curves may have to be used when a greater degree of precision is obtained in the experimental data.

III The Interpretation of Anomalous Lead Results

A *Classification of anomalous leads*

Isotope results which are not in accord with the criteria given for ordinary leads are probably anomalous and must be analyzed with multi-stage equations. Anomalous leads are most frequently recognized by the linear relationship of the isotope ratios on a graph of ^{206}Pb/^{204}Pb against ^{207}Pb/^{204}Pb. The necessity for a minimum of two stages in the production of anomalous leads such as the Joplin leads has been recognized since Holmes's calculation of the age of the Earth. Various authors, Houtermans (1953), Geiss (1954), Russell and others (1954), Bate and others (1957) have suggested an equation similar to that used by Kanasewich (1962a) for a two-stage anomalous lead but none seem to have presented the model in such a manner that a quantitative determination could be made of the time of the second stage of mineralization.

Anomalous leads may be classified into the following types (Kanasewich, 1962b; Kanasewich and Farquhar, 1965).

(1) Mixture of two ordinary leads
(2) Two-stage anomalous lead
(3) Short period anomalous lead
(4) Simple three-stage anomalous lead
(5) Higher order multi-stage lead

A few examples will be presented to illustrate the distinctive characteristics of these leads.

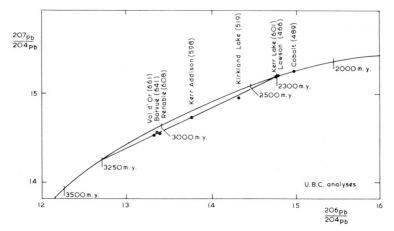

Figure 25 Graph of ^{207}Pb/^{204}Pb plotted against ^{206}Pb/^{204}Pb for galena samples from the Cobalt–Noranda area (Kanasewich and Farquhar, 1965). The results are intercompared with the Broken Hill No. 1 standard and have a precision of ± 0.1 %. Note their excellent fit to an anomalous lead line.

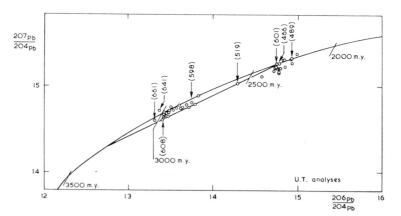

Figure 26 Graph of ^{207}Pb/^{204}Pb plotted against ^{206}Pb/^{204}Pb for the galena samples from the Cobalt–Noranda area (Kanasewich and Farquhar, 1965). Duplicate analyses have been averaged. The uncertainties due to random errors are less than $\pm 0.7\%$.

B *Mixture of two ordinary leads*

A mixture of two ordinary leads with different ages will yield a series of linearly related isotope ratios. The line should intersect the single-stage growth curve at two points and not extend beyond it. Examples of this type are to be found in eastern Canada and Finland.

Cobalt–Noranda area   The plots of the data obtained from the Cobalt–Noranda area, Canada, are shown in figures 25, 26 and 27. The precision of the results obtained by Kanasewich on a University of British Columbia mass spectrometer indicates that the results fit a straight line very closely. The older results obtained by Farquhar at the University of Toronto have some ^{204}Pb error and a consistent interlaboratory difference but they also fit a straight line within

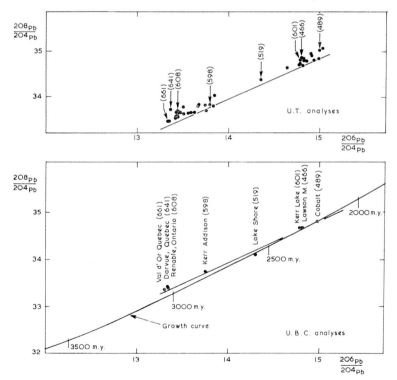

Figure 27   Graph of ^{208}Pb/^{204}Pb plotted against ^{206}Pb/^{204}Pb for the galena samples from the Cobalt–Noranda area. (After Kanasewich and Farquhar, 1965)

experimental error. The locations from which the samples were obtained are given in figure 28.

The linear trend through the samples from Kirkland Lake to Larder Lake also passes through the samples from Val D'Or, Barvue and Renabie. While these various samples can be classified into separate cogenetic groups, these linear isotopic relationships indicate that they cannot be explained individually sample by sample. Any attempt to date each sample individually on the single-stage model results in spurious relationships between the U/Pb and Th/U ratios in the source region and the time of mineralization.

Because of the relatively large number of samples analyzed, and the wide extent of the geographical area from which they were collected, it is felt that these limited and extremely linear isotopic distributions are significant. They suggest that the variations in isotopic ratios have been introduced by simple mixing of two single-stage leads having rather different isotopic compositions. The age of the two times of mineralization may be obtained from the intersection of the anomalous lead line with the single-stage growth curve. These intersections are at 3250 ± 150 m.y. and 2300 ± 150 m.y. (figure 25). The uncertainty of 150 m.y. includes analytic errors in the isotope ratios

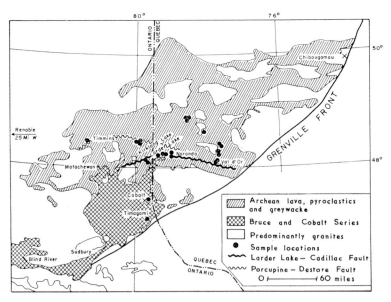

Figure 28 General geological features of the Cobalt–Noranda area.

and also probable errors in the primeval ratios, the age of the Earth and the decay constants of uranium.

The simplest interpretation of these results is as follows. The first period of mineralization occurred 3250 m.y. ago with the sulfides perhaps concentrated in the long faults which follow the regional strike. These faults undoubtedly extend to great depths and provided a channel for sulfide mineralization about 2300 m.y. ago. The lead must have come from a subsialic source since the isotope ratios do not have any of the characteristics of anomalous leads with development in a shallow crustal source region. The strongest evidence for this is that none of the analyses lie on an extension of the chord beyond the growth curve. The ^{208}Pb/^{204}Pb ratios (figure 27) support the conclusion that this is a simple mixture of two leads derived from a deep environment with a very constant U/Pb and Th/U ratio. The lack of contamination with crustal leads formed in a variable Th–U–Pb environment indicates that all the sulfide-bearing solutions traveled along faults extending to great depths and so had limited access to the country rock.

A large amount of effort has been expended to subdivide the Canadian Shield into suitable units of study. Early classifications (Wilson, 1939; Gill, 1948, 1949; Wilson, 1949) were based primarily upon stratigraphic, petrological and structural similarities. Features such as parallel fold belts, foliation trends, and truncation of ancient orogenic centers by younger mountain belts were invaluable aids in this work. Lately, the classification has been carried out with the aid of geochronological data (Wilson, 1949, 1952; Cumming and others, 1955; Farquhar and Russell, 1957a; Hurley and others, 1962; Goldich and others, 1961; Gastil, 1960; Stockwell, 1961; Burwash, Baadsgaard and Peterman, 1962). The classification is now based on the repeated recurrence of certain dates over widespread areas when K–Ar and Rb–Sr dating methods are used. These dates are related to periods of severe metamorphism during which minerals crystallize or recrystallize over large portions of the continent. On this evidence the oldest portions of the North American continent appeared to be 2400 to 2800 m.y. Portions of the continent with these ages include the Superior and Slave geologic province and are shown in figure 29.

Recently, the evidence from lead–uranium and model lead analyses has indicated that these portions of the continent are older than 3000 million years. Discordant zircon ages obtained by Catanzaro (1963) for southwestern Minnesota have been interpreted by him as originating from rock which is at least 3300 m.y. old. Work by Catanzaro

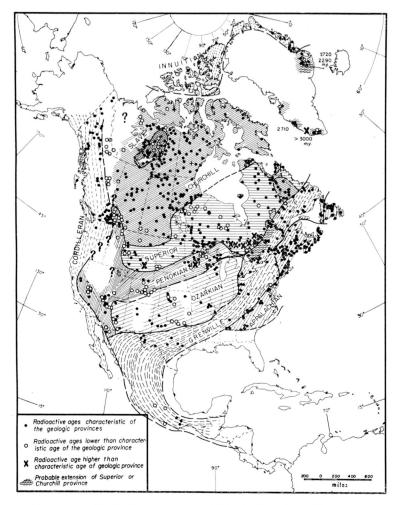

Figure 29 Geologic provinces of North America. The crosses mark areas with lead ages greater than 3000 m.y.

and Kulp (1964) on discordant zircons from Montana also yields a crystallization age greater than 3100 m.y. for that area.

The present work indicates that over an area with a length of more than 150 miles, from Timmins, Ontario, to Val D'Or, Quebec, the age of the crust is over 3250 m.y. An analysis of lead from Renabie indicates an extension to the west, and other lead isotope evidence from Manitouwadge, north of Lake Superior (Slawson and others,

1963), may be similarly interpreted. To the east, at Ivigtut, Greenland (Kanasewich and Slawson, 1964), the age of initial crustal deposition of these leads was over 3000 m.y. Thus both model lead and lead–uranium methods show that the Precambrian geology of North America extends beyond 3200 million years ago. Furthermore, these very old crustal rocks cover an extensive geographic area of North America (figure 29).

West Karelian zone A second example of a mixture of two ordinary leads probably occurs in the Karelian zone of the Baltic Shield. Analyses by Kouvo and Kulp (1961) suggest that an early period of ordinary lead mineralization occurred here about 2800 m.y. ago and that a second period of ordinary lead mineralization occurred about 1800 m.y. ago close to the boundary between the Fenno-Karelides and the Svecofennides. The data are shown in figure 30.

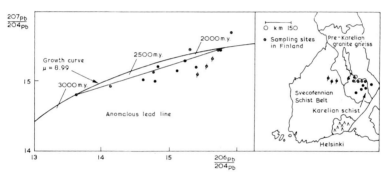

Figure 30 Lead isotope ratios from the West Karelian zone of Finland (Kouvo and Kulp, 1961). This suite may be interpreted as a mixture of two ordinary leads with ages about 2800 and 1800 m.y.

Wherever a mixture of two ordinary leads occurs, the isotope ratios cannot be interpreted point by point to yield a meaningful age. A large number of analyses must be made and the ages determined by the relationship of the lead isotope ratios to each other and to the growth curve. In particular, the single-stage lead models must not be applied indiscriminately to isolated analyses of galena.

C *Two-stage anomalous leads*

The principal characteristic of the $^{206}Pb/^{204}Pb$ and $^{207}Pb/^{204}Pb$ ratios for two-stage anomalous leads will be their exact fit to an anomalous lead line which intersects the single-stage growth curve at t_1, the time of ordinary lead mineralization, and at time t_2, a later period of mineralization. Most often some points fall on an extension of the line beyond t_2. If the crustal source region is very rich in uranium, the application of a single-stage equation will give negative apparent ages to some samples. The isotope ratios on a graph of $^{208}\text{Pb}/^{204}\text{Pb}$ against $^{206}\text{Pb}/^{204}\text{Pb}$ may fall on one or more straight lines, not necessarily passing through the ordinary lead ratios. The slope of each of these straight lines is related to the Th/U ratio of the chemically related parent atoms in the crustal source. If the straight line does not pass through the ordinary lead isotope abundances on this second graph, this is sufficient proof that each sample has developed in a two-stage growth system and is not a simple mixture of an ordinary lead and a very radiogenic component. The following example from Broken Hill will illustrate these points.

Broken Hill, Australia The main galena deposits at Broken Hill are conformable with the enclosing highly metamorphosed Willyama Series. Mass spectrometer analyses carried out at the University of British Columbia (Russell, Ulrych and Kollar, 1961) indicate that the lead isotope compositions in major mines are identical within $\pm 0.27\%$. These may be interpreted as ordinary leads and on the basis of a single-stage model their ages vary between 1613 and 1622 m.y. The uncertainty in the isotopic composition of the leads introduces only about 5 m.y. uncertainty in the age, but it must be remembered that the uncertainties in the primordial lead ratios and in the age of the Earth introduce an error which is much larger. Support for a major period of tectonic activity in southern central Australia at this time is obtained from uranium- and thorium-bearing minerals. Collins, Russell and Farquhar (1953) reported a $^{207}\text{Pb}/^{206}\text{Pb}$ age of 1510 m.y. on a davidite from Radium Hill. Greenhalgh and Jeffery (1959) quote an age value for Crocker's Well, also carried out on davidite. Complete uranium, thorium and lead analyses were made for this sample so that three independent age values could be calculated. They are 1628 m.y. ($^{206}\text{Pb}/^{238}\text{U}$), 1702 m.y. ($^{207}\text{Pb}/^{206}\text{Pb}$) and 1252 m.y. ($^{208}\text{Pb}/^{232}\text{Th}$). The very low value for the $^{208}\text{Pb}/^{232}\text{Th}$ age is very frequently observed, and does not detract from the rather good agreement between the uranium–lead ages.

In the same area there is a second mode of lead occurrence, known locally as the Thackaringa type because early examples were found near the Thackaringa fault zone. They occur in vein deposits and are thought to have been formed at a shallow depth and at lower temperatures than in the case of the conformable deposits (King and Thompson, 1953). The lead isotope ratios show large variations from vein to vein but the ^{206}Pb/^{204}Pb and ^{207}Pb/^{204}Pb ratios are linearly related with the isotopic ratios from the conformable deposits (figure 31). This clearly indicates that the Thackaringa leads are the product of growth in two distinct uranium systems.

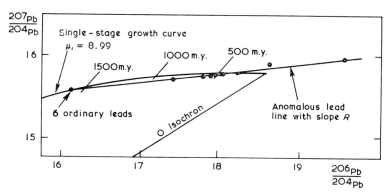

Figure 31 Graph showing the linear relationship between the ordinary leads and the anomalous leads at Broken Hill, Australia. Lead isotope data from Russell, Ulrych and Kollar (1961).

Kanasewich (1962a) has assumed that ordinary lead and uranium mineralization occurred contemporaneously in this area and used a two-stage model to obtain the age of a secondary period of tectonic activity to account for the anomalous Thackaringa lead isotope ratios. The age of secondary mineralization when some of the ordinary leads were remobilized and became contaminated with radiogenic lead to form vein deposits is found to be 510 ± 80 m.y. The uncertainty listed is only that due to the uncertainty of the measured isotopic abundances. The constants of Murthy and Patterson (1962) which are used in this paper decrease the estimated time of the second event to 485 ± 80 m.y. These times may be obtained by the intersection of the anomalous lead line with the single-stage growth curve as shown in figure 31. The growth curves in the second system are shown in figure 32 for all three isotope ratios.

It is useful to compare the limits on the ages using the Russell–Farquhar method for anomalous lead (equations A21 and A24 in appendix 1) with those obtained by the present model. Lead is found to be generated by uranium in the proportion given by the slope of the anomalous lead line, R, at a time t_{min} equal to 1110 m.y. ago. This is the oldest time at which the Thackaringa leads could have been deposited. A uranium mineral 1890 m.y. old ($=t_{max}$) would yield lead with this isotopic composition at the present time. From these calculations it is seen that the uranium and thorium could have been incorporated into the crust at any time between 1890 and 1110 m.y.

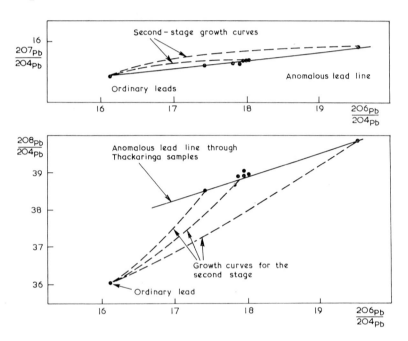

Figure 32 Graph showing the measured lead isotope ratios at Broken Hill and the growth curves along which the Thackaringa deposits developed between 1620 and 485 m.y. ago.

ago, while anomalous lead mineralization must have occurred at some time between 1110 and 0 m.y. ago. These data are summarized in figure 33 on the upper time line and compared with results from the two-stage model used by Kanasewich.

The time of secondary orogenesis is in the Ordovician or Upper Cambrian according to the Kulp (1961) time scale. Recently Richards and Pidgeon (1963) have reported age determinations on 24 mineral concentrates from the Broken Hill district. The results range from 430 to 1360 m.y. for K–Ar and from 460 to 550 m.y. for Rb–Sr and indicate that the last event of sufficient intensity to affect the biotites occurred in the Lower Paleozoic. It is possible that this may be correlated with the Tyennan orogeny which occurred in the Upper Cambrian and affected much of the area between Flinders Range and Tasmania. Thus the calculated time of mineralization of the anomalous leads is supported by K–Ar, Rb–Sr and by geologic evidence.

The ^{208}Pb/^{204}Pb and ^{206}Pb/^{204}Pb ratios are plotted in figure 32. There is a linear relationship between the lead isotope ratios from the Thackaringa vein deposits but the line does not pass through the

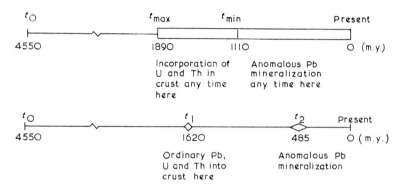

Figure 33 Chronological relationship between the Russell–Farquhar model for anomalous leads (top diagram) and the two-stage model (bottom diagram) as applied to Broken Hill.

Broken Hill ordinary leads. This is good evidence that this is primarily a growth system and not a simple mixture of a single ordinary and a single radiogenic lead. The ^{232}Th/^{204}Pb ratio, W_2, for the second stage of growth may be described by an equation of the form

$$W_2 = a + b\,\mu_2$$

The presence of the term, a, indicates that during the process in which uranium and thorium were transferred from a mantle to a crustal environment, thorium was locally enriched relative to uranium. The constant, b, is simply related to the slope of the line in figure 32. It represents the portion of uranium and thorium which underwent a similar chemical process in the crust. The calculated Th/U ratio for the seven Thackaringa-type leads varies between 3.8 and 6.6. This range is typical of the variation obtained from crustal granitic rocks (Rogers and Ragland, 1961).

Ozark Dome, Arkansas, Missouri and Oklahoma, U.S.A. The Ozark Dome area includes the Joplin leads which are in the Tri-State region, the Bonne Terre leads which are in southeastern Missouri and lead mines in northern Arkansas. A regression analysis was made on 30 samples reported by Bate and others (1957) (figure 34) to determine the slope of the straight line through the isotope ratios. The time of primary orogenesis, determined from two samples in Precambrian rocks, was found to be 1380 m.y. This compares with the

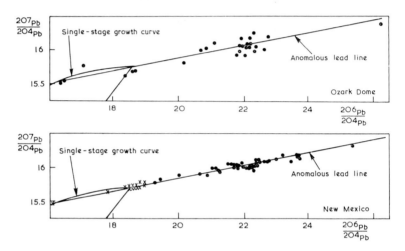

Figure 34 Graphs of lead isotope ratios from the Ozark Dome and New Mexico, U.S.A. Note the magnified scale of the $^{207}Pb/^{204}Pb$ axis. Measurements are by Bate and others (1957) and by Slawson and Austin (1960, 1962).

ages obtained by other researchers, as shown in table 4. All of these data strongly suggest that there was a period of major tectonic activity in this area between 1300 and 1400 m.y. ago and that the model lead age cannot be far from the absolute age of ordinary lead mineralization.

The secondary period of orogenesis is calculated to have occurred 90 m.y. ago with a very large standard deviation. This Cretaceous age correlates with statements by Philpott (1952) that there was regional uplift, folding, faulting and volcanism in Arkansas and Mississippi at the end of the Lower Cretaceous. As much as 7000 feet of section was truncated at the level of the Lower Cretaceous. In addition, diamonds occur in ultrabasic igneous intrusive rocks which penetrate Cretaceous sedimentary rocks near Murfreesboro in south-western Arkansas. This estimate agrees with the statement by Behre, Heyl and McKnight (1948) that 'the deposition of ores was contemporaneous with the last stages of the regional tectonic deformation'.

Table 4 Mineral ages from Missouri

Location	Mineral	$\dfrac{^{87}Rb}{^{87}Sr}$	$\dfrac{^{40}K}{^{40}Ar}$	$\dfrac{^{207}Pb}{^{206}Pb}$	$\dfrac{^{238}U}{^{206}Pb}$	$\dfrac{^{235}U}{^{207}Pb}$	$\dfrac{^{232}Th}{^{208}Pb}$
				(million years)			
Graniteville and Silver Mine, Mo.[a]	Muscovite			1200			
	Muscovite			1210			
Graniteville and Silver Mine, Mo.[a]	Muscovite			1210			
Silver Mine, Mo.[a]	Zinnwaldite books			1350			
Fredericktown, Mo.[b]	Muscovite	1350	1405				
Fredericktown, Mo.[b]	Zircon			1425	970	1120	1230
Graniteville, Mo.[b]	Biotite	1250	1280				
Graniteville, Mo.[b]	Feldspar	1230					
Decaturville, Mo.[b]	Muscovite	1360	1290				

[a] Allen, Hurley, Fairbairn and Pinson (1959).
[b] Tilton, Wetherill and Davis (1962).
Rubidium decay constant $= 1.47 \times 10^{-11}$ yr^{-1}.

New Mexico, U.S.A. Slawson and Austin (1960, 1962) have made a study of 63 samples of galena from deposits in West-Central New Mexico. The isotope ratios were all analyzed at the University of Toronto and lie close to a line with a slope of 0.0938 ± 0.0029 (figure 34). A sample from the Bosque del Apache site occurs in rocks classified as Precambrian and is calculated to have an age of 1500 m.y. Tilton and Davis (1959) have obtained Rb–Sr and K–Ar ages which vary from 1300 to 1490 m.y. near Mora, Dixon and Albuquerque, an area which lies northeast of the district under consideration.

The time of anomalous lead mineralization, according to the two-stage model (Kanasewich, 1962a), is found to be in the Tertiary, 50 m.y. ago. A Tertiary age has generally been assigned (Loughlin and Koschmann, 1942) to igneous rocks in the area; however, there is no direct stratigraphic correlation after the Permian.

A direct correlation has been found between lead isotopes and geologic structure in this area (Slawson and Austin, 1962). The group of samples marked with a cross in figure 34 coincides with a lineament which presumably formed a conduit for mineralizing solutions. A deeper crustal origin is suspected for the solutions since the average ^{238}U/^{204}Pb ($\mu_2 = 10$) ratio for this group of samples is only 10% higher than for ordinary leads. For the group of samples marked with a solid circle the average ^{238}U/^{204}Pb ratio is 170% higher than for ordinary leads. It is possible to contour both the ^{238}U/^{204}Pb ratio and the Th/U ratio for these samples to see the correlation with the lineament in the area.

D *Short period anomalous leads*

Many geologists such as King (1959) believe orogenic structures were built over long intervals by episodes of movement separated by even longer periods of crustal quiescence. We have taken advantage of this situation in our interpretation of two-stage anomalous leads wherein the evolution of lead in the second stage has taken over one billion years. This long quiescent period in a rich uranium crustal environment has contributed to the formation of extremely radiogenic leads such as those at Joplin.

Age determination studies indicate that the duration of an orogenic sequence may be as long as 200 to 400 m.y. Lead, uranium and thorium may be disseminated among a thick accumulation of volcanic and pyroclastic rocks during the early development of an eugeosyncline. If such a sequence is buried deeply, some of it may later become molten and be injected as a magma or by hydrothermal solutions into the upper part of the section. Leads which would be mineralized would have two stages of development with the second one being a relatively short period (less than 400 m.y. if the development of the Appalachian geosyncline is typical). The anomalous lead line in this case would be nearly tangent to the growth curve. The length of the line would depend upon the uranium/lead ratio, the length of time available in the second stage and the degree of homogenization before the lead and uranium were finally separated by mineralization.

Experimental uncertainties have created great difficulties in the identification of such theoretical examples of anomalous leads. The group of Hawaiian leads analyzed by Tatsumoto (figure 21) may be one example while Masuda's Japanese leads in volcanic rocks (figure 20) may be yet another. Both the examples given above are trace leads in extrusive and intrusive rocks and they have not been concentrated to form an ore. A possible example of a short period anomalous lead ore is the group of galenas from the Lake District in northern England which has been analyzed by Moorbath (1962). The results are shown in figure 35. The random error is too large to differentiate between the Hercynian or the Tertiary as the time of final mineralization.

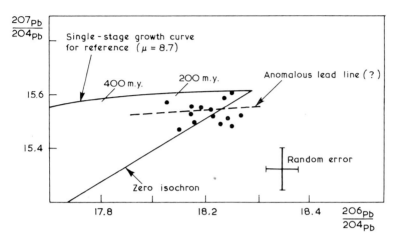

Figure 35 Lead isotope ratios from the Lake District, England. (Analyses by Moorbath, 1962)

E *Simple three-stage anomalous leads*

There are some special cases of a three-stage anomalous lead in which the $^{206}Pb/^{204}Pb$ and $^{207}Pb/^{204}Pb$ ratios lie exactly along an anomalous lead line. This occurs if the uranium and thorium are introduced into the crustal system either before or after the time of ordinary lead mineralization. These two hypothetical cases are illustrated in figure 36. For mathematical convenience the time of ordinary lead mineralization is always t_1, and, in this special case, μ_2 is zero or so small that it can be ignored when compared with the amount of uranium introduced into the third stage.

If a two-stage interpretation gives negative ages or ones which do not accord with independent information on the time of mineralization, a three-stage interpretation should be attempted even though it is generally impossible to solve the system completely without some further independent information. In any case, it is always possible to find useful limits for the various episodes in the development.

Examples of interpretation with a simple three-stage model are given by Kanasewich and Slawson (1964) for Ivigtut, Greenland; Ulrych and Russell (1964) for Sudbury, Ontario; Sinclair (1964) for the Kootenay Arc, British Columbia (figure 12). The Sudbury example will be discussed in greater detail since it contains some interesting Th/U ratios which have not been discussed previously.

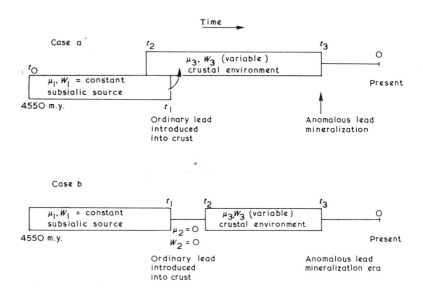

Figure 36 An illustration of the chronological development of two types of simple three-stage anomalous leads.

Sudbury, Ontario Although many potassium–argon and strontium–rubidium age determinations have been made on rocks and minerals from this region, the results have been generally discordant and confused. Perhaps the only convincing results have been whole rock strontium–rubidium data reported by Fairbairn, Hurley and Pinson (1960a) and Faure and others (1962). The strontium isochron plots are convincing and have been interpreted to give age values of 1650 m.y. for the Sudbury irruptive and 2160 m.y. for the Copper Cliff 'Rhyolite'. The reason for the complexity seems almost certainly related to the fact that Sudbury lies close to the boundary of at least two geologic provinces of vastly different age. Kanasewich (1962b) has suggested that near Sudbury there occurs the junction of four younger provinces (see figure 29) with the older Superior province.

The interpretation of ages determined for samples which have experienced more than one event is still ambiguous in most cases.

Isotopic analyses of galenas were reported for this region by Russell and others (1954) and by Russell and Farquhar (1960b). More recently, Ulrych and Russell (1964) have carried out isotopic analyses which are more precise than the earlier galena analyses.

As shown in figure 37, essentially the same straight line is obtained with the new analyses as with the combined new and old analyses when all have been corrected for pressure scattering of the peaks in the mass spectrometer. The analyses have been interpreted on the

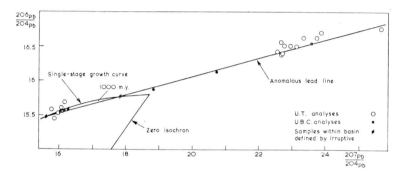

Figure 37 Lead isotope ratios which have evolved by radioactive decay of uranium in the Sudbury region. The ordinary and anomalous leads are linearly related and may be interpreted by a three-stage model. The analyses are from Russell and others (1954), Russell and Farquhar (1960b), Ulrych and Russell (1964) and Kanasewich and Farquhar (1965).

basis of a simple three-stage model (figure 36, case b), namely a stage between t_0 and t_1 during which an ordinary lead was produced, a stage between t_1 and t_2 during which there was no radiogenic addition of lead and a stage between t_2 and t_3 during which variable amounts of radiogenic lead were generated and added to the ordinary lead. A previous interpretation of the old Toronto analyses had been made by Kanasewich (1962a) on the basis of a two-stage model. Because of the uncertainties in the data, a mean of five ordinary leads was used to calculate an age of 1730 ± 150 m.y. for the ordinary leads. The time of secondary orogenesis was then found to be 870 ± 280 m.y. from the anomalous leads. Within limits of error this agreed with the time of the Grenville orogeny. For the new data, the apparent age of the least radiogenic lead is 1920 m.y. The maximum time at which the uranium could have been introduced into the crustal system is 2150 ± 50 m.y. and the minimum time is 1280 ± 50 m.y. This last figure is also the maximum age for the mineralization of the anomalous leads. In a three-stage model either t_2 or t_3 must be obtained from independent evidence, subject to the limits just discussed. In this case Ulrych and Russell assumed that t_3 was 950 m.y. obtained by averaging the values of strontium–rubidium ages by Fairbairn, Hurley and Pinson (1960b) for four gneisses in the nearby Grenville province. Using this value, an age of 1580 ± 50 m.y. was obtained for t_2, the time when uranium and thorium were introduced into the crust. This is close to the age of the Sudbury irruptive.

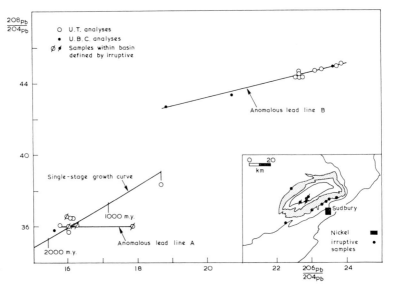

Figure 38 Lead isotope ratios from the Sudbury area together with a map indicating sample locations. The anomalous (third-stage) lead isotopes lie along lines not passing through the ordinary leads.

The ^{208}Pb/^{204}Pb ratios in the Sudbury region are very interesting because they illustrate the ordered mobility of thorium and uranium in near-surface crustal rocks (Kanasewich, 1964). Figure 38 shows that the lead isotope ratios have developed by growth in local country rock and suffered very little mixing during mineralization. The isotopes of ordinary leads are not linearly related to the anomalous leads even though they often occur in close proximity as in the Frood and McKim mines. The Th/U and Th/^{204}Pb ratios have characteristic values (indicated by the slope and positions of the lines through the anomalous leads in figure 38) which correlate with local geologic units. Since during the final period of mineralization very little mixing of lead with different isotopic composition has occurred, it is possible to calculate the Th/U and U/^{204}Pb ratios of the source regions.

Of particular interest is the fact that all the samples in the sediments of the Whitewater Group inside the basin have evolved in a thorium-free environment. These isotope ratios lie along line A which has zero slope, and therefore the Th/U ratios of the third-stage system must be zero. The line starts on the single-stage growth curve somewhere between 1600 and 1700 m.y. and this may be interpreted most simply by assuming that this is the time when uranium and thorium were mobilized in the third stage. The reasoning is strengthened by the fact that this corresponds with the previously calculated value for $t_2 = 1580 \pm 50$ m.y. Thus in this particular case it appears to be possible to calculate all three times for this three-stage model. The samples from the nickel mines have developed in a highly enriched thorium and uranium environment as indicated by line B in figure 38. It seems reasonable to suppose that some of this thorium and uranium was derived from the sediments in the Whitewater Group which had negligible thorium and uranium during the third stage of development.

G *Higher order multi-stage growth systems*

Growth systems with more than three consecutive stages of development have not been detected, possibly because they are rare, but most probably because of the imprecision of the analyses. Most of the gas-source mass spectrometer results from the University of British Columbia have a precision of one-tenth of one percent or better on the ^{206}Pb/^{204}Pb ratios. This has only just proved adequate for the interpretation of two-stage and simple three-stage models. Analyses by other laboratories, particularly if carried out by solid-source techniques, have five to ten times more random error and the interlaboratory comparisons on standard samples are usually inadequate. Thus it is difficult to justify more sophisticated models of interpretation at the present time. As was shown in section I.E, higher order growth systems may sometimes be treated as two-stage systems and this has been done by Chow and Patterson (1962) for oceanic leads.

H *Thorium/uranium ratios of anomalous leads*

Among the important numerical facts that can be derived rather precisely from anomalous lead isotope ratios is the Th/U ratio of the crustal system. The equations used for the calculations are given in the appendix and even if it is not certain whether a two- or three-stage model should be used, the values obtained will almost always be within 15% of the true value. This is about the magnitude of the uncertainties of such ratios as determined directly on rock samples by radiometric assays (Rogers and Ragland, 1961).

The calculations have been carried out by Russell, Kanasewich and Ostic (1964) on eleven suites of anomalous leads which have been analyzed by many research workers. The suites are from the Broken Hill area, Australia (Russell, Ulrych and Kollar, 1961); Sudbury, Canada (Ulrych and Russell, 1964); Goldfields, Canada (Russell and Farquhar, 1960a); Ozark Dome, U.S.A. (Bate and others, 1957); Thunder Bay, Canada (Farquhar and Russell, 1957a,b); Central Finnish Zone and the White Sea Region, U.S.S.R. (Vinogradov, Tarasov and Zykov, 1959); West-Central New Mexico (Slawson and Austin, 1960, 1962); Read Roseberry–Mount Farrell, Australia (Ostic, 1963a,b); Cobalt–Noranda area, Canada (Kanasewich and Farquhar, 1965). The results of these calculations are shown on a histogram in figure 39. The frequency has been plotted against the logarithm of the ratio because this form of plot for trace elements usually gives a more normal distribution and the shape of the distribution is then independent of constant errors such as could arise from uncertainties in the thorium decay constant. The shaded bar gives the distribution of Th/U for the 23 lead mines having approximately ordinary lead. For comparison purposes, figure 40 shows the published analyses of Th/U obtained by Evans and Goodman (1941); Keevil (1944), Whitfield, Rogers and Adams (1959), Rogers and Ragland (1961), Dimitriyev and Leonova (1962) and Leonova (1963) on acid and intermediate rocks.

It is clear that the distribution obtained for anomalous leads by the isotopic method is very similar to the distribution of Th/U obtained by direct analyses of crustal rocks. Figure 39 shows the very different distribution that is obtained for ordinary leads. The during mineralization are of great interest and have been closely studied by economic geologists and geochemists. However, the lead isotope data tell us very little about the processes taking place during mineralization since very often the lead ratios are not much changed at this time. What is obtained is an integrated geochemical history of the elements lead, uranium and thorium up to the time of final mineralization.

Lead isotopes, like strontium isotopes, may be used to trace the geochemical evolution of the continents. At the present time there are an insufficient number of precise determinations to decide if continental growth (Lawson, 1932; Wilson, 1952) has continued throughout geologic time from original nuclei or if substantially the entire continent formed very early in the Earth's history. Certainly, lead isotope data support other age determination studies which indicate that, by the time of the Grenville orogeny, the North American continent closely resembled its present dimensions (figure 29).

The hypothesis of multi-stage lead development is based on the episodic nature of orogenesis. It appears that a large proportion of lead isotope measurements may be interpreted on the basis of development in two or more uranium and thorium systems. The model assumes that the initial stage of lead isotope development occurred in a system having uniform uranium, thorium and lead. The calculated thorium/uranium ratios show that subsequent stages of development have occurred in a crustal sialic environment.

Appendix 1

Theory of Multi-Stage Models

For the purposes of geochronology it is convenient to let the present be $t = 0$ and count time positively into the past. Under this transformation the law of radioactive decay is

$$dN = +\lambda N \, dt \tag{A1}$$

where λ is the decay constant for a particular nucleus and dN is the number of nuclei out of a total N which are decaying in a short time dt. Letting N be the number of atoms at time t and N_0 be the number of radioactive atoms at the present time, the solution to equations (A1) is

$$N = N_0 \, e^{\lambda t} \tag{A2}$$

This equation is plotted in figure 1 for the radioactive atoms of ^{235}U, ^{238}U and ^{232}Th.

The rate of production of ^{206}Pb atoms at any time is proportional to the number of ^{238}U atoms $N(^{238}U)$ present in the system

$$\frac{dN}{dt}(^{206}Pb) = -\lambda N(^{238}U) \tag{A3}$$

The negative sign tells us that the amount of lead diminishes as we proceed into the past. Both sides of the equation may be divided by the number of ^{204}Pb atoms present and equation (2) may be substituted for the number of ^{238}U atoms present at any time t.

$$-d\frac{N(^{206}Pb)}{N(^{204}Pb)} = \lambda \left[\frac{N_0(^{238}U)}{N(^{204}Pb)} e^{\lambda t} \right] dt \tag{A4}$$

The quantity in brackets is the ratio of $^{238}U/^{204}Pb$ at any time t. In terms of symbols shown in table 1 equation (A4) can be integrated to yield the total amount of $^{206}Pb/^{204}Pb$ at any time t.

$$-\int_x^{a_0} dx = \int_t^{t_0} \lambda[\mu(t)e^{\lambda t}] \, dt \tag{A5}$$

Solving for x, the number of ^{206}Pb atoms relative to the number of ^{204}Pb atoms at any time t, one obtains the equation given by Russell and Allen (1956)

$$x = a_0 + \int_t^{t_0} \lambda[\mu(t)e^{\lambda t}] \, dt \tag{A6}$$

Similarly, the general equation for y, the ^{207}Pb/^{204}Pb ratio, and z, the ^{208}Pb/^{204}Pb ratio may be written

$$y = b_0 + \frac{1}{137.8} \int_t^{t_0} \lambda'[\mu(t)e^{\lambda' t}] \, dt \tag{A7}$$

$$z = c_0 + \int_t^{t_0} \lambda''[W(t)e^{\lambda'' t}] \, dt \tag{A8}$$

Uranium and thorium may be added or subtracted from the system at any time. Therefore μ and W, the ^{238}U/^{204}Pb and ^{232}Th/^{204}Pb ratios, reduced to the present time, are functions of time. The parameters μ and W may also vary from point to point within the Earth and are therefore three-dimensional variables of space.

Given x, y and z, the interpretation of lead isotopes involves the inverse solution of the three integral equations of the unknown four-dimensional variables μ and W. As indicated in the text, the analysis of a single sample of lead from a mine does not supply enough information and it is unwise to make any interpretation on isolated data even if the geological environment is well known.

The functions μ and W may vary from place to place but because metamorphic or orogenic events are episodic locally and brief in comparison to the half-lives of uranium and thorium it seems reasonable to treat them as constants for discrete intervals of time. The intervals may be as short as desired, so a general solution of equation (A6) has the form

$$x_m = a_0 + \mu_1(e^{\lambda t_0} - e^{\lambda t_1}) + \mu_2(e^{\lambda t_1} - e^{\lambda t_2}) \\ + \ldots + \mu_m(e^{\lambda t_{m-1}} - e^{\lambda t_m}) \tag{A9}$$

A lead may be specified by m, the number of uranium–thorium stages it has developed in, and the time of final mineralization as given by t_m. The lead is introduced into a new thorium–uranium–lead system characterized by parameters μ_1, W_1 at time t_0; μ_2, W_2 at time t_1; $\ldots$; μ_m, W_m at time t_{m-1}. Equation (9) may be written more compactly as

$$x_m = a_0 + \sum_{i=1}^{m} \mu_i(e^{\lambda t_{i-1}} - e^{\lambda t_i}) \tag{A10}$$

Similar equations may be written for y_m and z_m, the ^{207}Pb/^{204}Pb and ^{208}Pb/^{204}Pb ratios.

For a *single-stage* or *ordinary* lead the equations for the observed isotope ratios, x_1, y_1 and z_1 reduce to

$$x_1 = a_0 + \mu_1(e^{\lambda t_0} - e^{\lambda t_1}) \tag{A11}$$

$$y_1 = b_0 + \frac{\mu_1}{137.8}(e^{\lambda' t_0} - e^{\lambda' t_1}) \tag{A12}$$

$$z_1 = c_0 + W_1(e^{\lambda'' t_0} - e^{\lambda'' t_1}) \tag{A13}$$

Equations (A11) and (A12) may be solved for t_1, the time of mineralization, by eliminating μ_1 as noted by Houtermans (1946).

$$\frac{y_1 - b_0}{x_1 - a_0} = \phi = \frac{(e^{\lambda' t_0} - e^{\lambda' t_1})}{137.8(e^{\lambda t_0} - e^{\lambda t_1})} \tag{A14}$$

It must be emphasized that this equation should only be used for dating leads which have developed in a single closed lead–uranium–thorium system since the Earth formed. If a negative age is obtained the formula is incorrect for these leads and even the absolute value of t_1 is meaningless. A positive value for t_1 can be regarded as a *minimum* value of the first stage of lead isotope development in the area but must otherwise be interpreted with caution.

Leads which have developed in more than one distinct lead–uranium–thorium system are termed *anomalous*. A linear relationship between the $^{206}Pb/^{204}Pb$ and $^{207}/^{204}Pb$ ratios for a group of samples is proof of their anomalous nature and multi-stage equations must be used in the interpretation. A logical procedure is to consider that the isotope ratios have developed in a two-stage system ($m = 2$ in equations A9 and A10) and attempt an interpretation on this assumption. If the leads have developed in three or more lead–uranium–thorium systems a two-stage interpretation usually has some internal inconsistency. The precision of the experimental data is generally the limiting factor at this point and a more sophisticated model can seldom be justified.

For a two-stage interpretation consider the time t_1 to be a time of tectonic activity during which ordinary leads were differentiated from their source in the lower part of the crust or the upper part of the mantle and either concentrated to form a lead deposit at this time or disseminated throughout the host rock close to the surface. Contemporaneously, uranium- and thorium-bearing solutions were incorporated into the same environment. At time t_2, tectonic activity recurred in the area. Some of the ordinary lead may have been remobilized and become contaminated with radiogenic lead (which formed between times t_1 and t_2) to form anomalous leads. If the ordinary leads were disseminated at time t_1, they may be mixed with radiogenic lead and concentrated to form lead deposits at time t_2. In either case, the equation for the growth curve for a particular lead sample can be written in the form

$$x_2 = x_1 + \mu_2(e^{\lambda t_1} - e^{\lambda t_2}) \tag{A15}$$

$$y_2 = y_1 + \frac{\mu_2}{137.8}(e^{\lambda' t_1} - e^{\lambda' t_2}) \tag{A16}$$

$$z_2 = z_1 + W_2(e^{\lambda'' t_1} - e^{\lambda'' t_2}) \tag{A17}$$

The isotope ratios of the leads formed at time t_1 are denoted by x_1, y_1, z_1 as given in equations (A11) to (A13). If the second stage is in an upper crustal environment μ_2 and W_2 will vary considerably from sample to sample. However, μ_2 can be eliminated between equations (A15) and (A16) to obtain a result analogous to equation (A14).

$$\frac{y_2 - y_1}{x_2 - x_1} = R = \frac{(e^{\lambda' t_1} - e^{\lambda' t_2})}{137.8(e^{\lambda t_1} - e^{\lambda t_2})} \tag{A18}$$

According to equation (A18) there is a linear relationship between the ^{206}Pb/^{204}Pb and ^{207}Pb/^{204}Pb ratios. The experimental data will lie on a line which can be called simply an *anomalous lead line* and its slope depends only on the time, t_1, when uranium was introduced into the system, and the time, t_2, when uranium and radiogenic lead were separated in a mineralizing process. The two-stage model (Kanasewich, 1962a) makes the assumption, as indicated by the left-hand side of equation (A18), that ordinary lead mineralization and incorporation of uranium and thorium into the crust occurred contemporaneously during an early period, t_1, of tectonic activity.

The simplest form of a two-stage lead is a mixture of two ordinary leads (Kanasewich and Farquhar, 1965). The mathematical development is similar to that given above except that the second stage of development proceeds in a system identical to the first ($\mu_1 = \mu_2$; $W_1 = W_2$).

The equation for a complete three-stage model has the following format:

$$x_3 = a_0 + \mu_1(e^{\lambda t_0} - e^{\lambda t_1}) + \mu_2(e^{\lambda t_1} - e^{\lambda t_2}) + \mu_3(e^{\lambda t_2} - e^{\lambda t_3}) \tag{A19}$$

Similar equations may be written for y_3 and z_3. If the uranium concentration, μ_2 and μ_3, in the source varies from place to place with a

Gaussian distribution and the same mean, the isotope ratios will scatter about an anomalous lead line with the slope determined by times t_1 and t_3 (figure 15).

The isotope ratios may lie exactly along an anomalous lead line under certain special cases of a three-stage model. The most important is one in which μ_2 is zero. Geologically this means that there is a gap between times t_1 and t_2 during which the single-stage system is in a relatively uranium-free environment. At time t_2 uranium and thorium are introduced into the crustal system and the radiogenic leads produced are mixed with the ordinary leads at time t_3 when recrystallization occurs. The slope of the anomalous lead line will then be

$$R = \frac{y_3 - y_1}{x_3 - x_1} = \frac{(e^{\lambda' t_2} - e^{\lambda' t_3})}{137.8(e^{\lambda t_2} - e^{\lambda t_3})} \tag{A20}$$

This equation is equivalent to the one given by Farquhar and Russell (1957b).

It is impossible to obtain a unique solution for a three-stage lead from the isotope data alone, as the Sudbury case history illustrates. However, it is possible to obtain upper and lower limits on t_2 and t_3. By setting $t_3 = 0$ one finds the maximum value, t_{max}, of the time when uranium was incorporated into the crustal sequence.

$$R = \frac{(e^{\lambda' t_{max}} - 1)}{137.8(e^{\lambda t_{max}} - 1)} \tag{A21}$$

Such an equation was first used by Gerling (1942) to obtain the maximum difference in age of the crust at Ivigtut, Greenland and a group of widespread ore leads of Carboniferous age. The value he obtained for the maximum age of the crust at Ivigtut was 3230 m.y., and he interpreted this as a minimum value for the age of the Earth.

The time, t_{min}, when the uranium isotopes were generating lead isotopes in the ratio R may be obtained from equation (A4). This equation states that the rate at which $^{206}Pb/^{204}Pb$ is changing at any time such as t_m is

$$\frac{dx}{dt} = -\lambda \mu_3 \, e^{\lambda t_{min}} \tag{A22}$$

Similarly the rate at which $^{207}Pb/^{204}Pb$ is changing is

$$\frac{dy}{dt} = \frac{-\lambda' \mu_3}{137.8} e^{\lambda' t_{min}} \tag{A23}$$

The ratio is the slope of the anomalous lead line, R.

$$R = \frac{\lambda'}{137.8\lambda} \frac{e^{\lambda' t_{min}}}{e^{\lambda t_{min}}} \qquad (A24)$$

Equation (A24) was used by Russell and others (1954) on a group of galenas from Sudbury. The value t_{min} is the minimum time for t_2 when uranium was introduced into the crustal system. It can also be regarded as the maximum value for t_3, the time of lead mineralization. The solution for t_{min} is

$$t_{min} = \frac{1}{(\lambda' - \lambda)} \log_e \left(137.8 \frac{\lambda}{\lambda'} R\right) \qquad (A25)$$

The determination of the thorium/uranium ratio for the crustal source region follows immediately from the equations of either a two- or a three-stage model. For the simple three-stage model used in equation (A20) the ratio is

$$\frac{Th}{U} = \frac{137.8}{138.8} \frac{(z_3 - z_1)}{(x_3 - x_1)} \frac{(e^{\lambda t_2} - e^{\lambda t_3})}{(e^{\lambda'' t_2} - e^{\lambda'' t_3})} \qquad (A26)$$

The choice of a pair of values for t_2 and t_3 is relatively unimportant provided that the pair chosen is consistent with the slope relationship given by equation (A20). In fact, the use of the limiting value t_{min} will yield values within about 1% of the values given by equation (A26). The appropriate equation to use for the Th/U ratio in the crustal source region is then

$$\frac{Th}{U} = \frac{137.8}{138.8} \frac{(z_3 - z_1)}{(x_3 - x_1)} \frac{\lambda}{\lambda''} \frac{e^{\lambda t_{min}}}{e^{\lambda'' t_{min}}} \qquad (A27)$$

x_3 and z_3 are the isotopic ratios of the anomalous lead sample for which the calculation is being made, while x_1 and z_1 are the isotopic ratios of the ordinary lead. A suitable value for x_1 and z_1 is not always unambiguous. However, a value can usually be obtained within a few percent and the Th/U ratios will almost always be obtained within 15% which is about the magnitude of the uncertainty of such ratios as determined directly on rock samples by radiometric assays (Rogers and Ragland, 1961).

Appendix 2

A Table for Plotting Single-Stage Growth Curves

Constants—See table 1 $\mu_1 = 8.99 \pm 0.07$ Th/U $= 3.92 \pm 0.09$

	Growth curve $\mu_1 = 8.99$ $W_1 = 35.55$			Upper limit of growth band $\mu_1 = 9.06$ $W_1 = 34.96$			Lower limit of growth band $\mu_1 = 8.92$ $W_1 = 36.14$		
t_1(m.y.)	x	y	z	x	y	z	x	y	z
4550	9.56	10.42	30.00	9.56	10.42	30.00	9.56	10.42	30.00
4400	9.97	11.16	30.33	9.98	11.16	30.32	9.97	11.15	30.34
4200	10.51	11.99	30.77	10.51	12.00	30.76	10.50	11.98	30.78
4000	11.03	12.67	31.20	11.04	12.69	31.18	11.01	12.65	31.33
3700	11.77	13.48	31.85	11.79	13.50	31.82	11.76	13.45	31.88
3500	12.25	13.90	32.27	12.28	13.93	32.23	12.23	13.87	32.31
3000	13.39	14.65	33.32	13.43	14.69	33.28	13.37	14.62	33.37
2500	14.45	15.12	34.34	14.49	15.16	34.26	14.41	15.08	34.41
2000	15.43	15.40	35.33	15.47	15.44	35.24	15.38	15.36	35.42
1500	16.33	15.58	36.30	16.38	15.62	36.19	16.28	15.54	36.40
1000	17.17	15.68	37.24	17.23	15.73	37.12	17.11	15.65	37.36
500	17.94	15.75	38.16	18.01	15.80	38.02	17.88	15.71	38.29
400	18.09	15.76	38.34	18.16	15.80	38.20	18.02	15.72	38.48
300	18.24	15.77	38.52	18.30	15.81	38.38	18.17	15.73	38.66
200	18.38	15.77	38.70	18.45	15.82	38.56	18.31	15.74	38.85
100	18.52	15.78	38.88	18.59	15.83	38.73	18.45	15.75	39.03
0	18.66	15.79	39.06	18.73	15.84	38.91	18.59	15.75	39.21

Appendix 3

Lead Isotope Data on Deposits Used For Defining a Single-Stage Growth Curve

$c_0 = 30.00 \qquad t_0 = 4550$ m.y.

Mine area	$\frac{^{206}Pb}{^{204}Pb}$	$\frac{^{207}Pb}{^{204}Pb}$	$\frac{^{208}Pb}{^{204}Pb}$	No. of samples	Model lead age[b]	$\frac{^{238}U^{b}}{^{204}Pb}$	$\frac{^{232}Th^{b}}{^{204}Pb}$	$\frac{Th^{b}}{U}$	Source[a]
Hall's Peak, N.S.W., Australia	18.49	15.74	38.77	6	80	8.93	34.97	3.89	1
Bathurst, N.B., Canada	18.29	15.78	38.53	3	270	9.00	35.35	3.90	1
Cobar, N.S.W., Australia									
G.S.A. Mine, Cobar	18.22	15.78	38.57	6	320	9.00	35.89	3.96	1
Tharsis Mine	18.18	15.74	38.47	1	310	8.94	35.37	3.93	1
Great Cobar Mine	18.19	15.76	38.50	1	320	8.97	35.59	3.94	1
Silver Peak Mine	18.19	15.77	38.54	1	330	8.99	35.85	3.96	1
Queen Bee Mine	18.18	15.77	38.51	1	340	8.99	35.80	3.95	1
Captain's Flat, N.S.W., Australia	18.18	15.77	38.56	7	340	8.98	35.98	3.98	1
Sullivan Mine, B.C., Canada	16.63	15.64	36.58	2	1340	9.02	35.41	3.90	2
Mt. Isa, Queensland, Australia, Mine	16.23	15.60	36.26	4	1600	9.08	36.40	3.98	3
Black Star Mine	16.19	15.58	36.18	1	1600	9.04	35.99	3.95	3
Broken Hill, N.S.W., Australia Main Lode:	16.11	15.54	36.04	2	1620	8.98	35.38	3.91	3
Pinnacles Mine	16.11	15.53	36.01	1	1610	8.95	35.11	3.89	3
Little Broken Hill	16.13	15.55	36.12	1	1610	8.99	35.79	3.95	3
Globe Mine	16.11	15.53	36.03	1	1610	8.96	35.24	3.91	3
Centennial Mine	16.12	15.54	36.10	1	1620	8.98	35.78	3.95	3
Metsamouttu Aijala Kisko, Finland	15.77	15.47	35.61	1	1810	8.97	34.93	3.86	4
Korsnäs, Finland	15.81	15.47	35.65	1	1780	8.95	34.83	3.87	4, 5
Attu, Finland	15.80	15.48	35.64	1	1790	8.98	34.97	3.87	4
Orijärvi, Finland	15.82	15.50	35.67	1	1800	9.01	35.20	3.88	4
Pakila, Finland	15.76	15.48	35.63	1	1820	9.00	35.25	3.89	4
Errington Mine, Sudbury, Canada	15.60	15.46	35.75(?)	2	1920	9.02	?	—	6
Errington Mine	16.01	15.54	35.99	1	1700	9.03	35.95	3.95	6
Cobalt, Ontario, Canada	14.97	15.30	34.82	1	2250	9.02	35.39	3.90	7
						8.99 ± 0.07	35.55 ± 0.59	3.92 ± 0.09	

[a] 1. Ostic (1963a,b); 2. Sinclair (1964); 3. Russell, Ulrych and Kollar (1961); Kollar, Russell and Ulrych (1960); 4. Whittles (1962); 5. Kanasewich (1962b); 6. Ulrych and Russell (1964); 7. Kanasewich and Farquhar (1965).
[b] Calculated using the constants in table 1.

References

Aldrich, L. T. and Wetherill, G. W. (1960). Rb-Sr and K-A ages of rocks in Ontario and Northern Minnesota. *J. Geophys. Res.*, **65**, 337.
Aldrich, L. T., Wetherill, G. W., Bass, M. N., Compston, W., Davis, G. L. and Tilton, G. R. (1959). *Carnegie Inst. Wash. Yearbook*, **59**, 234. (Ann. Rept. Dept. Terr. Mag.)
Allen, V. T., Hurley, P. M., Fairbairn, H. W. and Pinson, W. H. (1959). Age of Precambrian igneous rocks of Missouri. Abstract. *Geol. Soc. Am.*, **2A**.
Alpher, R. A. and Herman, R. C. (1951). The primeval lead isotopic abundances and the age of the Earth's crust. *Phys. Rev.*, **84**, 1111.
Bate, G. L., Gast, P. W., Kulp, J. L. and Miller, D. S. (1957). Lead isotope abundances in lead minerals. Manuscript, Lamont Geological Observatory, Columbia Univ., New York.
Behre, C. H., Heyl, A. V. and McKnight, E. T. (1948). Zinc and lead deposits of the Mississippi Valley. *Intern. Geol. Congr., 18th, Great Britain, 1948*, Part 7, 51.
Boltwood, B. B. (1907). On the ultimate disintegration products of the radioactive elements. *Am. J. Sci.*, **23**, 77.
Burwash, R. A., Baadsgaard, H. and Peterman, Z. E. (1962). Precambrian K-Ar dates from the western Canada sedimentary basin. *J. Geophys. Res.*, **67**, 1617.
Carruthers, D. S. and Pratten, R. D. (1961). The stratigraphic succession and structure in the Zinc Corporation Ltd. and New Broken Hill Consolidated Ltd., Broken Hill, New South Wales. *Econ. Geol.*, **56**, 1088.
Cantanzaro, E. J. (1963). Zircon ages in southwestern Minnesota. *J. Geophys. Res.*, **68**, 2045.
Catanzaro, E. J. and Gast, P. W. (1960). Isotopic composition of lead in pegmatitic feldspars. *Geochim. Cosmochim. Acta*, **19**, 113.
Catanzaro, E. J. and Kulp, J. L. (1964). Discordant zircons from the Little Belt (Montana) Beartooth (Montana) and Santa Catalina (Arizona) Mountains. *Geochim. Cosmochim. Acta*, **28**, 87.
Chow, T. J. and Johnstone, M. S. (1963). Lead isotopes in the sediments of Hudson Bay and Baltic Sea. Abstract. *Trans. Am. Geophys. Union*, **44**, 890.
Chow, T. J. and Patterson, C. C. (1959). Lead isotopes in manganese nodules. *Geochim. Cosmochim. Acta*, **17**, 21.
Chow, T. J. and Patterson, C. C. (1962). The occurrence and significance of lead isotopes in pelagic sediments. *Geochim. Cosmochim. Acta*, **26**, 263.
Collins, C. B., Russell, R. D. and Farquhar, R. M. (1953). The maximum age of the elements and the age of the earth's crust. *Can. J. Phys.*, **31**, 402.
Cumming, G. L., Wilson, J. T., Farquhar, R. M. and Russell, R. D. (1955). Some dates and subdivisions of the Canadian Shield. *Proc. Geol. Assoc. Can.*, **7**, Part 2, 27.
David, Sir T. W. E. and Browne, W. R. (1950). *The Geology of the Commonwealth of Australia*, Vol. I. Arnold. London.
Dimitriyev, L. V. and Leonova, L. L. (1962). Uranium and thorium in the granitoids of the Kaib Massif (Central Kazakhstan). *Geochemistry (USSR) (English Transl.)*, 769.

Doe, B. R. (1962). Relationships of lead isotopes among granites, pegmatites and sulfide ores near Balmat, New York. *J. Geophys. Res.*, **67**, 2895.

Doe, B. R. and Hart, S. R. (1963). The effect of contact metamorphism on lead in potassium feldspars near the Eldora Stock, Colorado. *J. Geophys. Res.*, **68**, 3521.

Doe, B. R., Tilton, G. R. and Hopson, C. A. (1965). Lead isotopes in feldspars from selected granitic rocks associated with regional metamorphism. *J. Geophys. Res.*, **70**, 1947.

Eaton, J. P. (1962). Crustal structure and volcanism in Hawaii. *The Crust of the Pacific Basin.* Geophys. Monograph, No. 6. Natl. Acad. Sci.–Natl. Res. Council. p. 13.

Eberhardt, P., Geiss, J. and Houtermans, F. G. (1955). Isotopic ratios of ordinary lead and their significance. *Z. Physik*, **141**, 91.

Edwards, G. and Hess, D. C. (1956). Isolation and isotopic analysis of lead in meteorites and rocks. *Nuclear Processes in Geologic Settings*, Publ. 400. Natl. Acad. Sci.–Natl. Res. Council. p. 100.

Evans, R. D. and Goodman, C. (1941). Radioactivity of rocks. *Bull. Geol. Soc. Am.*, **52**, 459.

Evernden, J. F. and Richards, J. R. (1962). Potassium–argon ages in eastern Australia. *J. Geol. Soc. Australia*, **9**, 1.

Fairbairn, H. W., Hurley, P. M. and Pinson, W. H. (1960a). Mineral and Rock Ages at Sudbury-Blind River, Ontario. *Eighth Annual Progress Report.* Dept. Geol. and Geophys., Massachusetts Institute of Technology. p. 7.

Fairbairn, H. W., Hurley, P. M. and Pinson, W. H. (1960b). Mineral and rock ages at Sudbury-Blind River, Ontario. *Proc. Geol. Assoc. Can.*, **12**, 41.

Farquhar, R. M. and Russell, R. D. (1957a). Dating the Proterozoic in Canada. *Roy. Soc. Can., Spec. Publ.*, No. **2**, 28.

Farquhar, R. M. and Russell, R. D. (1957b). Anomalous leads from the Upper Great Lakes Region of Ontario. *Trans. Am. Geophys. Union*, **38**, 552.

Faure, G., Fairbairn, H. W., Hurley, P. M. and Pinson, W. H., Jr. (1962). The age of the nickel irruptive at Sudbury, Ontario. *Tenth Annual Progress Report.* Dept. Geol. and Geophys., Massachusetts Institute of Technology. p. 1.

Fleming, G. H., Jr., Ghiorso, A. and Cunningham, B. B. (1962). Specific alpha activities and half lives of U234, U235 and U236. *Phys. Rev.*, **88**, 642.

Flynn, K. F., and Glendenin, L. E. (1959). Half-life and beta spectrum of ^{87}Rb. *Phys. Rev.*, **116**, 744.

Gast, P. W. (1962). The isotopic composition of strontium and the age of stone meteorites—I. *Geochim. Cosmochim. Acta*, **26**, 927.

Gast, P. W., Tilton, G. R. and Hedge, C. (1964). Isotopic composition of lead and strontium from Ascension and Gough Islands. *Science*, **145**, 1181.

Gastil, G. (1960). The distribution of mineral dates in space and time. *Am. J. Sci.*, **258**, 1.

Geiss, J. (1954). Isotopic analysis of ordinary lead. *Z. Naturforsch.*, **9A**, 218.

Gerling, E. K. (1942). Age of the earth according to radioactivity data. *Dokl. Akad. Nauk SSSR*, **34**, 259.

Gerling, E. K. and Polkanov, A. A. (1958). The absolute age determination of the Precambrian of the Baltic Shield. *Geochemistry (USSR) (English Transl.)*, 867.

Gill, J. E. (1948). Mountain building in the Canadian Pre-Cambrian Shield. *Intern. Geol. Congr., 18th, Great Britain, 1948*, Part 13, 97.

Gill, J. E. (1949). Natural divisions of the Canadian Shield. *Roy. Soc. Can., Ser. 3*, **43**, 61.

Goldich, S. S., Nier, A. O., Baadsgaard, H., Hoffman, J. H. and Kreuger, H. W. (1961). The Precambrian geology and geochronology of Minnesota. *Minn. Geol. Surv., Bull.*, **41**.

Golubchina, M. N. and Rabinovich, A. V. (1957). Criteria of relationship between mineralization and magmatism based on the isotopic analysis of lead in the country rock and in the ore. *Geochemistry (USSR) (English Transl.)*, 238.

Greenhalgh, D. and Jeffery, P. M. (1959). A contribution to the Precambrian chronology of Australia. *Geochim. Cosmochim. Acta*, **16**, 39.

Hess, D. C. and Marshall, R. R. (1960). The isotopic compositions and concentrations of lead in some chondritic stone meteorites. *Geochim. Cosmochim. Acta*, **20**, 284.

Holmes, A. (1946). An estimate of the age of the earth. *Nature*, **157**, 680.

Holmes, A. (1947). A revised estimate of the age of the earth. *Nature*, **163**, 453.

Houtermans, F. G. (1946). The isotope ratios in natural lead and the age of uranium. *Naturwissenschaften*, **33**, 185. (See also addendum, 219.)

Houtermans, F. G. (1953). Determination of the age of the earth from the isotopic composition of meteoritic lead. *Nuovo Cimento*, **10**, 1623.

Houtermans, F. G., Eberhardt, A. and Ferrara, G. (1964). Lead of volcanic origin. In Craig, H., Miller, S. L. and Wasserburg, G. J. (Eds.), *Isotopic and Cosmic Chemistry*. North-Holland, Amsterdam. p. 233.

Hunt, G. H. (1961). The Purcell eruptive rocks. *Ph.D. Thesis*. Univ. Alberta, Canada.

Hurley, P. M., Hughes, H., Faure, G., Fairbairn, H. W. and Pinson, W. H. (1962). Radiogenic strontium-87 model of continent formation. *J. Geophys. Res.*, **67**, 5315.

Inghram, H. (1947). Manhattan project, Technical Series, Division II, 14, Chap. 5. McGraw-Hill, New York. p. 35.

Kanasewich, E. R. (1962a). Approximate age of tectonic activity using anomalous lead isotopes. *Geophys. J.*, **7**, 158.

Kanasewich, E. R. (1962b). Quantitative interpretations of anomalous lead isotope abundances. *Ph.D. Thesis*. Univ. British Columbia, Vancouver.

Kanasewich, E. R. (1964). Use of lead isotope data in tracing characteristic trends in Th/U and Th/^{204}Pb ratios of crustal rocks. Abstract. *Phys. Can., Bull. Can. Assoc. Physicists*, **20**, 45.

Kanasewich, E. R. and Farquhar, R. M. (1965). Lead isotope ratios from the Cobalt–Noranda area, Canada. *Can. J. Earth Sci.*, **2**, 361.

Kanasewich, E. R. and Slawson, W. F. (1964). Precision intercomparisons of lead isotope ratios; Ivigtut, Greenland. *Geochim. Cosmochim. Acta*, **28**, 541.

Keevil, N. B. (1944). Thorium/uranium ratios in rocks and minerals. *Am. J. Sci.*, **242**, 309.

King, H. F. and Thompson, B. P. (1953). The geology of the Broken Hill districts. *Empire Mining Met. Congr., 5th, Australia New Zealand, 1953, Publ.*, **1**, 533.

King, P. B. (1959). *The Evolution of North America.* Princeton Univ. Press, Princeton, U.S.A.

Kollar, F., Russell, R. D. and Ulrych, T. J. (1960). Precision intercomparison of lead isotope ratios; Broken Hill and Mount Isa. *Nature,* **187**, 754.

Kouvo, O. (1958). Radioactive ages of some Finnish Precambrian minerals. *Bull. Comm. Geol. Finlande,* **182**.

Kouvo, O. and Kulp, J. L. (1961). Isotopic composition of Finnish galenas. *Ann. N.Y. Acad. Sci.,* **91**, 476.

Kovarik, A. F. and Adams, N. E., Jr. (1955). Redetermination of the disintegration constant of ^{238}U. *Phys. Rev.,* **98**, 46.

Kulp, J. L. (1961). Geologic time scale. *Science,* **133**, 1105.

Lawson, A. C. (1932). Insular arcs, foredeeps and geosynclinal seas of the Asiatic Coast. *Bull. Geol. Soc. Am.,* **43**, 353.

Leonova, L. L. (1963). Uranium and thorium in hydrothermally altered rocks of the Kzyl-Ompul mountains (Northern Tien Shan). *Geochemistry (USSR) (English Transl.),* **6**, 566.

Loughlin, G. F. and Koschmann, A. H. (1942). Geology and ore deposits of the Magdalena mining district, New Mexico. *U.S. Geol. Surv., Profess. Papers,* **200**.

Lovering, J. F. and Morgan, J. W. (1964). Uranium and thorium abundances in stony meteorites. *J. Geophys. Res.,* **69**, 1979.

Lowdon, J. A. (1960). Age determinations by the Geological Survey of Canada. *Geol. Surv. Can. Paper,* **60-17**.

Lowdon, J. A., Stockwell, C. H., Tipper, H. W. and Wanless, R. K. (1963). Age determinations and geological studies. *Geol. Surv. Can. Paper,* **62-17**.

Marshall, R. R. (1960). The amounts and isotopic compositions of lead in eclogites from the Münchberg gneiss massif (Fithtelgebirge). *Intern. Geol. Congr., Rept. 21st Session, Norden, 1960,* Part 8, 404.

Marshall, R. R. (1962). Mass spectrometric study of the lead in carbonaceous chondrites. *J. Geophys. Res.,* **67**, 2001.

Marshall, R. R. and Hess, D. C. (1961). Lead from troilite of the Toluca iron meteorite. *Geochim. Cosmochim. Acta,* **21**, 161.

Martin, W. C. (1957). Errington and Vermillion Lake Mines. Structural Geology of Canadian Ore Deposits, II. *Commonwealth Mining Met. Congr., 6th, Vancouver, 1957,* 363.

Masuda, A. (1964). Lead isotope composition in volcanic rocks of Japan. *Geochim. Cosmochim. Acta,* **28**, 291.

Mohler, F. L. (1960). Isotopic abundance ratios reported for reference samples stocked by the National Bureau of Standards. *U.S. Dept. Comm., Tech. Note,* **51**.

Moorbath, S. (1962). Lead isotope abundance studies on mineral occurrences in the British Isles and their geological significance. *Phil. Trans. Roy. Soc. London, Ser. A,* **254**, 295.

Moorbath, S. and Pauly, H. (1962). Rubidium–strontium and lead isotope studies on intrusive rocks from Ivigtut, South Greenland. *Tenth Annual Progress Report.* Dept. Geol. and Geophys., Massachusetts Institute of Technology. p. 99.

Morgan, J. W. and Lovering, J. F. (1964). Uranium and thorium abundances in stony meteorites. *J. Geophys. Res.,* **69**, 1989.

Morgan, J. W. and Lovering, J. F. (1965). Uranium and thorium in the Nuevo Laredo Achondrite. *J. Geophys. Res.*, **70**, 2002.
Murthy, V. R. (1964). Stable isotope studies of some heavy elements in meteorites. In Craig, H., Miller, S. L. and Wasserburg, C. J. (Eds.), *Isotopic and Cosmic Chemistry*. North-Holland, Amsterdam. p. 488.
Murthy, V. R. and Patterson, C. C. (1961). Lead isotopes in ores and rocks of Butte, Montana. *Econ. Geol.*, **56**, 59.
Murthy, V. R. and Patterson, C. C. (1962). Primary isochron of zero age for meteorites and the earth. *J. Geophys. Res.*, **67**, 1161.
Ostic, R. G. (1963a). Isotopic investigation of conformable lead deposits. *Ph.D. Thesis*. Univ. British Columbia, Vancouver.
Ostic, R. G. (1963b). Recent isotopic investigations of conformable lead deposits. Abstract. *Intern. Union Geod. Geophys.*, *7th, Berkeley*, Part 7, 36.
Patterson, C. C. (1953). The isotopic composition of meteoritic, basaltic and oceanic leads and the age of the earth. *Nuclear Processes in Geological Settings*, Publ. 400. Natl. Acad. Sci.–Natl. Res. Council. p. 36.
Patterson, C. C. (1955). The $^{207}Pb/^{206}Pb$ ages of some stone meteorites. *Geochim. Cosmochim. Acta*, **7**, 151.
Patterson, C. C. (1956). Age of meteorites and the earth. *Geochim. Cosmochim. Acta*, **10**, 230.
Philpott, T. H. (1952). Louisiana–Arkansas region may yield additional oil. *World Oil*, **135**, 108.
Picciotto, E. and Wilgain, S. (1956). Confirmation of the period of thorium-232. *Nuovo Cimento*, **4**, 1525.
Polkanov, A. A. and Gerling, E. K. (1961). Problems of Geochronology and Geology. Research of the laboratory of Precambrian Geology. *Dokl. Akad. Nauk SSSR*, **12**, 1.
Rice, H. M. A. (1941). Nelson Map-Area, East Half, British Columbia. *Geol. Surv. Can. Mem.*, **228**, 55.
Richards, J. R. (1962). Age of the Earth's crust and lead model ages. *Nature*, **195**, 65.
Richards, J. R. and Pidgeon, R. T. (1963). Some age measurements on micas from Broken Hill, Australia. *J. Geol. Soc. Australia*, **10**, 243.
Richards, J. R., Webb, A. W. and Cooper, J. A. (1963). Potassium–argon ages on micas from the Precambrian region of North Western Queensland. *J. Geol. Soc. Australia*, **10**, 299.
Rogers, J. J. W. and Ragland, P. C. (1961). Variation of thorium and uranium in selected granitic rocks. *Geochim. Cosmochim. Acta*, **25**, 99.
Russell, R. D. (1956). Interpretation of lead isotope abundances. *Nuclear Processes in Geological Settings*, Publ. 400. Natl. Acad. Sci.–Natl. Res. Council. p. 68.
Russell, R. D. (1963). Some recent researches on lead isotope abundances. In Geiss, G. and Goldberg, E. D. (Eds.), *Earth Science and Meteoritics*. North-Holland, Amsterdam. p. 44.
Russell, R. D. and Allan, D. W. (1956). The age of the earth from lead isotope abundances. *Roy. Astron. Soc. (Geophys. Suppl.)*, **7**, 80.
Russell, R. D. and Farquhar, R. M. (1960a). *Lead Isotopes in Geology*. Interscience, New York.
Russell, R. D. and Farquhar, R. M. (1960b). Dating galenas by means of their isotopic constitution II. *Geochim. Cosmochim. Acta*, **19**, 41.

Russell, R. D., Farquhar, R. M., Cumming, G. L. and Wilson, J. T. (1954). Dating galenas by means of their isotopic constitutions. *Trans. Am. Geophys. Union*, **35**, 301.

Russell, R. D., Kanasewich, E. R. and Ostic, R. G. (1964). Quantitative interpretation of anomalous lead isotope abundances. *Chemistry of the Earth's Crust*. Akad. Nauk SSSR, Inst. Geochem. Anal. Chem. p. 638.

Russell, R. D., Ulrych, J. T. and Kollar, F. (1961). Anomalous leads from Broken Hill, Australia. *J. Geophys. Res.*, **66**, 1495.

Russell, R. D., Kanasewich, E. R. and Ozard, J. M. (1966). Isotropic abundance of lead from 'frequently-mixed' source. *Earth Planetary Sci. Letters*, **1**, 85.

Schütze, W. (1962). On the dating of Pb-minerals. *Geochim. Cosmochim. Acta*, **26**, 617.

Shaw, D. M. (1957). Comments on the geochemical implications of lead-isotope dating of galena deposits. *Econ. Geol.*, **52**, 570.

Shields, R. M. (1964). The ^{87}Rb–^{86}Sr age of stony meteorites. *Twelfth Annual Progress Report*. Dept. Geol. and Geophys., Massachusetts Institute of Technology. p. 3.

Sinclair, A. J. (1964). A lead isotope study of mineral deposits in the Kootenay Arc. *Ph.D. Thesis*. Univ. British Columbia, Vancouver.

Slawson, W. F. and Austin, C. F. (1960). Anomalous leads from a selected geological environment in West Central New Mexico. *Nature*, **187**, 400.

Slawson, W. F. and Austin, C. F. (1962). A lead isotope study defines a geological structure. *Econ. Geol.*, **57**, 21.

Slawson, W. F., Kanasewich, E. R., Ostic, R. G. and Farquhar, R. M. (1963). Age of the North American crust. *Nature*, **200**, 413.

Smith, C. H. and Skinner, R. (1958). Geology of the Bathurst–Newcastle Mineral District, New Brunswick. *Can. Mining Met. Bull.*, **51**, 150.

Soddy, F. (1911). *Ann. Rept. Progr. Chem. Chem. Soc. London*, **99**, 72.

Soddy, F. (1914). *Chemistry of the Radio-Elements*. Longmans, Green, London.

Stanton, R. L. (1960). General features of the conformable 'pyritic' orebodies. *Trans. Can. Inst. Mining Met.*, **63**, 22.

Stanton, R. L. and Russell, R. D. (1959). Anomalous leads and the emplacement of lead sulfide ores. *Econ. Geol.*, **54**, 588.

Staryk, E. E., Shatz, M. M. and Sobotovich, A. V. (1958). The age of meteorites. *Dokl. Akad. Nauk SSSR*, **123**, 424.

Staryk, E. E., Sobotovich, A. V., Lovtzyus, G. P., Shatz, M. M. and Lovtzyus, A. V. (1960). Lead and its isotopic composition in iron meteorites. *Dokl. Akad. Nauk SSSR*, **134**, 555.

Stockwell, C. H. (1961). Structural provinces, orogenies and time classification of rocks of the Canadian Precambrian Shield. *Geol. Surv. Can. Paper*, **61-17**, 108.

Swanson, G. O. and Gunning, H. C. (1948). Sullivan Mine. *Structural Geology of Canadian Ore Deposits*. Can. Inst. Mining Met. Publ. 219.

Tatsumoto, M. (1964). Isotopic composition of lead in volcanic rocks from Japan, Iwo Jima, and Hawaii. Abstract. *Trans. Am. Geophys. Union*, **45**, 109.

Tatsumoto, M. and Patterson, C. C. (1963). The concentration of common lead in sea water. In Geiss, J. and Goldberg, E. D. (Eds.), *Earth Science and Meteoritics*. North-Holland, Amsterdam. p. 74.

Thompson, B. P. (1953). Geology and ore occurrence in the Cobar District. Geology of Australian Ore Deposits. *Empire Mining Met. Congr.*, *5th, Australia New Zealand, 1953, Publ.*, **1**, 863.
Thomson, R. (1957). Cobalt and Larder Lake. *Structural Geology of Canadian Ore Deposits.* Can. Inst. Mining Met. Publ. 377.
Thurber, D. L. (1964). The uranium content of sea water. Abstract. *Trans. Am. Geophys. Union*, **45**, 119.
Tilton, G. R. and Davis, G. L. (1959). In Abelson, P. H. (Ed.), *Researches in Geochemistry.* Wiley, New York, pp. 190–216.
Tilton, G. R. and Patterson, C. C. (1956). The isotopic composition of lead in ultramafic rocks. *Trans. Am. Geophys. Union*, **37**, 361.
Tilton, G. R., Patterson, C. C., Brown, H., Inghram, M., Hayden, R., Hess, D. and Larsen, E. S., Jr. (1955). Isotopic composition and distribution of lead, uranium, and thorium in a Precambrian granite. *Bull. Geol. Soc. Am.*, **66**, 1131.
Tilton, G. R., Patterson, C. C. and Davis, G. L. (1954). Isotopic composition of lead in olivine bombs. *Bull. Geol. Soc. Am.*, **65**, 1314.
Tilton, G. R. and Reed, G. W. (1963). Radioactive heat production in eclogite and some ultramafic rocks. In Geiss, J. and Goldberg, E. D. (Eds.), *Earth Science and Meteoritics.* North-Holland, Amsterdam. p. 31.
Tilton, G. R., Wetherill, G. W. and Davis, G. L. (1962). Mineral ages from the Wichita and Arbuckle Mountains, Oklahoma and the St. Francis Mountains, Missouri. *J. Geophys. Res.*, **67**, 4011.
Tupper, W. M. (1959). Intrusive granites in New Brunswick. *Seventh Annual Progress Report.* Dept. Geol. and Geophys., Massachusetts Institute of Technology. p. 187.
Ulrych, T. J. (1964). The anomalous nature of Ivigtut lead. *Geochim. Cosmochim. Acta*, **28**, 1389.
Ulrych, T. J. and Russell, R. D. (1964). Gas source mass spectrometry of trace leads from Sudbury, Ontario. *Geochim. Cosmochim. Acta*, **28**, 455.
Vaasjoki, O. and Kouvo, O. (1959). A comparison between the common lead isotopic composition and minor base metal contents of some Finnish galenas. *Econ. Geol.*, **54**, 301.
Vinogradov, A. P. (1955). Lead isotopes and their geochemical meaning. *Peaceful Utilization of Atomic Energy.* Sess. Chem. Sci., Akad. Nauk SSSR. p. 320.
Vinogradov, A. P., Tarasov, L. S. and Zykov, S. I. (1959). Isotopic composition of leads from the ores of the Baltic Shield. *Geochemistry (USSR) (English Transl.)*, 689.
Vinogradov, A. P., Tarasov, L. S. and Zykov, S. I. (1960). Isotopic composition of leads from pyrite ore deposits of the Urals. *Geochemistry (USSR) (English Transl.)*, 565.
Vinogradov, A. P., Zykov, S. I. and Tarasov, L. S. (1958). The isotopic composition of admixtures of lead in ores and minerals as indication of their origin and time of their formation. *Geochemistry (USSR) (English Transl.)*, 653.
Wampler, J. M. and Kulp, J. L. (1964). An isotopic study of lead in sedimentary pyrite. *Geochim. Cosmochim. Acta*, **28**, 1419.
Wetherill, G. W., Oouvo, O., Tilton, G. R. and Gast, P. W. (1962). Age measurements on rocks from the Finnish Precambrian. *J. Geol.*, **70**, 74.

Whitfield, J. M., Rogers, J. J. W. and Adams, J. A. S. (1959). Relation between the petrology and the thorium and uranium contents of some granitic rocks. *Geochim. Cosmochim. Acta*, **17**, 248.

Whittles, A. B. L. (1962). The elusive isochron. Abstract. *Trans. Am. Geophys. Union*, **43**, 449.

Wickman, F. E., Blomquist, N. G., Geijer, P., Parwel, A., Ubisch, H. V. and Welin, E. (1963). Isotopic constitution of ore lead in Sweden. *Arkiv Mineral. Geol.*, **3**, 193.

Wilson, J. T. (1949). Some major structures of the Canadian Shield. *Trans. Can. Inst. Mining Met.*, **52**, 231.

Wilson, J. T. (1952). Orogenies as the fundamental geologic process. *Trans. Am. Geophys. Union*, **33**, 444.

Wilson, M. E. (1939). The Canadian Shield. *Geologie der Erde*. In Krenkle, E. (Ed.), *Geology of North America*, Vol. 1. Springer, Berlin.

Zartman, R. E. (1965). The isotopic composition of lead in microclines from the Llano uplift, Texas. *J. Geophys. Res.*, **70**, 935.

The interpretation of zircon ages

E. J. CATANZARO*

I *Introduction*, 225. II *Zircon*, 226. III *The discordancy problem*, 229.
IV *Geologic meaning of zircon ages*, 254. *References*, 255.

I Introduction

The uranium (thorium)–lead system in zircon is the only radioactive parent–daughter system which is frequently unaffected or only partially affected by a metamorphism of the enclosing rock. This general resistance to the complete rejuvenation of its uranium–lead system makes zircon unique among the common accessory minerals in its ability to retain some age evidence of an event which preceded the last significant metamorphic event in an area.

On the other hand, some zircon samples are known to have been completely rejuvenated (i.e. to have lost all of their previously accumulated radiogenic lead) during a metamorphic event; many others are known to have suffered partial rejuvenation during a metamorphic event; and, paradoxically, almost all zircon samples appear to be susceptible to some low-temperature process which generally causes measurable amounts of lead loss. If complete rejuvenation occurs, the sample will lose all age evidence of any previous igneous or metamorphic event. However, if the uranium–lead system is disturbed in any way short of complete rejuvenation, no age evidence will be completely obliterated but there will be internal discordance between the three isotopic ages (^{238}U–^{206}Pb age, ^{235}U–^{207}Pb age and ^{207}Pb/^{206}Pb age) obtained on the sample.

The combination of zircon's susceptibility to low-temperature lead loss and its resistance, but not general immunity to partial or complete

* Washington, D.C., U.S.A.

rejuvenation during metamorphic events generally makes the interpretation of zircon age results a twofold problem. First, if the ages are discordant, which is usual, the cause or causes of the discordances must be found and the discordant ages must be interpreted to yield true ages. Second, if the history of the rock or area under investigation is complex, as will generally be the case when 'relict' ages are found, the geologic meaning of the true ages must be deciphered. The first problem may be somewhat generalized into a choice between the basic hypotheses put forth to explain zircon discordances. The second problem cannot be generalized and its solution in any particular area will usually require detailed knowledge of the geology of that area.

The analytical aspects of zircon age determination are described in a number of papers (e.g. Tilton and others, 1955; Catanzaro and Kulp, 1964). In brief, these consist of decomposing (with the aid of a flux) and dissolving the sample; measuring the uranium and lead contents (by isotope dilution); and measuring the lead isotope composition (by mass spectrometry). Analytical errors usually attached to the resulting ages are of the order of 2 to 3%. The half-life of ^{238}U is known to better than 1% but that of ^{235}U has an uncertainty of $\pm 2\%$. However, discordant ages cannot be generally explained by either the analytical errors or the uncertainty in the half-lives.

Since zircon always contains thorium as well as uranium, a ^{232}Th–^{208}Pb age may be obtained, if a thorium analysis is performed. However, this age generally does not add significant information and, for analytical exigencies, is not even measured in some laboratories. (Thorium ages will not be generally discussed in this paper.)

II Zircon

Zircon is one of the most common accessory minerals, occurring in various amounts in almost all igneous, sedimentary and metamorphic rocks. In basic rocks, zircon usually crystallizes late and occurs as crystallites with irregular form, or as small rounded grains (Poldervaart, 1956). In intermediate and granitic rocks, zircon may crystallize early or late and usually forms sizeable euhedral crystals. In any igneous rock, the presence of rounded zircons, other than very minute grains, would suggest that the magma was contaminated by sediments, or that the igneous-looking rock is metasomatic. Sedimentary rocks usually contain rounded zircons while the shapes of zircons in metamorphic rocks usually reflect their premetamorphic nature, i.e.

metasedimentary rocks usually contain rounded zircons and meta-igneous rocks usually have euhedral zircons. However, zircon shapes may sometimes be affected by metamorphism. Harker (1932) has described detrital zircons whose rounded shape persisted even during ultrametamorphism, but Poldervaart and von Backstrom (1950) have shown that zircons in some rocks have recrystallized during intense metamorphism. The not uncommon presence of overgrowths shows that zircon shapes can be modified by later events.

The use of zircon characteristics in distinguishing generic and spatial relations has met with a number of successes. Larsen and others (1938) used zircons to establish mixing of magmas. Vitanage (1957) used the well-rounded shape and mixed assemblage of zircons to show that the Polonnarnenke charnockites (Ceylon) are metasedimentary. Taubeneck (1957) was able to distinguish three younger intrusives from the main body of the Bald Mountain Batholith (Oregon) on the basis of differences in the reduced major axis of their zircon suites, a technique developed by Larsen and Poldervaart (1957). Color has been used as a criterion for distinguishing zircons, but special caution is necessary because the color may be the result of metamictization which is a function of both the uranium and the thorium content and age.

The theoretical composition of zircon is $ZrSiO_4$ but many elements substitute for the Zr^{4+}, especially hafnium, strontium, sulfur, scandium, tantalum, niobium, yttrium, the rare earths, uranium and thorium (Rankama and Sahama, 1950). The uranium content of accessory zircon is quite variable but generally ranges between 100 and 2000 ppm. The thorium content is generally one-half to one-quarter that of the uranium content but in some samples thorium is more abundant than uranium. Pegmatitic zircon may contain as much as 25,000 ppm uranium and 7500 ppm thorium.

The occurrence of overgrowths on zircon grains is not uncommon (Butterfield, 1936; Smithson, 1937; Poldervaart and Eckelmann, 1955; Karakeda, 1961). Smithson (1937) found euhedral overgrowth rims on detrital zircon grains in Middle Jurassic sedimentary rocks and concluded that the overgrowths were formed *in situ*; presumably at low temperatures. However, in most cases, overgrowth formation occurs during a high-temperature rock-forming event such as the case of euhedral rims forming on detrital zircons incorporated in a magma (Karakeda, 1961), or the formation of anhedral to euhedral rims during granitization of sediments (Poldervaart and Eckelmann, 1955; Eckelmann and Poldervaart, 1957).

The importance of overgrowths in age determination studies is obvious. The age difference between the original zircon grains and the overgrowths will cause the zircons to yield discordant age results, unless the cores suffered total radiogenic lead loss at the time the overgrowths were formed. The recognition of rounded cores inside otherwise subhedral to euhedral zircons is very strong evidence that the original grains have been through a sedimentary cycle, a fact which has obvious implications for the geochronology as well as the geology.

Many natural zircons are partially or completely metamict. The metamictization is due to the energy released in the radioactive decay of uranium and thorium atoms present in the zircon lattice. Holland and Gottfried (1955) have estimated that about 1500 atomic displacements are produced during each alpha emission. Since young samples with high radioactivity fit the same radiation damage curve as old samples with low radioactivity, the self-annealing rate of zircon is probably negligible under ordinary natural conditions (Hurley and Fairbairn, 1952). However, annealing of metamict zircons may take place at high temperatures; it has been accomplished under hydrothermal conditions in the laboratory (Mumpton and Roy, 1961).

Metamictization is important in age determination work because of its obvious effect on the physical and chemical stability of zircon. In a number of multi-sample studies, it has been established that the degree of discordance of the zircon ages, i.e. the percent apparent lead loss, increases with increasing uranium content.

The general chemical and physical stability of zircon is well authenticated. Zircon heads the list in the order of persistence of minerals found in arenaceous sediments (Pettijohn, 1941). However, there is no doubt that zircon, especially if metamict, can be readily attacked under suitable natural conditions. Carroll (1953) observed that zircons in lateritic soil profiles are frequently strongly corroded. She ascribes the corrosion to alkaline solutions. Strock (1941) has shown that calcium bicarbonate solutions can remove considerable amounts of zirconium from zircon, and Maurice (1949) has shown that zircon has a wide range of stability in acid solutions but only a limited range in alkaline solutions. Frondel (1953) found that zircons which were especially susceptible to chemical attack always contained uranium, thorium and lead and characteristically had 10 to 12% water and a deficiency of silica. Mumpton and Roy (1961) studied the hydrothermal stability of the zircon–thorite group and concluded that the water in zircon is molecular water which enters the lattice only after a sufficient degree of metamictization has been attained.

The interpretation of zircon ages

They suggest that the silica deficiencies associated with hydrated and metamict zircons may be the result of leaching of the silica from zircons in a hydrothermal environment.

III   The Discordancy Problem

A   *Introduction*

If the two isotopic ages derived from the $^{206}Pb/^{238}U$ and $^{207}Pb/^{235}U$ ratios in a mineral differ by more than 5% the results are said to be 'discordant'. Differences less than 5% may be analytical, but differences greater than 5% are probably real and indicate that the mineral sample has not been a closed system throughout its history. Any temporal variation in the content of uranium, lead, or intermediate decay products (for an extended period of time), other than that due to radioactive decay, will lead to discordant age results.

Although a number of concordant zircons or zircon samples have been found, most samples are generally discordant. The age pattern exhibited by discordant zircons is almost always the same: ^{238}U–^{206}Pb age $<$ ^{235}U–^{207}Pb age $<$ $^{207}Pb/^{206}Pb$ age. Only a few reversals of this pattern have been found. Mathematically the usual pattern can result from either addition of uranium, loss of lead, or prolonged loss of one or more of the intermediate decay products in the ^{238}U chain. Naturally, for the first two processes, the mobility may include both daughter and parent atoms but the relative amounts would have to be consistent with the resulting pattern. The hypothesis of ^{238}U intermediate daughter loss can be eliminated as a major cause of zircon discordances by a number of considerations, the most important of which is that concordia plots (see below) show no evidence of this phenomenon. Of the two remaining processes, uranium addition and lead loss, evidence from leaching studies (Tilton, 1956) and U/Th ratio studies (Tilton and others, 1957), as well as general consideration of chemical potentials, suggest that lead loss is the more likely mechanism.

B   *Development of lead loss hypotheses*

The principal hypotheses put forth to account for discordant U–Pb ages were originally formulated on data from a variety of uranium-bearing minerals. Although different minerals may show different discordant age patterns (Kulp and Eckelmann, 1957), many show the same, most common, zircon pattern. It is possible that the cause of this type of discordance, which is the only type under consideration here, is basically the same in all of the different minerals. However, since the different minerals show varying susceptibilities to various geological phenomena, the practice of using other discordant minerals to explain discordant zircon results may complicate the situation.

The two hypotheses of lead loss presently used are episodic lead loss and continuous diffusion lead loss. Both hypotheses have been expanded somewhat from their initial conceptions and their historical development is, in essence, a synopsis of the accumulation of pertinent data.

In a study of the discordant uranium–lead ages of old monazites and uraninites from Rhodesia, Malagasy (formerly Madagascar) and Manitoba, Ahrens (1955a,b) noted that the results were linearly related when plotted on a graph of $\log t$ (^{238}U–^{206}Pb) vs. t (^{235}U–^{207}Pb). He concluded that the regularity of the results suggested that the lead loss was controlled by physical processes.

Wetherill (1956a,b) showed that the regularity noted by Ahrens could be explained by a simple geochemical process. On a plot of ^{206}Pb/^{238}U vs. ^{207}Pb/^{235}U atomic ratios, a 'concordia' curve can be drawn connecting all points at which $t(^{238}$U–^{206}Pb$) = t(^{235}$U–^{207}Pb$)$. Wetherill proved that all samples of the same age (t_1) which experienced a single episode of lead loss at the same time (t_2) will fall on a chord connecting the points t_1 and t_2 on the concordia curve (figure 1); the positions of the points along the chord will depend only on the percent lead loss for each sample. If two episodes of lead loss occur (at times t_2 and t_3) all of the discordant results will necessarily fall within, or on the lines of the triangle formed by the straight lines connecting the three points on the concordia curve (figure 2).

Russell and Ahrens (1957) made concordia plots of discordant uranium minerals (no zircons) of a number of ages. Their plots again showed the linear relationship between discordant minerals of the same age, but also suggested that the hypothetical times of lead loss (Wetherill's t_2) appeared to be related to the true ages of the minerals. That is, suites of minerals with probable true ages of 2680, 2020 and 1800 m.y. would hypothetically have had to suffer lead loss at times of 500, 300 and 150 m.y., respectively. They concluded that such a relationship was highly improbable if a chemical process was involved and suggested that the results supported Ahrens' (1955b) original contention that a physical process which favored removal of ^{207}Pb,

The interpretation of zircon ages

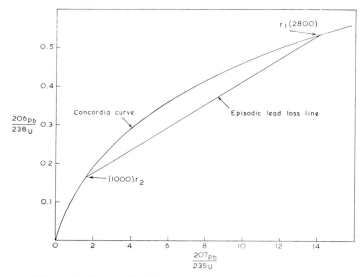

Figure 1 Concordia diagram with episodic lead loss line.

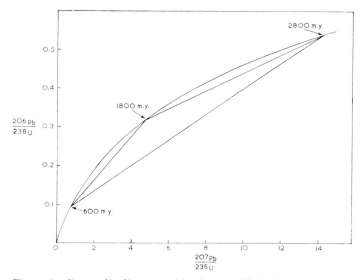

Figure 2 Concordia diagram with triangle of loci of results for two periods of episodic lead loss.

relative to ^{206}Pb, was involved. However, they were not able to correlate the data with any particular feature of the radioactive decay schemes, such as recoil distance, or rare-gas loss. The correlation of t_1 and t_2 noted by Russell and Ahrens has not been substantiated by further work but much of this work has been done in complicated areas where such a relationship could be masked.

Tilton (1960), making use of some calculations by Nicolaysen (1957), proposed that lead loss might be governed entirely by continuous diffusion of lead out of the uranium mineral during its entire

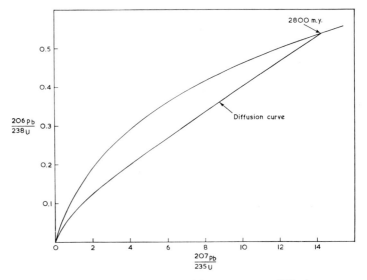

Figure 3 Concordia diagram with 2800 m.y. diffusion curve (constant D/a^2 model).

history. In this hypothesis the rate of diffusion is governed by a constant (for a given sample) diffusion coefficient, D, the effective radius of the mineral grains (also constant for a given sample), a, and the concentration gradient. Tilton showed that, for each true age, substitution of various values of D/a^2 into the diffusion equation (originally derived by Wasserburg, 1954) gives rise to a smooth curve on a concordia plot (figure 3). These curves are virtually straight lines for much of their length and begin to curve perceptibly only near the origin. Multiple samples of a mineral or minerals of the same age which have suffered less than about 50% lead loss by continuous diffusion will form a linear array on a concordia plot. Thus, with the continuous diffusion hypothesis, the general type of linear array of discordant results would be explained without the need for postulating an event at the lower end of the 'chord'. Moreover, the characteristics of the diffusion curves are such that there is a definite correlation between the age of a curve and a hypothetical time of lead loss arrived at by extrapolating the linear portion of the diffusion curve to the concordia curve. Approximate extrapolated intercept values for some diffusion curves are (2800,600), (1800,400) and (1100,300). The

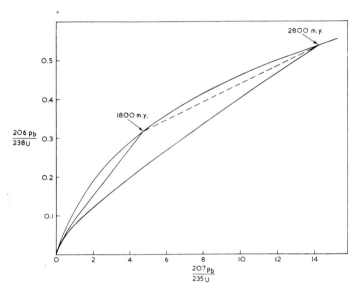

Figure 4 Concordia diagram with loci of results for continuous diffusion (constant D/a^2 model) lead loss with superimposed episodic lead loss (at 1800 m.y.).

continuous diffusion hypothesis could therefore explain (qualitatively) the apparent relationship noted by Russell and Ahrens (1957).

Tilton also recognized the possibility that a metamorphic event might cause episodic lead loss to be superimposed on the continuous diffusion lead loss and showed that such an episode would generally lead to a nonlinear array of points on a concordia plot; the points would necessarily fall on or between the diffusion curves for the original age and the age of the episode of lead loss (figure 4).

The simplicity of the continuous diffusion hypothesis and the fact that it appeared to explain all aspects of the data at that time made it very appealing. However, further work showed what appeared to be discrepancies or inconsistencies between new data and the continuous diffusion interpretations of the data. Silver and Deutsch (1961, 1963), Silver (1962) and Catanzaro and Kulp (1964) showed that, in some areas, discordant results formed linear arrays which extended well into the region of the concordia plot in which the diffusion curves are severely nonlinear. In addition they showed that there was frequently a direct relationship between the uranium contents of the samples and the amounts of lead loss. This fact suggested that there is a relationship between the metamictization of a sample and its susceptibility to lead loss, a relationship which would be incompatible with the idea of continuous diffusion lead loss governed by a constant diffusion coefficient.

The major criticism of the episodic lead loss hypothesis was generally held to be the fact that, in a number of areas, K–Ar and Rb–Sr ages did not indicate that any event had taken place at the time the discordant zircon samples were supposed to have suffered lead loss. Although Wetherill's original development of the hypothesis does not in any way depend on the cause of the lead loss, except to restrict it to a relatively short period of time, the episode was generally assumed to reflect a metamorphic event, which would be readily indicated by K–Ar and Rb–Sr ages of mica and feldspar. Catanzaro and Kulp (1964), however, suggested that metamict zircons may lose lead through leaching by groundwater solutions. If the leaching takes place over a relatively short period of time, the end result will be an 'episode' of lead loss identical in every respect to a metamorphic 'episode' save that the leaching episode might not seriously affect K–Ar and Rb–Sr ages in the rock.

Wasserburg (1963), recognizing the fact that uranium content and metamictization appeared to be a factor in zircon lead loss, extended the continuous diffusion hypothesis to include a model in which the diffusion coefficient is an arbitrary function of time. He showed that a radiation damage model which relates the diffusion coefficient, $D(t)$, to the integrated radiation damage and to the uranium and thorium contents yields curves which are very similar to those obtained by Tilton (1960) for a constant diffusion coefficient, D_0. The slopes of the curves are somewhat greater and the linear regions are more extended (figure 5). In order to achieve a simple curve, the radiation damage model must assume that the initial diffusion coefficient, D_0, is negligible, and that saturation of radiation damage does not occur. If D_0 is not negligible then $D(t)$ will equal $D_0 + D_1 t$ and the discordant points will fall within an area bounded by the curves for $D(t) = D_0$ and $D(t) = D_1 t$. If saturation occurs, then the points will fall between the curves for $D(t) = D_1 t$ and $D(t) = D_0$ (for saturation).

At present, then, the lead loss hypotheses are as follows. The diffusion hypotheses assume that the lead loss occurs by continuous diffusion of lead out of the mineral throughout its history. The constant D/a^2 model assumes that the values of the diffusion coefficient and effective radius are constant throughout the history of the mineral, except perhaps if recrystallization or annealing occur, and are independent of the effects of metamictization. The radiation

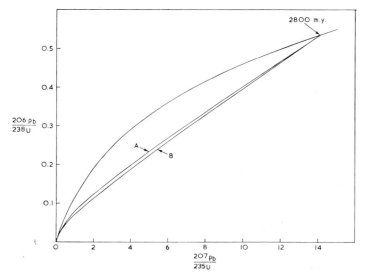

Figure 5 Concordia diagram comparing diffusion curves for the constant D/a^2 model (A) and the radiation damage model (B).

damage model (D_0 negligible, and no saturation) assumes that the diffusion coefficient is a function only of the integrated radiation damage and increases throughout the history of the mineral. The effective radius is presumably constant.

Both diffusion models yield curves which have extended linear portions and both can therefore explain the simple linear relationship in which the array of points has a slope extending toward the <700 m.y. portion of the concordia curve. This is the only general linear array possible under the continuous diffusion hypotheses. Both models recognize the possibility of metamorphic episodic lead loss being superimposed on the continuous diffusion lead loss. However, unless such an episode causes total loss of radiogenic lead in all samples, it will disrupt the linear relationship of the points and spread them over a quasi-triangular area on a concordia plot (figure 4).

The episodic lead loss hypothesis assumes that lead loss is not an intrinsic characteristic of zircons but occurs only when specific environmental factors affect the zircons. The lead loss may occur during a metamorphic event which causes annealing or partial recrystallization of the zircons, with preferential expulsion of the ill-fitting lead atoms, or during a nonmetamorphic event in which the lead is leached from metamict zircons by groundwater solutions. The nonmetamorphic event may be an episode of earth movement and new rock formation which does not drastically affect the K–Ar and Rb–Sr ages of the older rocks (e.g. the Laramide Revolution in western U.S.A.) or the 'event' may simply be the contact of alkaline groundwater solutions with sufficiently metamict zircons.

The episodic hypothesis can explain linear relationships no matter what their slopes, provided they remain within the bounds of the concordia plot, with intercepts >0 but <4500 m.y. If more than one episode of partial lead loss occurs, the linear relationship will be disrupted and the points will fall within a triangular (figure 2) or polygonal (more than two episodes) area on a concordia plot.

C *Case studies of discordant zircons*

Figures 6 and 8 to 12 show the results of a number of zircon age determination studies plotted on concordia diagrams. In each case two alternative explanations are given for the data: the first (figures a) uses episodic lead loss, recognizing both metamorphic and nonmetamorphic events; the second (figures b) uses continuous diffusion lead loss with, where indicated by nonzircon ages, metamorphic episodic lead loss. The continuous diffusion model used is the radiation damage model (D_0 negligible, no saturation) developed by Wasserburg (1963); his tabulated calculations were used in drafting the curves. This choice was made because in all cases given here this model fits the data better than, or at least as well as the constant diffusion coefficient model. Also, interpretations following the constant diffusion coefficient model can usually be found in the original papers.

The interpretations given here are not unique but are generally the simplest or most elucidating in each case. The specific cases given below are not an exhaustive survey of all zircon results, but represent virtually the complete range of complexity.

The size of the circles around each analytical point in the figures is an estimate of the uncertainty, where given by the original workers. The use of circles as error estimates is not strictly valid, since errors in uranium/lead isotopic ratios will move the points along a line through the points from the origin, and errors in $^{207}\text{Pb}/^{206}\text{Pb}$ ratios will move the points more or less perpendicular to these lines. However, the error statements are only estimates, not statistically significant analyses of the uncertainties, and the use of precise error boundaries is of little value.

Figure 6 is a concordia plot of the results of nine discordant zircon samples from the Little Belt Mountains, Montana (Catanzaro and Kulp, 1964). The area is a complex of metamorphic rocks (figure 7) some of which are definitely metaigneous and others of which are probably metasedimentary. The complex is overlain in the south by rocks of the Belt Series and in the north by Cambrian and younger sedimentary rocks. Felsic porphyry stocks and dikes (Laramide) intrude both the metamorphic rocks and the younger sedimentary rocks. The metamorphic rocks include migmatite, gneiss, granitic augen gneiss, metadiorite, phyllite and schist, and the structure in the rocks suggests two Precambrian metamorphic events.

Nonzircon age determinations establish two ages. The last metamorphism is dated at 1920 m.y. by a concordant monazite result ($^{238}\text{U}-^{206}\text{Pb}$ age = 1860, $^{235}\text{U}-^{207}\text{Pb}$ age = 1890, $^{207}\text{Pb}/^{206}\text{Pb}$ age = 1920 m.y.). Potassium–argon ages of biotite and hornblende from the metamorphic rocks range from 1500 to 1820 m.y. but these ages were probably lowered by argon loss during deep burial after the 1920 m.y. event or by the general heating of the rocks during the Laramide revolution. A potassium–argon age determination on hornblende from a porphyry dike in the middle of the metamorphic complex dates the intrusion at 120 m.y.

The episodic lead loss hypothesis (figure 6a) may explain the data in the following manner. All* of the zircons were initially formed or completely rejuvenated ⩾2470 m.y. ago. During the 1920 m.y. event samples 11, 13, 31, 34 and 41 lost all of their radiogenic lead (i.e. were completely rejuvenated), sample 14 lost a large portion of its lead

* Catanzaro (1967) has suggested that some of the zircons (those from the granitic augen gneiss) may have been initially formed during the 1920 m.y. event but there is no doubt that all of the other zircons were originally older than 2470 m.y.

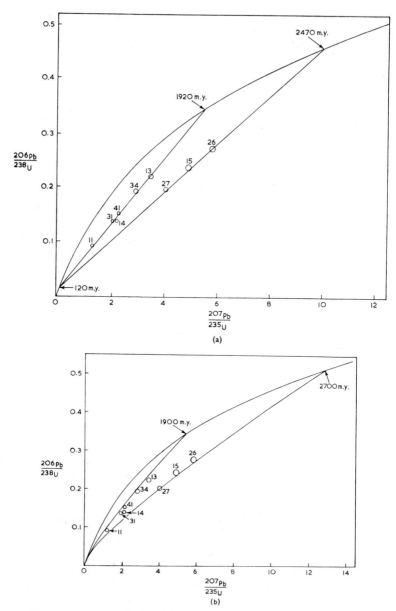

Figure 6 (a) Little Belt Mountains (data from Catanzaro and Kulp, 1964)—episodic lead loss interpretation; (b) Little Belt Mountains (data from Catanzaro and Kulp, 1964)—continuous diffusion interpretation.

and samples 15, 26 and 27 suffered little or no lead loss. All of the samples then experienced another episode of partial lead loss 120 m.y. ago, during the event which included the intrusion of the porphyry dike.

The episodic lead loss interpretation explains the data very well. It not only correlates the linear patterns with the independent age results, but it is also able to correlate the varying amounts of lead loss with other geological and geochemical factors. The variable effects of the 1920 m.y. event are explained on a geographical basis (figure 7). The completely rejuvenated zircons are found south of the sample 14 location and the unaffected (?) zircons are found

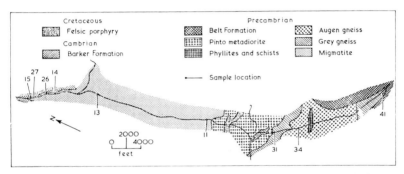

Figure 7 Outline geologic map of part of the Little Belt Mountains, showing zircon sample locations. (After Catanzaro and Kulp, 1964)

north of this location. This would suggest that the 1920 m.y. event was stronger in the southern portion of the area and would be in agreement with the recent suggestion (Catanzaro, 1967) that the granitic augen gneiss, which outcrops in the southern part of the area, may be syntectonic and may have been emplaced during the early stages of the 1920 m.y. event. The uniqueness of sample 14 can be explained by the chemical data, in addition to the geography. While morphologically related to the 'old' zircons (15, 26 and 27) (Catanzaro, 1962), sample 14 has a significantly higher uranium content (table 1) than these samples. Presumably it was more metamict at the time of the 1920 m.y. event and therefore more susceptible to lead loss. The variable amounts of lead loss suffered by the different samples during the 120 m.y. event is readily attributable to the varying degrees of metamictization of the samples. In both linear groups there is a direct correlation between the amount of lead loss (i.e. the relative positions on the straight lines) and the uranium content (table 1) (Catanzaro and Kulp, 1964).

In the episodic interpretation, the 2470 m.y. age must be considered a minimum age, since it is possible that all three of the samples lost some of their lead during the 1920 m.y. event.

Figure 6(b) shows a radiation damage model continuous diffusion interpretation of the Little Belt data. The 1900 m.y. diffusion curve was chosen because this is the age of the last metamorphism (Wasserburg's tabulations do not include those for a 1920 m.y. curve, but this small difference is of no consequence); the 2700 m.y. curve was chosen because this is a minimum age for at least some, if not all, of the samples. The data can be explained in the following manner. All of the zircons were initially formed or completely rejuvenated $\geqslant 2700$ m.y. ago. They lost lead by continuous diffusion both before and after the 1920 m.y. event, but during the event samples 11, 13, 14, 31, 34 and 41 suffered significant amounts of episodic lead loss and samples 15, 26, and 27 suffered little or no lead loss.

Although this interpretation, and the constant D/a^2 continuous diffusion interpretation (Catanzaro and Kulp, 1964), are plausible, both show some inconsistencies. There is no ready explanation, neither geographic nor geochemical, for the lack of correlation between the degree of lead loss suffered during the 1920 m.y. event and the presently indicated D/a^2 values of the different samples. For example, sample 11 has obviously suffered more total lead loss than sample 13, has more than four times as much uranium as sample 13 and is located south of sample 13, yet according to the diffusion hypotheses it suffered less lead loss during the 1920 m.y. event than did sample 13. Other such 'inversions' occur in both the 'young' and 'old' zircon groups and are obvious from the data.

Finally, the fact that the five results from the southern portion of the complex fall on the independently determined 120–1920 m.y. straight line must be considered a coincidence if any lead loss is presumed to have occurred by continuous diffusion. The fact that these points do not fall on any 1920 m.y. diffusion curve means that this particular linearity cannot be considered a result of the quasi-linear nature of continuous diffusion curves.

No matter which hypothesis is correct for the low-temperature lead loss, the discordant zircon data really add only one new age, since the 1920 m.y. event is solidly based on the concordant monazite result. The unique information obtained from the zircons is the minimum age of 2470 or 2700 m.y. for some event which affected the zircons prior to the 1920 m.y. event. Catanzaro (1967) has suggested that the true age of this event is probably 2700 m.y. and that it correlates with the definitely established 2700 m.y. event in the Beartooth Mountains of Montana and other areas in Wyoming.

The 1920 m.y. event was undoubtedly a metamorphic event. Certainly most, if not all, of the zircons dated were initially formed and present in the area long before that time. However, the geologic meaning of the 2700 m.y. age is not so definite. The three 'old' zircons are from migmatite (figure 7). They are very similar in physical characteristics (Catanzaro, 1962) and have similar uranium contents (table 1); they are probably related in origin. Three alternative expla-

Table 1 Uranium content of zircons from the Little Belt Mountains (data from Catanzaro and Kulp, 1964)

Sample number	U content (ppm)
11	2227
14	1417
31	1047
41	710
27	590
34	553
13	549
15	375
26	361

nations of the 2700 m.y. age are possible: (1) If the rock is metaigneous, this age could reflect the time of intrusion of this rock (*in situ*) and the time of the initial crystallization of the zircons. (2) If the rock is metasedimentary, the age could reflect the time of crystallization (or subsequent rejuvenation) of the zircons in another area. In this case the 2700 m.y. age might not be related to any particular event in the present area. (3) The rock may be either metaigneous or metasedimentary but the age might reflect a previous metamorphic event, in the Little Belt Mountain area, which caused complete radiogenic lead loss. Since the geology of the area suggests two metamorphic events, and since there is no age evidence for an event between 2700 and 1900 m.y., the last alternative seems most likely.

Figure 8 is a concordia plot of zircon age results from the Beartooth Mountains (Catanzaro and Kulp, 1964). Most of the samples were

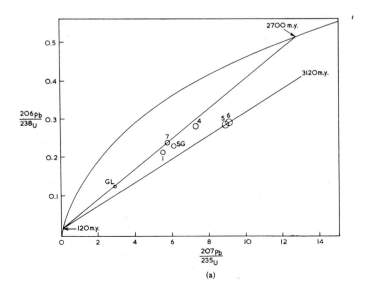

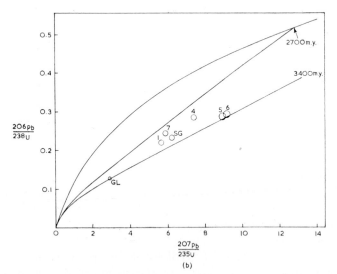

Figure 8 (a) Beartooth Mountains (data from Catanzaro and Kulp, 1964)—episodic lead loss interpetation; (b) Beartooth Mountains (data from Catanzaro and Kulp, 1964)—continuous diffusion interpretation.

taken from the Quad Creek area, which has been mapped and studied in detail by Eckelmann and Poldervaart (1957). The major rock units in this area are Precambrian granite gneiss, migmatite, metasedimentary rocks (principally paraamphibolites and quartzites) and metaigneous mafic dikes and stocks. Felsic Laramide porphyry dikes are also present. According to Eckelmann and Poldervaart the observable sequence of events in the Quad Creek area is: (1) deposition of ancient sediments, (2) folding, (3) metamorphism and granitization, (4) uplift, (5) peneplanation, (6) deposition of Paleozoic and younger sediments and (7) uplift, thrusting and emplacement of felsic porphyry dikes.

Zircon overgrowths are common, but the cores and unovergrown zircons are rounded and obviously detrital. Detailed studies (Poldervaart and Eckelmann, 1955; Eckelmann and Poldervaart, 1957) have shown that the frequency and completeness (i.e. attainment of euhedral form) of the overgrowths are greatest in the granite gneiss and least in the quartzite. These results, together with the overwhelming structural evidence, leave no doubt that the granitic rocks were formed *in situ* from sediments.

The period of metamorphism and granitization has been dated at 2700 m.y. by Rb–Sr and K–Ar dating of mica and feldspar and by a concordant uraninite result (Gast, Kulp and Long, 1958).

All of the zircon results are discordant. Samples GL, SG, 1 and 6 are from granitic gneiss, sample 5 is from migmatite, sample 4 is from amphibolite and sample 7 is from an anatectic granite near the Stillwater complex. The episodic lead loss interpretation (figure 8a) is as follows. The zircons were formed prior to 3120 m.y. ago (they may, of course, represent a number of ages), in the source area(s) of the Beartooth sediments. They lost lead and gained new zircon material (overgrowths) during the 2700 m.y. metamorphism and granitization and then again lost varying amounts of lead approximately 120 m.y. ago, presumably during the Laramide Revolution. The felsic porphyries in this area have not been dated; the 120 m.y. date is used by analogy with the age of the felsic porphyry dated in the Little Belt Mountains (noted above). The exact age of this event is not critical to the interpretation.

The detrital nature of the zircons, and the overgrowths, make it impossible to generally evaluate the amounts of lead lost by each sample during the 2700 m.y. event, or to correlate this with any other parameter. However, in this interpretation, samples GL and 7 are the only samples which would have lost all of their lead during the

2700 m.y. event and the uniqueness of these samples is reasonably plausible. Sample GL obviously suffered the greatest amount of total lead loss during the 120 m.y. event and, by analogy, might have been most susceptible to the 2700 m.y. event. Sample 7 is the only sample taken from an intrusive (anatectic) granite (R. Butler, cited in Catanzaro and Kulp, 1964) and was obviously in an ultrametamorphic environment during the 2700 m.y. event.

Figure 8(b) presents a possible continuous diffusion (radiation damage model) interpretation of the results. The 2700 m.y. curve was chosen because it represents the time of last metamorphism and new zircon formation. The 3400 m.y. curve was chosen because it represents a minimum age for at least some of the zircons; i.e. it is the youngest curve which can be drawn so that all points fall on or between the two curves. If the constant D/a^2 model is used, the youngest curve that can be put through point GL is approximately 4000 m.y. (Catanzaro and Kulp, 1964), an impossible value since at least some of the material in this sample is only 2700 m.y. old (overgrowths).

The continuous diffusion hypothesis can explain the data in many ways, the simplest being that the detrital zircons may be of more than one initial age and could thus not be expected to fall on any one or two particular curves. Even if they were initially of the same age the varying amounts of lead loss and overgrowths formed during the 2700 m.y. event would cause the results to scatter between two curves. Thus the scattering of the results does not preclude continuous diffusion lead loss. However, sample GL does make the constant D/a^2 continuous diffusion interpretation unlikely, or at least incomplete, and presents a problem for the radiation damage model too. Since the sample contains a large amount of overgrowth material (it is from granite gneiss) much of the zircon is 2700 m.y. old and the core material must therefore be much older than the supposed continuous diffusion minimum age of 3400 m.y. Even assuming that this necessarily extremely old age is possible, the continuous diffusion interpretation also presents the anomaly of having the sample with the highest D/a^2 value (GL) being the one least affected by any possible lead loss during the 2700 m.y. event. Finally, it is possibly significant that no points fall in the area on the concordia plot between the 120–2700 m.y. straight line in figure 8(a) and the 2700 m.y. diffusion curve in figure 8(b). This area is a possible locus of points in the continuous diffusion interpretation but is a 'forbidden' area for the given episodic lead loss interpretation.

Since the 2700 m.y. event is well established by nonzircon age determinations, the only new information given by the zircon results is the minimum age of 3120 or 3400 m.y. for the time of crystallization of the zircons, in the area or areas which were the sources of the Beartooth sediments. The location and true age(s) of the source area(s) are not known, but it is hoped that further study of the metasediments may suggest a source direction, and further zircon age determinations, especially if samples without overgrowths can be found, may lead to a more precise estimate of the true age or ages. Zircons with a minimum age of 3300 or 3550 m.y. have been found in Minnesota (see below) so a fairly extensive >3300 m.y. terrane seems very likely.

Figure 9 is a concordia plot of discordant zircon results from southwestern Minnesota (Catanzaro, 1963). Samples 1 and 2 are from the Morton gneiss and sample 3 is from the Montevideo gneiss (at Granite Falls). Two Precambrian events have been dated in these rocks: one at 2500 ± 100 m.y. and the other at 1800 ± 100 m.y. (Goldich and others, 1961; Goldich and Hedge, 1962). The older event dates a severe regional metamorphism and is reflected in the Rb–Sr ages of feldspar from both gneisses and in the Rb–Sr and K–Ar ages of biotite from the Morton gneiss. Rb–Sr and K–Ar ages of biotite from the Montevideo gneiss range from 1700 to 1800 m.y. A number of small igneous plutons have also been dated at approximately 1800 m.y. Presumably the younger event was a period of igneous activity which only locally affected the age systems of the gneisses.

The episodic lead loss interpretation (figure 9a) would say that the zircons are 3550 m.y. old and suffered partial but significant lead loss during the 1800 m.y. event. This interpretation assumes that the zircons were unaffected by the 2500 m.y. event, an assumption which superficially appears unlikely since the 2500 m.y. event was apparently stronger than the 1800 m.y. event. However, a lack of correlation between the ability of an event to reset K–Ar and Rb–Sr ages of mica and feldspar and its effect on the U–Pb system in zircon has been noted before (Tilton, 1960; Catanzaro and Kulp, 1964). In addition, since the degree of metamictization appears to be a significant parameter in zircon age discordances, it would appear likely that younger events may in general be more effective in causing lead loss (Catanzaro, 1963).

Figure 9(b) shows a possible continuous diffusion (radiation damage model) interpretation of the results. No diffusion curve of any model can be put through all three points; the 3300 m.y. curve simply

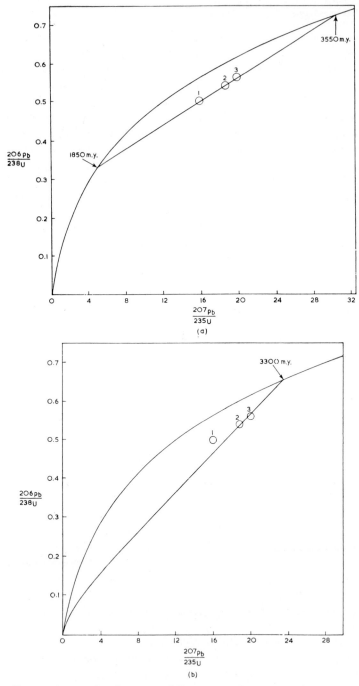

Figure 9 (a) Southwestern Minnesota (data from Catanzaro, 1963)—episodic lead loss interpretation; (b) Southwestern Minnesota (data from Catanzaro, 1963)—continuous diffusion interpretation.

represents the minimum age of at least some of the zircons, in the diffusion interpretation. The fact that the points do not fit any one curve may be explained as being due to: (1) the zircons are detrital and of different ages, or (2) they are of the same initial age but lost different amounts of lead during the 2500 and/or 1800 m.y. events. In either case, the linearity of the results and the 1850 m.y. intercept of the best-fit straight line (which agrees with an independently dated event) would have to be considered coincidences. This seems extremely unlikely, especially in view of the fact that Stern (1964) and Stern, Goldich and Newell (1965) report five more discordant zircon results, from these rocks, which fall on virtually an identical line; their best-fit line is 1900–3600 m.y.

The close similarity between the established ages of events in Minnesota: >3300 (or exactly 3600), 2500 and 1800 m.y.; and those in Montana: Little Belt Mountains, >2470 and 1900 m.y.; Beartooth Mountains, >3100 and 2700 m.y., is striking. Catanzaro (1967), on the basis of concordant U–Pb age results (monazite, Little Belt Mountains: 1920 m.y.; uraninite, Beartooth Mountains: 2700 m.y.), discordant zircon results and the general pattern of the Rb–Sr and K–Ar ages, which scatter downward from 2700 and 1900 m.y., has suggested the possibility that only three significant older Precambrian events are reflected in the rocks of Montana and Wyoming: >3100 m.y., 2700 m.y. and 1900 m.y. He suggests that the Rb–Sr and K–Ar ages which cluster below 2700 and 1900 m.y. are due to small amounts of radiogenic daughter loss and must be considered minimum ages which approximate the true ages as given by the concordant U–Pb results. If these ideas have any merit, they can obviously be extended to include the Minnesota results. This would suggest that the events reflected by the 2500 and 1800 m.y. groups of Rb–Sr and K–Ar ages in Minnesota have true ages of 2700 and 1900 m.y., respectively, and were contemporary with the identically dated events in Montana and Wyoming. The analogy might be carried further, to suggest that the detrital zircons in the Beartooth Mountain rocks may be 3600 m.y. old and may be temporally related to the Minnesota zircons.

Figure 10 is a concordia plot of some discordant zircon results from Arizona (Catanzaro and Kulp, 1964). The four bunched points represent results from the Johnny Lyon granodiorite (Silver and Deutsch, 1961) and the lone point at the lower left is from the Catalina gneiss (Catanzaro and Kulp, 1964). The Johnny Lyon points represent analyses of different layers successively stripped off the zircon grains from one sample, by partial fusions of the sample.

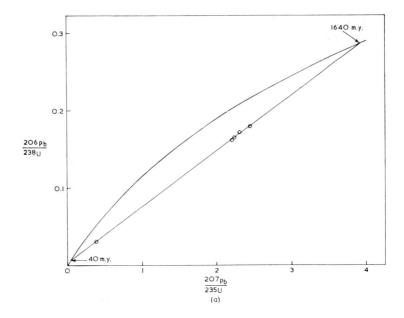

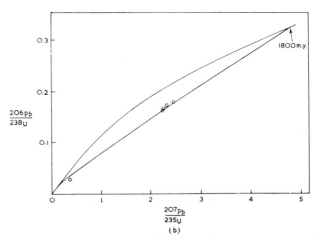

Figure 10 (a) Arizona samples. Johnny Lyon granodiorite (data from Silver and Deutsch, 1961) and Catalina gneiss (data from Catanzaro and Kulp, 1964)—episodic lead loss interpretation; (b) Arizona samples. Johnny Lyon granodiorite (data from Silver and Deutsch, 1961) and Catalina gneiss (data from Catanzaro and Kulp, 1964)—continuous diffusion interpretation.

The Johnny Lyon granodiorite is a posttectonic intrusive emplaced at the end of the Mazatzal Revolution (Silver and Deutsch, 1963). Rb–Sr dates of rocks presumably affected by this event range from 1200 to 1550 m.y. (Giletti and Damon, 1961). The Catalina gneiss has been studied by DuBois (1959a,b). He noted two orientations of linear and planar structure in the rock and interpreted this as evidence of two periods of metamorphism: one in Precambrian time and the other in post-Cretaceous time. Catanzaro and Kulp (1964) report a K–Ar age of 32 m.y. for muscovite from this rock.

The lack of geologic control for the relationship, if any, between the two rocks makes it difficult to rigidly assess the integrity of this plot as a whole, but certain interesting facts are apparent from the figures. The Catalina gneiss result is such that it is impossible to put a reasonable-aged diffusion curve through the point (figure 10b), showing that continuous diffusion lead loss, if real, must have been upset by the 32 m.y. metamorphism. The lack of spread in the Johnny Lyon points makes them essentially a single point so that it is a simple matter to put a straight line through these points and the Catalina gneiss point. However, it is interesting that a best-fit straight line (figure 10a) gives a lower intercept of 40 m.y., in very good agreement with the 32 m.y. K–Ar age. Silver and Deutsch (1963) have obtained some more discordant points by acid wash experiments on the Johnny Lyon sample. All of the points for this rock fall on a straight line with intercepts of 90 and 1655 m.y.

Figure 11 is a concordia plot of discordant zircon results from the Appalachian Province (Davis, Tilton and Wetherill, 1962). The samples are from gneissic rocks whose age relations are very complicated. At least three metamorphic events appear to have occurred in this area, with probable age of 1000–1100 m.y. (Tilton and others, 1960) and 450–550 and 350 m.y. (Kulp and Eckelmann, 1961), and at least some of the zircons appear to be detrital (Davis, Tilton and Wetherill, 1962).

A possible episodic lead loss interpretation is given in figure 11(a), The 485–1050 m.y. line was chosen by reference to some of the age data, and the 485–1390 m.y. line was chosen so that all of the points would fall on or between the two lines. In this simple interpretation, the 1390 m.y. date represents a minimum age of crystallization or rejuvenation for at least some of the samples; the 1050 m.y. date represents a time of partial or complete lead loss for previously present zircons, and, possibly, a time of new zircon formation; and the 485 m.y. date represents a time of lead loss for all of the samples.

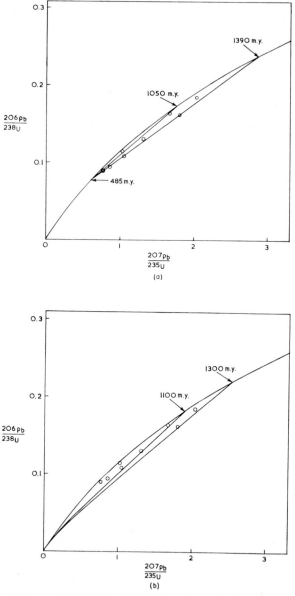

Figure 11 (a) Appalachian Province (data from Davis, Tilton and Wetherill, 1962)—episodic lead loss interpretation; (b) Appalachian Province (data from Davis, Tilton and Wetherill, 1962)—continuous diffusion interpretation.

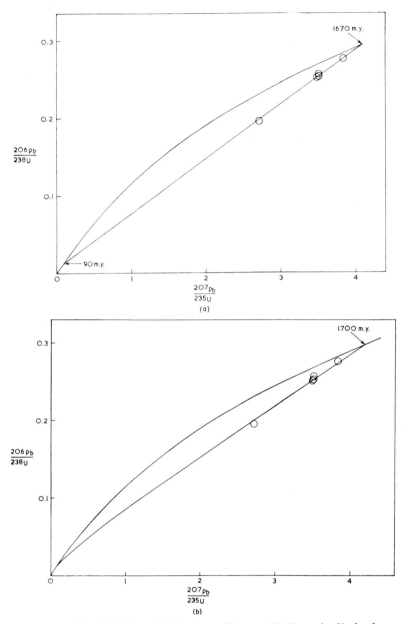

Figure 12 (a) Finland (data from Kouvo, 1958)—episodic lead loss interpretation;—(b) Finland (data from Kouvo, 1958)—continuous diffusion interpretation.

This interpretation does not preclude some lead loss during the 350 m.y. event, but it does not require it.

Figure 11(b) shows that the data cannot be accounted for by one or two continuous diffusion curves. The 1100 m.y. curve is used to show that some lead loss or new zircon formation must have occurred during the younger metamorphic events. The 1300 m.y. curve gives a minimum age for at least some of the zircons.

Figure 12 is a concordia plot of discordant zircon results from some Finnish granitic rocks (Kouvo, 1958). Three of the samples are from the Rapakivi granite and the other two are from the Bodom and Onas granites, all from the southeast coast of Finland. K–Ar and Rb–Sr ages on these and related rocks (postkinematic granites) range from 1550 to 1610 m.y. (Kouvo, 1958). The lack of spread in the plotted results and the lack of samples with large amounts of lead loss make it impossible to choose between the two alternative hypotheses (figures 12a and 12b), but both interpretations suggest that the zircons are 1700 m.y. old and have not been disturbed by any metamorphic event. There is no radioactive age evidence for a 90 m.y. 'event' (figure 12a) but the zircon data are not sufficient to preclude a zero age for the lead loss, by recent weathering.

D *Summary*

The studies noted above and other studies (Aldrich and others, 1958; Tilton, 1956; Tilton and others, 1957) lead to the following generally accepted conclusions:

(1) Zircon age discordances are primarily due to bulk radiogenic lead loss.

(2) In many cases, the results of multiple samples of discordant zircons of the same age and geologic history form linear arrays when plotted on concordia diagrams.

(3) In almost all cases of discordant zircon results there appears to be an excess loss of ^{207}Pb relative to ^{206}Pb (i.e. the ^{207}Pb/^{206}Pb age is a minimum).

(4) Lead loss from zircon may occur without significant radiogenic argon and strontium loss from mica and feldspar in the same rock.

(5) Metamorphism or heating sufficient to cause complete loss of radiogenic argon and strontium from mica and feldspar is not a sufficient condition for lead loss from zircon. However, a metamorphic event may cause partial or complete lead loss from zircon.

(6) Extensive metamictization is not a sufficient condition for

significant lead loss from zircon, but in a number of areas there is a direct correlation between the degree of radiation damage (based on uranium content) and the amount of lead loss.

(7) A low-temperature process exists which may cause extensive lead loss from zircons.

With specific regard to the relative merits of the episodic and continuous diffusion lead loss hypotheses, the following conclusions seem reasonable:

(1) Episodic lead loss during metamorphism is a well-authenticated phenomenon. However, whether or not the loss will occur in any particular case appears to depend on a number of parameters concerning both the metamorphism and the physical state of the zircons.

(2) Episodic lead loss during a nonmetamorphic event, i.e. an event which does not drastically affect K–Ar and Rb–Sr ages, appears to be a real phenomenon in at least certain areas, and groundwater leaching of lead from metamict zircons can conceivably explain all occurrences of low-temperature lead loss.

(3) The constant D/a^2 continuous diffusion lead loss hypothesis, with the corollary of possible superimposed metamorphic lead loss (Tilton, 1960), cannot explain all of the discordant age data in a number of areas. Moreover, it cannot explain the correlation between uranium content and lead loss observed in a number of cases where the zircon samples are from one rock or related rocks.

(4) The radiation damage continuous diffusion lead loss hypothesis, with possible metamorphic lead loss (Wasserburg, 1963), can formally explain all of the examples given here but yields unlikely results in at least one case (Beartooth Mountains) and shows inconsistencies in at least one other case (Little Belt Mountains). Moreover, in a number of cases where the purely episodic interpretation fits the independent age data very well, this agreement must be considered coincidental if any continuous diffusion hypothesis is presumed to be correct.

The radiation damage continuous diffusion model appears to be superior to the constant D/a^2 model in that it can explain the correlation of uranium content with lead loss seen in a number of cases (Silver and Deutsch, 1963; Silver, 1962; Catanzaro and Kulp, 1964). However, if radiation damage is itself a sufficient condition for lead loss, then the correlation between uranium content and lead loss should be more universal and should not appear to hold only in cases where the zircons are taken from one rock or a related set of rocks. Also, such an idea would preclude the occurrence of highly metamict but concordant zircons, which are rare but not unknown (Tilton and others, 1957; Stern, 1964).

In fact, the data suggest that although metamictization will increase a zircon's propensity for lead loss, it will not 'cause' the lead loss. Another step seems necessary. Under high-temperature conditions this step may be annealing or recrystallization; under low-temperature conditions it may be leaching by aqueous solutions.

The process of continuous diffusion of lead from zircons cannot be corroborated in the laboratory. In the 'field' this hypothesis is hampered by the fact that no suite of samples has been found which fits the curved portion of any diffusion curve. This would suggest that either continuous diffusion of lead from zircon is not a significant phenomenon or, it is significant but samples with the large diffusion coefficients necessary for so much lead loss may be too susceptible to lead loss by other means.

The possibility of lead loss by leaching of metamict zircon has been at least partially corroborated by laboratory experiments (Tilton, 1956; Mumpton and Roy, 1961) and it appears to be a real phenomenon. However, the substantiation of this phenomenon does not preclude the possibility of continuous diffusion.

As far as 'true' ages of discordant zircons are concerned, the cases noted here show that the two interpretations generally give quite similar results. In complicated areas, i.e. areas in which metamorphic episodic lead loss has occurred, the continuous diffusion models will generally give only minimum ages, while the episodic interpretation may give presumably accurate estimates of true ages (e.g. figure 9). Since it has been established that zircon discordances are almost always due to nonfractionating loss of radiogenic lead, it is almost always reasonable to accept the highest $^{207}Pb/^{206}Pb$ age as a minimum age. This minimum age would hold for all of the zircons if they are definitely cogenetic but need be correct for only some of the zircons if they are not cogenetic.

IV Geologic Meaning of Zircon Ages

Correlating zircon age results with the geologic history of the rocks in which the zircons occur is, of course, a major purpose of most zircon age determination studies. In any but the simplest of geologic cases this correlation requires a detailed knowledge of the geology of the area under study. Some of the specific data that must be known are: whether the zircons have been through a sedimentary cycle;

whether they are a mixture of different zircon suites; whether overgrowths are present; whether, in an igneous rock, the zircons are juvenile or assimilated; and whether, in a metamorphic rock, the material has been metamorphosed more than once. References to some of the case studies given in the previous section will show how the lack of certain mineralogic or geologic information makes it impossible to arrive at definite conclusions as to the meaning of some of the zircon ages.

Rb–Sr and K–Ar age determinations are always helpful in the interpretation of zircon results. They will tell immediately whether or not a zircon age is probably related to the last igneous or metamorphic event in an area and will, essentially, define any relict ages, but they will not generally give any information as to the significance of the relict ages. Whole rock and multi-mineral Rb–Sr age studies should prove very useful in this respect, but little has been done along this line.

The total value of any zircon age determination study will be realized only when it is correlated with a detailed geological study. In any complicated area, the 'geologist' will eventually have to make the final, most significant, interpretations.

References

Ahrens, L. H. (1955a). The convergent lead ages of the oldest monazites and uraninites (Rhodesia, Manitoba, Madagascar, and Transvaal). *Geochim. Cosmochim. Acta*, **7**, 294.

Ahrens, L. H. (1955b). Implications of the Rhodesia age pattern. *Geochim. Cosmochim. Acta.*, **8**, 1.

Aldrich, L. T., Wetherill, G. W., Davis, G. L. and Tilton, G. R. (1958). Radioactive ages of micas from granitic rocks by Rb–Sr and K–Ar methods. *Trans. Am. Geophys. Union*, **39**, 1124.

Butterfield, J. A. (1936). Outgrowths on zircon. *Geol. Mag.*, **73**, 511.

Carroll, D. (1953). Weatherability of zircon. *J. Sediment. Petrol.*, **23**, 106.

Catanzaro, E. J. (1962). A study of discordant zircons from the Little Belt (Montana), Beartooth (Montana), and Santa Catalina (Arizona) Mountains. *Ph.D. Thesis*. Columbia Univ., New York.

Catanzaro, E. J. (1963). Zircon ages in southwestern Minnesota. *J. Geophys. Res.*, **68**, 2045.

Catanzaro, E. J. (1967). Correlation of some Precambrian rocks and metamorphic events in parts of Wyoming and Montana. *The Mountain Geologist*, **4**, 9.

Catanzaro, E. J. and Kulp, J. L. (1964). Discordant zircons from the Little Belt (Montana), Beartooth (Montana) and Santa Catalina (Arizona) Mountains. *Geochim. Cosmochim. Acta*, **28**, 87.

Davis, G. L., Tilton, G. R. and Wetherill, G. W. (1962). Mineral ages from the Appalachian Province in North Carolina and Tennessee. *J. Geophys. Res.*, **67**, 1987.

DuBois, R. L. (1959a). Geology of the Santa Catalina Mountains. *Ariz. Geol. Soc., Southern Ariz. Guidebook*, **II**, 107.

DuBois, R. L. (1959b). Petrography and structure of a part of the gneissic complex of the Santa Catalina Mountains, Arizona. *Ariz. Geol. Soc., Southern Ariz. Guidebook*, **II**, 117.

Eckelmann, F. D. and Poldervaart, A. (1957). Geologic evolution of the Beartooth Mountains, Montana and Wyoming: Part I. Archean history of the Quad Creek area. *Bull. Geol. Soc. Am.*, **68**, 1225.

Frondel, C. (1953). Hydroxyl substitution in thorite and zircon. *Am. Mineralogist*, **38**, 1007.

Gast, P. W., Kulp, J. L. and Long, L. E. (1958). Absolute age of early Precambrian rocks in the Bighorn Basin of Wyoming and Montana, and southeastern Manitoba. *Trans. Am. Geophys. Union*, **39**, 322.

Giletti, B. J. and Damon, P. E. (1961). Rubidium–strontium ages of some basement rocks from Arizona and northeastern Mexico. *Bull. Geol. Soc. Am.*, **72**, 639.

Goldich, S. S. and Hedge, C. E. (1962). Dating of the Precambrian of the Minnesota River valley, Minnesota. Abstract. *J. Geophys. Res.*, **67**, 3561.

Goldich, S. S., Nier, A. O., Baadsgaard, H., Hoffman, J. H. and Krueger, H. W. (1961). The Precambrian geology and geochronology of Minnesota. *Minn. Geol. Surv., Bull.*, **41**.

Harker, A. (1932). *Metamorphism*. Methuen, London.

Holland, H. D. and Gottfried, D. (1955). The effect of nuclear radiation on the structure of zircon. *Acta Cryst.*, **8**, 291.

Hurley, P. M. and Fairbairn, H. W. (1952). Alpha-radiation damage in zircons. *J. Appl. Phys.*, **23**, 1408.

Karakeda, Y. (1961). Zircon overgrowths in the Ryoke metamorphic zone of the Yanai area, southwest Japan. *Mem. Fac. Sci., Kyushu Univ., Ser. D*, **10**, No. 2, 59.

Kouvo, O. (1958). Radioactive ages of some Finnish Precambrian minerals. *Bull. Comm. Geol. Finlande*, **182**.

Kulp, J. L. and Eckelmann, F. D. (1961). Potassium–argon isotopic ages on micas from the southern Appalachians. *Ann. N.Y. Acad. Sci.*, **91**, 408.

Kulp, J. L. and Eckelmann, W. R. (1957). Discordant U–Pb ages and mineral type. *Am. Mineralogist*, **42**, 154.

Larsen, D. S., Jr., Irving, J., Gonyer, F. A. and Larsen, E. S., 3rd. (1938). Petrologic results of a study of the minerals from the Tertiary volcanic rocks of the San Juan region, Colorado. *Am. Mineralogist*, **23**, 227.

Larsen, L. H. and Poldervaart, A. (1957). Measurement and distribution of zircons in some granitic rocks of magmatic origin. *Mineral. Mag.*, **238**, 544.

Maurice, O. D. (1949). Transport and deposition of the non-sulphide vein minerals. Part 5, Zirconium minerals. *Econ. Geol.*, **44**, 721.

Mumpton, F. A. and Roy, R. (1961). Hydrothermal stability studies of the zircon–thorite group. *Geochim. Cosmochim. Acta*, **23**, 217.

Nicolaysen, L. O. (1957). Solid diffusion in radioactive minerals and the measurement of absolute age. *Geochim. Cosmochim. Acta*, **11**, 41.

Pettijohn, F. J. (1941). Persistence of heavy minerals and geologic age. *J. Geol.*, **49**, 610.
Poldervaart, A. (1956). Zircon in rocks. 2. Igneous rocks. *Am. J. Sci.*, **254**, 521.
Poldervaart, A. and von Backstrom, J. W. (1950). A study of an area at Kakamas (Cape Province). *Trans. Geol. Soc. S. Africa*, **52**, 433.
Poldervaart, A. and Eckelmann, F. D. (1955). Growth phenomena in zircon of autochthonous granites. *Bull. Geol. Soc. Am.*, **66**, 947.
Rankama, K. and Sahama, T. G. (1950). *Geochemistry*. Univ. Chicago Press, Chicago.
Russell, R. D. and Ahrens, L. H. (1957). Additional regularities among discordant lead–uranium ages. *Geochim. Cosmochim. Acta*, **11**, 213.
Silver, L. T. (1962). The relation between radioactivity and discordance in zircons. *Nuclear Geophysics*, Publ. 1075. Natl. Acad. Sci.–Natl. Res. Council. p. 34.
Silver, L. T. and Deutsch, S. (1961). Uranium–lead method on zircons. *Ann. N.Y. Acad. Sci.*, **91**, 279.
Silver, L. T. and Deutsch, S. (1963). Uranium–lead isotopic variations in zircon: a case study. *J. Geol.*, **71**, 721.
Smithson, F. (1937). Outgrowths on zircon in the Middle Jurassic of Yorkshire. *Geol. Mag.*, **74**, 281.
Stern, T. W. (1964). Isotopic ages of zircon and allanite from the Minnesota River valley and La Sal Mountains, Utah. Abstract. *Trans. Am. Geophys. Union*, **45**, 116.
Stern, T. W., Goldich, S. S. and Newell, M. F. (1965). Effects of weathering on the U–Pb and Th–Pb systems in zircons. Abstract. *Trans. Am. Geophys. Union*, **46**, 164.
Strock, L. W. (1941). Geochemical data on Saratoga mineral waters, applied in deducing a new theory of their origin. *Am. J. Sci.*, **239**, 857.
Taubeneck, W. H. (1957). Geology of the Elkhorn Mountains, northeastern Oregon; Bald Mountain batholith. *Bull. Geol. Soc. Am.*, **68**, 181.
Tilton, G. R. (1956). The interpretation of lead-age discrepancies by acid-washing experiments. *Trans. Am. Geophys. Union*, **37**, 224.
Tilton, G. R. (1960). Volume diffusion as a mechanism for discordant lead ages. *J. Geophys. Res.*, **65**, 2933.
Tilton, G. R., Davis, G. L., Wetherill, G. W. and Aldrich, L. T. (1957). Isotopic ages of zircon from granites and pegmatites. *Trans. Am. Geophys. Union*, **38**, 360.
Tilton, G. R., Patterson, C., Brown, H., Inghram, M., Hayden, R., Hess, D. and Larsen, E. S., Jr. (1955). Isotopic composition and distribution of lead, uranium and thorium in a Precambrian granite. *Bull. Geol. Soc. Am.*, **66**, 1131.
Tilton, G. R., Wetherill, G. W., Davis, G. L. and Bass, M. N. (1960). 1000-million-year-old minerals from the eastern United States and Canada. *J. Geophys. Res.*, **65**, 4173.
Vitanage, P. W. (1957). Studies of zircons in Ceylon Precambrian complex. *J. Geol.*, **65**, 117.
Wasserburg, G. J. (1954). ^{40}A–^{40}K dating. In Faul, H. (Ed.), *Nuclear Geology*. Wiley, New York. p. 341.

Wasserburg, G. J. (1963). Diffusion processes in lead–uranium systems. *J. Geophys. Res.*, **68**, 4823.

Wetherill, G. W. (1956a). An interpretation of the Rhodesia and Witwatersrand age patterns. *Geochim. Cosmochim. Acta*, **9**, 290.

Wetherill, G. W. (1956b). Discordant uranium–lead ages, I. *Trans. Am. Geophys. Union*, **37**, 320.

Geochronological studies in Connemara and Murrisk, western Ireland

S. MOORBATH,* K. BELL,* B. E. LEAKE† and W. S. McKERROW*

I *Introduction*, 259. II *Sequence of events in the West of Ireland*, 266.
III *Experimental methods*, 269. IV *Results and discussion*, 270.
V *Concluding remarks*, 286. *Acknowledgements*, 290. *Appendix:
Locations and descriptions of the specimens studied*, 290.
References, 295.

I Introduction

The Dalradian Series of Scotland and Ireland is a major sedimentary–volcanic assemblage of late Precambrian, and possibly partly of Cambrian, age. Rough estimates of the total thickness vary from 25,000–40,000 feet. The outcrop extends from the Banffshire coast of northeastern Scotland through the Central and Southwest Highlands of Scotland and thence to the west coast of Ireland. During the Caledonian orogenic movements, the entire Dalradian Series underwent polyphase folding and polyphase regional metamorphism of varying intensity.

In contrast to the situation in Scotland, in the Connemara region of County Galway, western Ireland, there is a clear-cut, unequivocal relationship between metamorphosed Dalradian rocks (the Connemara Schists) and immediately overlying Ordovician and Silurian sediments, which are either unmetamorphosed or at a much lower metamorphic grade than the Dalradian. An outline of the geology of the area is shown in figure 1. Dalradian rocks occur both north and south of Lower Palaeozoic sediments; to the north the Silurian

---

* Department of Geology and Mineralogy, Oxford University.
† Department of Geology, Bristol University.

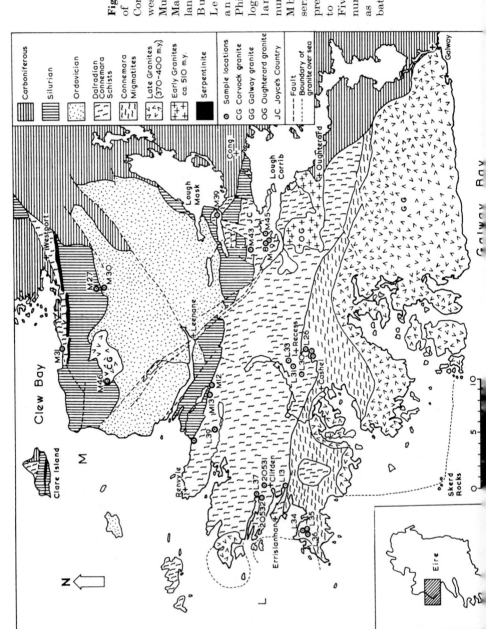

Figure 1 An outline of the geology of Connemara (north-west Galway) and Murrisk (southwest Mayo), western Ireland (after Anderson, Burke, Dewey, Leake, McKerrow and Campbell, Phillips, and the Geological Survey of Ireland). Specimen numbers prefixed by M belong to the BMM-series, whilst those prefixed by L belong to the BLM-series. Five-figure specimen numbers are the same as in Giletti, Moorbath and Lambert (1961).

is faulted down against Lower Dalradian rocks, but in the south, the Silurian (and at one locality the Ordovician) can be seen resting unconformably on the Connemara Schists. From a consideration of these observations, and on other grounds, there is virtually no doubt that the main metamorphism of the Connemara Schists occurred before the beginning of the Ordovician.

The majority of the K–Ar and Rb–Sr age measurements reported in this paper have been carried out on minerals from the Connemara Schists. It will become evident that the measured ages do not reflect the age of the main metamorphism. This fact may be of importance in the interpretation of isotopic ages from other regionally metamorphosed terrains.

In Scotland, the Dalradian Series can be divided into three stratigraphical groups (Knill, 1963):

(1) *Upper Dalradian.* Mainly greywackes, but including lavas, and generally with a limestone (the Loch Tay Limestone and its equivalents) at the base.

(2) *Middle Dalradian.* Varied clastic sediments, largely shelf deposits, but with greywackes in places, and some deposits that may be deltaic; the Schichallion Boulder Bed and its equivalents are at the base.

(3) *Lower Dalradian.* Varied shallow-water clastics and carbonates.

All three of these groups are represented in western Ireland; north of Croagh Patrick the Upper Dalradian is faulted against Silurian beds. In Connemara, both Middle and Lower Dalradian rocks (the Connemara Schists) occur south of the Lower Palaeozoics, but no Upper Dalradian is known. In general, the Dalradian rocks of Ireland and Scotland have so many features in common that there is virtually no doubt about their correlation, despite the absence of fossils.

The Dalradian rocks are intruded by granites; these include the pre-Arenig Oughterard granite, pebbles of which are present in the Lower Ordovician beds to the north, and the Galway granite of post-Wenlock age. The Oughterard granite is later than the main metamorphism and fold movements which affect the Connemara Schists, and thus provides a minimum age limit for these events. This is confirmed by other evidence, including the presence of boulders of schist in the Ordovician (Dewey, 1961) and the fact that most of the Lower Palaeozoic sediments (both Ordovician and Silurian) in the area are not metamorphosed above the chlorite grade.

The Ordovician sediments of Murrisk (i.e. the southwestern part of County Mayo) have been described by Stanton (1960) and Dewey

(1963). They consist of material largely derived from the surrounding Dalradian rocks.

The succession increases greatly in thickness towards the north of the present synclinal area of outcrop. The Lower Ordovician beds (Arenig and Llanvirn) range in thickness from 2000 to 21,000 feet; they consist of turbidites, slates, tuffs and boulder slides; the graptolites indicate continuous sedimentation from the *Didymograptus extensus* Zone to the *Didymograptus hirundo* Zone.

The Upper Ordovician is not so accurately dated by fossils; the few that have been found are consistent with a Llandeilo–Caradoc age. There is no sign of any major breaks in the Upper Ordovician; the thickness ranges from 3000 to 15,000 feet. These beds show a marked change in sedimentation from the Lower Ordovician in that they consist of coarse, deltaic, current bedded sandstones and conglomerates; beds of ignimbrite are also present and facilitate correlation of the two limbs of the syncline. This sudden change from dominantly fine-grained, deep-water beds to coarse, shallow-water beds will be discussed below in connexion with the interpretation of isotopic ages in the metamorphic rocks. Both the sedimentary evidence and the isotopic age work indicate a substantial elevation of the Dalradian massif immediately prior to the Caradocian.

The Silurian sediments range in age from Upper Llandovery to Wenlock (McKerrow and Campbell, 1960; Anderson, 1960; Dewey and Phillips, 1963). They occur in two synclinal areas to the north and south of the Ordovician outcrops; in both areas they tend to mask the relations between the Ordovician and Dalradian, but Dr. J. F. Dewey (personal communication) has recently discovered a section south of Kilbride Bay where black shales rest on Dalradian schists and where both are overlain by the basal conglomerate of the Silurian. These black shales have not yet yielded fossils. They are clearly pre-Silurian, but postdate the metamorphism of the Connemara Schists.

The Ordovician and Silurian sediments are chloritized wherever they are exposed in western Ireland. Muscovite is developed locally in both Ordovician and Silurian rocks in the region east of Cregganbaun and south of Croagh Patrick. This mild metamorphism of the Lower Palaeozoic rocks is post-Wenlock (the age of the youngest beds affected) and appears to be associated in time with the intrusion of the Corvock granite which cuts Ordovician and Silurian beds northeast of Cregganbaun.

Turning now to previous isotopic age data, Giletti, Moorbath and

Lambert (1961) reported Rb–Sr ages of micas from the Dalradian rocks of Scotland and western Ireland (Connemara) which suggested that these rocks were metamorphosed 475 ± 15 m.y. ago. At that time Rb–Sr ages of single minerals were calculated by assuming a fixed value for the ^{87}Sr abundance in the mineral at the time of metamorphism, conveniently expressed as the initial $^{87}Sr/^{86}Sr$ atom ratio. A value of 0.710 was commonly used, which is similar to that of strontium in sea water. It has now become clear to geochronologists that this assumption is unjustified for most metamorphic minerals. Recent advances in Rb–Sr geochronometry have shown that it is possible to calculate both the isotopic age and the initial $^{87}Sr/^{86}Sr$ of a mineral by analysis of the mineral and its corresponding whole rock (Compston, Jeffery and Riley, 1960; Nicolaysen, 1961; Fairbairn, Hurley and Pinson, 1961). Most of these earlier studies were concerned with metamorphosed igneous rocks. Recent Rb–Sr work by Bell (1964) on mica–whole rock pairs from Scottish Dalradian metasediments has shown that *single* mineral Rb–Sr ages, assuming an initial $^{87}Sr/^{86}Sr$ of 0.710, usually give anomalously high results, because of the unusually high enrichment in radiogenic strontium in the sediments at the time when metamorphism (or, more strictly, radiogenic strontium migration within the rock) ceased. It has been shown by Bell (1964), and in the present work, that initial $^{87}Sr/^{86}Sr$ values in the Dalradian sediments at this time were actually in the range of approximately 0.720–0.760, and vary from one rock to the next. It follows from the basic equations used in age calculations that the smaller the ratio Rb/Sr of a mineral from such a rock, the greater will be the difference between the single mineral age (assuming an initial $^{87}Sr/^{86}Sr$ of 0.710) and the corresponding mineral–whole rock age using the true $^{87}Sr/^{86}Sr$ ratio. This can be clearly seen by inspection of table 7 (p. 250) in the paper by Giletti, Moorbath and Lambert (1961), where samples with the lowest Rb/Sr ratios, in general, yield the highest apparent ages, and *vice versa*. These considerations suggest that the average figure of 475 ± 15 m.y. reported by Giletti and others for the termination of the main Dalradian metamorphism (in terms of cessation of radiogenic strontium mobility) is twenty to thirty million years too high. Rb–Sr mineral–whole rock ages from the Scottish Dalradian metasediments yield an average value of 440 ± 10 m.y. (Bell, 1964). Of course, this figure cannot be assumed without further evidence to represent the *true* age of metamorphism of the Dalradian Series, but only the time when diffusion of radiogenic ^{87}Sr out of the system ceased.

Recently, K–Ar ages on Scottish Dalradian rocks have been published by Harper (1964), Miller and Brown (1965) and Brown, York and others (1965). The average K–Ar age for twenty-three micas obtained by the last two groups of authors is 438 ± 10 m.y., although a few values scatter on either side of this range. Harper reports a mean K–Ar age of 429 ± 4 m.y. for forty muscovites, biotites and Middle Dalradian and Ballapel Foundation slates, whilst the mean K–Ar age for ten Upper Dalradian slates was significantly higher at 461 ± 8 m.y. Harper considers that the latter age, obtained on rocks from a high, stratigraphic level which were never heated to a high temperature and which probably cooled rapidly, may approximate more closely to the true age of the main Dalradian metamorphism in Scotland. The lower ages obtained on micas from the more deep-seated, high-grade rocks are interpreted by Harper in terms of a cooling history. Fitch, Miller and Brown (1964) have argued that the age patterns obtained from the Dalradian (and Moine) rocks of Scotland can best be explained as resulting from a combination of partial and complete overprinting of older metamorphic rocks (>720–765 m.y.) by a 420–410 ± 10 m.y. main Caledonian orogenic–magmatic–metamorphic event which was contemporaneous throughout the British Isles. This view differs in certain important respects from the conclusions of Bell (1964) and Harper (1964, 1965) who have been able to define a much more detailed isotopic chronology of the Scottish Dalradian on the basis of K–Ar and Rb–Sr determinations on many different rock types. Detailed discussion must be deferred until publication of the theses of Bell and Harper. In summary, however, it is found that the post-F_1 oldest granites (some of them foliated) in the Scottish Dalradian were emplaced about 530 m.y. ago, in Middle or Upper Cambrian times (Bell, 1964). This value sets a maximum limit for the age of the main metamorphic episodes. Furthermore, Harper's data strongly suggest that the main foldings and metamorphisms in the Scottish Dalradian occurred more than 480–490 m.y. ago, that is, in Arenig or pre-Arenig times. Further evidence for a pre-Arenig age of the Dalradian metamorphism in Scotland comes from K–Ar measurements on biotites from several posttectonic gabbros in northeastern Scotland, which yield ages in the range 460 to 498 m.y. (Brown, Miller, Grasty and Fraser, 1965).

The time relationships deduced *in the field* between Arenig rocks and the various episodes of folding in the Scottish Dalradian rocks are still uncertain (Shackleton, 1958; Johnson and Harris, 1965). In contrast, in Connemara it is virtually certain, as pointed out earlier, that the main foldings and metamorphisms were pre-Arenig in age.

In addition to the Connemara single mineral Rb–Sr ages of Giletti, Moorbath and Lambert (1961) already discussed above, four K–Ar mica ages from the Connemara Schists have been reported by Miller and Brown (1965). The four results are 420, 417, 414 and 384 m.y. and it is suggested by these workers and by Fitch, Miller and Brown (1964) that these dates provide clear evidence of complete overprinting of an earlier metamorphic complex by the main Caledonian event at 420–410 m.y.

The isotopic age for the base of the Arenig is not accurately known. A widely accepted value is 500 ± 15 m.y. (Holmes, 1960; Kulp, 1961; Harland, Smith and Wilcock, 1964), whilst Harris and others (1965) assign a *minimum* age of 475 m.y., and Evernden and Richards (1962) prefer a value of 480 m.y. for the base of the Arenig. The conclusion is inescapable, therefore, that the few isotopic mineral ages so far reported from the Connemara Schists (and probably the majority of the large number of published mineral ages from the Scottish Dalradian) have not preserved a record of the true age of the major metamorphism of the Dalradian Series.

This is also suggested by the work of Leggo, Compston and Leake (1966) who have shown that the Oughterard granite, which was intruded into the Connemara Schists post-F_4 in the fold sequence, gives a Rb–Sr–whole rock isochron whose youngest possible age is 510 ± 35 m.y. at the 95% confidence level, thus providing a minimum age for the early metamorphisms. A common lead model age of 580 ± 80 m.y. was reported by Moorbath (1962) for a period of galena mineralization associated with the Oughterard granite, and was originally interpreted as being grossly and anomalously high, as a result of the erroneous hypothesis that the Oughterard and Galway granite (ca. 380 m.y.) were contemporaneous. It now appears that the model lead age may actually represent the true age of mineralization and granite intrusion within the experimental error.

In contrast to the above, a potash feldspar–muscovite–whole rock age measured by Leggo, Compston and Leake (1966) on the Oughterard granite gives a 'metamorphic' age of 444 ± 7 m.y., a value close to that reported by the same authors for hornblende K–Ar ages separated from the Connemara Migmatites.

The geochronological evidence from western Ireland could be of great importance in Dalradian geochronology, particularly as the Connemara Schists have now been stratigraphically correlated with the Dalradian sequences in Donegal, Islay and Perthshire (Kilburn,

Pitcher and Shackleton, 1965). Accordingly, it seemed advisable to extend the reconnaissance work and obtain more isotopic ages from the Connemara Schists and the adjoining Ordovician and Silurian rocks to the north, which have themselves undergone a low-grade metamorphism (Dewey, 1963; Dewey and Phillips, 1963; Dewey and McManus, 1964). Measurements have been made on minerals, especially hornblendes, whose age of crystallization in the metamorphic and fold sequence is known.

II Sequence of Events in the West of Ireland

In western Ireland the deposition of the Dalradian rocks was followed by the intrusion of dolerite sills and then by sillimanite–garnet grade metamorphism (M_1) and isoclinal folding (F_1). After these events, basic and ultrabasic intrusions were emplaced in Connemara which extensively hornfelsed (M_2) their already regionally metamorphosed envelopes. It is certain that important folding and metamorphism preceded these intrusions because hornfelsed and partly dissolved pelitic xenoliths within the intrusions preserve folds and show an early mineral assemblage of sillimanite–garnet–biotite–oligoclase, which is destroyed by the hornfelsing and replaced by the assemblage green spinel–cordierite–magnetite $\pm$ orthopyroxene $\pm$ corundum. Furthermore, the envelopes to the intrusions show a progressive inward destruction of the regional metamorphic assemblages, whilst early fold hinges became detached and disoriented in mobilized hornfelses immediately next to the intrusions. Both the intrusions (Wager, 1932; Ingold, 1937; Rothstein, 1957; Leake, 1958a) and their hornfelses (Leake and Skirrow, 1960; Evans, 1964) have been fully described elsewhere.

Later, the intrusions were themselves strongly metamorphosed (M_3: amphibolite facies), injected and broken up into agmatites by quartz–andesine magma and folded (F_2) about folds with north–south trending hinges (Leake, 1964). These events partially or completely destroyed the original igneous assemblages. The replacing amphiboles vary from randomly arranged ones, formed during the early part of M_3, to north–south lineated ones produced during the later stages of M_3 when F_2 was acting. Likewise, variation from unlineated to north–south rodded quartz is seen in the quartz–andesine gneisses which form an important part of the Connemara Migmatites, thus dating their formation as about F_2. The final phase of the migmatization produced potash feldspar gneisses which are never lineated and, since they partly replace the earlier gneisses, it is deduced that their formation was post-F_2. It is possible that in the metasediments, M_2 and M_3 occurred as one continuous metamorphism.

The next recognizable phase is the tight east–west folding of F_3, accompanied by metamorphism (M_4) in north and central Connemara, sufficient to recrystallize hornblende, mica and feldspar. Throughout the 13 miles from south of Errislannan to the Skerd Rocks (figure 1) flat-lying metasediments, striped amphibolites and migmatized basic and ultrabasic rocks with their associated gneisses still preserve the F_2 north–south lineation and are therefore supposed not to have suffered sufficient metamorphism to have been recrystallized during F_3, although they have been ground and crushed by F_3 movements. Further north, the F_3 lineation becomes prominent and the F_2 lineation is preserved only as relicts. Much further north, at Clew Bay, a similar structural sequence is apparent in the Dalradian rocks (Dewey and Phillips, 1963) but the grade of metamorphism is generally much lower than in Connemara. Synchronous with the potash feldspar gneisses of Connemara are major folds at Cashel, which are F_3.

Closely following, or even overlapping, the later stages of F_3 there seems to have been another injection of basic magma which is now schistose amphibolite. This later magmatic phase is recognized because the intrusions occurred along the axial planes of the F_3 folds and cannot, therefore, predate F_3 (Edmunds and Thomas, 1966).

A late metamorphism, or metamorphisms, apparently unaccompanied by movements, has given rise to a patchy development of andalusite pegmatites and schists north and south of Lough Corrib and at Renvyle; to cordierite porphyroblasts in the Inagh Valley in central Connemara and to chloritoid porphyroblasts at Renvyle (Cruse, 1963). The age of this static metamorphism (M_5), or metamorphisms, is not known but is definitely post-F_3 and probably pre-F_4.

The F_4 folds produced the east–west Connemara antiform, and to the north a complementary synform (Connemara synform) and another antiform (Joyce's antiform), the latter two folds being best seen in Joyce's Country, north of Lough Corrib. South of the Connemara antiform the most obvious F_4 structure is that of the Delaney Dome, south of Clifden. This structure is accompanied by strain-slip and fracture cleavage, whereas in Joyce's Country the F_4 strain-slip cleavage was accompanied by sufficient metamorphism (M_5) to recrystallize biotite in the strain-slip cleavage planes. F_4 folds have also been recognized in the Dalradian rocks of Clew Bay.

The fifth period of folding (F_5) is most obvious south of Clifden where approximately north–south trending folds cross the east–southeast trending F_4 antiform, forming a dome structure. Usually, F_5 forms two sets of conjugate folds, whose axial planes strike NNE and NNW.

The Oughterard granite which has been dated at 510 ± 35 m.y., and is pre-Arenig, definitely postdates F_4 and may well postdate F_5, for it seems to be intruded along F_5 faults in some localities. The Oughterard granite has been affected by later fracture cleavages, shearing and movement along east–west faults as well as by various movements along NE–SW and NW–SE faults. The Galway granite (dated at 384 ± 2 m.y.) and the remaining granite intrusions of Connemara (Leggo, Compston and Leake, 1966) are all younger than the Oughterard granite. Even later are acid, intermediate and basic dykes and also various E–W, NNE and NNW faults, most of which appear to be pre-Carboniferous.

The stratigraphy of the Ordovician rocks of western Ireland has been described in detail by Stanton (1960) and Dewey (1963). The oldest rocks contain *Didymograptus extensus* and therefore belong to the Arenig, while the youngest Ordovician rocks are of uncertain age, possibly Caradoc or Llandeilo. The post-depositional sequence of events in the Ordovician and Silurian rocks of western Connaught has been summarized by Dewey and Phillips (1963). Uplift, folding and considerable erosion of Ordovician rocks occurred before the Silurian rocks were deposited; since no pre-Silurian cleavages have been recognized in the Ordovician rocks, it is thought that this folding was of concentric, rather than of similar type. A marked unconformity exists between the Silurian and the Ordovician rocks since the Upper Llandovery rests on Arenig rocks in several places in Murrisk. The Silurian rocks have been described by Anderson (1960) and McKerrow and Campbell (1960). The oldest Silurian is of Upper Llandovery age and is overlain, usually unconformably, by Lower Wenlock rocks; no rocks of Ludlow age are known in northwestern Galway or Mayo. Two penetrative sets of cleavages affected the Silurian and Ordovician rocks, followed by two pairs of conjugate folds with strain-slip cleavages, a single strain-slip cleavage and various episodes of faulting. The folding and most of the faulting are pre-Carboniferous.

Of major importance is the correlation of the later structural and metamorphic phases in the Dalradian rocks with those in the Ordovician and Silurian rocks. Present indications are that the Connemara antiform (F_4) predates all structures in the younger rocks. The slaty cleavage imposed on the Silurian rocks (Upper Llandovery and Wenlock) is represented by brittle strain-slip cleavages, fracture cleavages, faults and cold reworked schistosity planes in the Connemara Schists. At least some of the strain-slip cleavages and faults in the Silurian rocks occur as weak, very late fracture cleavages and faults in the basement. It is possible that F_5 may be related to the earliest movements on the Maam Valley faults which are mid-Ordovician or older. There are, however, still several very important uncertainties in the history of the Connemara Schists. Thus the ubiquitous chloritization of the biotite, sericitization and saussuritization of the plagioclase and other retrogressive changes, must belong to several metamorphisms. However, it appears certain that the metamorphisms M_1 to M_6 in Connemara predate the deposition of the Arenig rocks and that the F_4 folds in the Connemara Schists are earlier than the deposition of the Silurian rocks. Since these folds are of similar type and have a cleavage, they are almost certainly pre-Arenig also.

III Experimental Methods

Argon was extracted in a bakeable glass vacuum apparatus and determined by the isotope-dilution method, with ^{38}Ar as tracer ('spike'), in a Reynolds-type glass mass spectrometer. Potassium in micas and slates was determined by direct comparison with standard solutions using an EEL flame photometer coupled to a chart recorder. In the case of hornblende, the addition–dilution method proposed by Grimaldi (1960) was used, in order to correct for suppression of the potassium flame by other ions. Each potassium analysis reported in table 1 represents the mean of three determinations on separate mineral or rock sample aliquots. The experimental errors in measuring the K/Ar ratio are small enough to give a maximum error of $\pm 4\%$ on mica and slate ages and $\pm 6\%$ on hornblende ages. The decay constant for ^{40}K and $\lambda_e = 0.584 \times 10^{-10}$ yr^{-1}, and $\lambda_\beta = 4.72 \times 10^{-10}$ yr^{-1}. The isotopic abundance of ^{40}K was taken as 0.0119 atomic per cent.

Rubidium and strontium were determined by the isotope-dilution method, using either a single ^{86}Sr tracer or a mixed $^{84}Sr + ^{86}Sr$ tracer. Isotope measurements were carried out with a A.E.I., MS–2 mass spectrometer with a 60°, 6 inch analyser tube and an electron multiplier as an ion beam detector; a correction for the square root of the mass was applied to all observed isotope abundance ratios to correct for discrimination by the multiplier. A single tantalum filament surface ionization technique was used. Further experimental details have been described elsewhere (Giletti, Moorbath and Lambert, 1961; Bell, 1964; Moorbath and Bell, 1965). For micas with relatively high Rb/Sr ratios, the present-day ^{87}Sr/^{86}Sr ratios (table 3, column 7) were calculated directly from isotope-dilution runs. In the case of whole rocks, however, the ^{87}Sr/^{86}Sr ratio was measured directly on unspiked strontium. In order to eliminate unpredictable isotope fractionation effects, all measured ^{87}Sr/^{86}Sr ratios were normalized on the assumption that ^{86}Sr/^{88}Sr = 0.1194, in accordance with the procedure adopted by other workers (Faure and Hurley, 1963; Hedge and Walthall, 1963).

Rb–Sr ages have been calculated with a decay constant for ^{87}Rb of 1.47×10^{-11} yr^{-1} (Flynn and Glendenin, 1959). This is 6% greater than the decay constant of 1.39×10^{-11} yr^{-1} proposed by Aldrich and others (1956). All of the Rb–Sr ages reported in this paper would be increased by 6% if the smaller decay constant were used. The natural abundance of ^{87}Rb is taken as 27.85 atom per cent.

The final errors quoted for the Rb–Sr ages were calculated by changing each of the parameters one at a time by their experimental errors and finding the square root of the sums of the squares of the residuals. Computations were carried out on the C.D.C. 1604 digital computer at the University of Texas. Errors in the normal isotopic abundances of both Rb and Sr, and in the decay constant of ^{87}Rb, were ignored. The spike concentrations of Rb and Sr were considered to be accurate to $\pm 1\%$. Rb–Sr isochrons were calculated by means of a least squares analysis in which both the x and y coordinates were taken into consideration. Errors were calculated for the latter by altering each of the parameters by its experimental error, one at a time, and finding the square root of the sums of the squares of the sequences of the residuals in the usual statistical manner.

IV Results and Discussion

A *K–Ar mineral ages*

The K–Ar results are given in table 1, whilst figure 2 is a histogram of measured ages. It is evident that on the basis of the currently accepted time scale (Holmes, 1960; Kulp, 1961; Harland, Smith and Wilcock, 1964) none of the ages are pre-Arenig and consequently that they do not relate directly to the main Connemara metamorphisms M_1 to M_6. There appears to be no correlation between the type of lineation and the measured age. Thus F_2-lineated hornblendes do not record an older age than F_3-lineated ones, while petrographically early micas do not retain older ages than later overgrowths. In fact, no less than 22 out of the 25 Connemara K–Ar mineral ages lie between 435 and 465 m.y., which is regarded as almost within the experimental error spread of the individual measurements. The

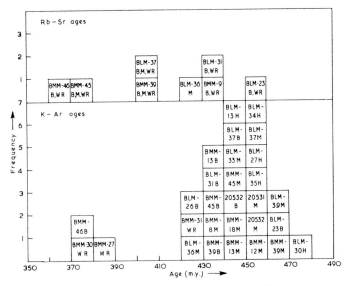

Figure 2 Histogram of K–Ar and Rb–Sr ages. Specimen numbering as in tables 1 and 3. B, biotite; M, muscovite; H, hornblende; WR, whole rock.

average for all 25 Connemara K–Ar ages is 447 ± 8 m.y. It is postulated that some event occurred at this time and that radiogenic argon has been quantitatively retained in the minerals since then. The exact nature of this event is uncertain. The following speculations are based upon the very strong evidence, referred to previously, that (a) metamorphisms M_1 to M_6 of the Connemara Schists are pre-Arenig (Dewey, 1961), (b) the base of the Arenig is at least 480 m.y. old (Evernden and Richards, 1962; Harris and others, 1965). Two principal possibilities suggest themselves:

(1) Uplift of the Connemara massif occurred about 440–460 m.y. ago, possibly coinciding with the folding of the Ordovician rocks to the north at the end of the Ordovician or beginning of the Silurian

Table 1 K–Ar results

Sample number	Description	Mineral	K content (%)	^{40}Ar, rad. (scc/g $\times 10^{-4}$)	^{40}Ar, rad. / ^{40}Ar, total (%)	Age[a] (m.y.)
BMM-8	Andalusite–muscovite pegmatite, Claggan, Connemara	Muscovite	8.29	1.616	98.4	435
BMM-12	Mica schist, Owenduff Bridge, Connemara	Muscovite	4.94	1.020	97.6	457
BMM-13	Mica schist, Lecknavarna, Connemara	Biotite	3.95	0.770	69.9	435
		Muscovite	8.51	1.718	97.4	446
BMM-18	Mica schist, Lettergesh, Connemara	Muscovite	6.94	1.379	98.6	442
BMM-39	Sillimanite–biotite–cordierite schist, Ferry Bridge, Connemara	Biotite	4.44	0.868	95.0	435
		Muscovite	5.76	1.205	98.1	462
BMM-45	Mica schist, Cornamona, Connemara	Biotite	7.03	1.393	97.8	440
		Biotite	7.03	1.376	97.5	435
		Muscovite	8.25	1.660	98.9	446
		Muscovite	8.25	1.634	98.5	440
BLM-13	F$_3$ amphibolite with relict F$_2$. Ardbear Bridge, Connemara	Hornblende	0.309	0.0612	87.8	440
BLM-23	Biotite–hornblende norite, Lough Wheelaun intrusion, Connemara	Biotite	5.83	1.232	98.6	466
		Biotite	5.77	1.210	96.5	463
BLM-26	Plagioclase–hornblende–biotite rock, Lough Wheelaun intrusion, Connemara	Biotite	5.62	1.076	96.6	427
BLM-27	Hornblende–plagioclase–pyroxene rock, Lough Wheelaun intrusion, Connemara	Hornblende	0.619	0.126	92.2	451
BLM-30	F$_3$ amphibolite, nr. Cashel, Connemara	Hornblende	0.207	0.0444	71.2	473
		Hornblende	0.207	0.0441	54.3	470

Sample	Description	Mineral				Age
BLM-31	F_3 mica schist, nr. Recess, Connemara	Biotite	6.07	1.120	98.5	439
BLM-33	Quartz–feldspar pegmatite vein, Recess, Connemara	Muscovite	8.42	1.701	98.4	448
BLM-34	Amphibolite lens in F_2-gneiss, Truska Lough, Connemara	Hornblende	0.586	0.120	93.0	454
BLM-35	F_2 hornblendite in migmatites, Truska Lough, Connemara	Hornblende	0.836	0.172	91.9	455
BLM-36	Pegmatite in migmatites, Truska Lough, Connemara	Muscovite	8.50	1.599	97.6	421
		Muscovite	8.11	1.528	96.4	421
BLM-37	F_3 mica schist, Streamstown Bay, Connemara	Biotite	7.45	1.503	99.1	448
		Muscovite	8.44	1.722	97.9	452
BLM-39	Post-F_2 pegmatite crossing ultrabasic intrusion, Currywongaun, Connemara	Muscovite	8.43	1.778	95.8	465
20531	Pegmatite, nr. Clifden, Connemara	Muscovite	8.63	1.783	88.9	457[b]
20532	Mica schist, Streamstown, Connemara	Biotite	7.23	1.448	99.0	447[b]
		Muscovite	8.76	1.797	99.7	454[b]
BMM-31	Slate in Upper Dalradian Westport Grit, Clew Bay, nr. Louisburgh, Murrisk, Co. Mayo	Whole rock	6.32	1.211	99.0	427
BMM-27	Slate in Oughty Group of the Wenlock. Letterbrock, Murrisk, Co. Mayo	Whole rock	2.79	0.479	98.7	387
BMM-30	Slate in Letterbrock Group of the Arenig, Letterbrock, Murrisk, Co. Mayo	Whole rock	1.48	0.246	97.7	377
BMM-46	Corvock granite, Murrisk, Co. Mayo	Biotite	3.97	0.661	97.9	376

[a] Maximum error for mica ages is ±4%, and for hornblende ages ±6%.
[b] Single mineral Rb–Sr ages for these samples were reported by Giletti, Moorbath and Lambert (1961) assuming an initial $^{87}Sr/^{86}Sr$ of 0.710. The ages (same order as above) were 463 ± 10, 460 ± 12 and 493 ± 15 m.y. The last sample has the lowest Rb/Sr ratio and hence shows a significant age discordance (see text).

(Dewey, 1963; Dewey and Phillips, 1963). If, prior to this uplift, the Connemara rocks were sufficiently deeply buried for their temperature to remain above that at which ^{40}Ar could diffuse out of the crystal lattices of minerals, then the uplift may have accelerated the cooling of the massif and resulted in the temperature falling too low for this diffusion to continue. According to this hypothesis the rocks would have had to remain above the threshold temperature of argon diffusion for at least 30–40 m.y. before uplift. Geologists differ a good deal in their views on the feasibility of such a lengthy period of burial, with a correspondingly slow cooling rate, after cessation of the main metamorphic–orogenic activity. It is not possible to enter into all the arguments here, but it may be relevant to recall the very low K–Ar and Rb–Sr ages (5 m.y.) obtained from recently uplifted biotite gneisses of probable Jurassic age (ca. 150 m.y.) in the central part of the uplift zone of the New Zealand Alps (Hurley and others, 1962). This example is, of course, much more extreme than the situation in Connemara.

The actual threshold temperatures for argon diffusion are not known with any certainty, but probably lie in the general range 150–300°C. This topic has recently been reviewed by Lambert (1964). However, it is known that the threshold temperature for hornblende exceeds that of muscovite, which in turn exceeds that of biotite (Hart, 1964). The mean K–Ar ages of relevant groups of Connemara minerals are shown in table 2. Whilst it appears that the average age

Table 2 Average K–Ar ages of minerals from the Connemara Series[a]

Mineral	Age (m.y.)
All biotites (8)	442 ± 6
All muscovites (12)	448 ± 5
Schist biotites (6)	441 ± 7
Schist muscovites (7)	451 ± 7
Coexisting biotite (5)	441 ± 8
Coexisting muscovite (5)	451 ± 8
Hornblendes—this work (5)	456 ± 12
Hornblendes—this work and Leggo, Compston and Leake (1966) (8)	447 ± 10

[a] Figures in parentheses indicate total number of specimens.

of hornblendes and muscovites is a few million years higher than that of biotites, it is not really possible to make any significant distinctions on account of the experimental error, the spread of the individual results and the comparatively small number of measured ages. It may be concluded, however, that if uplift and cooling were the main factors in producing the observed ages then (a) the time interval during which radiogenic argon was 'frozen in' (uplift?) did not exceed about 10 m.y. at the most, and (b) observed ages are closely related to the time of maximum sedimentation which occurred in Upper Ordovician times and gave rise to a great thickness of sediments in the area to the north of the Connemara massif (McKerrow and Campbell, 1960; Dewey, 1963). It is at about this time, too, that the Llanvirn Glenummera Shales were rapidly succeeded by the coarse conglomerates of the Mweelrea Group.

(2) A widespread retrogressive metamorphism occurred in Connemara about 440–460 m.y. ago. This has been suggested by Leggo, Compston and Leake (1966) as the mechanism producing a whole rock–muscovite–potash feldspar Rb–Sr age of 444 ± 7 m.y. for the Oughterard granite. This figure is remarkably close to the average K–Ar value obtained in the present work. This retrograde metamorphism is regarded as being responsible for the main episode of chloritization in the Oughterard granite and also elsewhere in Connemara.

The present evidence lends some limited support to this hypothesis. It may be significant that hornblende from the least chloritized and epidotized amphibolite (BLM–30, table 1) yields the oldest age of 472 ± 20 m.y., while those reported by Leggo, Compston and Leake (1966) which come from substantially chloritized rocks yielded K–Ar ages of 432, 434 and 437 m.y. respectively. One feature of all the rocks examined in the present work is the ubiquitous presence of chlorite and late sericite. It is, therefore, not unreasonable to correlate the mean K–Ar age with this late retrograde metamorphism which is impressed on the rocks irrespective of their earlier M_1–M_6 crystallizations. Whether such a low-grade metamorphism can really produce a more or less uniform K–Ar age on a regional scale in such diverse minerals as biotite, muscovite and hornblende with very variable grain sizes (e.g. from schists, pegmatites, amphibolites) is something that only future work can resolve.

Whole rock K–Ar ages of slates Two whole rock K–Ar ages (samples BMM–27, BMM–30) are reported in table 1 for Lower Palaeozoic sediments from central Murrisk, Co. Mayo, to the north of Connemara, where Ordovician and Silurian rocks together acquired a grade of metamorphism approaching the chlorite/biotite isograd (Anderson, 1960; Dewey, 1961, 1963; Dewey and Phillips, 1963). The Arenig slate yields an age of 377 ± 14 m.y., and the Wenlock slate gives 387 ± 14 m.y., both agreeing within experimental error. It has been demonstrated beyond any reasonable doubt that whole rock K–Ar measurements on well-cleaved slates can give a close approximation to the time at which the slaty cleavage was impressed on the rocks (Dodson, 1963; Harper, 1964, 1965). On the basis of the current time scale, this event in central Murrisk occurred in Lower Devonian times.

An Upper Dalradian slate (BMM–31, table 1) belonging to the Westport Grit Series from the area of relatively low-grade metamorphosed Dalradian rocks on the southern shore of Clew Bay, Co. Mayo, yielded an age of 427 ± 17 m.y. This is just within the range of the Connemara K–Ar mineral ages discussed previously. However, the apparent age could have been slightly lowered by the relative proximity of the post-Silurian event at about 380 m.y.

Turning back for a moment to the topic of the 440–460 m.y. Connemara K–Ar mineral ages, it seems most unlikely to the authors that these can be interpreted simply as due to overprinting of a pre-Arenig (pre-500 m.y.) event by the Lower Devonian (ca. 380 m.y.) event. It is believed that such a process would lead to a highly discordant age pattern between different minerals and rock types, with apparent ages probably along the whole interval between pre-Arenig and Lower Devonian. On the other hand, there may well have been overprinting of the main Connemara K–Ar event by the Lower Devonian event, giving apparent ages in the range of about 440–380 m.y. This could be the explanation for the four apparent ages in this range reported by Fitch, Miller and Brown (1964), which they interpret somewhat differently as overprinting of a pre-Arenig event by a main Caledonian event at 420–410 m.y. However, until further evidence comes to light, it rather looks as if the main Connemara K–Ar event at about 440–460 m.y., whatever its nature, was sufficiently strong to eradicate practically all preexisting ages and nearly all isotopic record of the pre-Arenig event.

Corvock granite K–Ar age   A biotite from the Corvock granite (BMM–46, table 1) yields a K–Ar age of 376 ± 14 m.y. The Corvock granite (figure 1) is a small, transgressive stock that intrudes the Ordovician and Silurian rocks of the Mayo trough, some four miles southwest of Croagh Patrick. Dewey and Phillips (1963) consider that the granite predates the penetrative schistosities and regional tight folding of the

Ordovician and Silurian rocks, which occurred in post-Wenlock times. The specimen was collected about 4–5 miles from the Arenig and Wenlock slate specimens (BMM–27, BMM–30, table 1) which yielded a closely similar K–Ar age to the Corvock granite.

B Rb–Sr age measurements

The Rb–Sr results are given in table 3, and are also shown on the histogram (figure 2). With one exception, the ages were obtained from biotite and/or muscovite and the corresponding whole rock, as discussed in the introduction. The ages may be calculated graphically (figures 3 and 4), or by solution of simultaneous equations. The Rb–Sr ages represent the time at which diffusion of radiogenic ^{87}Sr within the rock system ceased, and localized strontium isotope homogenization was no longer possible.

The only case in which a whole rock measurement was not possible was the pegmatitic muscovite BLM–36. However, in view of its high Rb/Sr ratio a small uncertainty in the initial ^{87}Sr/^{86}Sr makes little difference to the calculated age.

Rather surprisingly, the Rb–Sr ages show a greater scatter than the K–Ar ages. Table 4 shows a comparison of K–Ar and Rb–Sr ages where both are available for the same specimen. The first four pairs of K–Ar and Rb–Sr ages are in excellent agreement. However, samples BLM–37, BMM–39 and BMM–45, from the more northerly parts of the Connemara Schist outcrop, have significantly lower Rb–Sr ages than K–Ar ages. The biotite–muscovite–whole rock Rb–Sr ages of these three specimens are 408 ± 10, 409 ± 15 and 372 ± 10 m.y. respectively, where the errors are quoted at the 66% (1σ) confidence level. At the 95% (2σ) confidence level, the ages would overlap. Until further data become available we assume, possibly unjustifiably, that these three Rb–Sr ages refer essentially to the same event, with a mean value of 393 ± 13 m.y. at the 2σ level. This age, perhaps, very nearly reflects the post-Silurian event which was also recorded by K–Ar measurements on the Arenig and Wenlock slates of the Mayo trough to the north (table 1), as well as the intrusion of the post-Silurian granites generally. It appears that in some cases a relatively mild thermal event can bring about strontium isotope homogenization in a mineral–whole rock system (at least the size of a hand specimen) without the corresponding expulsion of argon from the mica lattice. In any case, the comparative data lend support to the hypothesis, already mentioned earlier, of partial-to-complete overprinting of the

Table 3 Rb–Sr results

Sample number	Description	Mineral	Rb (ppm)	Sr (ppm)	$^{87}Rb/^{86}Sr$	$^{87}Sr/^{86}Sr$ (observed)	Age (m.y.)	$^{87}Sr/^{86}Sr$ (initial)
BMM-9	Staurolite–garnet–mica schist, Claggan, Connemara	Biotite Whole rock	214 92	28.0 138	22.84 1.92	0.8673 0.7339	434 ±20	0.7127 ±0.0038
BMM-39	Sillimanite–biotite–cordierite schist, Ferry Bridge, Connemara	Biotite Muscovite Whole rock	228 149 98	18.9 67.6 141	35.78 6.44 2.02	0.9412 0.7650 0.7360	409 ±15	0.7249 ±0.0037
BMM-45	Mica schist, Cornamona, Connemara	Biotite Muscovite Whole rock	370 177 156	15.2 52.2 44.5	73.67 9.92 10.15	1.1493 0.8019 0.7996	375 ±10	0.7440 ±0.0045
BLM-31	Mica schist, near Recess, Connemara	Biotite Whole rock	381 226	81.9 409	13.58 1.60	0.8050 0.7288	430 ±38	0.7186 ±0.0047
BLM-37	F$_3$ mica schist, Streamstown Bay, Connemara	Biotite Muscovite Whole rock	727 434 163	38.3 21.5 133	57.00 60.65 3.54	1.0790 1.1169 0.7654	408 ±10	0.7433 ±0.0047
BLM-36	Pegmatite in migmatites, Truska Lough, Connemara	Muscovite	894	58.3	46.03	1.0019	427 ±15	0.7120 (assumed)
BLM-23	Biotite–hornblende norite, L. Wheelaun intrusion, Connemara	Biotite Whole rock	358 50	80.2 659	13.04 0.221	0.7965 0.7106	453 ±35	0.7091 ±0.0041
BMM-46	Corvock granite, Murrisk, Co. Mayo	Biotite Whole rock	280 117	23.9 117	34.56 1.91	0.8957 0.7196	367 ±18	0.7093 ±0.0035
BMM-12	Mica schist, Owenduff Bridge, Connemara	Whole rock	111	228	1.41	0.7356	—	—
BMM-18	Mica schist, Lettergesh, Connemara	Whole rock	155	95.4	4.70	0.7620	—	—

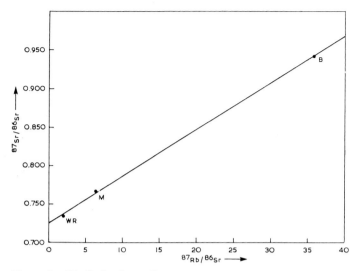

Figure 3 Rb–Sr isochron diagram for schist BMM–39. The slope corresponds to an age of 409 ± 15 m.y. and the intercept to an initial $^{87}Sr/^{86}Sr$ of 0.7249 ± 0.0037. B, biotite; M, muscovite; WR, whole rock.

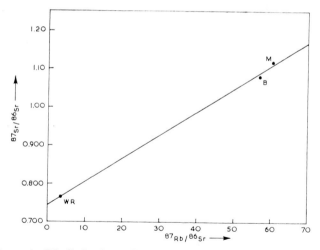

Figure 4 Rb–Sr isochron diagram for schist BLM–37. The slope corresponds to an age of 408 ± 10 m.y. and the intercept to an initial $^{87}Sr/^{86}Sr$ of 0.7433 ± 0.0047. B, biotite; M, muscovite; WR, whole rock.

Table 4 Comparison of K–Ar and Rb–Sr ages[a]

Sample number	Rock type	K–Ar age (m.y.)	Rb–Sr age (m.y.)
BMM–46	Corvock granite	376 ± 14 (B)	367 ± 18 (B, WR)
BLM–23	L. Wheelaun intrusion	465 ± 19 (B)	453 ± 35 (B, WR)
BLM–31	Schist	439 ± 18 (B)	430 ± 38 (B, WR)
BLM–36	Pegmatite	421 ± 17 (M)	427 ± 15 (M)
BLM–37	Schist	448 ± 18 (B) 452 ± 18 (M)	408 ± 10 (M, B, WR)
BMM–39	Schist	435 ± 18 (B) 462 ± 18 (M)	409 ± 15 (M, B, WR)
BMM–45	Schist	438 ± 18 (B) 443 ± 18 (M)	372 ± 10 (M, B, WR)

[a] M, muscovite; B, biotite; WR, whole rock.

widespread Connemara isotopic event (ca. 440–460 m.y.) by the Lower Devonian event or events (ca. 380–390 m.y.).

Instances where Rb–Sr ages are significantly lower than K–Ar ages on the same mineral or rock are not uncommon in metamorphic rocks. This problem has been reviewed by Kulp and Engels (1963), who also investigated the effect of base exchange on micas. They were mainly concerned with circulating ground waters in the surface zone, but it is possible that their findings could be extended to water involved in metamorphic or metasomatic processes. The main conclusions of Kulp and Engels were as follows. (a) The critical factors in base exchange are the structural conditions of the mica, its inherent base-exchange capacity, the concentration of divalent ions in the water and its flux; (b) Up to 50% of the potassium in a biotite may be removed by base exchange without affecting the K–Ar age. Even when 80% is removed, the K–Ar age will not be lowered by more than about 10%; (c) If sufficient rubidium is present in the water, rubidium will be added to the mica whilst radiogenic ^{87}Sr is removed, so that the Rb–Sr age will be lowered relative to the unchanged K–Ar age.

It is possible that some mechanism of this kind accounts for the discordant age pattern observed in the case of samples BLM–37, BMM–39 and BMM–45, and that the post-Silurian event affected *some* of the mica Rb–Sr ages, but not necessarily the K–Ar ages.

Kulp and Engels (1963) observed the above effect on biotites, but found that metamorphic muscovites usually gave concordant K–Ar and Rb–Sr ages. In the present case, of course, the Rb–Sr ages of BLM–37, BMM–39 and BMM–45 were calculated from biotite–muscovite–whole rock isochrons (figures 3 and 4).

It was pointed out earlier that a ^{87}Rb decay constant of 1.47×10^{-11} yr^{-1} has been used in this work (as in all previous work from the Oxford laboratory). It could be argued that the 1.39×10^{-11} yr^{-1} decay constant, used by a number of other laboratories, would increase the Rb–Sr ages by 6% and bring samples BLM–37 and BMM–39 (and to a lesser extent BMM–45) more into line with the K–Ar ages. In that case, however, the other four pairs in table 4 would become correspondingly discordant. It is quite fruitless to speculate along these lines until the ^{87}Rb decay constant has been accurately fixed once and for all. This remains one of the most urgent problems in geochronometry.

Table 3, column 9, gives the ^{87}Sr/^{86}Sr ratio at the time of the last isotopic event which homogenized the strontium isotopes within individual subunits (at least the size of hand specimens) of the rock. (This corresponds to the intercept of the isochron on the ^{87}Sr/^{86}Sr axis in figures 3 and 4.) It is evident that this ratio is very variable, with values in the range ca. 0.720–0.755 for the metasediments. As was pointed out in the introduction, the choice of a fixed initial ratio of about 0.710 is quite unjustified for such rocks and would lead to spuriously high isotopic ages. In Connemara, and elsewhere, this could in some cases lead to quite fortuitous agreement between the calculated 'metamorphic' age and the true (as defined by stratigraphy and time scale) age of metamorphism. Thus, in Connemara, the calculation of Rb–Sr mineral ages, assuming initial ^{87}Sr/^{86}Sr $= 0.710$, will yield variable (depending on the Rb/Sr ratio of the mineral) ages generally in the range of about 450–600 m.y., many of which would normally be interpreted as 'pre-Arenig'. However, as shown above, the use of quantitative methods of interpretation which are based upon the true behaviour of radiogenic strontium during geological events indicates that isotopic ages *may* reflect quite low-grade metamorphic events and/or cooling as a result of uplift, all of which may considerably postdate the main metamorphism and remove all isotopic record of its occurrence. Much the same reasoning applies to K–Ar ages and it is interesting to note that Rickard (1965) has obtained identical K–Ar ages on muscovites which on the basis of detailed petrographical and structural analysis indicate different ages of metamorphism.

The only relatively low initial ^{87}Sr/^{86}Sr (0.7091 ± 0.0041) was obtained from the biotite–whole rock pair of sample BLM–23 (table 2). This is a hornblende–biotite norite from the Lough Wheelaun intrusion. Such rocks are expected to have low initial $^{87}Sr/^{86}Sr$ values in the range 0.703–0.706 (Faure and Hurley, 1963). However, enrichment in radiogenic ^{87}Sr is not proved in this particular case, since the error on the initial ratio is relatively large.

Turning once again to the Corvock granite, a biotite–whole rock measurement (table 2) yields an age of 367 ± 18 m.y., in close agreement with the previously discussed K–Ar age of 376 ± 14 m.y. This age is identical, within experimental error, with the ages of the Newer Granites of Connemara, including the Galway granite (Giletti, Moorbath and Lambert, 1961; Leggo, Compston and Leake, 1966). An initial $^{87}Sr/^{86}Sr$ value of about 0.709 is characteristic for many granites.

C Age of deposition of the Connemara metasediments (Lower and Middle Dalradian)

Rb–Sr isotopic analyses of whole rock sedimentary and metasedimentary material has greatly extended the use of isotopic dating in determining the approximate age of deposition of unfossiliferous sedimentary successions. The dating of sedimentary material is complicated by the inheritance of radiogenic ^{87}Sr in detrital minerals such as micas. The total radiogenic ^{87}Sr content of a sediment can be regarded as consisting of two components—an inherited radiogenic component from the source region, plus an additional radiogenic component formed by decay of ^{87}Rb from the time of deposition to the present day. If the inherited radiogenic ^{87}Sr were completely absent, the parent-to-daughter ratio of ^{87}Sr to ^{87}Rb could be used to evaluate the true age of sedimentary deposition. Such conditions are rarely, if ever, fulfilled, so that ages obtained from single samples of whole rock sediments are normally greater than the true age of deposition.

Compston and Pidgeon (1962) and Whitney and Hurley (1964) concluded that if the amount of inherited radiogenic ^{87}Sr is small (i.e. a young source region), or if the strontium isotopic composition of the sediments was relatively uniform at the time of deposition, whole rock isochrons (i.e. a plot of $^{87}Rb/^{86}Sr$ vs. $^{87}Sr/^{86}Sr$) could yield ages which closely correspond to the true age of deposition. However, if these conditions are *not* fulfilled, or if there is very little variability in the Rb/Sr ratios, an apparent limiting age (t) can still be calculated from a single whole rock analysis using the following equation:

$$t = \frac{1}{\lambda} \log_e \left[1 + \frac{{}^{86}Sr}{{}^{87}Rb} (R_t - R_i) \right] \quad (1)$$

where, λ = decay constant of ^{87}Rb
$R_t = {}^{87}Sr/{}^{86}Sr$ measured today
$R_i = {}^{87}Sr/{}^{86}Sr$ initially

Since R_t and $^{86}Sr/^{87}Rb$ are experimentally determined quantities and λ is known, the only unknown parameter necessary for an age calculation is some estimate of R_i. Although R_i may vary from one sample to another, *maximum* ages can be set for deposition by assuming the lowest possible $^{87}Sr/^{86}Sr$ ratio for R_i (Compston and Pidgeon, 1962; Long and Lambert, 1963). Normally, $^{87}Sr/^{86}Sr$ ratios similar to those of oceanic basalts are substituted for R_i. The inherent assumptions in such an approach do not require the sediments to have had a uniform strontium isotopic composition at time of deposition, but only require that the whole rocks have remained closed chemical systems with respect to both Rb and Sr from the time of deposition to the present day. With regard to the latter point, there is obviously no the guarantee that the primary chemical composition of a sediment is retained during a regional metamorphism. However, detailed geochemical work by Shaw (1954) has indicated that certain regionally metamorphosed sediments have remained closed or nearly closed systems during metamorphism and that their chemistry reflects original sedimentary inhomogeneities. Similar conclusions have been drawn for rocks from the Dalradian Series (Pitcher and Sinha, 1958; Leake, 1958b; Mercy, 1963). It is, in any case, well known that the overall Rb–Sr characteristics of an *igneous* rock may remain unchanged during regional metamorphism, so that the original age of intrusion can be obtained. A good example of this kind is the Carn Chuinneag intrusion in the Moine Series of Scotland (Long, 1964).

Whole rock Rb–Sr isotopic data from eight samples of Connemara Schists are summarized in table 5. Ages calculated by assuming an initial $^{87}Sr/^{86}Sr$ ratio similar to oceanic basalts (i.e. 0.704) are given. It is emphasized that the maximum ages thus obtained do not necessarily bear any relation whatsoever to the stratigraphic position of the analysed sediment, but are only dependent on the physical and chemical processes of sedimentary differentiation and diagenesis, and on the assumed initial $^{87}Sr/^{86}Sr$ ratio. The simple theory indicates that the greater the increase in the Rb/Sr ratio at the time of deposition, the closer will be the approximation of the calculated maximum age to the true age of deposition. Favourably high Rb/Sr ratios may be brought about in a fine-grained sediment at the time of deposition by preferential incorporation of Rb from sea water and fixing as an interlayer cation in illitic and mixed clay structures on the surface of fine-grained particles. The overall decrease of computed maximum ages with increasing Rb/Sr ratios is clearly shown in table 5. With higher Rb/Sr and present-day $^{87}Sr/^{86}Sr$ ratios,

Table 5 Rb–Sr Whole rock maximum ages of Connemara metasediments

(Assumed minimum initial $^{87}Sr/^{86}Sr = 0.704$)

Sample number	$^{87}Rb/^{86}Sr$	Apparent maximum age (m.y.)[a,b]
BMM–45	10.15	640 ± 30
BMM–43	4.52	690 ± 70
BMM–18	4.70	830 ± 60
BLM–31	1.60	1050 ± 170
BMM–9	1.92	1070 ± 150
BMM–39	2.02	1070 ± 140
BLM–37	3.54	1170 ± 90
BMM–12	1.41	1510 ± 200

[a] Rounded off to nearest 10 m.y.
[b] The error includes analytical uncertainties only. Full analytical details are given in table 2.

the actual difference between the *assumed* initial $^{87}Sr/^{86}Sr$ ratio (0.704) and the *true* initial $^{87}Sr/^{86}Sr$ ratio becomes less critical in the age calculation (see equation 1).

From these considerations it follows that the youngest age (i.e. 640 ± 30 m.y. for sample BMM–45) should represent a maximum for the age of deposition of at least a part of the Dalradian Series of Connemara. In view of the small number of measurements and also the uncertainties with regard to the extent of Rb and Sr migration during metamorphism, a tentative maximum age estimate of about 700–750 m.y. for the Connemara Series is suggested. It must be borne in mind that the deposition of such a major sedimentary sequence may well have lasted 100 m.y., if not more. Strictly speaking, a series of Rb–Sr whole rock determinations should be carried out on each stratigraphic horizon and the stratigraphic position of an analysed sample within the sequence should be known. This is not easy in a structurally complex, regionally metamorphosed terrain, such as Connemara. However, the above maximum age estimate probably does not exceed the true age of deposition by more than 10–15% and it is hoped that analyses of further whole rock specimens with relatively high Rb/Sr ratios will give an even closer approximation to the true age of deposition.

Unfortunately the eight samples of table 5 do not fit at all closely to a straight line in an isochron plot of ^{87}Rb/^{86}Sr versus ^{87}Sr/^{86}Sr. There is little point in presenting the actual plot here, but it may be verified by any interested reader who cares to plot these parameters, which are given in tables 3 and 5. The scatter of points thus obtained reflects considerable differences in the amounts of inherited radiogenic ^{87}Sr in each of the individual samples, since all the metasediments are considered to be broadly of the same age. It simply shows that, on a large scale, the strontium isotopic composition of the sediments remained inhomogeneous at the time of deposition and consequently the *true* age of deposition cannot be determined from an isochron plot made up of these eight samples.

Much Rb–Sr whole rock work is currently in progress to define the ages of the principal Precambrian sedimentary sequences of the British Isles within reasonably narrow limits. This late Precambrian age of the Connemara Series is in reasonable agreement with Rb–Sr whole rock isotopic data from the Scottish Dalradian Series (Bell, 1964). Because of the relatively uniform initial strontium isotopic composition of the Dalradian sediments in the central and northerly parts of the main Dalradian belt of the Scottish Highlands, an isochron plot has yielded tentative limits of between about 650 and 750 m.y. for the deposition of these sediments. Long and Lambert (1963) were unable to obtain an isochron plot on six samples of Moinian whole rock metasediments from northern Scotland, but showed that the maximum age of deposition of the sediments was probably about 900–1000 m.y. Unfortunately, the Moinian samples investigated by Long and Lambert had considerably lower Rb/Sr ratios than the Dalradian samples, rendering the age determinations correspondingly uncertain, as pointed out earlier. In contrast, some of the fine-grained unmetamorphosed Torridonian sediments of northwestern Scotland have comparatively high Rb/Sr ratios and yield maximum ages of deposition in the range 800–850 m.y. (Moorbath and Stewart, unpublished work).

The Rb–Sr dating of whole rock sediments and metasediments offers bright prospects for future work. Many more of these still somewhat time-consuming measurements are required, because interpretation of the results is essentially statistical. Already there is strong isotopic evidence for major sedimentation in the British Isles between about 900–650 m.y. ago, possibly comprising the *whole* of the Moinian, Torridonian and Dalradian. Much work will be required, however, before finer correlations between these three major sedimentary sequences become isotopically feasible.

V Concluding Remarks

The primary aim of this work was to date the main metamorphism of the Connemara Schists. As the work progressed, it became evident that this objective was unlikely to be achieved. At present, the writers have no alternative but to accept (a) the very strong, circumstantial geologic evidence that the principal metamorphisms M_1–M_6 of the Connemara Dalradian are pre-Arenig (Dewey, 1961), (b) that the base of the Arenig is 500 ± 15 m.y. old (Holmes, 1960; Kulp, 1961; Harland, Smith and Wilcock, 1964), (c) that the age of the post-metamorphic Oughterard granite is at least 510 ± 35 m.y., as defined by a whole rock Rb–Sr isochron obtained by Leggo, Compston and Leake (1966).

If these points are ever amended or disproved, then the conclusions of this paper may have to be revised. It is of interest to note, for example, that T. G. Miller (1964) has questioned the stratigraphic evidence for the pre-Arenig metamorphism of the Connemara Schists. However, if the above points are accepted, it follows that the observed isotopic ages, none of which exceeds 470 m.y. and most of which are below this value, reflect relatively minor (with regard to temperature) geological events. They demonstrate the ability of such events to influence and determine the diffusion behaviour of radiogenic isotopes. It was suggested earlier in this paper that the concentration of isotopic ages in the range of about 440–460 m.y. may reflect either a widespread uplift and cooling event and/or a retrograde metamorphism with chloritization. A further group of ages is found in the range 370–390 m.y. and clearly reflects a low-grade metamorphic event which affected Ordovician and Silurian sediments in Co. Mayo, and parts of the northernmost Connemara Schists to the south. This is also the age of the Newer Granites in the area, including the Corvock, Galway and Roundstone granites, and others. Clearly, events occurred at this time which had a fringing effect on the Connemara Schists and which were sufficient in some cases to cause partial or complete overprinting of the earlier set of 440–460 m.y. isotopic ages. A summary of the sequence of events in western Ireland is presented in table 6.

There is little doubt that the situation in the Scottish Dalradian closely parallels that in Connemara. The majority of published K–Ar and Rb–Sr mineral ages from the Scottish Dalradian fall in the same general range (ca. 440–460 m.y.) as the main group of isotopic ages reported in the present paper from Connemara (Bell, 1964; Harper, 1964; Miller and Brown, 1965; Brown, York and others, 1965). Nevertheless, there is an increasing amount of evidence to suggest that the principal foldings and metamorphisms in the Scottish Dalradian actually occurred earlier than about 480–490 m.y. ago, almost certainly in pre-Arenig times. Of particular relevance are the whole rock slate K–Ar studies of Harper (1964, 1965) and four biotite K–Ar ages of posttectonic gabbros from Aberdeenshire which fall in the range 460–498 m.y. (Brown, Miller, Grasty and Fraser, 1965). A maximum age limit for the principal metamorphisms and foldings in the Scottish Dalradian is supplied by whole rock Rb–Sr data on pretectonic granites, with an average age of 530 ± 10 m.y. (Bell, 1964). In contrast, no undoubted pretectonic granites have yet been described from western Ireland.

The sequence of events outlined in table 6 may be compared and contrasted with tables prepared by Brown, Miller, Soper and York (1965), summarizing all available K–Ar age work on the Scottish Caledonides up to about the middle of 1964. These authors present the detailed pros and cons for the contrasted 'overprinting' and 'cooling' hypotheses which may be invoked to explain the observed age patterns. The present interpretation of the geochronology of Connemara incorporates features of both hypotheses, although the *major* isotopic event of ca. 440–460 m.y. is regarded as the consequence of cooling, and not of overprinting.

The present work has not yielded any really positive evidence regarding the true age of major folding and metamorphism in Connemara. Nevertheless, it may help to emphasize some of the current uncertainties in interpretative geochronology, and the difficulty in deriving unambiguous geological information from K–Ar and Rb–Sr ages on minerals (particularly micas) from regionally metamorphosed complexes. Such ages nowadays form a substantial proportion of published geochronological data. The situation is more

Table 6 Sequence of events in western Ireland

Isotopic age	Geological age	Sequence of geological events	Geochronometric details
Equal to, or younger than, ca. 700 m.y.	Late Precambrian, possibly ranging into the Cambrian	Deposition of Dalradian sediments in Connemara	Maximum age of deposition obtained from Rb–Sr measurements on whole rock metasediments
Greater than 510 ± 35 m.y.	Pre-Arenig	Foldings F_1–F_4 (and possibly F_5). Metamorphisms M_1–M_6. Basic and ultrabasic intrusions	Isotopic record of these events has largely been erased
510 ± 35 m.y.	Pre-Arenig, probably late Cambrian	Intrusion of Oughterard granite	Rb–Sr whole rock studies by Leggo, Compston and Leake (1966)
ca. 510–ca. 470 m.y.	Cambrian to Ordovician	No further metamorphism (*sensu stricto*), but rocks remain sufficiently deeply buried for continued radiogenic argon and strontium diffusion	Oldest age in this study is a hornblende K–Ar age of 470 ± 20 m.y.
ca. 470–ca. 430 m.y. (with maximum in the range ca. 460–ca. 440 m.y.)	Middle to Upper Ordovician	*Main isotopic event in Connemara.* Dalradian minerals become closed systems for radiogenic argon and in most cases for radiogenic strontium. Probably corresponds broadly to uplift and cooling contemporaneous with influx of coarse, thick sediments into the Mayo trough to the north of the rising Connemara massif. Accompanied by retrogressive metamorphism and chloritization	K–Ar on hornblende and micas Rb–Sr on micas and whole rock

| ca. 390–380 m.y. | Lower Devonian | Low-grade metamorphism of Arenig and Wenlock sediments with production of slates and expulsion of preexisting radiogenic argon in Murrisk, Co. Mayo. Partial overprinting of main Connemara isotopic event, yielding some 'mixed' K–Ar and Rb–Sr ages, particularly in northern Connemara | K–Ar on whole rock slates from Murrisk
K–Ar and Rb–Sr on Dalradian minerals and whole rocks from northern Connemara |
| ca. 400–370 m.y. | Lower Devonian | Intrusion of Newer Granites of Connemara and Murrisk, e.g. Galway, Roundstone, Corvock, etc. | K–Ar on micas. Rb–Sr on micas and whole rocks |

serious when, as in Connemara, the spread of events (and to a lesser extent the experimental error) corresponds to a time range which represents an appreciable fraction of the actual age. This type of situation will hardly arise in the case of ancient Precambrian complexes.

It is hoped that the present work has contributed to a closer appreciation of the potentialities and the limitations of isotopic age work. The geological significance and interpretation of 'isotopic events' provides a worthy topic for future research in geochronology.

Acknowledgements

We thank D. C. Rex, J. Simons and R. Goodwin for many of the potassium and argon determinations, Miss J. Dunsdon for help with mineral separations, and Mrs. H. McArdle for skilled technical assistance. Drs. C. T. Harper and N. J. Snelling provided valuable help and advice.

Appendix

Locations and Descriptions of the Specimens Studied

BMM–8 Andalusite–muscovite pegmatite pod in Connemara Schist. Claggan Quarry, between bench marks 252.5 and 197.4 on 6 in sheet Co. Galway 39.

Randomly arranged pink andalusite crystals averaging $2 \times 2 \times 5$ cm are intergrown with, and partly replaced by $0.5 \times 0.5 \times 0.1$ cm muscovites. Between the andalusites are clots of quartz and stellate aggregates of muscovite and chlorite enclosing abundant apatite, leucoxenized sphene and ilmenite.

BMM–9 Staurolite–garnet–biotite–muscovite–oligoclase schist. 2 yd from BMM–8. Connemara Schist.

A rock consisting of oligoclase, quartz, clouded garnet, staurolite, biotite, muscovite and rare tourmaline and leucoxenized ilmenite. The plagioclase, garnet and staurolite all preserve a distinct internal foliation and schistosity which is earlier than, and oblique to the present schistosity of the rock. A few late biotite flakes overgrow this later schistosity.

BMM–12 Quartz–plagioclase–chloritized biotite–muscovite schist. Roadside exposure of Connemara Schist, 450 yd west of Owenduff Bridge at bench mark 221.8 on 6 in sheet Co. Galway 11. 150 yd from the nearest outcrop of Llandovery conglomerate. The schist is composed of partly kaolinized oligoclase–andesine, quartz, partly chloritized biotite, muscovite and accessory iron ore.

BMM–13 Oligoclase–quartz–biotite–muscovite schist. Connemara Schist. ½ mile SE of Lecknavarna Cottage, south of the track, south of Lough Fee on 6 in sheet Co. Galway 11.

Partly kaolinized oligoclase with quartz, slightly chloritized biotite, muscovite and myrmekitic potash feldspar. Two generations of muscovite are present.

BMM–18 Muscovite–biotite–chlorite–quartz schist. On the shore, in the SE corner of Gortnaling Cove, Lettergesh, on 6 in sheet Co. Galway 10. 50 ft below the Upper Llandovery basal conglomerate in the Connemara Schist.

A few chloritized and sericitized garnets are set in muscovite, chloritized biotite and quartz. Pinitized cordierite and sericitized sillimanite may be present together with accessory apatite, pyrite and iron ore.

BMM–27 Calcareous slate from the Oughty Group of the Wenlock. ½ mile NW of Letterbrock village, on 6 in sheet Co. Mayo 97.

Thin (1 cm) bands of fine-grained pelite composed of chlorite, quartz, muscovite, feldspar and a little calcite alternate with calcite-rich, chlorite, quartz bands. Magnetite octahedra are scattered throughout both bands and there is clear indication of two periods of chlorite growth; one parallel to the muscovite and the slaty cleavage and the other oblique to it, but parallel to the preferred orientation of the calcite.

BMM–30 Arenaceous slate from the Letterbrock Group or possibly at the base of the Derrymore Group, but definitely Arenig. Quarry, west of the road at Letterbrock village, 1½ miles due east of Oughty Hill, 6 in sheet Co. Mayo 97.

Detrital grains of quartz and feldspar are set in a micaceous matrix with small flakes of new biotite, chlorite and muscovite. Some detrital epidote and sphene, partly altered to leucoxene are also present.

BMM–31 Fine pelitic slate band in the Upper Dalradian Westport Grit. Shore, 3¾ mile ENE of Louisburgh on 6 in sheet Co. Mayo 86.

A fine pelitic slate with a few chlorite laminae and thin calcareous bands.

BMM-39 Retrograded sillimanite–biotite–cordierite schist. Connemara Schist. 100 yd west of the Ferry Bridge on the Finny–Clonbur road, on 6 in sheet Co. Mayo 120A.

A high-grade mobilized pelite with a primary assemblage of plagioclase–sillimanite–biotite–cordierite with secondary muscovite replacing sillimanite and biotite, pinite replacing cordierite, chlorite with opaque needles replacing biotite, and some quartz replacing now sericitized plagioclase.

BMM-43 Quartz–plagioclase–muscovite–chlorite schist. Connemara Schist. 220 yd downstream from the stone bridge at Crumlin East, i.e. 180 yd SSE of bench mark 208.8 on 6 in sheet Co. Galway 26.

A strongly sheared schist with quartz, muscovite, altered plagioclase, chlorite and a little iron ore.

BMM-45 Quartz–oligoclase–biotite–muscovite schist. Connemara Schist. Quarry, $1\frac{1}{2}$ miles SW of Cornamona, on the south side of the road to Maam Bridge, 6 in sheet Co. Galway 39.

A very fresh semi-pelite with two generations of biotite and muscovite; one grows parallel to the schistosity, the other across it.

BMM-46 Corvock granite. On road from Cregganbaun to Louisburgh, about $\frac{3}{4}$ mile north of Cregganbaun. 6 in sheet Co. Mayo 96.

Quartz-basic oligoclase–microcline–biotite granite with accessory orthite, sphene, apatite and magnetite. The biotite has been partly chloritized and partly replaced by muscovite, the plagioclase contains randomly arranged flakes of replacing muscovite, epidote grains and kaolinite alteration while the sphene has been partly altered to leucoxene.

BLM-13 Striped amphibolite in the Connemara Schists. 120 yd SSE of Ardbear Bridge, on the Clifden–Ballyconneely road, 6 in sheet Co. Galway 35.

A reticular fabric of fairly fresh hornblendes around sericitized and saussuritized plagioclase with a little ilmenite and sphene. The hornblendes grew in a lineated fashion during F_3, but rare F_2-lineated hornblendes have resisted recrystallization.

BLM-23 Biotitic hornblende norite. 70 yd south of the southern point of Loughanillaun in the northeast portion of 6 in sheet Co. Galway 51. Ultrabasic rock in the Lough Wheelaun intrusion in the Connemara Schists.

The rock consists of basic bytownite, bronzite and a little amphibolized augite with late biotite flakes, a little iron ore and apatite. The plagioclase is considerably saussuritized and sericitized while the bronzite is partly altered to amphibole and partly altered to a chloritic mass. A metamorphic crystallization, producing hornblende, is clearly superimposed on the original igneous crystallization.

BLM–26   Basic bytownite–hornblende–biotite rock from the Lough Wheelaun intrusion into the Connemara Schists. 250 yd NE of the east point of Loughanillaun in the northeast portion of 6 in sheet Co. Galway 51.

Biotite porphyroblasts overgrow a rock composed of basic bytownite and hornblende formed from pyroxenes. Small ($10 \times 10 \times 3$ mm) desilicated pelite xenoliths composed of completely saussuritized plagioclase, magnetite, green spinel, corundum and högbomite are scattered throughout the rock. Late quartz has been introduced. Clearly the rock is a metamorphosed, contaminated ultrabasic rock.

BLM–27   Poikiloblastic hornblende-basic bytownite–orthopyroxene rock of the Lough Wheelaun intrusion into the Connemara Schists. 60 yd ESE of the south point of Lough Wheelaun on 6 in sheet Co. Galway 51.

An original noritic rock with basic bytownite in which hornblende poikiloblasts have replaced plagioclase, augite and bronzite and at a later date biotite porphyroblasts have partly replaced the hornblende. Late quartz has been introduced, the plagioclase is considerably sericitized and saussuritized and clots of iron ore are present. A metamorphosed ultrabasic rock.

BLM–30   Striped amphibolite in the Connemara Schists. 330 yd SW of the Galway–Clifden–Cashel road junction; on the sheet boundary between 6 in sheets Co. Galway 37 and 51.

Exceptionally fresh hornblendes in a perfectly schistose, reticular fabric enclosing slightly sericitized andesine and a little ilmenite. The hornblendes are perfectly lineated parallel to the hinges of F_3 folds.

BLM–31   Plagioclase–quartz–biotite schist. Connemara Schist. West side of the Galway–Clifden road, west of Glendalough, Recess, on 6 in sheet Co. Galway 37.

Kaolinized and sericitized oligoclase–andesine, quartz, partly chloritized biotite schist with accessory apatite and potash feldspar. The mineral growth appears to be para-F_3.

BLM–33 Quartz–plagioclase–microcline vein crossing biotite schist. Connemara Schist. Road-cutting, about 60 yd NE of Recess station on 6 in sheet Co. Galway 37.

A patchy textured quartz, altered oligoclase, microcline rock with epidote, chlorite and a little muscovite.

BLM–34 Amphibolite lens in quartz–andesine–labradorite–hornblende gneiss of the Connemara Migmatites. 240 yd WSW of the north point of Truska Lough, 6 in sheet Co. Galway 49.

A partly chloritized and carbonatized hornblende-saussuritized and -sericitized plagioclase lens with biotite fringing and marginally replacing the hornblende. The lens grades into a quartz-sericitized and -saussuritized andesine–labradorite–hornblende–chlorite gneiss. The whole rock is badly altered by retrogressive changes but the hornblende is of F_2 age.

BLM–35 Hornblende-rich metagabbro in the Connemara Migmatites. 15 yd ENE of BLM–34.

A massive hornblende-rich rock with a little sericitized and saussuritized plagioclase and clinopyroxene relicts inside the hornblende, which is fairly fresh. A little apatite, pyrite, magnetite and interstitial quartz are present and the hornblende is believed to be of F_2 age.

BLM–36 Quartz–oligoclase–muscovite pegmatite in the Connemara Migmatites. 390 yd SW of the north point of Truska Lough, 6 in sheet Co. Galway 49.

A quartz-rich vein with oligoclase and muscovite. The oligoclase is partly replaced by smaller, randomly arranged muscovite flakes.

BLM–37 Muscovite–biotite–quartz–oligoclase schist. Connemara Schist. Roadside quarry on the Clifden–Letterfrack road, immediately east of Streamstown Bay, 300 yd SE of Streamstown school at southern end of 6 in sheet Co. Galway 22.

A perfectly schistose muscovite-rich quartz–oligoclase–microcline very slightly chloritized biotite rock with a little iron ore and apatite. The muscovite and biotite are F_3 lineated.

BLM–39 Muscovite-rich pegmatite crossing the Currywongaun ultrabasic norite intrusion into the Connemara Schists. 420 yd E20°S, from the northern end of Lough Anivan, 6 in sheet Co. Galway 10.

The vein is composed of partly sericitized oligoclase and potash feldspar, quartz and randomly grown muscovite plates with a little tourmaline. The quartz appears to corrode the plagioclase. The vein is post-F_2.

20531 Muscovite from a mica–microcline pegmatite dyke in the Connemara Schists, 1000 yd north of Clifden, Co. Galway (Giletti, Moorbath and Lambert, 1961).

20532 Semi-pelitic Connemara Schist. Roadside quarry $\frac{1}{4}$ mile SSE of Streamstown school, near Clifden, Co. Galway (Giletti, Moorbath and Lambert, 1961).

A fine-grained fresh biotite–muscovite–oligoclase–quartz schist with minor microcline.

References

Aldrich, L. T., Wetherill, G. W., Tilton, G. R. and Davis, G. L. (1956). Half-life of ^{87}Rb. *Phys. Rev.*, **103**, 1045.

Anderson, J. G. C. (1960). The Wenlock strata of South Mayo. *Geol. Mag.*, **97**, 265.

Bell, K. (1964). A geochronological and related isotopic study of the rocks of the Central and Northern Highlands of Scotland. *D. Phil. Thesis.* Oxford Univ.

Brown, P. E., Miller, J. A., Grasty, R. L. and Fraser, W. E. (1965). Potassium–argon ages of some Aberdeenshire granites and gabbros. *Nature*, **207**, 1287.

Brown, P. E., Miller, J. A., Soper, N. J. and York, D. (1965). Potassium–argon age pattern of the British Caledonides. *Proc. Yorkshire Geol. Soc.*, **35**, 103.

Brown, P. E., York, D., Soper, N. J., Miller, J. A., MacIntyre, R. M. and Farrar, E. (1965). Potassium–argon ages of some Dalradian, Moine and related Scottish rocks. *Scot. J. Geol.*, **1**, 144.

Compston, W., Jeffery, P. M. and Riley, G. H. (1960). Age of emplacement of granites. *Nature*, **186**, 702.

Compston, W. and Pidgeon, R. T. (1962). Rubidium–strontium dating of shales by the total-rock method. *J. Geophys. Res.*, **67**, 3493.

Cruse, M. A. J. B. (1963). The geology of Renvyle, Inishbofin and Inishshark, north-west Connemara, Co. Galway, Eire. *Ph.D. Thesis.* Univ. Bristol.

Dewey, J. F. (1961). A note concerning the age of metamorphism of the Dalradian rocks of Western Ireland. *Geol. Mag.*, **98**, 399.

Dewey, J. F. (1963). The Lower Palaeozoic stratigraphy of central Murrisk, County Mayo, Ireland, and the evolution of the South Mayo trough. *Quart. J. Geol. Soc. London*, **119**, 313.

Dewey, J. F. and McManus, J. (1964). Superposed folding in the Silurian rocks of Co. Mayo, Eire. *Geol. J.*, **4**, 61.

Dewey, J. F. and Phillips, W. E. A. (1963). A tectonic profile across the Caledonides of South Mayo. *Liverpool Manchester Geol. J.*, **3**, 237.

Dodson, M. H. (1963). Isotopic ages from southwest England. *D.Phil. Thesis.* Oxford Univ.

Edmunds, W. M. and Thomas, P. R. (1966). The stratigraphy and structure of the Dalradian rocks north of Recess Connemara, Co. Galway. *Proc. Roy. Irish Acad., Sect. B*, **64**, 527.

Evans, B. W. (1964). Fractionation of elements in the pelitic hornfelses of the Cashel–Lough Wheelaun intrusion, Connemara, Eire. *Geochim. Cosmochim. Acta*, **28**, 127.

Evernden, J. F. and Richards, J. R. (1962). Potassium–argon ages in eastern Australia. *J. Geol. Soc. Australia*, **9**, 1.

Fairbairn, H. W., Hurley, P. M. and Pinson, W. H. (1961). The relation of discordant Rb–Sr mineral and whole rock ages in an igneous rock to its time of crystallisation and to the time of subsequent $^{87}Sr/^{86}Sr$ metamorphism. *Geochim. Cosmochim. Acta*, **23**, 135.

Faure, G. and Hurley, P. M. (1963). The isotopic composition of strontium in oceanic and continental basalts: application to the origin of igneous rocks. *J. Petrol.*, **4**, 31.

Fitch, J. F., Miller, J. A. and Brown, P. E. (1964). The age of the Caledonian orogeny and metamorphism in Britain. *Nature*, **203**, 275.

Flynn, K. F. and Glendenin, L. E. (1959). Half-life and beta spectrum of ^{87}Rb. *Phys. Rev.*, **116**, 744.

Giletti, B. J., Moorbath, S. and Lambert, R. St. J. (1961). A geochronological study of the metamorphic complexes of the Scottish Highlands. *Quart. J. Geol. Soc. London*, **117**, 233.

Grimaldi, F. S. (1960). Dilution–addition method for flame spectrophotometry. *U.S. Geol. Surv., Profess. Papers*, **400-B**, 494.

Harland, W. B., Smith, A. G. and Wilcock, B. (1964). The Phanerozoic time-scale. *Quart. J. Geol. Soc. London, Symp.*, **120S**, 458.

Harper, C. T. (1964). Potassium–argon ages of slates and their geological significance. *Nature*, **203**, 468.

Harper, C. T. (1965). A potassium–argon isotopic age study of the British Caledonides. *D. Phil. Thesis.* Oxford Univ.

Harris, P. M., Farrar, E., MacIntyre, R. M., York, D. and Miller, J. A. (1965). Potassium–argon age measurements on two igneous rocks from the Ordovician system of Scotland. *Nature*, **205**, 352.

Hart, S. R. (1964). The petrology and isotopic-mineral age relations of a contact zone in the Front Range, Colorado. *J. Geol.*, **72**, 493.

Hedge, C. E. and Walthall, F. G. (1963). Radiogenic strontium-87 as an index of geologic processes. *Science*, **140**, 1214.

Holmes, A. (1960). A revised geological time-scale. *Trans. Edinburgh Geol. Soc.*, **17**, 183.

Hurley, P. M., Hughes, H., Pinson, W. H. and Fairbairn, H. W. (1962). Radiogenic argon and strontium diffusion parameters in biotite at low temperatures obtained from Alpine fault uplift in New Zealand. *Geochim. Cosmochim. Acta*, **26**, 67.

Ingold, L. M. (1937). The geology of the Currywongaun–Doughruagh area, Co. Galway. *Proc. Roy. Irish Acad., Sect. B*, **43**, 135.

Johnson, M. R. W. and Harris, A. L. (1965). Is the Tay Nappe post-Arenig? *Scot. J. Geol.*, **1**, 217.

Kilburn, C., Pitcher, W. S. and Shackleton, R. M. (1965). The stratigraphy and origin of the Portaskaig Boulder Bed Series (Dalradian). *Geol. J.*, **4**, 343.

Kilroe, J. R. (1907). The Silurian and metamorphic rocks of Mayo and North Galway. *Proc. Roy. IrishAcad., Sect. B*, **26**, 129.

Knill, J. L. (1963). A Sedimentary History of the Dalradian Series. In *The British Caledonides*. Oliver and Boyd, Edinburgh. p. 99.

Kulp, J. L. (1961). Geological time scale. *Science*, **133**, 1105.

Kulp, J. L. and Engels, J. (1963). Discordances in K–Ar and Rb–Sr isotopic ages. In *Radioactive Dating*. Intern. Atomic Energy Agency, Vienna. p. 219.

Lambert, R. St. J. L. (1964). The relationship between radiometric ages obtained from plutonic complexes and stratigraphical time. *Quart. J. Geol. Soc. London*, **120S**, 43.

Leake, B. E. (1958a). The Cashel–Lough Wheelaun Intrusion, Co. Galway. *Proc. Roy. Irish Acad., Sect. B*, **59**, 155.

Leake, B. E. (1958b). Composition of pelites from Connemara, Co. Galway, Ireland. *Geol. Mag.*, **95**, 281.

Leake, B. E. (1964). New Light on the Dawros Peridotite, Connemara, Ireland. *Geol. Mag.*, **101**, 63.

Leake, B. E. and Skirrow, G. (1960). The pelitic hornfelses of the Cashel–Lough Wheelaun intrusion, County Galway, Eire. *J. Geol.*, **68**, 23.

Leggo, P., Compston, W. and Leake, B. E. (1966). The geochronology of the Connemara granites and its bearing on the antiquity of the Dalradian Series. *Quart. J. Geol. Soc. London*, **122**, 91.

Long, L. E. (1964). Rb–Sr chronology of the Carn Chuinneag intrusion, Rossshire, Scotland. *J. Geophys. Res.*, **69**, 1589.

Long, L. E. and Lambert, R. St. J. (1963). Rb–Sr isotopic ages from the Moine Series. In *The British Caledonides*. Oliver and Boyd, Edinburgh. p. 217.

McKerrow, W. S. and Campbell, C. J. (1960). The stratigraphy and structure of the Lower Palaeozoic rocks of north-west Galway. *Sci. Proc. Roy. Dublin Soc., Sect. A*, **1**, 27.

Mercy, E. L. P. (1963). The geochemistry of some Caledonian granitic and metasedimentary rocks. In *The British Caledonides*. Oliver and Boyd, Edinburgh. p. 189.

Miller, J. A. and Brown, P. E. (1965). Potassium–argon age studies in Scotland. *Geol. Mag.*, **102**, 106.

Miller, T. G. (1964). Age of Caledonian orogeny and metamorphism in Britain. *Nature*, **204**, 358.

Moorbath, S. (1962). Lead isotope abundance studies on mineral occurrences in the British Isles and their geological significance. *Phil. Trans. Roy. Soc. London, Ser. A*, **254**, 295.

Moorbath, S. and Bell, J. D. (1965). Strontium isotope abundance studies and Rb–Sr age determinations on Tertiary igneous rocks from the Isle of Skye, northwest Scotland. *J. Petrol.*, **6**, 37.

Nicolaysen, L. O. (1961). Graphic interpretation of discordant age measurements on metamorphic rocks. In Kulp, J. L. (Ed.), Geochronology of Rock Systems. *Ann. N.Y. Acad. Sci.*, **91**, 198.

Pitcher, W. S. and Sinha, R. C. (1958). The petrochemistry of the Ardara aureole. *Quart. J. Geol. Soc. London*, **113**, 393.

Rickard, M. J. (1965). Taconic orogeny in the Western Appalachians: Experimental application of microtextural studies to isotopic dating. *Bull. Geol. Soc. Am.*, **76**, 523.

Rothstein, A. T. V. (1957). The Dawros Peridotite, Connemara, Eire. *Quart. J. Geol. Soc. London*, **113**, 1.

Shackleton, R. M. (1958). Downward-facing structures of the Highland Border. *Quart. J. Geol. Soc. London*, **113**, 361.

Shaw, D. M. (1954). Trace elements in pelitic rocks. *Bull. Geol. Soc. Am.*, **65**, 1151.

Stanton, W. I. (1960). The Lower Palaeozoic rocks of south-west Murrisk, Ireland. *Quart. J. Geol. Soc. London*, **116**, 269.

Sutton, J. (1963). Some events in the evolution of the Caledonides. In *The British Caledonides*. Oliver and Boyd, Edinburgh. p. 249.

Wager, L. R. (1932). The geology of the Roundstone district, Co. Galway. *Proc. Roy. Irish Acad., Sect. B*, **41**, 46.

Whitney, P. R. and Hurley, P. M. (1964). The problem of inherited radiogenic strontium in sedimentary age determinations. *Geochim. Cosmochim. Acta*, **28**, 425.

Radiometric dating and the pre-Silurian geology of Africa

T. N. CLIFFORD*

I *Introduction*, 299. II *Radiometric ages in the 400–700 m.y. range*, 306. III *Radiometric ages in the 800–1300 m.y. range*, 337. IV *Radiometric ages in the 1300–2100 m.y. range*, 347. V *Radiometric ages greater than 2100 m.y.*, 370. VI *Conclusions*, 392. *References*, 409.

I Introduction

The geology of Africa can be broadly subdivided into: the Precambrian and early Palaeozoic history for which diagnostic fossils are not available; and the Phanerozoic history datable by fossil means (figure 1). Although radiometric dating has an important role to play in fixing precise boundaries within the Phanerozoic, by far its greatest contribution lies in unravelling the complexities of the Precambrian. For this reason, the discussions which follow will be oriented towards an analysis of the Precambrian-to-early Palaeozoic history of the African continent.

From a geological approach, work by government geological surveys, mining companies, other institutions and private individuals has provided a basic understanding of the stratigraphy and structure of large parts of the continent. Summary works by Krenkel (1925, 1957), Du Toit (1954), Furon (1963), Haughton (1963) and others, clearly testify to the success of these programmes which are a prerequisite for the application of more modern geophysical and geochemical studies. However, the vastness of Africa, the complexity of

* Research Institute of African Geology, University of Leeds, England.

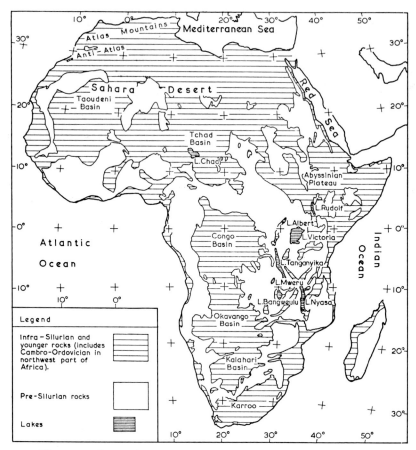

Figure 1 Generalized map showing the pre-Silurian 'basement' in relation to the cover rocks of Africa.

Precambrian geology, and the obscuring effect of the great interior spreads of younger sediment (e.g. the Congo and Kalahari Basins), have led to innumerable correlations over long distances, based purely on similarity of lithology, degree and style of deformation, metamorphism, etc. In recent years this approach to structural and stratigraphic correlation has been supplemented by the application of quantitative age measurements. Perhaps the greatest pioneer effort in this field was the work of Holmes (1951) who, on the basis of the limited number of age measurements then available, presented a 'Provisional map of Pre-Cambrian orogenic belts in South and Central Africa' (figure 2). Even after the intensive work of the past 15 years, a surprisingly high proportion of Holmes' map is, in essence, correct; for this reason, this chapter is dedicated to his memory.

Subsequently Holmes and Cahen (1955, 1957) presented invaluable

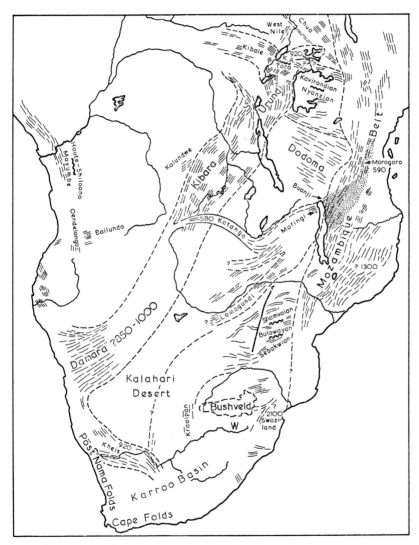

Figure 2 Holmes' (1951) provisional map of Precambrian orogenic belts in South and Central Africa. The figures represent the ages of radioactive minerals in millions of years.
K.A. = Karagwe Ankolean; W = Witwatersrand.

lists of African radiometric ages, and discussed their reliability and significance. These earlier works resulted in the recognition of ten geological cycles in Precambrian–Palaeozoic times as follows: 3200–3400 m.y. cycle (unnamed); 2800–3100 m.y. cycle (unnamed); 2650 m.y. cycle (Shamvaian); 2000 m.y. cycle (Limpopo); 1040 m.y. cycle (Karagwe Ankolean); 875–900 m.y. cycle (unnamed); 630 m.y. cycle (Katanguian); 485 m.y. cycle (L. Cambrian); 400 m.y. cycle (Ordovician); and 230–255 m.y. cycle (Hercynian). Many of these earlier geochronological observations were utilized by Furon and others (1959) in the preparation of the 'Esquisse structurale provisoire de l'Afrique'.

Most recently, two important review papers on the geochronology of parts of Africa have been published: (1) on middle and northern Africa, by Cahen (1961); and (2) on southern Africa, by Nicolaysen (1962).* Together these papers present clear expositions of the advances made in the understanding of African stratigraphy and structure through the use of radiometric dating techniques.

In the discussion which follows, these earlier works are supplemented and amplified in the light of most recent determinations and hypotheses. Because of the complexity in the interpretation of common lead ages, only Rb–Sr†, K–Ar and U–Th–Pb ages are considered, except where common lead ages in a critical interval of African history represent the only available data. At the moment some 800–900 such ages are available for the 'African Shield'. In addition, a considerable number of ages for Middle Silurian and younger events have been published, particularly for igneous episodes.

When the Precambrian-to-early Palaeozoic radiometric ages are considered in histogram form (figure 3), the frequency of the major age patterns emerges as follows: (I) 400–700 m.y.; (II) 850–1050 m.y.; and (III) 1600–2050 m.y. That these peaks *reflect* orogenesis is now well known, but it is also clear that these patterns include ages which resulted from a wide variety of 'thermal causes', and with a 'time-

---

* Since this work was presented for publication, a valuable study of the 'Geochronology of Equatorial Africa' has been presented by L. Cahen and N. J. Snelling (1966) (North-Holland, Amsterdam).

† All Rb–Sr ages are recalculated using the ^{87}Rb decay constant of 1.47×10^{-11} yr^{-1}; this constant is chosen because in recent years, with the exception of South African laboratories, almost all geochronologists working on African material have quoted their ages based on this constant. The decay constant of 1.39×10^{-11} yr^{-1} utilized by South African geochronologists yields an age approximately 6% higher.

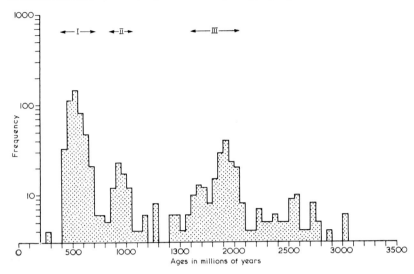

Figure 3 Histogram of African K–Ar, Rb–Sr and U–Th–Pb ages older than 200 m.y., showing the principal peaks of concentration (I, II and III).

lag' between the cause and the recorded age in many instances. Nevertheless, the age concentrations and other data reviewed herein will provide a sequence of generalized bench marks in the subdivision of the Precambrian (see table 1).

There are, of course, a number of drawbacks to the use of histograms to demonstrate thermal events, notably: that significant earlier events may not be represented due to eradication by later events; and that undue emphasis may be given to results of detailed studies of only local significance. Although the peaks provide a framework for discussion, therefore, to avoid the greatest of these drawbacks, African ages and their significance are discussed here under four headings: ages in the 400–700 m.y. range; ages in 800–1300 m.y. range; ages in the 1300–2100 m.y. range; and ages greater than 2100 m.y.

From conisderations of geology, thermal events within these age ranges may represent either orogenic (pre-, syn- and postorogenic) or anorogenic effects. Although in Africa it is not always possible to draw precise time boundaries between orogenic and anorogenic effects, both types of process reflect crustal (or mantle) instability. The role of radiometric dating lies in the determination of thermal anomalies which have accompanied the effects of this instability with its attendant deformation, igneous activity and mineralization.

Despite statements to the contrary (Brock, 1956, p. 153) there is no reason to suppose that orogenic deformation and mountain building processes which affected Africa during the latter half of the Precambrian and the early part of the Phanerozoic (500–2000 m.y. ago) differed significantly from the processes of orogenesis in the type orogen of the Appalachians and other similar Phanerozoic orogenic zones. Indeed, the interpretation of an 'ancient orogenic belt' given by Handley (1956) and modified in figure 4, although oversimplified, embodies most of the accepted fundamental aspects of orogenic belts of greater modernity. Figure 4 also serves to illustrate the general differences in the manner of reaction of (1) 'cover' rocks and (2)

Table 1 Pre-Silurian bench marks of Africa, and two types of subdivision (I and II) of the Precambrian

Dated events	Stratigraphic subdivision	
	I	II
450–680 m.y.	—Includes Damaran and Katangan episodes	
	LOWER PALAEOZOIC AND UPPER PROTEROZOIC	LOWER PALAEOZOIC AND UPPER PRECAMBRIAN
1100 ± 200 m.y.	—Includes Kibaran episode	
	MIDDLE PROTEROZOIC	MIDDLE PRECAMBRIAN
1850 ± 250 m.y.	—Includes Eburnian episode	
	LOWER PROTEROZOIC	LOWER PRECAMBRIAN
ca. 2600 m.y.	—Includes Shamvaian episode	
	ARCHAEAN	
ca. 3000 m.y.		
	KATARCHAEAN	

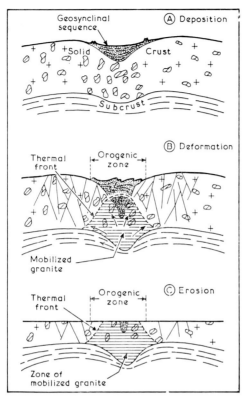

Figure 4 Idealized sections of a fold belt (modified after Handley, 1956) showing: A. the depositional phase of the mobile zone; B. the deformed geosynclinal volcanic–sedimentary pile (geosynclinal facet of orogenesis); and C. the imprint of tectonothermal effects on the older basement floor (vestigial facet of orogenesis) [see text].

'basement floor', during orogenesis; depth zoning, particularly depth-controlled changes in P_{H_2O}, marginal thrusting and the involvement of remnants of more than one previous orogenic belt in the 'basement floor' rocks of the geosyncline, are some additional complications which must be borne in mind in the interpretation of orogenic belts and their thermal imprints. The significance of these factors, and the special position of Africa, is clear when it is realized that the continent was not only subjected to lengthy periods of Precambrian and Phanerozoic erosion, but that it has also remained largely continental since Lower Palaeozoic times; as a result, a variety of levels in Precambrian-to-early Palaeozoic orogenic zones are exposed, and play an important role in efforts to understand the varied aspects of orogenic expression. In this regard, figure 4 focuses attention on the fact that orogenesis may be recorded in one of two ways: (a) a *geosynclinal facet* of deformed cover; and (b) a *vestigial facet* (vestigeosyncline) of rejuvenated floor rocks (basement with respect to the geosynclinal sediments) from which the cover rocks have been locally, or regionally, removed or on which, for some reason, geosynclinal sediments were never deposited (see figure 4). In addition, a third facet must also be recognized: (c) *rejuvenated facet*: (a) or (b) subsequently affected by a more youthful orogenic event.

In contrast to the extensive orogenic events of Africa, anorogenic events occur on a variety of scales. In general, although anorogenic events may be related in time to some important orogenic process, they are not related to it in space; or, alternatively, if related in space to some previous orogeny, they are separated from that orogeny by a significant time gap.

It is clear, therefore, that the thermal imprints of structural instability of either orogenic or anorogenic type are diverse and caution must be exercised in interpreting ages, particularly where they relate to poorly documented processes.

II Radiometric Ages in the 400–700 m.y. Range

A General statement

By far the greatest concentration of ages in Africa lie in the broad range of 400–700 m.y. (see figure 3); these are mainly recorded in zones of late Precambrian-to-early Palaeozoic orogenesis; the most significant concentration lies within the more limited interval of 450–650 m.y. (see figure 5). When the K–Ar, Rb–Sr and U–Th–Pb ages from this interval are considered separately it is seen (figure 6) that: (i) the majority of K–Ar ages lie within the 450–570 m.y. range, with a minor peak at 420 ± 10 m.y., and a mild plateau at 590–670 m.y.; (ii) the histogram based on Rb–Sr ages shows a distinct positive skewness covering a range from 430 m.y. to 650 m.y., with a peak at 500 m.y.; and (iii) U–Th–Pb ages show concentrations at ca. 520 m.y. and 630 ± 20 m.y.

The older portion of this age range (ca. 600 m.y.) has been interpreted as reflecting the Katangan (Kundelunguan, or Lufilian) orogeny in Katanga Province, and the West Congolian (or Post-Schisto-Gréseux) orogeny of the Lower Congo (Cahen, 1963; Cahen

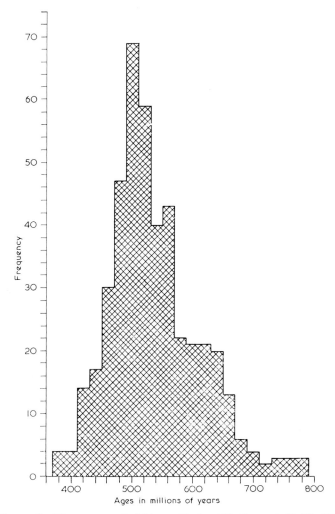

Figure 5 Histogram of African K–Ar, Rb–Sr and U–Th–Pb ages in the 400–800 m.y. range.

and others, 1963). The younger portion of the range (ca. 500 m.y.) has been regarded as reflecting orogenesis in Malagasy Republic (Emberger, 1958); in southern Africa (Clifford, 1963); in Nigeria (Jacobson and others, 1964); and in West Africa (Kennedy, 1964).

In southern Africa it has been suggested that the term *Katangan episode* be reserved for orogenesis reflected by ages in the 580–680

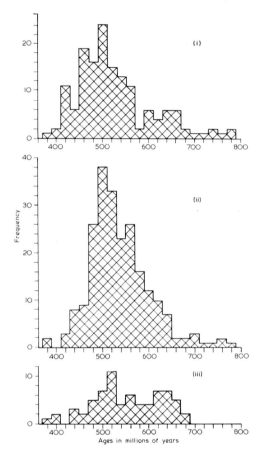

Figure 6 Histograms of African ages in the 400–800 m.y. range: (i) K–Ar ages; (ii) Rb–Sr ages; and (iii) U–Th–Pb ages.

m.y. range, whereas the term *Damaran episode* has been utilized for orogenesis reflected by ages in the 450–580 m.y. range (Clifford and others, 1962; Clifford, 1967); more recently, ages of ±500 m.y. have been cumulatively referred to as the Pan-African orogeny (Kennedy, 1965).

The distribution of 400–700 m.y. ages in Africa is shown in figure 7, the most important feature of which is that when these geochronological data are combined with geological data, a clear pattern of zones of late Precambrian–early Palaeozoic crustal instability is evident. This pattern of mobile orogenic zones and stable cratons

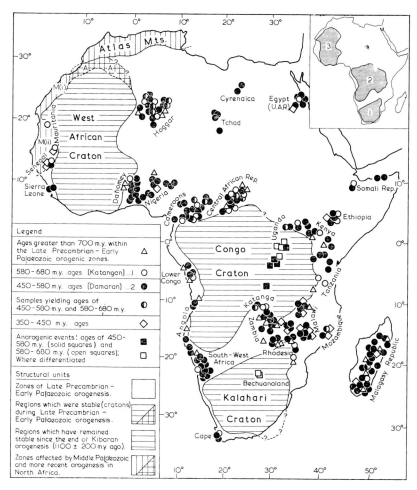

Figure 7 Distribution of 450–680 m.y. ages shown in relation to the late Precambrian–early Palaeozoic structural units, and other selected elements of African structure.
A–A. Anti-Atlas; M(i). Mauritanide zone (northern sector); and M(ii). Mauritanide zone (southern sector).

Inset: a simplified map of the late Precambrian–early Palaeozoic structural units, showing the stable blocks (1. Kalahari Craton; 2. Congo Craton; 3. West African Craton) and the orogenic zones (including the Mozambique Belt) (see also Kennedy, 1964).

has been discussed elsewhere (Clifford, 1963, 1967; Kennedy, 1964); suffice it, therefore, to say that these zones now consist of: (a) a *geosynclinal facet* represented by orogenically deformed Upper Precambrian and older rocks in Central Africa, South-West Africa, the Cape Province, the Lower Congo, Egypt, Ethiopia (?) and the Somali Republic (?); and (b) a *vestigial facet* represented by belts of rejuvenated basement rocks in Central and East Africa (the Mozambique and Zambesi Belts), Malagasy Republic, the Cameroons, the Central African Republic, Nigeria, the Hoggar of the Sahara and elsewhere. In addition, the *rejuvenated facet* of this orogenic activity is preserved along the southern tip of Africa within the late Palaeozoic-to-early Mesozoic Cape Fold Belt, and probably in West Africa within the Mauritanide zone (see figure 7).

The similarity of the age patterns of these three facets, together with their obvious geographical and structural continuity, indicate that these zones represent different structural expressions of the same major orogenic zone, now represented by different structural levels through a sinuous mountain chain. From the point of view of description, the ages reflecting orogenic effects within this chain are most conveniently described under the following regional headings: (a) Central Africa, South-West Africa, South Africa and the Lower Congo; (b) the Mozambique and Zambesi Belts; (c) Ethiopia, Somali Republic and northeastern Africa; (d) Cameroons and Central African Republic; (e) Malagasy Republic; and (f) West Africa.

In contrast to the extensive 400–700 m.y. ages reflecting *orogenic effects*, it is clear from figure 7 that ages reflecting truly *anorogenic effects* are uncommon.

B *Orogenic effects*

Central Africa, South-West Africa, South Africa and the Lower Congo region Western Zambia, southern Katanga and the northwestern part of Rhodesia (Central Africa), South-West Africa, western and southern South Africa and the western borders of Africa to the north and south of the Lower Congo, all merit joint description in that they contain the largest development of Upper Precambrian rocks in Africa. Therefore these regions offer an opportunity to consider age measurements in relation to tectonic processes (folding, regional metamorphism, granite emplacement, etc.) in geosynclinal zones of late Precambrian–early Palaeozoic orogenesis.

The distribution of deformed and cratonic rocks of probable Upper Precambrian age in southern Africa is shown in figure 8, together with certain problematical rocks whose stratigraphic position is still in doubt. Of these sequences, the Katanga System of Central Africa represents the most accurately dated succession since it is younger than the Kibaran deformation (ca. 1100 ± 200 m.y.) and older than 500–650 m.y.

In *Katanga Province*, these rocks pass from an undeformed sequence in the north to highly folded rocks in the south (figure 8) (see Cahen, 1954). Furthermore, southwards into Zambia, towards the Zambesi Valley, the rocks of this succession become progressively regionally metamorphosed (Macgregor, 1951). A mirror image of this thermal increase is seen in the Lomagundi–Piriwiri–Deweras sequence of Rhodesia, where regional metamorphism increased broadly from south to north.

Recently, Cahen and others (1961) have reviewed available age data from Katanga, and recognize two periods of uranium mineralization in the rocks of the Katanga System as follows: (1) mineralization at 620 ± 20 m.y. (at Shinkolobwe, Kalongwe and Luishya); and (2) mineralization at 520 ± 20 m.y. (at Kolwezi and Kamoto). From these data, Cahen (1963, p. B189) concludes that the principal phase of Katangan orogenic deformation '. . . est liée, ou un peu antérieure, à la date de 620 ± 20 m.a.'

In *Zambia*, to the south, there is little evidence of the 620 m.y. event (see figure 7). In the Zambian Copperbelt, uraninite and brannerite ages of 522 ± 15 m.y. and 503 ± 15 m.y.* have been obtained (Darnley and others, 1961; Cahen and others, 1961). Within this region of low-grade regional metamorphism, K–Ar and Rb–Sr ages of mica and actinolite from the Katanga metasediments and the underlying basement demonstrate a uniform thermal imprint reflected by ages in the 455–530 m.y. range (Snelling and others, 1964b; Snelling, 1965). Moreover, although a number of more ancient mineral ages are preserved in the pre-Katanga System basement, the pre-Katanga granite at Nchanga has yielded data which indicate homogenization of its whole rock strontium isotope composition 570 m.y. ago (Snelling and others, 1964b); the tectonic significance of this effect is, however, still uncertain.

To the south, in the higher grade portion of the Katanga System

---

* Secondary pitchblendes from Luanshya and Nkana have also given ages of 365 ± 40 m.y. and 235 ± 35 m.y. respectively (Cahen and others, 1961).

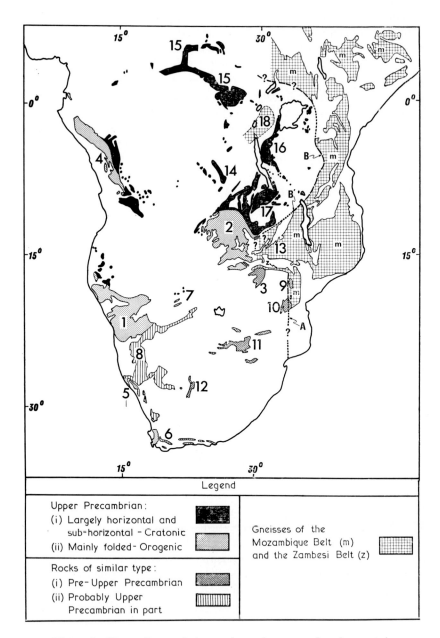

Figure 8 Upper Precambrian rocks and some rocks of uncertain age in southern Africa.

in Zambia, two granitic bodies have also been dated; they are the Hook Batholith and the Lusaka granite. Both have yielded Rb–Sr isochron ages of 500 ± 20 m.y. based on whole rock and biotite samples, whereas the constituent biotites for these two granites have given biotite K–Ar ages of 465 m.y. and 540 m.y. respectively (Snelling and others, 1964b). Since the emplacement of these two bodies is intimately related to the orogenesis of the Katanga System, these dates give a measure of the time of deformation (Phillips, 1959; De Swardt and others, 1964; Simpson and others, 1963). Similar ages have been obtained for minerals from post-Katanga System pegmatites, though two K–Ar pegmatite muscovite ages fall within the somewhat lower range of 420–430 m.y. (Snelling, 1965).

In northwestern *Rhodesia* a similar picture is emerging (figure 7). The Lomagundi–Piriwiri–Deweras sequence (figure 8) shows a broad

---

1. Otavi and Damara Facies of the Outjo System of South-West Africa.
2. Katanga System of Katanga Province, northeastern Angola and Zambia.
3. Lomagundi–Piriwiri–Deweras sequence of Rhodesia.
4. Western Congo System of the Lower Congo.
5. Gariep System of the Sperrgebiet and the Richtersveld.
6. Malmesbury Beds of the Cape Province.
7. Tsodilo quartzites, Kihabe dolomite and the Aha carbonates of northwestern Botswana.
8. Nama–Tsumis–Auborus sequence of South-West Africa and the Ghanzi Beds of Botswana.
9. Frontier–Gairezi System of eastern Rhodesia and western Mozambique.
10. Umkondo System of southeastern Rhodesia.
11. Waterberg–Loskop Systems of South Africa and eastern Botswana.
12. Matsap System of South Africa and southern Botswana.
13. The Zambia extension of the Mozambique Belt.
14. Bushimaie System of the Republic of the Congo.
15. Lindi and Ubangi Groups, the Basse–Kotto System and the Bobassa Series of the Republic of the Congo and the Central African Republic.
16. Bukoban System of East Africa.
17. Plateau Series of Central Africa.
18. Thermal 'node' of Damaran–Katangan activity in Rwanda and southwestern Uganda.

A. a portion of the western boundary of the Mozambique Belt (after Holmes, 1951); B. the western boundary of the Mozambique Belt in East Africa; and B'. the probable western boundary of the Mozambique Belt in Zambia.

northward increase in regional metamorphism, culminating in a pegmatite–granite province of high-grade metamorphism around Miami (Wiles, 1961). Within this high-grade region, muscovite and biotite from late orogenic pegmatites have yielded K–Ar ages of 470–530 m.y. and one Rb–Sr age of 433 m.y.; less satisfactory U–Th–Pb ages of 400–440 m.y. have also been obtained on betafite and columbo-tantalite from pegmatites (Wiles, 1961; Clifford, 1963; Clifford, Rex and Snelling, 1967; Holmes and Cahen, 1957). Moreover, biotite from one of the granites at Miami has given Rb–Sr and K–Ar ages of 500 m.y. and 480 m.y. respectively, whereas the whole rock and feldspar ages from the same sample indicate that the granite is, in fact, very much older (Clifford, Rex and Snelling, 1967).

To the west, across the Kalahari, another extensive sequence of supposed Upper Precambrian in *South-West Africa* (see figure 8) presents a similar age picture to that in Central Africa (see figure 7). In essence, the Upper Precambrian consists of two correlative facies, the Damara and Otavi Facies, both of which are strongly folded. These differ in that the Otavi Facies is largely an unmetamorphosed miogeosynclinal sequence, whereas the Damara Facies is a regionally metamorphosed eugeosynclinal sequence; in the latter, the regional metamorphism culminates in a high-grade central zone profusely intruded by granite and pegmatite.

Published radiometric ages from the slightly metamorphosed portion of the regime of the Otavi Facies include Rb–Sr ages of 526 ± 30 m.y. yielded by biotite from pre-Otavi granite which was affected by the low-grade (biotite–chlorite) effects of regional metamorphism (Clifford and others, 1962); and K–Ar ages of 486 ± 10 m.y. and 861 ± 10 m.y. for muscovites from meta-limestone and 'granitoid' rock respectively, and 340 m.y. for biotite from dolomite, all from the Otavi Facies (Clifford, Grasty and Miller, unpublished work).

Within the Damara Facies published ages include: Rb–Sr and U–Th–Pb ages of 510 ± 60 m.y. for pegmatite and granite minerals (Nicolaysen and Burger, in Clifford and others, 1962); seven K–Ar ages of 450–550 m.y. for biotites from metasediments from a wide variety of regional metamorphic grades (Clifford, 1963); U–Th–Pb ages of monazite and sphene of 500 ± 20 m.y. and 600 ± 25 m.y. respectively from the pegmatite zone (Burger and others, 1965); and K–Ar ages of 454 and 470 m.y. from granitic and dioritic plutons of the orogenic granite suite (Clifford, Grasty and Miller, unpublished work).

In South-West Africa, therefore, there is a consistent age pattern of 450–570 m.y. for a wide variety of minerals of a wide variety of geneses from this orogenically deformed Upper Precambrian terrain. It was on the basis of ages within this range that the *Damaran orogenic episode* was defined (Clifford and others, 1962).

Southwards from South-West Africa, the Damaran orogenic zone is considered to be represented along the coast by the folded Gariep System (figure 8) around the mouth of the Orange River, and by the folded Malmesbury System in the Cape Province of *South Africa*, both of which have been correlated with the Damara Facies. Although no ages are available from Gariep terrain, a series of mineral and whole rock ages are available for the Cape granite, one of a suite of plutons intruded into the Malmesbury sediments, reputedly during the waning stages of Malmesbury orogenesis (Scholtz, 1946). Published age data for the Cape granite are as follows: zircon, 530 ± 40 m.y.; biotite (K–Ar), 505 ± 25 m.y.; biotite and feldspar (Rb–Sr), 485–520 m.y.; and whole rock–feldspar (Rb–Sr isochron age), 520 ± 8 m.y. (Aldrich and others, 1958; Allsopp and Kolbe, 1965). In addition, Allsopp and Kolbe (1965) carried out Rb–Sr isochron studies on samples of sediment from the Malmesbury System and concluded that the age of the system cannot greatly exceed 560 ± 45 m.y.

In addition to this southern extension of the Damaran orogenic belt, a branch of the folded Damara and Otavi Facies trends northwards from South-West Africa, into *Angola* (figures 7 and 8). Although folded representatives of the Upper Precambrian are rare in the southern part of that territory, Mendes (1964) obtained Rb–Sr ages in the range 490 ± 8 m.y. to 513 ± 59 m.y. for biotite from gneisses and granites along the coastal region; in addition, certain Rb–Sr biotite ages in the 900–1400 m.y. range demonstrate the antiquity of this terrain and indicate that, in the western part of Angola, the 500 m.y. event is preserved in 'rejuvenated' pre-Upper Precambrian rocks (Mendes, 1964).

To the north, in the *Lower Congo* (figure 8), the correlatives of the Katanga System—the Western Congo System—are widely represented and strongly folded on the west where they abut against gneisses representing the northward extension of the coastal zone studied by Mendes in Angola. In contrast, in the eastern part of the outcrop of the Western Congo System, the rocks are weakly folded or unfolded.

The only dates available from the Western Congo System itself are common lead galena ages of 740 m.y. (Cahen and others, 1963) and are of uncertain significance due to possible remobilization (Cahen, 1961); the principal thermal imprint is preserved in the gneisses to the west of this deformed, principally sedimentary sequence. Two pre-Western Congo System orogenies have been dated at ca. 2760 ± 500 m.y., and ca. 1480–1800 m.y., while Cahen and others (1963) recognize two post-Western Congo System thermal events superimposed on this ancient crystalline terrain:

(a) Regional metamorphism and granite emplacement—the effects of the West Congolian orogeny; dated by Rb–Sr microcline and muscovite ages of 590 ± 30 to 660 ± 66 m.y.

(b) Pegmatite and vein emplacement reflected by Rb–Sr and K–Ar ages of biotite, muscovite and microcline in the 440–550 m.y. range.

These gneisses represent the geanticlinal (or vestigial) interior of the orogenic zone, and the difficulties of interpreting such ages have already been mentioned elsewhere (Clifford, 1967); suffice it to say that the older ages (580–660 m.y.) may reflect the *Katangan episode* whilst the younger ages are within the range of the *Damaran episode*.

The Mozambique and Zambesi Belts of eastern Africa Outside these areas of deformed geosynclinal rocks in southern Africa, important late Precambrian–early Palaeozoic tectonothermal activity undoubtedly took place in certain extensive regions where Upper Precambrian rocks are poorly represented or are absent. The most important regions of southern Africa for which radiometric data are available are: the Mozambique and Zambesi Belts (figure 8), the Cameroons and the Central African Republic (figure 7).

The *Mozambique Belt* represents one of the fundamental features of Africa. It was first defined and delineated by Holmes (1951); more recently, Cahen (1961) has outlined some of its salient features (see figure 8). The main points established by these published data are:

(1) The belt trends approximately north–south from Rhodesia and Mozambique northwards through Malawi (Nyasaland), Tanzania, Kenya and Uganda (Holmes, 1951; Cahen, 1961).

(2) The belt consists of high-grade metamorphic rocks with associated granites and pegmatites, and is considered to be 'the deep interior of part of an orogenic belt which has been ... uplifted and ... deeply eroded' (Holmes, 1951, p. 256).

(3) The belt is characterized by ages of 400 to 700 m.y.; Cahen (1961, p. 560) notes, however, that 'despite its "young" age determinations, the Mozambique belt may be a multicyclic belt in

which ancient rocks belonging to older cycles have been re-involved in a "young" end-Pre-Cambrian to early Palaeozoic activity.'

Although a generally N-S oriented belt, in Rhodesia the Mozambique Belt swings to an E-W trend to become what is locally called the *Zambesi Belt* (Snelling and others, 1964a), a belt which is thermally and structurally contiguous with the northern part of the folded Lomagundi-Piriwiri-Deweras sequence of northwestern Rhodesia (see figure 8). To the north, the structural continuity of the Zambian sector of the Mozambique Belt with the folded Katanga System is clearly implied by the works of Holmes (1951) and Cahen (1961, p. 561).

Rb-Sr, K-Ar and U-Th-Pb ages are available from more than 50 localities in the Mozambique and Zambesi Belts in Rhodesia, Zambia, Mozambique, Malawi, Tanzania, Kenya and Uganda. Almost all of these ages fall within the 400-650 m.y. range. Moreover, of the 20 K-Ar and Rb-Sr ages of metamorphic micas and hornblende, 75% fall within the 440-570 m.y. range, while four ages fall in the 615-650 m.y. range (Snelling, 1962, 1963, 1964, 1965; Snelling and others, 1964a).

Twenty K-Ar and Rb-Sr ages are also available for biotite and hornblende from plutonic rocks, and granite and hornblendite dykes, and all are less than 600 m.y.; of these, 13 are in the 450-560 m.y. range and include granite, syenite and pyroxenite mineral ages; in addition, whole rock samples of granite dykes cutting the basement in the Uganda segment of the Mozambique Belt have yielded an Rb-Sr age of 565 ± 20 m.y. The majority of the remainder of the data are for minerals from intrusives yielding ages of less than ca. 420 m.y., and include granite from Zambia, nepheline syenite and perthosite from Malawi and the granites from Mozambique (Snelling, 1962, 1963, 1964, 1965; Mendes, 1961).

K-Ar and Rb-Sr ages of pegmatite micas and feldspars in the Mozambique and Zambesi Belts have given a variety of ages. Muscovite from the Kenailmet pegmatite of Kenya has given a K-Ar age of 615 m.y., while muscovite and lepidolite from an ancient rejuvenated pegmatite affected by the thermal activity of the Zambesi Belt in northern Rhodesia has yielded K-Ar ages of 450 and 965 m.y. respectively. Lepidolite from a pegmatite in Mozambique has been dated by the Rb-Sr method at 410 m.y.; similar 'low' K-Ar ages of 420 and 430 m.y. have been obtained for muscovite and biotite respectively from the Serenji district of Zambia. Other silicate mineral ages from pegmatites include: feldspar K-Ar ages of 410-490 m.y. and a single Rb–Sr feldspar age of 585 ± 20 m.y. from Kenya; and a biotite Rb–Sr age of 470 m.y. from Zambia (Holmes and Cahen, 1957; Snelling, 1963, 1964, 1965; Snelling and others, 1964b; Cahen, 1961).

U–Th–Pb ages of samarskite, betafite, uraninite and monazite from eight pegmatite localities within the Mozambique and Zambesi Belts in Rhodesia, Zambia, Tanzania and Kenya have yielded ages concentrated in two ranges: (1) 485–545 m.y.; and (2) 615–650 m.y. (Snelling and others, 1964b; Burger, 1960; Darnley and others, 1961; Holmes and Cahen, 1957).

Finally, three U–Th–Pb ages have been published for minerals from rather rarer rock types as follows: zircon from nepheline syenite gneiss, ca. 540 m.y.; monazite from albite-aegirine rock, 542 m.y.; and davidite from norite–anorthosite, 578 m.y. (Holmes and Cahen, 1957; Darnley and others, 1961).

These age data show an overwhelming concentration of mineral and whole rock ages, from a wide variety of geneses, in the 450–570 m.y. range of the Damaran episode; only 15% (10 ages) fall within the 580–680 m.y. range and even that concentration is fortuitous since four of the ages are from pegmatites within a mile of each other (Darnley and others, 1961; Snelling, 1963). Notwithstanding these youthful ages, there are few representatives of Upper Precambrian sequences within the Mozambique Belt. Along with its undoubted continuation from Uganda westwards into the Central African Republic and the Cameroons, this portion of the late Precambrian–early Palaeozoic orogenic belt represents an example *par excellence* of the vestigial facet of orogeny; the Malagasy Republic is yet another fine example.

Ethiopia, Somali Republic and northeastern Africa The Precambrian of Ethiopia (Boje Series) and the Somali Republic (Basement Complex) consists largely of metamorphic rocks including schists, quartzites, marbles, amphibolites, paragneisses and granitic bodies; the ages of these rocks are still uncertain (Furon, 1963). However, Greenwood (1961) has suggested that in Somali Republic the Basement Complex may be the westward extension of the Inda Ad Series, a sequence of slates, quartzites, slaty sandstones and limestones in the eastern part of the territory. This thesis has been amplified by Rogers and others (1965) who consider that the whole of the Horn of Africa, from Ethiopia to the Somali Republic 'formed part of a north–south . . . trough of late Pre-Cambrian sedimentation

... subjected to an intense regional metamorphism ... followed by light dynamic metamorphism.'

K–Ar ages from the crystalline rocks in *Ethiopia* fall into two groups: (i) 615–660 m.y. for hornblende from gneisses, for amphibolite (whole rock) and for granite samples; and (ii) 450–520 m.y. for phyllite (whole rock) and for micas from quartz diorite, augen gneiss, muscovite gneiss and granite. In addition, the following ages have been obtained: pyroxenite (whole rock), 330 ± 180 m.y.; chlorite from gneiss, 303 ± 184 m.y.; muscovite from pegmatite, 569 ± 35 m.y.; and biotite from hornblende gneiss, 739 ± 14 m.y. (Rogers and others, 1965; Snelling, 1964).

From their data, Rogers and others (1965) deduce an event of katametamorphism at ca. 600 m.y. and an event of epimetamorphism at ca. 500 m.y., supporting the view that the region is the northward extension of the Mozambique Belt.

Similar K–Ar ages have been obtained for the Basement Complex and the Inda Ad Series of the *Somali Republic*. The ages of biotites from granodiorite gneiss, synkinematic nepheline syenite gneiss, and from plutonic rocks intrusive into the Inda Ad Series all fall within the 460–515 (± 25) m.y. range. In addition, a sample of metamorphosed argillaceous sandstone from the Inda Ad Series gave a K–Ar age of 500 ± 20 m.y., whereas a mudstone from the same series has yielded an age of 380 ± 40 m.y. Finally, a K–Ar age of 600 ± 25 m.y. has been recorded for muscovite from a gneiss in the Basement Complex (Snelling, 1963; Snelling, in Gellatly, 1963).

In *Egypt*, the pre-Carboniferous rocks are represented by a mixed sequence of rocks which are, from oldest to youngest, as follows: (i) Fundamental Complex (migmatites); (ii) Mitiq Series (highly metamorphosed schists, gneisses and marbles); (iii) Atalla Series (quartzites, itabirites, rhyolites and andesites); (iv) Old Paraschist Series (metasediments and andesites); (v) Shadli Series (schists, lavas, tuffs and breccias); (vi) Dokhan Series (dioritic flows and some sediments); and (vii) Hammamat Series (sandstones, conglomerates and some lavas). Two granite series, the Shaitian and the Gattarian, intrude this sequence; the Gattarian postdates the Hammamat Series (Schürmann, 1961; Gheith, 1961). Moreover, Schürmann (1961) considers that units (ii)–(vii) represent a thick (up to 35,000 ft) sequence of rocks of Riphean ('young' Precambrian) age.

A number of age measurements have been carried out by the K–Ar method on this sequence (Gheith, 1961; Higazy and El-Ramly, 1960) and have given the following results: biotite from gneisses of the Fundamental Complex, 456 and 510 m.y.; chloritized biotite from the Mitiq–Atalla sequence, 308 m.y.; whole rock gneissose Shaitian granite (?), 590 m.y.; biotites from the early phase of the Gattarian granite, 435 and 520 m.y.; biotites from latest phase of Gattarian granite, 462 and 540 m.y.; whole rock samples of Gattarian granite, 410, 460 and 470 m.y.; whole rock samples of aplite and xenolith associated with Gattarian granite, 410 and 420 m.y.; muscovite and microcline from pegmatite, 600 and 470 m.y. respectively; and whole rock trondjemite gneiss and gneissose granite, 450 m.y. and 430 m.y. respectively. Similar results showing a principal concentration of 450–550 m.y. (with a few exceptions greater than 600 m.y.) have been published by Schürmann (1964) for Rb–Sr, K–Ar and some U–Th–Pb ages of mineral fractions from gneiss, gneissic granite, granite and pegmatite from the northern part of the Eastern Desert of Egypt.

Other ages from northeastern Africa are few. Snelling (1963, 1964) has reported two K–Ar ages of 535 ± 25 m.y. and 560 ± 25 m.y. for biotites from quartz–biotite granulite and granite from bore holes in Cyrenaica respectively. These data suggest an extension of late Precambrian–early Palaeozoic tectonothermal activity far to the west of the Egyptian basement outcrop.

Clearly, the consistency of the 450–650 m.y. age pattern throughout eastern Africa supports the view that the Precambrian of Somali Republic, Ethiopia and eastern Egypt, represents the northward continuation of the Mozambique Belt. However, even though the balance of present evidence favours a late Precambrian (or even Lower Palaeozoic) age for the Inda Ad Series of Somali Republic and for part of the sequence in eastern Egypt, the age of the widespread gneisses of Somali Republic and Ethiopia is still in doubt. For this reason, it is still uncertain whether this region is the geosynclinal or vestigial facet of late Precambrian–early Palaeozoic orogenesis.

Cameroons, the Central African Republic and Tchad On the western side of equatorial Africa, work by Lasserre (1964) has shown that a uniform pattern of ages is present over a large part of the *Cameroons* in rocks which have long been regarded as 'Lower' and 'Middle' Precambrian (see figure 7). These rocks consist of a variety of metasediments, granulites and migmatites intruded by granite, synkinematic dunite and gabbro (Haughton, 1963).

Lasserre (1964) studied a number of mineral and whole rock samples from these crystalline rocks. Nine samples from the crystalline schists and gneisses (Middle Precambrian (?)) of the Yaoundé district of southern Cameroons yielded Rb–Sr biotite ages of 546–585 m.y., with an average age of 565 m.y. for the metamorphism; in contrast, two biotites from central Cameroons yielded an average of 615 m.y. More significantly, perhaps, four biotites from old syntectonic granites in southern Cameroons, and two from central Cameroons, gave ages of 560–585 m.y. with a single whole rock age of 683 ± 110 m.y., but Lasserre is cautious in his interpretation of these data, and notes (p. 999) 'Il n'est pour l'instant pas possible de savoir si ces âges correspondent réellement à l'âge du métamorphisme et à celui des granites syntectoniques anciens, ou à un rajeunissement de ces formations aux environs de 570–600 M.A.'

In addition to these ages, Lasserre obtained Rb–Sr ages of 512 ± 10 and 538 ± 10 m.y. for biotites, and 534 ± 110 m.y. for a whole rock sample, from the late syntectonic granites of the central and northern Cameroons; he feels that these ages, at least, correspond to a younger (Cambrian) episode of granitization which was followed by posttectonic granites giving biotite ages of 495 ± 20 m.y.

Similarly, in the *Central African Republic*, a consistent pattern of 500–650 m.y. ages has been obtained from terrain believed to be very antique (Haughton, 1963; Roubault and others, 1965).

Ages from granite massifs and their schistose host rocks have been determined for samples from three geologically distinguishable groups: (I) migmatitic and metamorphic basement; (II) heterogeneous metasomatic granites; and (III) latest granites (*granites ultimes*). From the first of these groups (I), biotites from gneiss, charnockite and migmatite yielded Rb–Sr ages mainly in the range 575–655 m.y. with an average age of 602 m.y. Rb–Sr ages of biotites from group II fall within the much broader range of 530–640 m.y. with an average age of 575 m.y., whilst similar Rb–Sr biotite determinations from group III fall within the range 515–565 m.y., with an average age of 533 m.y. (Roubault and others, 1965). From the same region, Vachette (1964a) reports Rb–Sr ages of 535–560 m.y. for biotite and muscovite samples from anatectic granite and migmatite, and a Rb–Sr age of 550 ± 9 m.y. for biotite from a granite (group III (?)) which yielded a zircon Pb/α age of 1350 m.y.

In addition to this very uniform age pattern, certain mineral samples, particularly of microcline, from these three rock groups have yielded Rb–Sr ages greater than 800 m.y., with one age of 1950 ± 30 m.y. These ages are discussed later, but in those cases where these samples have been dated by the K–Ar method, the ages are in the 490–600 m.y. range (Roubault and others, 1965). The antiquity of the terrain and the absence of Upper Precambrian sediments, class this region as a vestigial facet of the late Precambrian–early Palaeozoic activity. To the south, however, in the region of flat-lying supposed Upper Precambrian of the Basse–Kotto System (see figure 8), the only ages available are Rb–Sr ages of biotite from the underlying granite and gneiss giving dates of 970 ± 30 and 890 ± 30 (Roubault and others, 1965).

From *Tchad*, to the north (see figure 7), Vachette (1964a) obtained Rb–Sr ages of 514–560 (± 12) m.y. for biotite and muscovite samples from three posttectonic granites, and an age of 520 ± 8 m.y. for pegmatite phlogopite.

Malagasy Republic The geology of the *Malagasy Republic* is, for all intents and purposes, part of Africa, and the studies of the pre-Karroo of the island have a distinct bearing on the analysis of the history of the African continent. The most important recent studies have been carried out by H. Besairie and his colleagues (see Haughton, 1963); their work has resulted in the recognition of the following sequences in the pre-Karroo crystalline rocks (from oldest to youngest): (1) Androyan System; (2) Graphite System; (3) Vohibory System; (4) Sahatany System (including the Cipolin Series); and (5) Quartzite Series. Of these, systems 1–3 make up the major part of the crystalline rocks of the island (see Furon and others, 1959).

Early common lead ages suggested that the Vohibory System is older than ca. 2140 m.y. (Holmes and Cahen, 1957, p. 83), whilst it was claimed that other galenas yielding ages of 1110–1140 m.y. placed a minimum age limit on the Cipolin Series (Holmes and Cahen, 1957, p. 56). Recent extensive radiometric dating by Delbos (1964) has partially confirmed these data since a number of whole rock and mineral ages from the Androyan and Vohibory Systems have given ages greater than 2000 m.y. old; these ages are discussed in a later section of this chapter.

Although these ancient formations comprise almost the whole of the pre-Karroo terrain of Malagasy, the major number of Rb–Sr biotite ages from the Androyan and Vohibory Systems fall within the 440–510 m.y. range (Delbos, 1964). Moreover, muscovite and sericite from the overlying Quartzite Series have yielded Rb–Sr ages of 570 and 630 m.y., whilst stratified granites and migmatites which affect a wide variety of levels throughout the crystalline sequence of the island have yielded monazite and zircon ages of

510–580 m.y., biotite Rb–Sr ages of 470–510 m.y. and a whole rock Rb–Sr age of 540 ± 15 m.y. In addition, pegmatites almost everywhere yield mineral ages of ca. 500 m.y., whilst phlogopite and thorianite from pyroxenites and peridotites, and from pegmatites associated with pyroxenites, have given ages of 515–550 m.y. (phlogopite) and 480–490 (thorianite) (Delbos, 1964; Holmes and Cahen, 1957); a few uncorrected, and less reliable, zircon ages fall in the 390–460 m.y. range (Holmes and Cahen, 1957, p. 29).

From these data, Delbos (1964, p. 1855) concludes that a generally weak metamorphism affected the Quartzite Series 600 m.y. ago, but that the widespread ages in the 460–550 m.y. range represent an event of major importance in which three interrelated phenomena are recognizable:

550 m.y.   Granitization giving rise to the stratified granites.
500 m.y.   Pegmatite mineralization.
460 m.y.   Rejuvenation of biotites throughout the whole crystalline complex.

That these complex events of the 500 m.y. cycle reflect the imprint of an orogeny is beyond doubt (Emberger, 1958). However, if this region ever contained geosynclinal sediments related to that orogeny then they have been removed, with the possible exception of the locally preserved Quartzite Series in the centre of the island (the Itremo massif). Malagasy Republic thus represents the vestigial facet of activity of the Damaran orogenic episode (450–580 m.y.) largely preserved in rocks of much greater antiquity (ca. 2500 m.y. (?)) (Holmes and Cahen, 1957; Cahen, 1961).

West Africa In West Africa, two regions of late Precambrian–early Palaeozoic activity have been recognized and quantitatively defined by radiometric dating as follows: (1) an eastern region extending westwards and northwards from the Cameroons (see p. 320), through Nigeria, Dahomey, Togo, Niger Republic, Mali and into the Hoggar (see figure 7); and (2) a western region extending from Sierra Leone northwards through Senegal and Mauritania (see figure 7). The two zones are separated by the ancient stable region of the West African Craton (Kennedy, 1964).*

* In his synthesis of West African geology, Kennedy (1964, 1965) recognized the presence of an ancient stable block (the West African Craton) partially encircled by more youthful *circumstructural* zones. The details of Professor Kennedy's views are presently in preparation for publication; in the meantime, I am greatly indebted to him for permission to include his data, and the broad elements of his geological interpretation, in this section (pp. 323–331).

In the southern part of the *eastern region*, in Dahomey, Ghana, Togo and Nigeria, three distinctive members are recognized from west to east as follows (Haughton, 1963; Kennedy, in preparation):

(1) An *outer zone* of folded and non-metamorphic sediments and lavas of the Buem Formation; a miogeosynclinal sequence overfolded and thrust towards the ancient stable foreland to the west.

(2) An *intermediate zone* occupied by the Togo (Atacorian) Series of quartzites and schists becoming increasingly metamorphosed towards its thrust contact with the Dahomeyan on the east.

(3) An extensive *eastern zone* occupied by ortho- and paragneisses, granulites and granites of the Dahomeyan.

Of these zones, the most extensive is the Dahomeyan. In the east, for example, at least half of Nigeria is underlain by the crystalline basement attributed to this system and consisting essentially of granites, gneisses, migmatites and metasediments. In particular, this basement contains widespread *older granites** representing a complex plutonic suite with associated tin and columbite–tantalite pegmatites (Jacobson and others, 1964). The emplacement of this suite is believed to be related to a major orogenic cycle, and all known occurrences of older granite are synkinematic.

Northwards from Togo, Dahomey and Nigeria, these rocks disappear beneath the Mesozoic and Tertiary sequence of the Gao trench, but reappear again in Mali in the great bend of the Niger River where the *Bourré gneiss*, consisting of muscovite and amphibole gneisses, is believed to represent the Dahomeyan. The *Ansongo Series* of quartzites, quartzitic granulites and schists has been correlated with the Atacorian, and is believed to postdate the Bourré gneiss; and the *Labbezenga Series* of mica schists and quartzites has been correlated with the Buem to the south (Haughton, 1963).

Further to the north, the supposed correlatives of some of these sequences outcrop extensively in the classic region of the Hoggar (see figure 7) where the following tectonostratigraphic units have been recognized; (i) Suggarian ($\equiv$ Dahomeyan (?)); (ii) Pharusian; and (iii) Nigritian (late Precambrian (?)). Of these, the *Suggarian* includes a strongly metamorphosed sequence of biotite and amphibole gneisses, marbles and quartzites, partially migmatized, and intruded by granites. It is also of particular interest to note that it includes

---

* In contrast, Nigeria is well known for its later suite of younger granites of Jurassic age (Jacobson and others, 1964).

a major charnockite zone, known as the In Ouzzal Facies, extending in a N–S belt some 400 miles long and 12 miles wide through the Adrar des Iforas (Furon, 1963). The *Pharusian* is largely conglomerates, quartzites, marbles and volcanic rocks and is considered to be younger than the Suggarian (Furon, 1963); this sequence is metamorphosed and is intruded by batholithic and subvolcanic igneous complexes with associated rhyolites. The *Nigritian* rests unconformably on the Pharusian and is primarily a sandstone and conglomerate sequence.

All of these sequences are folded and, from a broad structural point of view, the Hoggar region is characterized by major tectonic 'horsts' (oriented north–south) of Suggarian, separated by complementary zones of the Pharusian (see Furon, 1963, p. 91). That the Suggarian suffered an earlier (pre-Pharusian) orogenic event has long been known, and it is in fact unconformably overlain by the Pharusian (Black, 1966). It has, moreover, been suggested (Black, 1966) that during late Precambrian–early Palaeozoic mountain building, movement took place along the N–S thrust faults which delimit the Suggarian and Pharusian compartments, and that the life of this portion of the orogenic belt came to an end with the deposition of the Nigritian molasse and related sequences.

A number of detailed radiometric dating studies have been published for this complex region of Togo, Dahomey, Nigeria, Mali and the Hoggar; the majority of ages fall consistently in the 450–650 m.y. range (see figure 7). For example, six biotites from Dahomeyan gneisses and migmatites from *Dahomey* yielded Rb–Sr ages of 500 ± 13 m.y. to 523 ± 9 m.y. (Bonhomme, 1962). Four concordant granites from the same region gave biotite Rb–Sr ages of 492 ± 10 m.y. to 506 ± 11 m.y. whereas the Sinendé intrusive granite gave whole rock and biotite ages of 507 ± 34 m.y. and 506 ± 6 m.y. respectively. In contrast, the Kouandé orthogneiss which is intrusive into part of the Dahomeyan gneisses, yielded biotite and whole rock ages of 592 ± 9 m.y. and 1650 ± 220 m.y. respectively; these data suggest that at least part of the Dahomeyan may be older than 1650 m.y. (Bonhomme, 1962, pp. 34–35). A similar antique Rb–Sr age of 1860 ± 90 m.y. was obtained for a whole rock sample of granitized mica schist of the Atacorian, while the biotite from this sample gave an age of 291 ± 9 m.y. (Bonhomme, 1962); the significance of these ages is still uncertain.

In the extension of the Dahomeyan basement in *Nigeria*, the most detailed whole rock–mineral study has been carried out on the Kusheriki *older granite* which gave the following ages: biotite

(Rb–Sr), 479 ± 10 m.y.; biotite (K–Ar), 490 ± 18 m.y.; and whole rock (Rb–Sr), 530 ± 140 m.y. (Jacobson and others, 1964). In addition, biotite from the older granite at Kagara has given a K–Ar age of 540 ± 20 m.y., whereas K–Ar determinations on biotite, and Rb–Sr determinations on muscovite and lepidolite, from pegmatites have yielded ages of ca. 500 ± 20 m.y. (Jacobson and others, 1964). A high K–Ar age of 780 ± 40 m.y. given by hornblende from syenite may be due to the presence of relict argon, whereas a low age of 190 ± 10 m.y. yielded by biotite from one of the older granites may reflect the effects of the Jurassic younger granites (Snelling, 1964). Finally, one discordant U–Th–Pb age for monazite gave a $^{207}Pb/^{206}Pb$ date of 485 m.y. (Jacobson and others, 1964).

Of the metamorphic rocks of the Basement Complex in Nigeria, the following K–Ar ages have been obtained: hornblende and biotite from charnockite, 535 ± 25 m.y.; biotite samples from granulite, migmatite and charnockitic hypersthene diorite, 480 ± 20 to 510 ± 20 m.y.; muscovite from kyanite schist, 495 ± 20 m.y.; and muscovite from marble, 435 ± 16 m.y. (Snelling 1964, 1965).

These older granites and associated metamorphic rocks can be traced eastwards into the Cameroons (see p. 320) and westwards into Dahomey. Since the Kusheriki granite and other older granites are synkinematic or late synkinematic, it has been suggested that they represent a plutonic episode related to the orogenic conditions in a Lower Palaeozoic orogenic belt.

The folded sequence of Bourré gneiss–Ansongo–Labbezenga of *Mali* has been correlated with the Dahomeyan–Atacorian–Buem to the south. From that region, Vachette (in Reichelt, 1966) reports muscovite and biotite Rb–Sr ages of 562 ± 41 and 606 ± 22 m.y. respectively for samples from the Bourré gneiss, and a whole rock Rb–Sr age of 586 ± 95 m.y. for a sample of mica schist from the Takamba Series ($\equiv$ Dahomeyan (?); in Haughton, 1963).

Of all the areas studied of this eastern region of late Precambrian–early Palaeozoic activity in West Africa, the most extensive work has been carried out in the Suggarian and Pharusian rocks of the *Hoggar* (see figure 7). In the western part of the region, in the Central Sahara pegmatites cutting the In Ouzzal Facies of the Suggarian have given biotite K–Ar and Rb–Sr ages of 1820 ± 50 m.y. and 1730 ± 70 m.y. respectively, suggesting that an old orogenic cycle (perhaps the Eburnian, see p. 350) affected these rocks. Similarly, migmatitic granite of Ouallen in the same region has yielded muscovite and biotite Rb–Sr ages of 1795 ± 50 and 745 ± 35 m.y. respectively, and a zircon ^{207}Pb/^{206}Pb age of 1885 ± 80 m.y. (Lay and Ledent, 1963; Lay and others, 1965); although originally considered to be a 'Pharusian' granite, it has recently been suggested that the Ouallen granite may be 'un panneau suggarien en pays pharusien' (Lay and others, 1965).

These ancient ages are, however, exceptional in the Hoggar region. Within the *Suggarian* 'compartments', the following Rb–Sr ages have been obtained for mineral and whole rock samples of metamorphic rocks: muscovite and biotite from mica schists, 570 ± 30 and 555 ± 25 m.y. respectively; biotite and muscovite from migmatite and gneiss, 575 to 630 m.y.; whole rock migmatite, 968 ± 100 m.y.; whole rock and muscovite samples from quartzites, 1157 ± 114 and 1210 ± 110 m.y. respectively; and biotite from migmatite, 506 ± 16 m.y. Although the interpretation of ages of ca. 1000–1200 m.y. is uncertain, the ages around 570–630 m.y. are taken to indicate metamorphism around that time, with subsequent rejuvenation at ca. 500 m.y. (Lay and Ledent, 1963; Boissonnas and others, 1964).

In addition to these metamorphic rocks, granites from Suggarian terrain have given a variety of Rb–Sr ages mainly in the 420–660 m.y. range. Lay and Ledent (1963), for example, found that biotites from four Suggarian granites range from 445 ± 30 m.y. to 700 ± 35 m.y., with one muscovite age of 540 ± 40 m.y. Zircon U–Pb ages from these samples showed only one concordant U–Pb age of 595–635 m.y.; the other samples yielded ^{207}Pb/^{206}Pb ages of 630–665 m.y., but ^{206}Pb/^{238}U and ^{207}Pb/^{235}U ages in the 505–525 m.y. range (Lay and others, 1965). In addition to these data, other granite bodies have been dated by Boissonnas and others (1964) who found that concordant Rb–Sr whole rock and mica samples indicated the presence of granite emplacement events at ca. 520 m.y. and ca. 480 m.y. ago; other biotite ages of ca. 420 m.y. may also indicate plutonism around that time.

In contrast to the rather wide range of Rb–Sr ages for biotites from 'Suggarian' granites, biotite and zinnwaldite from '*Pharusian*' granites have given consistent Rb–Sr ages in the range of 470 ± 15 m.y. to 530 ± 35 m.y., with one biotite age of 620 ± 30 m.y. The zircons, however, have all yielded discordant ages with ^{207}Pb/^{206}Pb ages of 590–650 m.y. and ^{206}Pb/^{238}U and ^{207}Pb/^{235}U dates falling in two ranges of 515–565 m.y. and 355–440 m.y. (Lay and others, 1965). As an example of the mineral–whole rock age pattern, the following data may be noted from the Tin Touafa granite: biotite (Rb–Sr age), 525 ± 15 m.y.; whole rock (Rb–Sr age), 510 ± 100 m.y.; zircon (^{207}Pb/^{206}Pb age), 590 ± 40 m.y.; zircon (^{206}Pb/^{238}U age), 355 ± 15 m.y.; and zircon (^{207}Pb/^{235}U age), 390 ± 20 m.y. (Lay and Ledent, 1963; Lay and others, 1965). The Tin Touafa granite is typical of the syntectonic 'Pharusian' granites and, on the basis of these age data, it has been concluded that this and similar granites were emplaced during the Pharusian tectonic cycle 550–650 m.y. ago, and that the lower Pb/U ages resulted from subsequent episodic lead loss (Lay and Ledent, 1963; Lay and others, 1965).

Few ages are available for Pharusian metamorphic rocks: a Rb–Sr age of 505 ± 60 m.y. has been obtained for a whole rock quartzite sample, and 960 ± 95 m.y. for muscovite from mica schist.

Summarizing these data from the Hoggar, the following general conclusions have been reached: (1) the 1700–1800 m.y. ages from the Suggarian of the Central Sahara reflect an ancient tectonic cycle, to which the Ouallen granite may also belong; (2) the syntectonic 'Pharusian' granites were emplaced some 550–650 m.y. ago during an important tectonic episode, and the major metamorphism of the Suggarian (dated at ca. 600 m.y.) may belong to the same event; and (3) later events include the outpouring of lavas (based on a single whole rock rhyolite Rb–Sr age of 550 m.y.), emplacement of granites ca. 520 m.y. ago and ca. 480 m.y. ago, emplacement of pegmatites 470–510 m.y. ago and the rejuvenation of biotites yielding ages as low as 450 m.y. (Lay and Ledent, 1963; Lay and others, 1965; Boissonnas and others, 1964).

In contrast to the eastern region of late Precambrian–early Palaeozoic orogenesis in West Africa, the *western region* is less well defined by radiometric age determinations (see figure 7); moreover, it is characterized by a more extended Phanerozoic orogenic history. In Sierra Leone, it forms a well-defined orogenic belt extending across the country from southeast to northwest truncating and deflecting the structures within the older stable shield to the east (Kennedy, in preparation). This marginal belt consists of three well-defined units which are, from east to west, as follows:

(1) An *outer zone*, immediately adjacent to the shield, consisting of folded, mainly non-metamorphic sediments and volcanics of the Rokel River Series which rest unconformably on the granites and gneisses of the shield.

(2) A *central zone* of schists (the Marampa schists) with associated granitic rocks and migmatites.

(3) An *inner zone* of high-grade ortho- and paragneisses and granulites (the Kasila gneisses), overlapped on the seaward side by Tertiary and Recent sediments.

This marginal orogenic zone extends northwards into Guinea where the representatives of the Rokel River Series, the Marampa schists and the Kasila gneiss are overlain, unconformably, by horizontal sandstones (*grés horizontaux*) that pass upwards without visible break, into strata containing Silurian fossils; the pre-Silurian age of the marginal orogenic belt is, consequently, firmly established.

To the north of this intervening spread of definite Phanerozoic rocks, similar sequences to those of the marginal orogenic belt of Sierra Leone and Guinea, reappear from beneath the *grés horizontaux* (see Furon and others, 1959), and form part of the great 'Mauritanide Zone' which extends northwards along the western margin of the continent from Senegal and Portuguese Guinea in the south, through Mauritania and the Spanish Sahara, and into Morocco (Sougy, 1962). This major zone has been ascribed to the Hercynian Revolution (Sougy, 1962) and consists of two sectors: (i) a *southern sector* (Senegal–Mauritania) comprising mainly metamorphic rocks, and folded (unfossiliferous) sediments of uncertain age; and, (ii) a *northern sector* (Spanish Sahara–northwest Mauritania) in which fossiliferous Devonian rocks are involved.

In the southern sector of the Mauritanide zone two main rock formations are concerned: a narrow eastern zone of folded non-metamorphic rocks of the Falemian ($\equiv$ Rokel River Series (?)), marginal to the ancient stable shield to the east; and, a western zone of regionally metamorphosed epimetamorphic schists of the Bakel–Akjoujt Series ($\equiv$ Marampa schists (?)), with associated granites. The age of these two sequences in Senegal and to the north, is uncertain; in recent years, it has been suggested that Bakel–Akjoujt and Falemian may be correlative, and equivalent to the flat-lying rocks (infra-Cambrian and younger) which rest on the Precambrian of the ancient shield to the east (Bassot and others, 1963). In Senegal the Bakel–Akjoujt Series is complicated by the fact that it forms two branches; one running N–S along the line of strike of the Marampa schists of Sierra Leone; the other (the Koulountou Series) running WSW–ENE. The two branches are separated by a 'wedge' of rocks of the Youkounkoun Series, the whole sequence being separated from the shield to the east by the Falemian Series.

The northern sector of the Mauritanide zone has been discussed by Sougy (1962). In essence, in Spanish Sahara and northwestern Mauritania, the fossiliferous Devonian and underlying rocks are folded, and thrust eastwards over the stable region of interior West Africa. The post-Devonian age of deformation in the northern sector is thus established, but the connexion between it and the southern sector is less certain; if the two join, the Bakel–Akjoujt Series of the southern sector must suffer an abrupt swing of some 300 kilometres to the west to join the northern sector; alternatively, the northern and southern sectors may be separate orogenic belts of differing age.

Radiometric ages from this orogenic zone in the western marginal region of West Africa are sparse. The contrast between this zone and the stable shield to the east is, however, clearly defined by radiometric dates in Sierra Leone, Senegal and Mauritania where the Precambrian of the 'shield' has given undisturbed ages of >1800 m.y., in contrast to ages of ca. 600 m.y. and younger in the marginal orogenic belt (Kennedy, in preparation).

In *Sierra Leone*, the marginal belt has yielded K–Ar ages of 486 ± 20 m.y. and 552 ± 20 m.y. for biotite from granulitic Kasila gneiss and muscovite from the Marampa schists respectively (Kennedy, in preparation). To the north, Rb–Sr age determinations on whole rock and separated mineral fractions from *Senegal* have been published by Bassot and others (1963). Biotite from the Niokolo–Koba granite which predates the Youkounkoun Series gave an age of 645 ± 39 m.y. suggesting a Phanerozoic age for that series, a view which is perhaps supported by an age of 560 ± 50 m.y. yielded by the argillaceous fraction of one of the constituent pelites. In addition, the Koulountou Series, considered to be the metamorphic equivalent of the Youkounkoun (Bassot and others, 1963) has yielded Rb–Sr ages of 441 ± 17 m.y. for a whole rock rhyolite sample, and 868 ± 370 m.y. and 433 ± 25 m.y. for whole rock and sericite–chlorite samples respectively from a pelite.

On existing maps, the Koulountou Series passes northwards along the strike into metamorphic Bakel Series which Bassot and others (1963) consider to be equivalent to the Falemian Series to the east. Age data for mica schists from the Bakel Series in Senegal show Rb–Sr ages of 357 ± 21 m.y. and 358 ± 33 m.y. for muscovite samples, and 206 ± 6 m.y. for a biotite sample (Bassot and others, 1963). On the present evidence, therefore, these data suggest the following sequence of events: (1) emplacement of the Niokolo–Koba granite (ca. 650 m.y.); (2) deposition of the Youkounkoun Series ($\equiv$ Koulountou Series $\equiv$ Bakel Series $\equiv$ Falemian Series (?)); (3) metamorphism of the Koulountou Series (435 m.y.); (4) metamorphism I of the Bakel Series (355 m.y.); and (5) metamorphism II of the Bakel Series (205 m.y.).

In *southern Mauritania*, to the north, the Bakel Series is represented by schists and quartzites resting on the Guidimaka crystalline basement; the contact is marked by mylonites (Dars and Sougy,

1964). This pre-Bakel basement contains acid plutonic rocks which have yielded Rb–Sr ages as follows: biotite from granodiorite, 297 ± 14 m.y.; and muscovite from granite, 600 ± 24 m.y. (Vachette, 1964b). A very similar pattern has been obtained from the Akjoujt region to the north, where the El Kleouat granite, cropping out in the region of the Akjoujt Series, has yielded Rb–Sr whole rock and biotite ages of 600 ± 62 m.y. and 243 ± 6 m.y. respectively (Vachette, 1964b); in addition, biotite from the Hajakh Dkhem granite in the same region has given a Rb–Sr age of 327 ± 6 m.y. (Bonhomme, 1962). The data from Akjoujt are consistent with the view expressed by Giraudon and Sougy (1963), that these granites represent part of the Precambrian which has been thrust over the Akjoujt Series during the Mauritanian (Hercynian) orogeny (cf. ages of 330 and 240 m.y.). However, it has also been suggested that these granites intrude the Akjoujt Series; if this is the case, and Vachette (1964b) is correct in taking the whole rock age of 600 m.y. as the emplacement age of the granite, the Akjoujt Series must be a Precambrian sequence subsequently reinvolved in Mauritanian tectonism.

These data from the western region of West Africa indicate an important late Precambrian–Phanerozoic orogenic belt in which important tectonothermal pulses in Middle and Upper Palaeozoic times are recorded in the Mauritanide zone by ages in the 300–350 m.y. and 200–250 m.y. ranges; these two events accord with the Acadian and Hercynian orogenies respectively (Kennedy, in preparation).

Having regard to the consistent ca. 600 m.y. ages sporadically recorded in Senegal and Mauritania, Kennedy (in preparation) postulates that the Mauritanide zone 'must coincide with the late Precambrian orogenic belt.' Such an analysis integrates, at least structurally, the Mauritanide zone with the marginal orogenic zone identified along the coastal margin of Sierra Leone and regions to the north, and characterized by ca. 450–600 m.y. ages.

C *Anorogenic effects*

A number of anorogenic events reflected by ages within the 450–660 m.y. range, or slightly older, have now been recorded from Africa; they include igneous and more extensive thermal activity. The igneous activity may include the outpouring of the Kisii basalt in Kenya, which has yielded a whole rock K–Ar age of 674 ± 27 m.y. (Snelling, 1962). The basalt lies within the essentially flat-lying (cratonic) Upper Precambrian Bukoban System which rests on basement presumed to be the northward extension of the Tanzania nucleus (>ca. 2000 m.y.). Similar K–Ar ages of 590 ± 20 m.y. and 660 ± 20 m.y. have been obtained from plagioclase and pyroxene samples from the Shushong sill in eastern Botswana (Snelling, 1965). However, although this sill postdates the gently dipping rocks of the Waterberg System (of age >1700 m.y.), the significance of these ages is, as yet, unknown.

Biotite from perthosite of the Mbozi syenite–gabbro complex of southern Tanzania has given a K–Ar age of 745 ± 30 m.y. (Snelling, 1963); this complex is intruded into the rocks of the Ubendian Belt which had a lengthy and varied tectonothermal history indicated by ages concentrated around ca. 1800 m.y. and 1000 m.y. (Snelling, 1964, p. 34).

Finally, phlogopite from metasomatized country rocks around the Nkumbwa Hill carbonatite of northern Zambia has given a K–Ar age of 680 ± 25 m.y. (Snelling, 1962). The country rocks are biotite and hornblende gneisses and granites of unknown age.

In addition to these examples of discrete igneous outpourings and intrusives, thermal activity of anorogenic character may well be the cause of a number of 500–580 m.y. ages in southwestern Uganda, Rwanda and southern Kivu, in terrain of demonstrably greater antiquity. In southwestern Uganda, for example, although whole rock samples have yielded Rb–Sr ages of 1300–1700 m.y., their constituent biotites have given K–Ar and Rb–Sr ages largely in the range 490–570 m.y. (Snelling, 1962, 1963, 1964). In addition, in Rwanda and southern Kivu, Monteyne-Poulaert and others (1962a,b) obtained local ages of 500–580 m.y. for microcline, biotite, microlite and fergusonite mainly from pegmatites within the Kibaran Belt which itself has yielded abundant mineral, and some whole rock, ages in the 800–1300 m.y. range. Together, these regions are, for convenience, referred to as a thermal 'node' of Damaran–Katangan activity (Clifford, 1967). Further work is required on this important area, but available data suggest the existence of largely Damaran activity of an entirely thermal nature, in contrast to the tectono-thermal activity of this and the Katangan episode elsewhere in Africa.

D *Summary*

Geological and geochronological data from Africa clearly demonstrate the presence of important zones of crustal mobility and

deformation in late Precambrian–early Palaeozoic times (see figure 7). The tectonothermal activity within these zones is now expressed as three differing facets:
(1) A *geosynclinal facet* in which Upper Precambrian geosynclinal sediments are widely represented; for example, Central Africa (the Katanga System), South-West Africa (the Damara and Otavi Facies of the Outjo System), and the Lower Congo region (the Western Congo System).
(2) A *vestigial facet* of rejuvenated basement rocks from which the geosynclinal cover rocks have been extensively removed or on which, for some reason, they were never deposited; for example, the southern Cameroons and the Central African Republic, the Mozambique Belt, the Malagasy Republic, and perhaps the Dahomeyan of West Africa.
(3) A *rejuvenated facet* in which the zones have been involved in later orogenesis; for example, the folded Malmesbury Beds of South Africa subsequently involved in the Hercynian movements in the Cape Fold Belt, and the pre-Bakel–Akjoujt basement of West Africa (e.g. Guidimaka granitic rocks) subsequently involved in the Acadian and the Hercynian orogenies in the Mauritanide zone.

There are, of course, parts of the orogenic belt which include both (1) and (2); for example: in the Lower Congo, an important zone of basement rocks occupies the interior of the orogenic zone to the west of the geosynclinal sequence of the Western Congo System; in the Hoggar, the Suggarian compartments of ancient basement alternate with north–south oriented compartments of Pharusian rocks (which may, in part, be the geosynclinal Upper Precambrian); and in Kenya supposed Upper Precambrian rocks (the Turoka System) crop out in the Mozambique Belt and may, in part, be correlative with some of the higher grade gneisses within the belt, as has been suggested for the relationship between the Inda Ad Series and the Basement Complex in the Somali Republic.

Within these differing facets, the late Precambrian–early Palaeozoic orogenic activity is reflected by ages of 450–680 m.y., and a number of less precise terms have been used for events in this range; for example, the Mozambiquian orogeny in the Mozambique Belt (Holmes, 1965), and the Riphean orogeny in the Hoggar (Black, 1966). Discussing this time range, however, Cahen (1963) maintains that the ca. 600 m.y. ages reflect the principal orogenesis in the Katanga Province and the Lower Congo, and that the ca. 500 m.y. ages represent posttectonic events; to the former the name *Katangan* (or West Congolian) orogeny or orogenic episode has been given, and is apparently synonymous with such terms as 'Kundelunguan' and 'Post-Schisto-Gréseux' orogeny.

There is no doubt that the 450–680 m.y. age range shows two distinct concentrations at 580–680 m.y. and at 450–580 m.y. in southern Africa, and Clifford (1963, 1967) has argued that these reflect two orogenic episodes: (a) the *Katangan episode* (ages of 580–680 m.y.); and (b) the *Damaran episode* (ages of 450–580 m.y.). It has also been noted that the greatest preponderance of ages in the late Precambrian–early Palaeozoic zone fall within the Damaran episode range, and that this episode can probably be correlated with the Sardian phase of the Caledonian orogeny of Europe (Clifford, 1963). Moreover, on the basis of geochronological data from West Africa, this subdivision has been supported by Kennedy (in preparation) who recognizes two orogenic pulses, at ca. 600 m.y. ago (early phase) and ca. 500 m.y. ago (main phase), constituting the Pan-African thermotectonic episode (Kennedy, 1964).

Of these two age concentrations the Damaran range includes Rb–Sr, K–Ar and U–Th–Pb ages yielded by: minerals from a wide range of metamorphic rocks, and from pegmatites and uranium-bearing mineral deposits; and mineral and whole rock samples of granite, including synorogenic types. These samples represent a great diversity of conditions of formation, and the consistent nature of the age pattern over a wide area indicates that the recorded thermal imprint reflects fundamental crustal activity. That the activity was orogenic is most conclusively demonstrated in the geosynclinal facets of the orogeny where deformation of the geosynclinal pile was accompanied by: (a) emplacement of synorogenic granites yielding Rb–Sr whole rock isochron and mineral ages of ca. 500 m.y. (e.g. Zambia, and Cape Province of South Africa); (b) late and synorogenic pegmatization yielding 470–550 m.y. ages (e.g. South-West Africa and Rhodesia); (c) regional metamorphism; and (d) uranium mineralization giving ages of ca. 500–520 m.y. (Katanga and Zambia). In this regard, the data from the Cape granite are of particular importance in that the granite is late orogenic, and has yielded Rb–Sr whole rock isochron and mineral ages of ca. 520 m.y., and intrudes strongly folded Malmesbury sediments which, on Rb–Sr isochron evidence, are considered to be not significantly older than ca. 560 ± 45 m.y. (Allsopp and Kolbe, 1965).

In contrast to these geosynclinal facets, the status of the 450–580 m.y. ages in the vestigial facets of late Precambrian–early Palaeozoic orogenesis cannot be so directly tested since the imprint affects older basement; as, for example, in the Mayumbe Belt of the Lower Congo region and the western part of Angola, the Mozambique Belt, the pre-Karroo of the Malagasy Republic, etc. In these regions, ages in the 450–580 m.y. range, have, however, been attributed to orogenesis in Nigeria (Jacobson and others, 1964) and Malagasy Republic; elsewhere in the late Precambrian–early Palaeozoic orogenic zone they are considered to reflect the Damaran episode (main phase) because of the analogy with the geosynclinal facets.

Age data reflecting the Katangan episode (580–680 m.y.) are less widespread than for the Damaran episode (see figure 7). However, there are scattered ages in this range from: (i) South-West Africa, (ii) the Lower Congo, (iii) Senegal, (iv) Mauritania, (v) Ethiopia and (vi) the Mozambique Belt; and more widespread ages from (vii) the Cameroons, (viii) Central African Republic, (ix) the Hoggar and (x) Katanga. Many of these dates (for example, ii, vi, vii, viii and ix) are from the vestigial facets of orogenesis and it is as difficult to interpret their structural significance as it would be to interpret the vestigial ages in the Damaran range *without* the evidence from the geosynclinal facet of orogeny. However, uraninite ages from Katanga, and Rb–Sr data from the Mayumbe Belt, the Cameroons and Central African Republic strongly support the view that the 580–680 m.y. ages within the late Precambrian–early Palaeozoic orogenic zone reflect an orogenic pulse (Katangan episode or early phase) earlier than the Damaran episode.*

The zones affected by Damaran–Katangan (Pan-African) orogenic activity are extensive and they separate and, in part, surround three regions which acted as stable cratons during Upper Precambrian–early Palaeozoic times;† these have been termed the *Congo* and *Kala-*

---

* It should, of course, be noted that the recognition of the orogenic status of these episodes does not imply that ages in the 450–580 m.y. and 580–680 m.y. ranges always reflect orogeny; for example, the 500–580 m.y. ages of the 'node' of Uganda, Rwanda and southern Kivu apparently reflect an imprint of an entirely thermal nature.

† Rocci (1965) has suggested that a fourth craton (the Nile Craton) may have existed in northeastern Africa. However, there is no support for this suggestion, and figure 7 indicates that if it did exist, it was very much smaller than the other three.

hari Cratons (Clifford, 1963) and the *West African* Craton (Kennedy, 1964).

The *Congo Craton* is bounded by the Mozambique Belt on the east, the northeast trending ('transcontinental') segment of the Damaran–Katangan (Pan-African) chain on the south, the coastal folded zone of rocks of the Western Congo System on the west and the vestigial zone of the Cameroons and the Central African Republic on the north. Within this cratonic area, the crystalline basement is of considerable antiquity, and undeformed or weakly deformed correlatives of the Katanga System are widespread (see figures 7 and 8).

The *Kalahari Craton* lies to the south of the Congo Craton and is bounded by the Mozambique Belt on the east, the 'transcontinental' segment of the Damaran–Katangan chain on the north, the folded coastal zone including the Gariep System on the west, and the orogenic zone of the Malmesbury Beds on the south (see figure 7). Within this craton, indisputable correlatives of the Upper Precambrian are rare. However, the major part of the outcrop of the Nama System in South-West Africa (see figure 8) is flat-lying or weakly folded, even though its northern portion was involved in the folding of the Damara Facies. Elsewhere, the presence of even older sequences which have not suffered Upper Precambrian-to-Lower Palaeozoic orogenic activity, demonstrates the prolonged stability of the region; such sequences include the Transvaal System (older than 2000 m.y.), the Waterberg–Loskop sequence (older than 1700–1800 (?) m.y.), and the Umkondo System which has traditionally been correlated with the Waterberg System (see figure 8).

The *West African Craton* (see figure 7) has been defined by Kennedy (in preparation) as an ancient stable region extending from the Anti-Atlas southwards to the Ivory Coast, and from the western margin of the Hoggar westwards to the Mauritanide zone. It represents a long established rigid area within which-flat lying sediments of infra-Silurian, Palaeozoic and later age rest on Precambrian crystalline basement that emerges in the south, and over an extensive area in northern and northwestern Mauritania. This basement has been dated at 1850 ± 250 m.y., and stands in marked contrast to the younger marginal belts to the east and west, including: (i) on the east, the Nigeria–Dahomey–Hoggar zone of deformation yielding ages of 450–680 m.y.; and (ii) on the west, the Mauritanide zone of late Precambrian (?)-to-Upper Palaeozoic deformation, and the ca. 500 m.y. belt in Sierra Leone.

III Radiometric Ages in the 800–1300 m.y. Range

A *General statement*

In southern Africa, a significant number of ages between 800–1300 m.y. reflect the effects of the Kibaran orogenic episode. The undisturbed effects of this episode are best preserved in the regions which were not subsequently affected by orogenesis; viz. the Congo and Kalahari Cratons (see figure 7); in contrast, this important event has not, thus far, been recognized in the West African Craton. Within these regions, ages reflecting orogenic effects occur principally in two zones: (i) the Kibara–Karagwe Ankolean Belt of Central and East Africa; and (ii) the Namaqualand (Cape Province)–Natal Belt of South Africa (see figure 9). In addition, a number of 'Kibaran' ages have been recognized within the zones of more youthful orogenesis; in most cases, however, the effects which produced these apparent 'Kibaran' ages are uncertain as, for example, in the case of scattered 800–1300 m.y. ages in the Mozambique Belt, the Central African Republic, northern South-West Africa, the Lower Congo, Angola, the Hoggar, Senegal and elsewhere (see figure 9).

Apart from the ages reflecting orogenic effects, and the smaller number of 'relict' ages in more youthful zones of deformation, anorogenic effects of similar radiometric date include igneous intrusions in the Transvaal, Botswana and Rhodesia where they intrude undeformed sedimentary sequences.

In the discussion which follows, ages for orogenic effects, anorogenic effects and effects recorded in more youthful orogenic zones, are treated separately.

B *Orogenic effects*

The principal examples of orogenic effects reflected by 800–1300 m.y. ages are: the Kibara–Karagwe Ankolean Belt; and the Namaqualand–Natal Belt. In addition, the Muva–Irumide Belt has been correlated with the former, whereas the Ubendian–Ruzizi Belt has many points in common with the latter (see figure 9); although, therefore, the significance of the 800–1300 m.y. event is still uncertain in these two belts, they are most logically treated in this section.

The Kibara–Karagwe Ankolean Belt This belt consists of 30,000 feet of geosynclinal sediments referred to as the Kibara, Urundi and Karagwe Ankolean Groups (Haughton, 1963). For the most part,

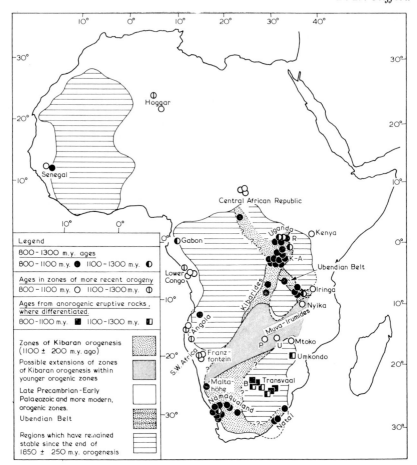

Figure 9 Distribution of 800–1300 m.y. ages shown in relation to certain established, and suggested, elements of African structure.

B. Botswana; U. Urungwe; P. Phoenix, southern Zambia; K–A. Karagwe Ankolean region, Uganda; R. Ruwenzori, Uganda.

these rocks are argillites, quartzites, grits and conglomerates, with volcanics and some carbonate rocks. This sequence can be traced from Katanga Province, where it appears from beneath the rocks of the Katanga System, northeastwards into Uganda, where it swings to a westerly trend in the Congo and disappears beneath the flat-lying Upper Precambrian rocks which skirt the great Congo Basin (see Furon and others, 1959).

A wide variety of ages are available for the mineralization in this belt. Pegmatite mineral ages include: muscovites and microcline from Rwanda, Burundi and Kivu which have yielded Rb–Sr ages ranging from 870 m.y. to 1130 m.y.; and uraninite and cyrtolite from Kivu with ^{207}Pb/^{206}Pb ages of 850 ± 20 m.y. (Monteyne-Poulaert and others, 1962a,b). To the SSW, in the Kibara System in the Congo, pegmatite lepidolites from Manono and Muika have given Rb–Sr ages of 920(945) and 900 m.y.; whereas their K–Ar ages are 840 and 915 m.y. respectively (Aldrich and others, 1958). In addition, yttrocrasite from tin-bearing veins at Mitwaba has given an age of 1030–1130 m.y. (Ledent and others, 1956), and biotite from a late tectonic granite at Mandwe in the Kibara Mountains has given a K–Ar age of 905 ± 40 m.y. (Snelling, 1964).

In the northward extension of this geosynclinal belt in southwestern Uganda, pegmatite cutting the Karagwe Ankolean sediments has given an unreliable age of 750 m.y. (Holmes and Cahen, 1957, p. 54; Cahen, 1961), whilst in western Tanzania muscovite from Karagwe Ankolean quartz–muscovite metasediment has given a K–Ar age of 900 ± 30 m.y. Furthermore, a whole rock K–Ar age of 1285 ± 50 m.y. has been obtained for a slate-mudstone sample from the Karagwe Ankolean Group at Masaka in Uganda (Snelling, 1963).

In addition to these ages from the regime of Kibara–Urundi–Karagwe Ankolean rocks, Cahen and others (1966) report: a Rb–Sr whole rock–microcline isochron age of 1310 ± 40 m.y. for pre- or early-tectonic granite in northern Katanga; Rb–Sr ages of 1240 ± 70 m.y. for whole rock and microcline samples from syntectonic granite; and Rb–Sr ages of 875 ± 26 to 1010 ± 50 m.y. for mineral and whole rock samples from posttectonic granites, pegmatites and veins. In this regard, it is perhaps worthy of note that Snelling (1964) reported whole rock Rb–Sr determinations on granite gneisses in the Ntungamo arena of southwestern Uganda which yielded an isochron indicative of strontium isotope homogenization at 1415 ± 35 m.y.; Snelling tentatively interpreted this as the age of the post-Karagwe Ankolean metamorphism.

From their NNE trend in the Congo, the folds of the Kibara (Karagwe Ankolean) Group swing westwards from southwestern Uganda and disappear beneath the younger sediments of the Congo Basin (see figure 9). The extent of the Kibara–Karagwe Ankolean Belt in that direction is, therefore, uncertain. However, in the Central African Republic, ages between 800–1300 m.y. have been recorded from crystalline rocks of the Ouango massif which underlie the supposed Upper Precambrian just north of the Oubangui River. The rocks are granites intrusive into supposed Lower Precambrian amphibolites (Gérard, 1956) and biotite from a granite and from a gneissic enclave has given Rb–Sr ages of 890 ± 30 and 970 ± 30 m.y. respectively (Roubault and others, 1965). Although these data are inconclusive, Roubault and others note that they accord with ages in the Kibara Chain to the southeast; they are, therefore, *tentatively* connected with that chain in figure 9.

Furthermore, it should be pointed out that although the Karagwe Ankolean sediments of southwestern Uganda swing westwards, and show no extensive development towards the north, a number of ages in the 800–1300 m.y. range have now been recorded from supposedly pre-Karagwe Ankolean rocks in western Uganda in the vicinity of the western branch of the rift valleys. For example, three biotites and a muscovite from schist, gneiss and biotite–chalcopyrite samples from the Ruwenzori Mountains and the Kilembe Mine have yielded K–Ar ages of 1030 ± 40 to 1265 ± 30 m.y. (figure 9). In addition, biotite from a pegmatite at Kilembe Mine has yielded a K–Ar age of 705 ± 30 m.y., whilst a whole rock phyllite associated with Toro quartzites of Butiti has yielded an age of 1200 ± 40 m.y. (Snelling, 1963, 1965). The rocks from which these 'Kibaran' ages were obtained have long been regarded as older than the Karagwe Ankolean sequence; Snelling (1965, p. 30) concludes, therefore, that they have presumably 'been thrust . . . during the post Karagwe Ankolean diastrophism' which is reflected by ages in the 1000–1250 m.y. range. The thermal influence of Kibaran deformation thus clearly extends northwards in the vicinity of the western rift valley.

The Namaqualand–Natal Belt Early work by Holmes (1950) on minerals from the Gordonia region of the northern Cape Province of South Africa revealed ages of ca. 1000 m.y. for pegmatite uraninite. On the basis of these data, Holmes linked the Gordonia (Namaqualand) Belt with the Kibara–Karagwe Ankolean Belt via the Damara orogen of South-West Africa (see figure 2). However, subsequent work has shown that orogenic activity in the Damara orogen took place at a much later date (the Damaran episode) some 450–550 m.y. ago.

More recent radiometric dating studies of samples from the northern and western parts of the Cape Province have substantiated the earlier results from that region (see figure 9) (Nicolaysen, 1962; Burger, 1962; Nicolaysen and Burger, 1965). The detailed radiometric data from the Orange River region and Namaqualand fall within the

900–1250 m.y. range, and include: K–Ar ages for muscovite and amphibole samples from schist and amphibolite respectively; Rb–Sr ages for muscovite and microcline samples from granite, pegmatite and gneiss; and U–Th–Pb ages for zircon, monazite and other rare earth minerals from a variety of source rocks including granite gneiss, pegmatite and sulphide ore bodies (Burger, 1962; Burger, 1964; Nicolaysen and Burger, 1965) These data are taken as support for a '1000 million year old orogeny in this region' (Burger, 1964); in more detail, it has been considered that the metamorphic strata were 'intensely metamorphosed and reconstructed when they moved into giant upwellings of basement "infrastructure" about 1000 m.y. ago' (Nicolaysen and Burger, 1965).

The extent of this event is shown in figure 9; however, its northward extension into South-West Africa is only confirmed by a single whole rock K–Ar age of 954 ± 34 m.y. for a pre-Nama System hornblende rock in the Maltahöhe Region of South-West Africa (Clifford, Grasty and Miller, unpublished work).

Although the widespread ages in the Namaqualand portion of the Cape Province are similar to those in the Kibara–Karagwe Ankolean Belt, no extensive geosynclinal sequences equivalent in age to the Kibara Group have been recognized. The country rocks are largely granite, granite gneiss, and remnants of the Kheis System previously correlated with the Archaean systems (older than 2600 m.y.); the latter include acid and basic lavas and a wide variety of sediments.

The problem of the Kheis System has been discussed by Nicolaysen (1962); present evidence seems to suggest that this region represents, in large part, the remobilized infrastructure of an ancient orogenic zone which is *now* largely devoid of recognizable geosynclinal sediments of Kibara Group age (see Nicolaysen and Burger, 1965).

To the south these crystalline rocks disappear beneath younger rocks, largely Karroo (see Furon and others, 1959). Further east, however, along the Natal coast, a group of crystalline rocks appears from beneath the younger Palaeozoic cover. In general these rocks have roughly E–W structural trends, and include granite and gneiss together with a variety of metasediments, lavas, hornblendic rocks and granulites (Haughton, 1963, p. 114). Radiometric data from four localities in the main granite–gneiss complex of Natal have given evidence of ca. 950–1150 m.y. gneissification and granite emplacement (Nicolaysen and Burger, 1965). Rb–Sr ages include: 855 m.y. for biotite from syenitic gneiss; 900–1070 m.y. for microcline and biotite from gneissic granite; 950 m.y. for biotite from charnockite; and

955 m.y. for phlogopite from marble. In addition, $^{207}Pb/^{206}Pb$ ages of 1100–1140 m.y. were obtained for zircon from gneiss. In view of the similarity of the age patterns of Natal and Namaqualand, Nicolaysen and Burger (1965) have suggested a structural link beneath the thick Karroo basin which presently separates them (see Furon and others, 1959); the relation shown in figure 9 is merely a modified version of their originally suggested pattern.

The Muva–Irumide Belt   In contrast to the Kibara–Karagwe Ankolean and Namaqualand–Natal Belts there are no definitive ages for the Muva–Irumide segment over the major portion of its postulated length (figure 9; see also Ackermann, 1950, 1960; Furon and others, 1959). This zone is occupied by quartzites, quartzite conglomerate and argillaceous rocks of the Muva System with widespread granite, forming a NE–SW trending chain approximately parallel to the Kibara–Karagwe Ankolean Belt to the west (Guernsey, 1950). This rock sequence has been tentatively correlated with the Kibara Group (Garlick, in Mendelsohn, 1961, p. 28); more recently, however, the 'Muva' has been found to consist of a number of sequences separated by unconformities (De Swardt and others, 1964, p. 8).

The problem of the stratigraphic and structural significance of the Muva–Irumide Belt is aggravated by the fact that a large part of it was apparently subsequently affected by the orogenic activity of the Damaran and possibly Katangan episodes (Cahen, 1961, p. 561; Clifford, 1965), with the result that ages greater than 500–600 m.y. are sparse. However, in the Southern Province of Zambia, pegmatite yielding consistent Rb–Sr and K–Ar mineral ages in the 1000–1100 m.y. range occurs in association with folding along NE trending axes in pre-Upper Precambrian rocks (figure 9); De Swardt and others (1964) note that the northeasterly trending fold phase may be comparable in age with the Kibara–Karagwe Ankolean Belt. In addition, to the south, in the Urungwe district of Rhodesia a Rb–Sr age of 1170 ± 40 m.y. has been obtained for muscovite from a granite which has yielded whole rock and perthite Rb–Sr ages of 2080 m.y. and 1610 m.y. respectively (Clifford, Rex and Snelling, 1967). In further support, to the WSW, common lead ages in the same general range have been obtained from the Wankie district of Rhodesia. The only other available age data in the ca. 1000 m.y. range from within the general trend of the Muva–Irumide Belt is a whole rock Rb–Sr age of 1000–1100 m.y. obtained from the Nyika granite in northern Malawi (Snelling, 1963; see figure 9), within the Mozambique Belt, which is characterized by radiometric ages in the 400–700 m.y. range (Cahen, 1961).

Clearly, evidence of the effects of the Kibaran orogenic episode in the Muva–Irumide zone is scant. The zone is, however, of considerable importance in East and Central Africa; as a zone of rejuvenation during the Damaran orogenic episode, its analysis and the nature and trend of its boundaries are critical to the understanding of the polycyclic Mozambique Belt.

The Ubendian–Ruzizi Belt The Ubendian Belt of Tanzania and its continuation, the Ruzizi Belt of the eastern Congo (see figure 9), form a linear NW trending orogenic zone composed largely of metamorphosed pelites and psammites with an abundance of biotite gneiss and migmatite (Quennell and Haldemann, 1960). This belt was originally tentatively assigned to the 2000 m.y. cycle by Holmes and Cahen (1957, p. 103); Snelling (1963) has suggested that the more recent age data set a younger limit of 1800 m.y. to the age of this zone; these data are discussed later in this work.

However, in addition to this older age pattern, it has long been known that this belt is also the regime of much younger ages of ca. 1000 m.y. (Cahen, 1961). These more youthful ages fall within the range of 840–1040 m.y. and include: K–Ar ages of biotites from granite (Iringa) and carbonatite (Chunya); K–Ar and Rb–Sr ages of muscovite and microcline respectively from the Tungwa Mine; and K–Ar and Rb–Sr ages of muscovite from the Karema Mine. In addition, a K–Ar age of 1300 m.y. has been obtained for hornblende from the Chimala granite (Aldrich and others, 1958; Snelling, 1964, 1965).

Although no extensive rock sequences of Kibara Group age have been identified within the Ubendian–Ruzizi Belt, these ages clearly reveal a thermal imprint coeval with that within the Kibara–Karagwe Ankolean Belt; whether this imprint reflects an orogeny, or merely a thermal disturbance, is still not known.

'Kibaran' ages preserved in younger orogenic zones An increasing number of ages in the 800–1300 m.y. range are being found 'preserved' within the late Precambrian–early Palaeozoic system of orogenic zones (see figure 9). Some of these have already been referred to; for example, the Muva–Irumide segment in central Africa. In other regions, however, an even greater degree of speculation surrounds similar ages.

In northern *South-West Africa*, basement granite and pegmatite underlying the folded Upper Precambrian (the Otavi Facies) have given whole rock (granite) and muscovite (pegmatite) Rb–Sr ages of 1185 ± 80 m.y., and a microcline (granite) Rb–Sr age of 900 ± 50 m.y. (Clifford and others, 1962; Simpson and Otto, 1960).

Similar ages of 920 ± 24 m.y. to 1281 ± 57 m.y. have been obtained on biotites from granites of the same area in association with the late Precambrian–early Palaeozoic orogenic zone along the coast of *Angola* (Mendes, 1964).

As has been noted earlier, the coastal zone of crystalline rocks in the *Lower Congo* represents a zone of great antiquity on which important Damaran and Katangan events have been imprinted (figure 9). Within that coastal Mayumbian Belt, Rb–Sr ages of 945 ± 380, 1090 ± 285 and 1420 ± 160 m.y. have been obtained for microcline from augen gneisses (Cahen and others, 1963). Furthermore, Pb/α zircon ages of 1180–1480 m.y. have been obtained from granites and diorites in the same belt (Cahen, 1961) whilst further north, in Gabon, a Pb/α age of 1292 m.y. was obtained for zircon from granitic rocks close to the apparent boundary of the 450–680 m.y. orogenic belt (Cahen, 1961; see figure 9).

In the *Central African Republic* ages in the 800–1300 m.y. range have been identified in the zone of basement crystalline rocks which were involved in the late Precambrian–early Palaeozoic orogenesis (see p. 321; and figure 9). Within this zone a number of Rb–Sr ages ranging from 755 ± 30 m.y. to 1040 ± 70 m.y. have been obtained from muscovite, biotite and microcline from granite, gneiss and gneissic enclave samples (Roubault and others, 1965). These are relict ages within 500–600 m.y. isochronic terrain; indeed, a microcline yielding a Rb–Sr age of 987 ± 300 m.y., gave a K–Ar age of 490 ± 20 m.y. (Roubault and others, 1965). These 'Kibaran' ages may represent the northward extension of the Kibaran Belt from the Ouango massif (see p. 339).

Other 'Kibaran' ages within this younger belt of orogeny elsewhere in Africa include the following: Rb–Sr ages of 960 ± 95 to 1210 ± 110 m.y. for muscovite from mica schist and quartzite, and whole rock quartzite in the Hoggar of West Africa; a K–Ar age of 855 ± 30 m.y. for hornblende from an amphibole inclusion in basement charnockite in Kenya; a lead–uranium age of 835 m.y. for monazite from a pegmatite at Baragoi, Kenya; and a K–Ar age of 965±40 m.y.for lepidolite (Rb–Sr age of 2370±55 m.y.) and a Rb–Sr age of 1090±70 m.y. for muscovite from the Benson Mine in Rhodesia (Lay and Ledent, 1963; Snelling, 1965; Cahen, in Snelling, 1963; Snelling and others, 1964a; Boissonnas and others, 1964).

Additional ages of 800 to 1300 m.y. have been obtained for sediments. Rb–Sr determinations of the argillaceous fraction of the Walidiala argillite, and of the whole rock Koulountou pelite of Senegal have given ages of 1022 ± 20 and 868 ± 370 m.y. respectively (Bassot and others, 1963).

The interpretation of all of these scattered ages is still in doubt. It is worth bearing in mind, however, that Holmes (1951) *originally* considered the possibility that a zone of Kibaran activity may have extended northwards from South-West Africa into the Lower Congo along the western coastal region of Africa. An extensive programme of whole rock analysis within the late Precambrian–early Palaeozoic orogenic zones in the eastern and western parts of southern Africa should be most revealing.

C *Anorogenic effects*

The principal thermal events of anorogenic character are igneous activities which particularly characterize eastern Botswana and the Transvaal (see figure 9). Although these events do not fall entirely within the 800–1300 m.y. period, they represent events which are not readily subdivided.

Of particular interest are the age measurements on samples from alkaline complexes in the Transvaal. These include: a K–Ar age of 1250 ± 60 m.y. for biotite from the foyaite contact zone in the Pilanesberg alkaline complex, and Rb–Sr ages of 1215 ± 180 m.y. on biotite–chlorite concentrates from the Robinson dyke, one of a suite of 'Pilanesberg dykes' (Schreiner and Van Niekerk, 1958; Snelling, 1963). Although a somewhat older age of 1420 ± 70 m.y. has been obtained for soda syenite from the Leeuwfontein alkali syenite complex northeast of Pretoria (Oosthuyzen and Burger, 1964), the results may be included here as this complex is probably consanguineous with the Pilanesberg.

More recently, Snelling (1965) has obtained K–Ar ages of 940 ± 30 to 1255 ± 40 m.y. for plagioclase–pyroxene samples from dolerite sills in eastern Botswana, and 865 ± 30 and 870 ± 30 m.y. for similar samples from the Pretoria–Loskop Dam region of the Transvaal. Moreover, a K–Ar whole rock age of 1140 m.y. has been obtained for dolerites cutting the Umkondo System in southeastern Rhodesia (see Johnson and Vail, 1965).

These ages seem to suggest a period of intrusion of alkali syenite complexes some 1200–1400 m.y. ago, and a period of dolerite emplacement at a rather younger (800–1100 m.y.) period. In addition, the dates are of stratigraphic significance in that these igneous rocks postdate the extensive sequence of sediments of the Waterberg–Loskop Systems (Du Toit, 1954; Snelling, 1965), and thus place a minimum of 1300–1400 m.y. on the age of that system; additional ages, discussed later in this work, indicate an even older age (greater than 1790 m.y.) for at least part of this system.

D Summary

The distribution of ages in the 800–1300 m.y. range in Africa is shown in figure 9. It should be borne in mind that Holmes (1951, p. 261) was of the opinion that because of the 'remarkable convergence of evidence it can be suggested with some confidence that the Karagwe Ankolean, Urundi, Kibara, Damara and Kheis Systems all belong to a single geosyncline which became an orogenic belt about 900 m.y. ago'. With the exception of the Damara, which is now known to belong to the late Precambrian–early Palaeozoic system of orogenic belts, this statement still accommodates a number of available facts. However, the relation shown in figure 9 between the Kibara–Karagwe Ankolean Belt in the north and the Namaqualand–(Kheis System)–Natal Belt in the south, cannot be regarded as firmly established; the precise nature of their connexion is obscured by the late Precambrian–early Palaeozoic orogenic zone, and by the extensive cover of Upper Palaeozoic-to-Recent sequences of the Kalahari Basin (see Furon and others, 1959). Moreover, the special difficulties of the Muva–Irumide segment and its association with this 'main stem' of orogeny have already been stressed, and focus attention on the problems of the recognition of the regional extent of older orogenic effects in more youthful belts. Despite these glaring problems, the pattern presented in figure 9 coordinates the available data into a unified picture which, at least, presents a target for future research.

Dating the actual events of orogenesis within the 800–1300 m.y. range is still under discussion. On the evidence from the Kibara–Karagwe Ankolean Belt, Cahen and others (1966) report syntectonic granite emplacement at ca. 1250 m.y. ago and maintain that ages younger than 1100 m.y. date posttectonic events. In contrast, largely on the basis of U–Th–Pb and Rb–Sr ages, tectonism involving massive upwelling of basement infrastructure around 1000 m.y. ago has been inferred for the Namaqualand–Natal Belt (Nicolaysen, 1962; Nicolaysen and Burger, 1965). It is, of course, possible that different orogenic phases affected these two belts, though this view seems unlikely because their age patterns are virtually identical. More radiometric data are required before this enigma can be resolved, but in the present state of our knowledge of the *subsequent* late Precambrian–early Palaeozoic orogenic activity, it is at least *less* likely that the 800–1300 m.y. thermal imprint *only* consisted of: (1) orogeny reflected by ages in the 1100–1300 m.y. range; and (2) posttectonic events reflected by ages in the 800–1100 m.y. range. For the time being it is considered that the Kibaran orogenic episode should be defined as the regional orogenesis reflected by ages in the 1100 ± 200 m.y. range.

From a consideration of stratigraphy, the Kibaran orogenic episode places a maximum age on the rocks of the Katanga System (see figure 8) and, with the minimum age of 500–600 m.y. for that system, clearly places the major part of the sequence in the Upper Precambrian. Similar ages of ca. 900–1200 m.y. ages set a *maximum* age limit on the following: the Otavi and Damara Facies and the Nama (and Auborus) System of South-West Africa; the Sijarira Beds which rest directly on the Urungwe granite of northern Rhodesia; and the flat-lying correlatives of the Katanga System in the Ouango massif of the Central African Republic (see figure 8).

Conversely, the ages from the Kibaran orogenic episodes place a *minimum* age on the Kibara, Karagwe Ankolean and Urundi Groups in the Congo, Uganda and Rwanda. In addition, anorogenic ages from igneous rocks in southern Africa place minimum ages on the Waterberg System (and presumably the Matsap System) of Botswana and South Africa, and on the Umkondo System of Rhodesia; additional, more definitive data on the ages of these systems indicate even greater antiquity (see p. 370).

IV Radiometric Ages in the 1300–2100 m.y. Range

A General statement

In the southern half of the African continent and in West Africa, there is widespread evidence of thermal and tectonothermal events reflected by ages in the 1300–2100 m.y. range. The evidence is most clearly preserved in those regions which have remained cratonic during Kibaran and more youthful orogenic episodes. These ancient stable regions include (see figure 10): a large part of West Africa;

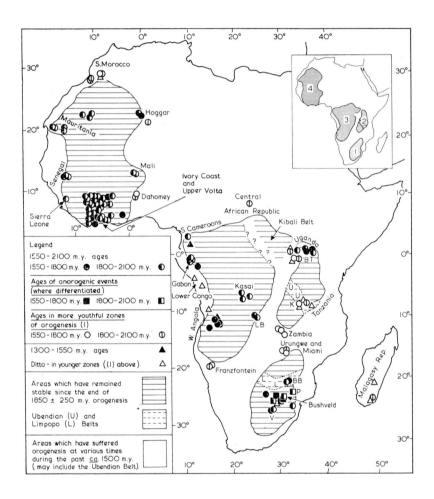

Figure 10 Distribution of 1300–2100 m.y. ages shown in relation to certain elements of African structure.
L. Limpopo Belt; LB. Lukoshian Belt; BB. Beit Bridge; K. Kate, Zambia; P. Palabora; U. Ubendian Belt; V. Vredefort Ring. BT. Buganda–Toro Belt of Uganda.

Inset: regions of Africa which have remained largely stable since the end of the 1850 ± 250 m.y. orogenesis as follows: 1. Rhodesia–Transvaal Craton; 2. Tanzania Craton; 3. Kasai–Angola Craton; 4. West African Craton.

southern Uganda, western Kenya and a major part of central Tanzania; the central region of Rhodesia and a large part of eastern South Africa; and large areas of western Angola, the Congo, Gabon, southern Cameroons and the Central African Republic.

Within these regions there is abundant evidence of orogenic and more limited evidence of anorogenic effects, yielding ages largely in the 1600–2100 m.y. range. In addition, a considerable number of ages within this range are preserved within belts of more youthful orogenesis, particularly the late Precambrian–early Palaeozoic mountain chain.

B *Orogenic effects*

Orogenic effects reflected by ages in the 1300–2100 m.y. range are known from a number of parts of Africa. Perhaps the most characteristic data have been obtained from West Africa where a uniform age pattern in the 1800–2050 m.y. range dates the *Eburnian orogenic episode*, particularly in the Ivory Coast and the Upper Volta region; similar ages have been noted from Niger Republic, Senegal, Mauritania and southern Morocco. From the same general region, there is evidence of a thermal pulse dated at 1600–1700 m.y.; although this may represent an orogeny in southern Morocco, its structural significance elsewhere in West Africa is uncertain.

In addition, although the present results are less conclusive, events in the Buganda–Toro Belt of Uganda, the Ubendian–Ruzizi Belt of Tanzania, Zambia and the Congo, the Limpopo Belt of Rhodesia and the Transvaal, the Lukoshian Belt of the Congo and the Luiza Series of Kasai, have all given ca. 1800–1900 m.y. ages suggestive of Eburnian orogenic activity. To a similar cause may be ascribed the radiometric ages from Gabon and the southern Cameroons, though the areal extent of the activity in that region is, as yet, uncertain. Equally uncertain is the structural significance of 1600–1700 m.y. ages from western Angola; they are included in this discussion of orogenic effects because they reflect a regional process in granitic rocks.

West Africa It has already been noted that the major part of West Africa represents a region of antique stability bounded by the more youthful zones of orogenesis of: the Atlas and Anti-Atlas on the north, the Mauritanides on the west, and the Dahomey–Hoggar Belt on the east. Within this region, rocks of infra-Silurian and younger age rest on Precambrian basement which emerges: in Sierra Leone, Liberia, Guinea, Upper Volta, Ivory Coast and Ghana in the south; over extensive areas of northern and northwestern Mauritania; in the Spanish Sahara; and locally in the Anti-Atlas region (see figure 1). These basement areas show the widespread influence of orogenic activity reflected by ages in the 1600–2100 m.y. range (figure 10); local evidence of even older events in Mauritania and elsewhere, are dealt with later in this work.

In the *Ivory Coast, Upper Volta* and surrounding regions a major part of the Precambrian consists of a strongly folded succession of schists with quartzites and extrusive greenstones referred to the Birrimian System (Haughton, 1963). These rocks are widely invaded by granites which have been subdivided into: *Baoulé* granites which are very extensive and concordant bodies; and *Bondoukou* granites, which are more local and discordant bodies.

Intensive Rb–Sr dating studies have been carried out by Bonhomme (1962) and Vachette (1964c) on mica and whole rock samples of the granites from the Ivory Coast and Upper Volta (see figure 10). Of the 70–80 granite ages obtained, approximately 15% fall in the 2000–2100 m.y. range, more than 50% fall in the 1900–2000 m.y. range, 20–25% fall in the 1800–1900 m.y. range and 6 ages fall in the 1600–1800 m.y. range (Bonhomme, 1962; Vachette, 1964c).

The most detailed study has been carried out on the Baoulé granite types, and analyses of whole rock, muscovite and biotite samples from various granite masses show a clear concordance in age (see table 2), and support Bonhomme's view that the granites were emplaced in the Birrimian during the time interval 1830 to 2030 m.y. ago. This important time event is the Eburnian episode, and the fact that the Baoulé granites are concordant and syntectonic suggests that it was an important orogenic event.

Other minerals yielding Rb–Sr ages in the 1830–2030 m.y. range include: muscovite, potash feldspar and biotite from pegmatite associated with the Baoulé granite; biotite from an orthogneiss 'boulder' in the Birrimian; and biotites from migmatites (Bonhomme, 1962; Vachette, 1964c). In addition, a U–Th–Pb age of ca. 1950 m.y. was obtained for uraninite from pegmatite cutting the Birrimian System. These data support the concept of the widespread influence of the Eburnian tectonothermal event.

Despite this uniform pattern, however, it should be noted that Pb/α ages of 2680 m.y. and 2310 m.y. have been obtained for zircon from the Man granite, and from post-Birrimian pegmatite respectively

Table 2 Rb–Sr ages of granites from Ivory Coast and Upper Volta

Granite body	Dated sample	Age (m.y.)	Reference[a]
Baoulé granite type			
Bouaké	Biotite	1988 ± 29	1
	Muscovite	2016 ± 25	1
	Whole rock	1930 ± 50	1
Niakaramandougou	Biotite	1932 ± 27	1
	Muscovite	1925 ± 27	1
	Whole rock	1930 ± 80	1
Dziguitiela	Biotite	1913 ± 27	2
	Muscovite	1931 ± 89	2
	Whole rock	1870 ± 157	2
Tovré	Biotite	1930 ± 12	2
	Whole rock	1914 ± 153	2
Agboville	Biotite	1850 ± 21	1, 2
	Muscovite	1995 ± 29	1, 2
	Whole rock	1937 ± 117	1, 2
Adzope	Biotite	1932 ± 20	1, 2
	Muscovite	1938 ± 48	1, 2
	Whole rock	1923 ± 135	1, 2
Bianao	Biotite	1898 ± 49	1, 2
	Muscovite	1948 ± 34	1, 2
	Whole rock	1927 ± 60	1, 2
Bondoukou granite type			
Ayame	Biotite	1805 ± 32	2
	Whole rock	2095 ± 58	2

[a] 1. Bonhomme (1962); 2. Vachette (1964c).

(Roques, 1959); however, biotite from charnockite associated with the former has given a Rb–Sr age of 2028 m.y. (Bonhomme, 1962).

In contrast to the Baoulé granites, the Bondoukou granites are discordant and smaller in size; they probably postdate the Baoulé granites. The mineral–whole rock age of the Bondoukou granites (table 2), together with 14 biotite ages in the 1850–2025 m.y. range, is taken as support for the view that the emplacement of these granites took place during the Eburnian event (Bonhomme, 1962).

Apart from these consistent ages, a number of Rb–Sr ages in the 1610–1770 m.y. range have been obtained as follows: muscovite and biotite from migmatites, 1697 ± 41 to 1721 ± 55 m.y.; biotite from the Tienlo granite of Baoulé type, 1707 ± 44 m.y. (whole rock age, 1947 ± 64 m.y.); biotites from two other Baoulé granites, 1610–1620 m.y.; and biotite from a fine-grained enclave in Bondoukou granite, 1683 ± 30 m.y. From the similarity of their age in the 1610–1770 m.y. range, Vachette (1964c) recognizes a relation between migmatization and the rejuvenation of certain granite biotites; the structural significance of this subsidiary event is, however, still uncertain.

Equally uncertain is the significance of ages obtained from granites which intrude the Kinkéné Series in the Ivory Coast; this series has been correlated with the Tarkwaian of Ghana, a sequence of conglomerates, schists and phyllites which were laid down after orogenesis and granite emplacement in the Birrimian. The post-Kinkéné granite has yielded whole rock and biotite Rb–Sr ages of 2004 ± 42 and 2020 ± 35 m.y. respectively. In the light of their similarity to Eburnian ages, the relationship of the Kinkéné and Birrimian sequences requires reappraisal. A previous K–Ar age determination on feldspar from pegmatite cutting type Tarkwaian east of Konongo in Ghana gave an age of ca. 1650 m.y. (Holmes and Cahen, 1957, p. 60).

In Niger Republic to the NNE of the Ivory Coast, the Birrimian System is considered to be represented by the Yatakala Series of amphibolites, schists and quartzites outcropping in the midst of the Liptako granites and granitized by them (Haughton, 1963). Rb–Sr ages of biotite from the Liptako granite and lepidolite from pegmatite are 1860 ± 110 m.y. and 1920 ± 60 m.y. respectively (Bonhomme, 1962).

In *Senegal*, the supposed equivalents of the rocks of the Birrimian System comprise a strongly folded sequence of: (i) metamorphosed volcanics (Mako Series); and (ii) metamorphosed pelites, greywackes and limestones (Dialé and Daléma Series) (Bassot and others, 1963). This succession is cut by syntectonic and posttectonic granites which have given biotite Rb–Sr ages of 2011 ± 39 m.y., 2037 ± 39 m.y. and 2086 ± 106 m.y. (Bassot and others, 1963); these data indicate that the Mako, Dialé and Daléma Series are older than 2000 m.y. and that the granites are part of the Eburnian episode of granite emplacement.

To the north, in *Mauritania*, two extensive metamorphic series, the Bakel–Akjoujt and the Amsaga, have been recognized; of these, the former has already been discussed (p. 328). The Amasaga Series lies to the east of the Bakel–Akjoujt outcrop and consists of a highly folded complex of gneisses with quartzites, mica schists and calc-silicate rocks; the whole sequence contains granites (both discordant and concordant) and shows strong regional metamorphism with the development of sillimanite, garnet, cordierite and hypersthene (Haughton, 1963).

A number of Rb–Sr biotite ages have been published for the granites in the Amsaga Series, and both concordant and discordant bodies have yielded similar ages in the 1950–2000 m.y. range (Bonhomme, 1962). These data offer support for the view that the granites were emplaced during the Eburnian episode, and that the Amsaga Series is older than ca. 2000 m.y. That the Eburnian episode also affected older rocks is demonstrated by ages of ca. 1810–1860 m.y. obtained on biotites from leucocratic and migmatitic granites associated with ca. 2600 m.y. pegmatites in the Sattle Ogmane complex in the same general region of Mauritania; even more youthful (1610–1670 m.y.) ages are reported from the nearby Tasiast Group (older than 2600 m.y.) (Bonhomme, 1962; Vachette, 1964b).

The exact age of the Amsaga Series is unknown; it may be correlated with the Tasiast Group (and therefore older than 2600 m.y.), or younger than that group (and therefore representative of the 2000–2600 m.y. time interval). It is, perhaps, worthy of note that the Amsaga Series has been correlated with the Dahomeyan (see Bonhomme, 1962, p. 37).

In *southern Morocco*, Precambrian rocks appear from beneath the Palaeozoic rocks at a number of points along the Anti-Atlas (see Furon and others, 1959). These have been subdivided into three series (from oldest to youngest): Precambrian I, the Zenaga Series; Precambrian II, the Kerdous Series; and Precambrian III, the Ouarzazate Series (Choubert, 1951).

Of these, the Zenaga Series has been tentatively correlated with the Birrimian System (Choubert, 1951), and is mainly a sequence of mica schists folded along generally N–S axes. It exhibits two granite events: the earlier (Azguemerzi granite) is associated with migmatites; the later (Tazenakht granite) is accompanied by pegmatites.

The Kerdous Series of the Kerdous massif, 125 miles east of the Zenaga Plains, consists of a variety of sediments and lavas folded along E–W or ENE axes, and intruded by granite, serpentine and gabbro. The Ouarzazate Series is a series of conglomerates, sandstones and lavas.

Age measurements for the Zenaga region have shown Rb–Sr ages of 1637 ± 100, 1669 ± 85 and 690 ± 35 m.y. for whole rock, muscovite and biotite respectively, from the Azguemerzi granite (see figure 10). In contrast, muscovite and whole rock samples from the Tazenakht granite yielded Rb–Sr ages of 1584 ± 65 and 1537 ± 90 m.y. respectively, whereas pegmatites from this region have given muscovite Rb–Sr and K–Ar ages of 1635 ± 50 m.y. and 1655 m.y. respectively, and a microcline Rb–Sr age of 1460 ± 45 m.y. (Cahen and others, 1964; Choubert, 1959). In view of the better precision of the measurements, the muscovite ages of 1670 ± 85 m.y. and 1580 ± 65 m.y. are adopted for the respective ages of these two granites (Cahen and others, 1964); the 690 ± 35 m.y. age obtained for biotite is regarded as a rejuvenation effect.

Radiometric studies of samples from the Kerdous massif in southern Morocco have given a whole rock granite age of 2600 ± 130 m.y. with a muscovite age of 1806 ± 55 m.y.; in addition, muscovite from a pebble of pegmatite in rocks at the base of Precambrian III has given an age of 1837 ± 55 m.y. (Cahen and others, 1964). Studies have shown that the granite predates Precambrian II, and the whole rock date has been taken to represent the age of the granite. The concordance between the pegmatite muscovite and granite muscovite ages indicates either complete rejuvenation due to orogenesis around 1820 m.y. ago (Eburnian?); or partial rejuvenation during a more recent event (e.g. 1600 m.y. ago) (Cahen and others, 1964).

These ages from southern Morocco serve to indicate the antiquity of orogenic events and focus attention on the problems of stratigraphic correlation in the region.

Kibali–Buganda Toro Belt In the northern part of the *Congo*, the rocks of the Kibali Group reputedly rest on an older metamorphic complex to the north, although 'the plane of contact between the two is in places a thrust plane' (Haughton, 1963, p. 55). This group includes sericite schists, graphitic schists, quartzites, conglomerates, carbonate rocks, tuffs, lavas and itabirites; the sequence is cut by post-Kibalian granites.

The Kibalian folding trends approximately WNW, and the fold belt stretches ESE towards the Buganda–Toro folding of Uganda (see figure 10). No K–Ar, Rb–Sr or U–Th–Pb ages are yet available for the Kibalian sector of this belt. However, galena from gold-reefs in the Kibali Group has given an average model age of 1840 ± 100 m.y. (Cahen, 1961, p. 546) suggesting that the Kibalian orogenic episode is older than 1850 m.y. (Cahen, 1963).

To the east, in *Uganda*, the pre-Karagwe Ankolean rocks of the southern, western and southwestern parts of the territory have largely been described in terms of the Toro System (including the Igara Schist Series and the Madi Quartzites), the Samia Series (Kavirondian System), the Bulugwe Series (Nyanzian System), intrusive granites of various ages, gneisses and migmatites.

McConnell (1959) attempted to clarify the stratigraphy and suggested that the largely sedimentary sequences of low metamorphic grade which underlie a large part of Buganda Province on the northern side of Lake Victoria, should be grouped as the Buganda Group. Extending the correlation to the west, McConnell (1959) included the following series of the Ruwenzori Mountain region in that group: the Stuhlmann Pass Series; the Stanley Volcanic Series; the Freshfield Pass Series; the Butahu Series; and the Kilembe Series. In addition, it was suggested that the Igara Schist Series (the upper part of the Toro System) of metasediments should be correlated with the Buganda Group, but that the Madi Quartzites (the lower part of the Toro System) are older. More recently the whole sequence has been classified as the Buganda–Toro System.

To the east of the main outcrop of the Buganda–Toro in Buganda Province, a similar sequence of rocks known as the Samia Series and the Bulugwe Series crops out in the southeastern corner of Uganda; the former is equated with the Kavirondian System of Kenya; whereas the latter is correlated with the Nyanzian System. Although these two series are correlated with the Buganda Group by McConnell (1959), it has been shown that an important period of granite intrusion and diastrophism separates the Nyanzian and the Kavirondian in Kenya (see Cahen, 1961, p. 545).

The region of the Buganda–Toro sequence of southern, western and southwestern Uganda has given consistent ages of ca. 1800 m.y. (see figure 10). Pegmatite in the Lunya district, for example, has given a Rb–Sr lepidolite age of 1860 ± 30 m.y., whereas others in the general Kampala–Mubende region have yielded Rb–Sr ages in the 1720–1770 m.y. range (Snelling, 1962; Holmes and Cahen, 1957). In addition, biotite from mica schists of the Stuhlmann Pass Series of Ruwenzori has given a K–Ar age of 1830 ± 80 m.y., whilst biotite from a cordierite–biotite gneiss at Mosozi has yielded a K–Ar age of 1723 ± 90 m.y. (Snelling, 1962, 1963). In contrast, K–Ar ages of 1345 ± 60 m.y. and 1260 ± 40 m.y. obtained for hornblende and muscovite respectively from Ruwenzori (Snelling, 1965), probably reflect the later imprint of the Kibaran orogeny along the line of the present Western Rift Valley (Snelling, 1965, p. 30).

Perhaps the most interesting age data for the Buganda–Toro rocks and their relationship to the overlying Karagwe Ankolean is from the Lwanda unconformity in the Masaka district of southwestern Uganda, where the Karagwe Ankolean consists of only slightly folded grits, quartzites and shales which rest unconformably on metamorphosed and highly folded schists probably belonging to the Buganda Group (McConnell, 1959). K–Ar age measurements from these two rock sequences support the view that the Buganda–Toro extends into southwestern Uganda and predates the Karagwe Ankolean (McConnell, 1959). The age data are as follows (Snelling, 1963).

	Age
Whole rock mudstone of the Karagwe Ankolean	1285 ± 50 m.y.[a]
---------------- UNCONFORMITY ----------------	
Muscovite from schist of the Buganda–Toro	1810 ± 80 m.y.

[a] This whole rock age is probably too low due to argon leakage but 'suggests a younger limit of 1300 million years for the . . . Karagwe Ankolean System' (Snelling, 1963).

To the west, within the outcrop area of the Karagwe Ankolean, whole rock Rb–Sr ages of 1600–1800 m.y. have been obtained for granite from the Masha arena, a granite mantled by, and close to the northern limit of, the Karagwe Ankolean outcrop and these ages suggest that it is pre-Karagwe Ankolean basement (Snelling, 1963, p. 33). The biotite age from this Masha arena is younger, reflecting the late Precambrian–early Palaeozoic *thermal* imprint in the region (see p. 328). Similar relationships have been identified in the Bwizibera granite to the east of Ruwenzori indicating the northward extension of the youthful node of southwestern Uganda and Rwanda (see p. 328).

Away from these later events, which seem to be concentrated broadly along the Western Rift, the Mubende granite gives a biotite K–Ar age of 1803 ± 80 m.y.; and a whole rock Rb–Sr age of 1920 ± 200 m.y. This granite is intrusive into the Buganda–Toro System in the western part of its main outcrop, and these data support an age of ca. 1800 m.y. for its emplacement (Snelling, 1962, 1964).

Similar ages have been obtained for granites from the eastern border region of Uganda (figure 10). For example, biotite from the Masaba granite mass has given K–Ar and Rb–Sr ages of 1870 ± 150 and

1850 ± 80 m.y., respectively; in addition, biotites from the Nagongera and Buteba granites in the same region have given K–Ar ages of 1749 ± 90 m.y. and 1879 ± 80 m.y., respectively. However, in the case of the Buteba granite, muscovite and potash feldspar have yielded Rb–Sr and K–Ar ages of 2175–2600 m.y., and a whole rock Rb–Sr age of 2515 ± 250 m.y. (Snelling, 1962, 1964). This granite has intruded the rocks of the Samia–Bulugwe Series and, in view of the mineral–whole rock age agreement, an age of ca. 2600 m.y. has been accepted as the date of emplacement of the granite; the discordance between this age and some of the mineral ages is taken to indicate the effect of the post-Buganda–Toro metamorphism ca. 1800–1900 m.y. ago.

The age data from eastern, western and southwestern Uganda indicate a well-defined belt of rocks characterized by ages in the 1800 ± 100 m.y. range. This belt extends from eastern Uganda, through Buganda Province, Ankole, Ruwenzori and into the Kibali Belt of the Republic of the Congo; the consistency of the ages 'suggests that this 1800 million year event marks an important episode of regional metamorphism and orogeny' (Snelling, 1962) and places a maximum age on the Karagwe Ankolean Group, and a minimum age on the Buganda–Toro sequence.

Ubendian–Ruzizi Belt The Ubendian Belt of southern Tanzania and northern Zambia, and the Ruzizi of the eastern Congo, form a continuous NW–SE trending zone largely occupied by crystalline rocks including metamorphosed pelites, psammites and limestones, with migmatites and biotite gneisses (Quennell and Haldemann, 1960); this belt was tentatively assigned to the 2000 m.y. cycle by Holmes and Cahen (1957, p. 103). More recently, a number of K–Ar and Rb–Sr determinations have been made on samples from granite and gneiss. To date, the most intensive study has been made on the Kate granite in northern Zambia, a pluton which intrudes a 1000 ft sequence of shales and sandstones of the so-called Lower Plateau Series, but which is unconformably overlain by the arenites, shales and cherts of the Upper Plateau Series of possible Katanga System age (Page, 1962, p. 29). The ages from this granite are: Rb–Sr whole rock age, 1460 ± 24 m.y.; biotite K–Ar age, 1725 ± 70 m.y.; and the K–Ar age of muscovite from a pegmatite adjacent to the granite, 1785 ± 75 m.y. K–Ar ages for micas from other rocks within the Ubendian Belt are as follows: muscovite from greisen in Ubendian granite at Chunya, Tanzania, 1800 ± 80 m.y.; biotites from the Ufipa gneiss of Zambia and from an intrusive granite at Chunya, 1420 ± 60 m.y.; and biotite from the Lumi gneissic granite of Zambia, 1590 ± 70 m.y. The K–Ar hornblende age of 1300 ± 50 m.y. from the Chimala granite of Tanzania may also belong to this group, although it has been grouped with the later 'Kibaran' ages which also characterize this belt (see p. 343) (Snelling, 1963, 1964, 1965).

Discussing these radiometric data, Snelling (1963) provisionally concludes that they set a younger limit of about 1800 m.y. to the Ubendian Belt, and attributes the lower mica ages to 'slight argon loss.' He notes, moreover, that the lower whole rock Rb–Sr age of the Kate granite can be 'attributed to subsequent chemical alteration.' Exactly how much older than ca. 1800 m.y. the Ubendian orogenesis really was is still not known; for the time being this figure may be taken as a rough, though minimum, approximation to the date of orogenesis. The deposition of the Lower Plateau Series must be older than that date.

Kasai and Lukoshi Regions In the Kasai Province, in the southern part of the Congo, a number of crystalline rock units of pre-Kibara Group age have a general ENE structural trend (Cahen, 1963), and three zones have been recognized as follows (from north to south): (i) Dibaya basement; (ii) Intermediate zone; and (iii) Luiza basement. From a study of the Rb–Sr ages of mineral samples in this region, Ledent and others (1962) concluded: that the Luiza basement has a minimum age of 3300 m.y.; that an age of 2700 m.y. dates the granitization of the Dibaya basement; and finally that the metamorphism and orogenesis of the Luiza Series of sediments in the Intermediate zone took place around 2100 m.y. ago, an event which also affected the Dibaya basement.

Of these events, only the 2100 m.y. imprint of the Intermediate zone concerns us at this juncture. The Luiza Series in this zone is a metamorphic sequence of quartzites, sericite schists, itabirites, and micaceous gneisses. Rb–Sr ages of 2100 and 2110 m.y. for muscovite from schists, place a general age on the metamorphism and orogenesis; a Rb–Sr age of 2010 m.y. on a microcline from pegmatite from the underlying Dibaya basement to the WNW is taken to indicate the extensive activity of this episode of deformation and thermal imprint (Ledent and others, 1962). To the WSW, in northeastern *Angola*, a similar Rb–Sr age of 1852 ± 62 m.y. has been obtained for biotite from a gneiss on the line of strike of the Intermediate zone (Mendes, 1964).

To the south of the Kasai region a belt of folding affects the rocks of the Lukoshi Formation. This formation consists of pelitic and psammitic sediments with conglomerates, greenstones and economic manganese deposits; this sequence rests on migmatitic basement and is folded with N 10°W trends in the east, and east–west trends in the west (Haughton, 1963, p. 78; Ledent and others, 1962; Cahen, 1963). The dating of this orogenic episode is important stratigraphically in that the Lukoshi Formation is unconformably overlain by the rocks of the Kibara Group in the east (Cahen, 1963). Muscovite from a pegmatite cutting the manganese member of the Lukoshi Formation has yielded a Rb–Sr age of 1845 m.y. (Ledent and others, 1962) and provides a younger limit for the Lukoshian orogenesis. From the same general region, but 40 miles to the southeast, in Angola, Mendes (1964) has obtained an age of 1869 $\pm$ 40 m.y. for biotite from a granite; however, the structural relationship of this granite mass to the Lukoshi Formation is unknown.

Limpopo Belt Bordering the Limpopo River in the southern part of Rhodesia and the northern part of South Africa, an ENE trending belt of crystalline rocks stretches from Botswana eastwards almost to the Mozambique border; this belt has been termed the Limpopo Orogenic Belt (see figure 10).

Although the very general structural pattern of this belt is known (Cox and others, 1965), little detailed work has been carried out over the major part of the zone. However, in the northern part of South Africa, a number of detailed studies are available particularly in the vicinity of the economic district of Messina (Söhnge and others, 1948; Van Eeden and others, 1955). These more intensive studies have shown that the crystalline rocks of the Limpopo Orogenic Belt consist of (1) metasediments of the Messina Formation; (2) basic and ultrabasic rocks; and (3) granitic rocks, gneisses and pegmatite. Deformation strongly affected the Messina Formation, the basic and ultrabasic rocks, and certain of the granitic rocks.

Few radiometric ages are presently available for this belt. However, a radioactive concentrate (mainly sphene) from a sphene-rich schlieren at Beit Bridge has yielded a concordant U–Th–Pb age of 1940 $\pm$ 60 m.y. (Holmes and Cahen, 1957; Nicolaysen, 1962); Nicolaysen considers that this age 'probably dates an important period of metamorphism of the Messina Formation.' In addition, microcline from a pegmatite at Beit Bridge gave a K–Ar age of 1660 m.y. (Holmes and Cahen, 1957). Recently, Van Breemen and others (1966) have reported Rb–Sr ages for four biotites from gneisses which support a ca. 2000 m.y. isotopic event in the belt; they consider, however, that an event older than 2500 m.y. was responsible for some of the high-grade metamorphism.

Gabon and southern Cameroons In Gabon, a series of pegmatites cutting the 'Lower Precambrian' Ogooué gneiss has given muscovite Rb–Sr ages of 1890 ± 46 to 1930 ± 52 m.y. (Vachette, 1964a) (see figure 10). These gneisses are part of the Ogooué System and they have been correlated with the Mayumbe System of the Lower Congo (Cahen, 1961). The ages are consistent with the Eburnian orogenic episode of the Ivory Coast, and are not inconsistent with the Mayumbian orogeny (minimum age 1480–1800 m.y.) of the Lower Congo.

A similar age of 1980 ± 70 m.y. was obtained by Lasserre (1964) for biotite from the Ebolowa granite in the southern Cameroons; the significance is uncertain since this granite has given other ages of 2200–2500 m.y. (see p. 385). Equally difficult to interpret is the Pb/α age of 1785 m.y. yielded by alluvial zircon from the Du Chaillu massif, to the southeast of the Ogooué region, and the zircon Pb/α ages of 1466 m.y. and 1292 m.y. obtained for zircon from orthogneiss and granite north of that region (Bessoles and Roques, 1959).

Western Angola The crystalline rocks which occupy the major part of western Angola have long been considered to belong to the oldest Precambrian, with some younger sediments of the Oendolongo System (see Furon, 1963; Furon and others, 1959). The former, Basement Complex, consists largely of granitic rocks and gneisses which have been compared with the *pre*-Mayumbian (older than ca. 2500 m.y.) of the Lower Congo.

Published Rb–Sr determinations from western Angola show that, excluding the youthful ages of the coastal zone, biotite ages from granite samples fall broadly within the range 1585–1695 (± 40) m.y.; with two biotites from a granite and a gneiss sample yielding ages of 1463 ± 30 m.y. and 1531 ± 36 m.y. respectively (Mendes, 1964). This age pattern in the 1580–1700 m.y. range is consistent with the Huabian granite of northern South-West Africa (see p. 362).

C *1300–2100 m.y. ages preserved in younger zones*

A number of ages between 1300 and 2100 m.y., are 'preserved' in the zones of more youthful orogenesis, particularly in the late

Precambrian–early Palaeozoic orogenic belts in Malagasy Republic, Zambia, Rhodesia, South-West Africa, Lower Congo, Central African Republic, the Hoggar and Dahomey (see figure 10). In addition, ages from the Precambrian terrains of southern Morocco should logically be included here since they occur in the region of the Anti-Atlas subsequently affected by Hercynian deformation; however, since these later movements left no recognizable thermal imprint, the ages from this region have been dealt with along with other undisturbed orogenic effects.

Malagasy Republic As has been previously noted, the main part of the pre-Karroo of the Malagasy Republic shows a uniform thermal imprint reflected by widespread mineral ages in the ca. 500 m.y. region. However, a number of samples from the Androyan System (see p. 322) have yielded older Rb–Sr ages as follows: whole rock granite and leptynite, 2160 ± 23 and 2057 ± 20 m.y. respectively; and whole rock gneiss, 1315 ± 10 m.y. In addition, a Pb/α age of 1515 m.y. has been obtained for zircon from pegmatite cutting the Vohibory System (Delbos, 1964; Holmes and Cahen, 1957). Although these ages suggest ancient activity within this crystalline zone, their actual significance is still uncertain.

Zambia In the general region of the Zambian Copperbelt a number of relict ages in the 1500–1650 m.y. range have been obtained from rocks of pre-Katanga System age (see figure 10). These rocks include an older sequence of schists and gneisses of the Lufubu System, and a younger sequence of quartzite, schist and conglomerate of the Muva System, in addition to widespread granite and gneiss; the two systems are separated by an unconformity (Mendelsohn, 1961).

A Rb–Sr age of 1610 ± 150 m.y. obtained for potash feldspar from pre-Katanga rocks at Roan Antelope is comparable with the whole rock Rb–Sr age of 1625 ± 50 m.y. yielded by aplite from Mtuga to the southeast. The latter is related to granites generally accepted as being older than the Muva System; the age thus represents a provisional *maximum* age for the Muva ($=$ Kibara Group?). In addition, muscovite from a segregation pegmatite in pre-Katanga System gneisses has given a K–Ar age of 1550 ± 50 m.y. even though biotite from the country rock gneisses has yielded an age of 560 ± 20 m.y. (Snelling and others, 1964b; Snelling, 1965).

The evidence suggests the existence of a ca. 1600 m.y. metamorphic–igneous episode in this region of Africa (Snelling and others, 1964b; Snelling, 1965).

Rhodesia In the Urungwe district of northwestern Rhodesia, granite intrusive into the Piriwiri System has given microcline and whole rock Rb–Sr ages of 1610 ± 90 m.y. and 2080 ± 125 m.y. respectively (Clifford, Rex and Snelling, 1967); a 'Kibaran' age of 1170 ± 40 m.y. was obtained for the constituent muscovite by the same method. To the northeast, in the regionally metamorphosed Piriwiri System characterized by ca. 500 m.y. ages in the Miami region, granite has yielded a whole rock Rb–Sr age of 2800 ± 110 m.y.; a microcline age of 2894 ± 110 m.y.; and biotite K–Ar and Rb–Sr ages of 480–500 m.y.

These ages seem to indicate that ancient rocks were involved in much later orogenic activity; the mica ages are consistent with rejuvenation at 1100 m.y. and 500 m.y. respectively for these two granites.

South-West Africa In northern South-West Africa, a large mass of pre-Otavi granite forms the core of a regional anticline, mantled by the unconformably overlying folded sediments of the Otavi Facies ($\equiv$Katanga System) (Clifford and others, 1962). This mass represents *at least* one major plutonic event of batholithic dimensions; that event has been termed the *Huabian episode*.

The granite has yielded a number of Rb–Sr ages in the 500–600 m.y. and 900–1200 m.y. ranges (see p. 314 and p. 344). In contrast, U–Pb zircon ages have given a slightly discordant pattern dating a period of crystallization 1700 ± 70 m.y. ago (Clifford and others, 1962, p. 269). The dated granite sample is probably a product of liquid $\rightleftharpoons$ crystal equilibrium; this fact, coupled with the idiomorphic form of the zircons, indicates that the rock is probably the product of magmatic crystallization. The age of 1700 ± 70 m.y. is, therefore, provisionally taken as the date of emplacement of the granite.

Lower Congo Within the coastal crystalline Mayumbian Belt of the Lower Congo, in a zone now widely characterized by ages in the 500–600 m.y. range, Cahen and others (1963) have noted Rb–Sr ages of 1420 ± 160 m.y. from microcline in augen gneisses in the Boma district (see figure 10). In this region, an orogeny which predates the Western Congo System ($\equiv$Katanga System?) was originally recognized on stratigraphic grounds, and named the *Mayumbian orogeny*. The Lufu granite and its associates in the Lower Congo were intruded around the time of that orogeny and have given zircon Pb/α ages of 1182 and 1483 m.y. respectively (Cahen and others, 1962, p. B258); farther north, in the Republic of the Congo, quartzitic diorite of Saras intruding rocks believed to be correlative with those of the

Mayumbe Belt has given a zircon Pb/α age of ca. 1250–1375 m.y. Coexisting biotite from this diorite has given a Rb–Sr age of 515 ± 14 m.y. indicating that the zircon is a 'relict' age within the zone of coastal late Precambrian–early Palaeozoic activity (Vachette, 1964a). However, because of the problems of lead loss from zircons, and the possibility of absorption of radiogenic strontium by plagioclase coexisting with the dated microcline at Boma, the ages of 1480 and 1420 m.y. are regarded as *minimum* ages for the Mayumbian orogeny which probably took place sometime in the 1480–1800 m.y. interval (Cahen and others, 1963, p. B263).

Central African Republic In the discussion of the geological subdivision in the Central African Republic, it has been noted that ancient rocks have been reinvolved in a much younger event reflected by consistent ages in the 450–650 m.y. range. A few Kibaran ages (ca. 1000 m.y.) have been recognized and, in addition, one microcline Rb–Sr age of 1950 ± 70 m.y. has been obtained. The significance of this latter isolated age is not known, but it suggests that some rocks of the Kibali Belt age may have been involved in the late Precambrian–early Palaeozoic orogenesis (Roubault and others, 1965).

The Hoggar, Central Sahara and Dahomey The Hoggar is a classic region for Precambrian geology which has been subdivided into: (1) the Suggarian; (2) the Pharusian; and (3) the Nigritian. Structurally the region is characterized by major tectonic 'horsts' (oriented N–S) of Suggarian separated by complementary zones of Pharusian rocks (see Furon, 1963, p. 91).

A considerable number of ages are now available from the Hoggar and the majority fall broadly in the 500–650 m.y. range. In the west, however, outside the influence of this later activity, the migmatitic granite of Ouallen occurring close to the eastern edge of the West African Craton (see figure 10), has been ascribed to the Pharusian, and has given the following radiometric data: muscovite and biotite Rb–Sr ages of 1795 ± 50 m.y. and 745 ± 35 m.y. respectively; and discordant U–Pb ages for zircon with a $^{207}Pb/^{206}Pb$ age of 1885 ± 80 m.y. (Lay and Ledent, 1963; Lay and others, 1965).

To the southeast, in the Silet region of the Hoggar, the N–S trending charnockite zone of the In Ouzzal Facies (Suggarian) has given K–Ar and Rb–S rages of 1820 ± 50 and 1730 ± 70 m.y. respectively for biotite from pegmatite (Eberhardt and others, 1963). Very recently, Ferrara and Gravelle (1966) have reported additional age data from the northern extension of the Adrar des Iforas. In particular, they have obtained Rb–Sr ages of 1750–1840 m.y. and K–Ar ages of 1690–1795 m.y. for biotites from granite, biotite pyroxenite and biotite gneiss samples. In contrast, however, Rb–Sr whole rock data for granite, syenite and gneiss samples gave an isochron age of 2860 m.y. Although the Adrar des Iforas appears to be structurally part of the Nigeria–Hoggar segment of the late Precambrian–early Palaeozoic orogenic zone, the *thermal* imprint of ca. 500–650 m.y. age is absent.

In Dahomey, in the southward extension of this late Precambrian–early Palaeozoic orogenic zone, the Kouandé orthogneiss and a sample of granitized Atacorian mica schist have yielded antique Rb–Sr whole rock ages of 1650 ± 200 m.y. and 1860 ± 90 m.y. respectively, but lower biotite ages of 592 ± 9 m.y. and 291 ± 9 m.y. respectively (Bonhomme, 1962).

D   *Anorogenic and related effects*

The eastern part of South Africa and Botswana represents a critical region in discussions of ages in the 1600–2100 m.y. range. In essence, the Precambrian stratigraphy is (from oldest to youngest): (1) Crystalline Basement with schist belts; (2) Witwatersrand and Dominion Reef Systems; (3) Ventersdorp System; (4) Transvaal System; (5) Bushveld complex; (6) Waterberg and Loskop Systems; and (7) Pilanesberg and other alkaline complexes (see figure 11).

Although broadly protected from widespread linear orogenesis between 1300–2100 m.y., much of the crustal activity during this period does not fall strictly into 'anorogenic' events; for example, the Vredefort Ring. Other events, however, such as the emplacement of the Bushveld igneous complex and Palabora intrusion are, in all respects, anorogenic.

Bushveld igneous complex   This complex is a large 'lopolithic' body occupying an elongated area almost 300 miles in length in the Transvaal (see figure 11). It intrudes the rocks of the Transvaal System but is unconformably overlain by the Waterberg–Loskop sequence, thus providing a minimum age for the former, and a maximum age for the latter. The igneous activity was initiated by the emplacement of basic lavas, tuffs and felsites, and this early phase was followed by the intrusion of the main phase of plutonic

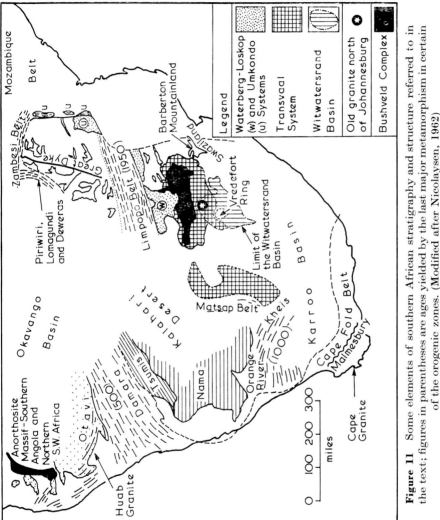

Figure 11 Some elements of southern African stratigraphy and structure referred to in the text; figures in parentheses are ages yielded by the last major metamorphism in certain of the orogenic zones. (Modified after Nicolaysen, 1962)

basic and ultrabasic activity. The latest phase of igneous activity was the emplacement of granite (Hall, 1932; Du Toit, 1954). At the present level of erosion, these various igneous rocks cover a total exposed and concealed area of more than 23,000 square miles.

Early Rb–Sr age determinations on whole rock and feldspar fractions of the granite were carried out by Schreiner (1958) and yielded ages of ca. 1800 m.y. In addition, work by Nicolaysen and others (1958) has shown that the Rb–Sr ages of biotites from granite and pyroxenitic gabbro samples fall within the 1780 ± 60 m.y. to 1930 ± 50 m.y. range, with one granite biotite age of 1600 ± 90 m.y. In addition, U–Th–Pb determinations on six monazites strongly support an age of crystallization of 2010 ± 80 m.y., whereas determinations on three zircons from the granite indicate an age close to 1900 m.y. for the formation of this rock (Nicolaysen and others, 1958).

These data suggest an age of 1950 ± 200 m.y. for the intrusion of the Bushveld complex. The Transvaal System must, therefore, be older than this figure, whereas the Waterberg–Loskop Systems are younger.

Palabora complex The Palabora complex in the northeastern Transvaal is composed largely of pyroxenite, syenite, olivine–diopside–phlogopite pegmatoid, fenite, carbonatite and serpentinized dunite (Hanekom and others, 1965); it is well known for its economic concentrations of apatite, vermiculite and copper sulphides. Thorianite present in the carbonatite has given a concordant U–Pb age of ca. 2000 m.y. (Holmes and Cahen, 1957, p. 75), suggesting that this is one of the oldest carbonatite complexes in Africa.

Vredefort Ring In the northern part of the Orange Free State and the southern part of the Transvaal, a remarkable ring of strata occurs upturned around the Vredefort granite mass (see figure 11). The mantling strata largely belong to the Witwatersrand, Ventersdorp and Transvaal Systems which have been tilted to high angles and, 'over the north-western half of the periphery they have been strongly overturned . . . producing an asymmetrical mushroom-like structure' (Du Toit, 1954, p. 198). The ring is now partly concealed by Karroo beds, but is considered to be roughly pear shaped (Du Toit, 1954). Alkali granites and enstatite-bearing basic granophyre dykes intrude the folded sediments of the Witwatersrand and Transvaal Systems.

In recent years, 'shatter cones' have been found in all elements of this sequence except the central granite itself and the basic granophyre dykes. Their orientation appeared to 'indicate that the cones originated by an intense shock-wave radiating outwards from the present centre of the ring' (Hargraves, 1961, p. 152). Although these cones could be explained by a meteor-impact theory, Hargraves was cautious in his interpretation of them.

Most recently, radiometric dating studies by Nicolaysen and others (1963) have shown that the ring structure was formed *before* 1900 m.y. ago. Although the age data from this important study have not yet been released for publication, they do indicate: (i) the near identity of the Vredefort ages with those of the Bushveld complex; and (ii) the improbability of the meteor-impact hypothesis (Nicolaysen and others, 1963). In the light of these data, Nicolaysen and others (1963) favour a terrestrial theory and suggest that the formation of the ring structure is probably related to an unusual regime of crustal mechanics prevailing in the southern and central Transvaal during the time of emplacement of the Bushveld complex.

Other regions To the *north of Johannesburg*, a mass of granitic and other crystalline rocks is mantled by rocks of the Witwatersrand and Ventersdorp Systems on its southern side; the whole is unconformably overlain by the Transvaal System on the east, west and north. Petrologically this 'Old Granite' has many features in common with the granitic rocks in the core of the Vredefort Ring.

Rb–Sr determinations on whole rock samples from this body indicate an age of emplacement of 3010 ± 60 m.y. for the granite (Allsopp, 1961). However, the apparent ages of the separated minerals vary widely and this discordance is regarded as the result of the diffusion of radiogenic strontium from mineral to mineral during a later reheating. Indeed, biotite from the granite mass has yielded a Rb–Sr age of ca. 2000 m.y., and it has been concluded that the reheating occurred around that time; this age is consistent with the ages obtained for the emplacement of the Bushveld complex (Allsopp, 1961, p. 1507).

A somewhat similar pattern is recorded in conglomerates of the Dominion Reef, Witwatersrand and Ventersdorp Systems in which oval and spherical uraninite grains giving a crystallization age of 2950 ± 150 m.y. show evidence of intense chemical alteration about 2050–2150 m.y. ago (Nicolaysen, 1962); and the Gaberones granite of eastern Botswana which has given whole rock Rb–Sr ages of ca. 2600–2800 m.y., but K–Ar ages of hornblende and biotite of 2225 ± 100 m.y. and 1630 ± 65 m.y. respectively (Snelling, 1962, 1965).

E *Summary*

Ages in the 1850 ± 250 m.y. range are widespread in southern and West Africa (see figure 10), and clearly reflect at least one important orogenic event. Within this range, the most definitive dates are for the Eburnian episode of orogenesis (1850–2050 m.y. ago) characterized thermally by widespread synorogenic granite emplacement and metamorphism. The type region for this event is the Ivory Coast and the Upper Volta, but correlative tectonothermal activity has been recognized in Senegal and Mauritania. Moreover, on the basis of presently available data, orogenic activity reflected by ca. 1800–2000 m.y. ages in the following regions may also be tentatively ascribed to the Eburnian orogenic episode: (1) southern Cameroons; (2) the belt of the Luiza Series of Kasai; (3) the Lukoshi Belt of the Congo; (4) the Ubendian–Ruzizi Belt of East Africa; (5) the Kibali and Buganda–Toro Belts of Congo and Uganda; and (6) the Limpopo Belt of Rhodesia, Botswana and South Africa. It is perhaps worthy of note that (2)–(6) inclusive are linear orogenic belts and represent the oldest known developments of *extensive* linear orogenic belts in the African structural record. Whether the deformed Birrimian and related sequences in West Africa occupy similar elongate belts is not known; present evidence suggests that the deformation which accompanied granite emplacement in those regions is 'spatial' and has more in common with the patterns of pre-2100 m.y. orogenesis in Africa.

The Eburnian effects in Africa provide *minimum* ages of 1800–2000 m.y. for the following sequences: (1) the Birrimian, Yatakala, Amsaga and Mako–Dialé–Daléma successions of West Africa; (2) the Ogooué gneiss (= Mayumbe System) of the Lower Congo region and Gabon; (3) the Luiza Series of Kasai; (4) the Lukoshi Formation of the Congo; (5) the Ubendian System and Ruzizi Group of the Congo and East Africa; (6) the Kibali Group of the Congo; (7) the Buganda–Toro System of Uganda; (8) the Messina Formation of Rhodesia, Botswana and South Africa; and (9) the Suggarian (In Ouzzal Facies) of the Hoggar. Moreover, these dated orogenic effects provide *maximum* ages for a number of unconformably overlying groups including

(a) the Kibara, Karagwe Ankolean and Urundi Groups; (b) the Upper Plateau Series (Zambia); (c) the Umkondo System (resting on the eroded roots of the Limpopo Belt) of Rhodesia;* (d) the Western Congo System (resting on the Mayumbe System); and (e) the Pharusian rocks of the Hoggar.

The significance, within the 1850 ± 250 m.y. range, of the 1600–1700 m.y. age concentration (see figure 3) is uncertain. It has been suggested that it may reflect an orogenic phase related to, but later than, the 1850–2050 m.y. event (Kennedy, in preparation). Certainly, it is widely represented in Mauritania, southern Morocco, Ivory Coast, Dahomey, western Angola, Zambia, Rhodesia and South-West Africa; it is, therefore, no mere thermal accident.

The stratigraphic position of granites yielding ca. 1600 m.y. ages sets a *minimum* age on the Zenaga Series (supposed Precambrian I) of southern Morocco; a similar age obtained for the Kouandé orthogneiss of Dahomey is considered to support the view that the Dahomeyan is older than ca. 1650 m.y. (Bonhomme, 1962). Furthermore, acid igneous rocks of this age are believed to *predate* the Muva System of Zambia and, since this system is unconformably overlain by the Katanga System, it occupies a comparable time interval to the Kibara Group of the Congo.

In contrast to the above effects, one of the most striking anorogenic features of African history around 1850 ± 250 m.y. times, was the emplacement of the Bushveld complex ca. 1950 ± 200 m.y. ago. This major event sets a *minimum* age limit on the Transvaal, Ventersdorp, Witwatersrand and Dominion Reef Systems. However, similar ages of ca. 2000 m.y. have also been obtained from basement granite, from uranium minerals in the Witwatersrand System, and from intrusive rocks associated with the Vredefort Ring. These data suggest that the emplacement of the Bushveld may have had a profound effect not only in the reheating of underlying rocks, but also in crustal tectonics of local significance but orogenic intensity, as in the Vredefort Ring. Nor are these widespread effects so surprising when it is considered that even in its present form, the Bushveld complex represents approximately 100,000 cubic miles of

---

* A K–Ar age of 1140 m.y. from a basic igneous rock cutting the Umkondo System sets a minimum age limit on that system. Moreover, on palaeomagnetic evidence it has been suggested that the Umkondo is *older* than ca. 1600 m.y. (see Jones and McElhinny, 1966); it has been correlated with the Waterberg System of South Africa (Du Toit, 1954) (see p. 370).

magma evacuated from the crust or upper mantle; a volume sufficient to represent a circular mass of rock 1–2 miles in thickness and with a radius of 140 miles.

Finally, the ages from the Bushveld complex provide a *maximum* limit for the unconformably overlying Waterberg–Loskop Systems (see figure 11), a sequence of some 5000 feet of brown, red and purple sandstones, conglomerates, shales and volcanic rocks (Du Toit, 1954, p. 211). Oosthuyzen and Burger (1964) have obtained a lead isotope age of 1790 ± 70 m.y. for granophyre which intrudes sediment correlated with the Loskop sequence (the lower part of the Waterberg–Loskop succession). Although the most definitive data show that the Waterberg sediments are older than 1420 m.y., Oosthuyzen and Burger (1964) consider that they are also 'probably at least 1790 m.y. old.' In view of the data from the Bushveld, this suggests maximum and minimum limits of 1950 ± 50 m.y. and 1790 ± 70 m.y. respectively for the age of the Waterberg–Loskop Systems.

V Radiometric Ages Greater Than 2100 m.y.

A General statement

Radiometric dating studies have shown that rocks yielding ages greater than ca. 2100 m.y. occur in parts of West, Central, East and southern Africa. Within these areas a number of quantitative structural and stratigraphic bench marks have been recognized, and give an insight into the early geological history of the continent. These ages, and the events which they record, are best preserved in those regions which have not been affected by later thermal or tectonothermal activity. Figure 12 shows the main concentration of ancient ages in those regions of Africa which have not been orogenically deformed since at least ca. 1500 m.y. ago.

In certain cases, it is possible to place a rather more precise outline around these ancient nuclei; for example, the Transvaal, Rhodesia and Dodoma–Nyanza nuclei (see figure 13). In other cases, however, the areal extent of the nuclei is uncertain; as, for example, in the Kasai, Gabon–Cameroons and Mauritania nuclei (see figure 12).

These nuclei are all characterized by very old ages. However, in certain examples, they also show the subsequent influence of 'Eburnian' events (ca. 1850–2050 m.y.) whose structural significance is, in most cases, uncertain. The nuclei of this type include (i) the Transvaal nucleus with its imprint of the effects of the emplacement of the

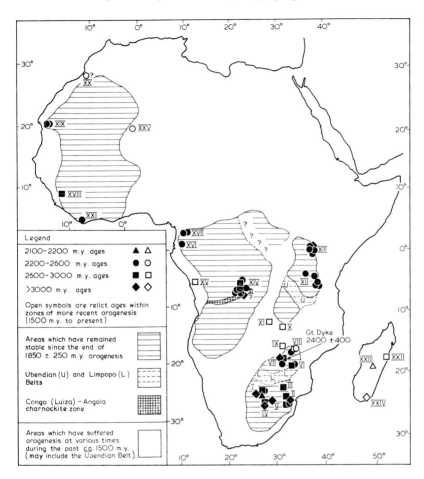

Figure 12 Distribution of ages >2100 m.y., shown in relation to certain elements of African structure.

I. Botswana granites; II. Swaziland and eastern Transvaal; III. Letaba, Transvaal; IV. Witwatersrand; V. Johannesburg; VI. Bikita, Rhodesia; VII. central Rhodesia; VIII. Mtoko, Rhodesia; IX. Miami, Rhodesia; X. Irumi, Zambia; XI. Copperbelt, Zambia; XII. Tanzania; XIII. eastern Uganda and western Kenya; XIV. Kasai, Congo and northeastern Angola; XV. Lower Congo; XVI. Gabon; XVII. southern Cameroons; XVIII. Sula Mts, Sierra Leone; XIX. Sattle Ogmane and Tasiast, Mauritani ; XX. southern Morocco; XXI. Ivory Coast; XXII. Ivoloina, Malagasy Rep.; XXIII. Vohimena, Malagasy Rep.; XXIV. Behara, Malagasy Rep.; XXV. Hoggar.

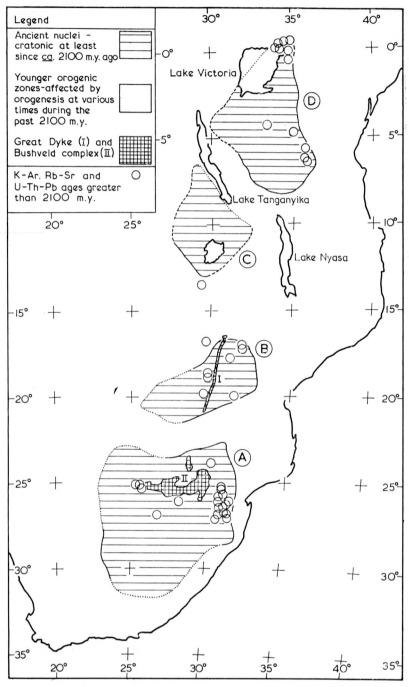

Figure 13 Ancient nuclei of eastern and southern Africa: A. Transvaal nucleus; B. Rhodesia nucleus; C. Zambia nucleus; D. Dodoma–Nyanza nucleus.

Bushveld complex and associated crustal phenomena (see p. 369); (ii) the Kasai nucleus; (iii) the Gabon–Cameroons nucleus; (iv) the ancient terrain of southern Morocco; and (v) the Mauritania nucleus.

In addition to these regions, a number of ages greater than ca. 2100 m.y. are preserved in the late Precambrian–early Palaeozoic mountain chain (see figure 12). These data clearly show that part, at least, of the basement floor of this chain was composed of ancient crustal materials; for example, in Rhodesia, Malagasy Republic, the Lower Congo and the Hoggar.

B *Ancient nuclei*

Transvaal nucleus The Transvaal nucleus is bounded by the Limpopo Belt on the north and the Lebombo monocline on the east. A large part of its western and southern boundary is obscured by Phanerozoic rocks except in the southern part of Botswana and the northern Cape Province where its western boundary is defined by the eastern 'front' of folding (ca. 1000 m.y.) of the Matsap System (see figure 11). Within this nucleus (see figure 13), the ancient Precambrian sequence consists of (from oldest to youngest): (1) Crystalline Basement and schist belts; (2) Dominion Reef System; (3) Witwatersrand System; (4) Ventersdorp System; and (5) Transvaal System (see Du Toit, 1954). All predate the Bushveld complex and are, therefore, older than 1950 ± 200 m.y.

The Crystalline Basement and schist belts are particularly well developed in the eastern and northern Transvaal and in Swaziland. Although large masses of granitic rock predominate, good examples of 'schist belts' (with orogenic fold patterns) occur in the Murchison Range, and the Barberton Mountainland. As a result of detailed field work in the latter region, the following schist belt sequence was recognized (Visser and others, 1956): (1) a lower *Swaziland System* composed of a basal *Onverwacht Series* of basic-to-acid lavas, and an overlying *Fig Tree Series* of predominantly pelitic rocks with cherts and ironstones in the lower portion and clastic sediments in the upper part; and (2) an upper *Moodies System* consisting largely of psammitic rocks with a well-developed basal conglomerate. These systems are intruded by basic and ultrabasic rocks, and by extensive granitic bodies which occur widely around the margins.

Most of the age measurements from the vicinity of the Barberton Mountainland are from the extensive granite terrain which surrounds the schist belt. Intensive radiometric dating on the granites in Swaziland has been carried out by Allsopp and others (1962) who determined the Rb–Sr ages of whole rock samples from granites belonging to three different granite series designated, G1, G4 and G5. The results are tabulated in table 3; the G1 granite gave a whole rock age of 3240 ± 300 m.y.; the ages of four samples of G4 granite gave a satisfactory mean of 2900 ± 60 m.y.; and three samples of G5 granite gave whole rock ages of 2070 ± 50 m.y., 2400 ± 50 m.y. and 2720 ± 340 m.y. Mineral fractions separated from some of these granites have yielded discordant ages interpreted as due to diffusion of ^{87}Sr from mineral to mineral (see table 3).

Table 3 Rb–Sr ages of whole rock and mineral samples of Swaziland granites. (After Allsopp and others, 1962)

Granite	Dated sample	Age (m.y.)
G1	Whole rock	3240 ± 300
	Microcline	2595 ± 90
	Plagioclase	5660 ± 1000
G4	Whole rock	2900 ± 60
	Biotite	2330 ± 50
	Muscovite	3800 ± 100
	Microcline	2670 ± 30
G5	Whole rock	2070 ± 50
	Whole rock	2400 ± 50
	Whole rock	2720 ± 340
	Feldspar	2375 ± 70

It has been considered that these granites developed during the post-Swaziland System orogeny; this *suggests* that the Swaziland System is older than 3240 ± 300 m.y. However, the best minimum age for the Moodies System (and underlying sequences) is provided by the G4 granite age of 2900 ± 60 m.y. since this granite has been mapped as intrusive into that system (Nicolaysen, 1962). It is worthy of note that the Kubuta pegmatite, considered to be related to the G4 granite, has given lepidolite ages of 2760 and 3020 (Rb–Sr), and 2660 m.y. (K–Ar) (Nicolaysen, 1962; Aldrich and others, 1958); a similar Rb–Sr age of 2650 m.y. has, moreover, been recorded for lepidolite from pegmatite at Letaba (Holmes and Cahen, 1957), in the northward extension of this ancient rock province. The minimum age of 2900 m.y. for the schist belt sequence is consistent with these and other forthcoming data (Nicolaysen 1962; Nicolaysen and others, 1965).

Further data for this time event have been obtained from whole rock samples of Old Granite outcropping north of Johannesburg; these have yielded concordant Rb–Sr results indicating an emplacement age of 3020 ± 65 m.y. (Allsopp, 1961). This granite intrudes amphibolites and serpentines which have been correlated with the Swaziland System, and is mantled by the rocks of the Witwatersrand and Ventersdorp Systems.

The even greater extent of these ancient rocks is indicated by an Rb–Sr age of 2740 ± 150 m.y. obtained for muscovite from pegmatitic granite in the basement *beneath* the Dominion Reef System south of Johannesburg (Nicolaysen and others, 1962). In this context, it is particularly worthy of note that isotopic age measurements relating to uraninite and monazite from auriferous conglomerates of the Dominion Reef System have given ages of 3000–3100 (± 100) m.y. (Nicolaysen and others, 1962); on the basis of these data, and bearing in mind the unquestionably detrital origin of the monazite, the deposition of the Dominion Reef System (and overlying systems) must have taken place *less* than 3050 ± 100 m.y. ago (Nicolaysen, 1962).

These sedimentary and igneous sequences of the Transvaal extend westwards into eastern Botswana (Du Toit, 1954; Boocock and Van Straten, 1962), from which a number of valuable ancient ages have been obtained. For example, the Dominion Reef System, which is a sequence of acid lavas including felsites and feldspar porphyries, and tuffs in Botswana, has given Rb–Sr whole rock ages of 3060 ± 170 m.y. for felsite samples (Snelling, 1965); these ages are reasonably consistent with the maximum age of Dominion Reef sedimentation given by Nicolaysen (1962). Intruded into the felsites, the Gaberones-type granite plutons have yielded Rb–Sr whole rock ages of 2650 and 2820 (± 300) m.y.; hornblende and biotite from these granites have given rather younger ages (see table 4) indicating a later disturbance (see p. 368) (Snelling, 1965).

These varied age data are consistent with the stratigraphic sequence for the Transvaal and eastern Botswana given in table 5.

Rhodesia nucleus Ancient rocks affected by events yielding ages

Table 4 Age determinations from Gaberones-type granite, eastern Botswana. (After Snelling, 1962, 1965)

Dated sample	Locality	Age (m.y.)	
		K–Ar	Rb–Sr
Whole rock	Lobatsi pluton	—	2820 ± 300
Hornblende	Lobatsi pluton	2185 ± 100	—
Whole rock	Gaberones pluton	—	2665 ± 300
Hornblende	Gaberones pluton	2225 ± 100	—
Biotite	Gaberones granite	1629 ± 65	—

Table 5 Stratigraphic sequence for the Transvaal and eastern Botswana

Loskop and Waterberg Systems
1950 ± 200 m.y.
Transvaal System Ventersdorp System Witwatersrand System Dominion Reef System
ca. 3000 m.y.
Moodies System Swaziland System

greater than 2100 m.y. are, from a structural and stratigraphic point of view, particularly well represented in Rhodesia. This nucleus is bounded by the Limpopo Belt on the south, the Mozambique Belt on the east, and the Zambesi Belt on the north; on the west, however, the boundary is largely obscured by younger cover sediments (figures 11 and 13).

For the most part, the rocks of this nucleus consist of a folded sequence of sedimentary rocks and lavas of the 'schist belts,' and widespread granitic plutons considered to represent gregarious batholiths intruding the sediment–lava sequence (see figure 14) (Macgregor, 1951).

In terms of stratigraphy, the sequences of the schist belts have been subdivided into three systems: the Sebakwian, the Bulawayan and the Shamvaian. The oldest of these, the Sebakwian System, consists largely of metasedimentary rocks including rocks of granulitic texture ranging from quartzites to hornblende schists, biotite and chlorite schists, banded ironstones, metagreywackes and some limestones. Metamorphosed lavas are locally important. This whole sequence has been intruded by serpentinized peridotite and invaded by granite which has produced partial granitization.

An unconformity separates this system from the overlying Bulawayan System, which forms the principal formation of all the 'schist belts'. This system is composed mainly of basic and intermediate lavas and breccias, with intrusive serpentines; rhyolites are, however, rare (Macgregor, 1951). A variety of sedimentary rock types are interbedded with these extrusives, including banded ironstone, greywacke, conglomerate, phyllite, quartzite and algal limestone.

The Shamvaian System forms the upper member of this sequence and it consists chiefly of arkose, greywacke, conglomerate, phyllite, slate, dolomite and some ironstone.

Correlation of this succession with the schist belts of South Africa is, as yet, uncertain; Du Toit (1954), however, suggested the following provisional correlation.

South Africa	Rhodesia
Moodies	Shamvaian
...........................Unconformity	
	Bulawayan
Fig Tree	
	..............Unconformity
Onverwacht	
	Sebakwian

Age data from these ancient rocks of Rhodesia have recently been summarized by Nicolaysen (1962). A whole rock K–Ar age of 3100 m.y. for a granite boulder from the base of the Bulawayan System suggests the previous existence of very old granites in the region, and

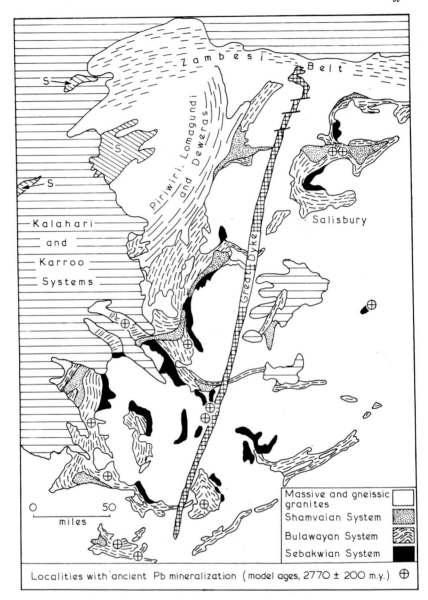

Figure 14 Tectonic sketch map of part of Rhodesia, with localities of ancient lead mineralization showing 'model' ages of 2770 ± 200 m.y. (Modified after Nicolaysen, 1962)

S. Sijarira Beds of late Precambrian or early Palaeozoic age.

gives a rather uncertain maximum age to the Bulawayan System (Nicolaysen, 1962; Wasserburg and others, 1956).

From a study of post-Shamvaian pegmatites it has been found that Rb–Sr and K–Ar ages for micas (largely lepidolites), and U–Th–Pb ages for monazite and microlite, fall within the general range of 2450–2650 m.y. (Nicolaysen, 1962). This broad range is supported by 'model' ages of 2770 ± 200 m.y. yielded by galenas from 26 localities within the central part of Rhodesia (figure 14), and together these data place a minimum age of ca. 2600–2700 m.y. on the schist belt sequences. However, it should be borne in mind that, if these rocks can be correlated with those of the Transvaal nucleus, this lava–sediment succession may be older than ca. 2900–3000 m.y. Moreover, if the intimate relationship of deformation, gregarious batholith emplacement, and some of the pegmatite emplacement, envisaged by Macgregor (1951), is accepted, the dates of 2500–2700 m.y. may *approximate* to the *last* important episode of orogenesis in the Rhodesia nucleus.

Intruding these older rocks and running for over 300 miles in a NNE direction is the Great Dyke (see figure 14). This impressive feature is some 3–7 miles wide and exhibits a pseudo-stratification and a synclinal cross-section. It is largely basic and ultrabasic in composition (Du Toit, 1954; Worst, 1960). Recent dating studies have been carried out by Allsopp (1965) using picrite samples from the dyke, and these results are presented in table 6. Rb–Sr ages are recalculated to $\lambda_\beta = 1.474 \times 10^{-11}$ yr^{-1}. On the basis of the K–Ar age of plagioclase from the Wedza complex of the dyke, Allsopp (1965) concluded that the age of emplacement of the rocks could be as great as 2800 m.y. Moreover, on the basis of agreement between K–Ar and Rb–Sr (using $\lambda_\beta = 1.39 \times 10^{-11}$ yr^{-1}) ages for biotite, he considered that the minimum age of the dyke is 2530 ± 30 m.y.; this agreement is lost, however, when the Rb–Sr ages are recalculated with the $\lambda_\beta = 1.474 \times 10^{-11}$ yr^{-1}. Using the whole rock isochron approach, Allsopp determined an age of 2400 ± 400 m.y., if converted to the latter decay constant; which is consistent with the major number of biotite, plagioclase and pyroxene ages (see table 6).

Earlier whole rock Rb–Sr dating by Faure and others (1963) of two samples (quartz gabbro and norite) from the Wedza complex gave an age of 2110 ± 350 m.y. and was regarded as support for the conclusion that the Great Dyke and the Bushveld complex (1950 ± 200 m.y.) were emplaced during the same magmatic cycle. However, from his new and more detailed investigation, Allsopp (1965)

Table 6 Age determinations from the Great Dyke of Rhodesia

		Age (m.y.)	
Rock type	Dated sample	Rb–Sr	K–Ar
Picrite[a]	Biotite	2385 ± 25	2520 ± 50
	Plagioclase	—	1720 ± 50
	Pyroxene	—	3500 ± 200
Picrite[b]	Biotite	2380 ± 15	2520 ± 50
	Plagioclase	—	2730 ± 80
	Pyroxene	—	2490 ± 100
Picrite[a,b,c]	Whole rock	2400 ± 400	—
Quartz gabbro and norite[b]	Whole rock	2110 ± 350[d]	

[a] Hartley complex.
[b] Wedza complex.
[c] Selukwe complex.
[d] After Faure and others (1963); all other ages from Allsopp (1965).

concluded that the Great Dyke is significantly older than the Bushveld complex.

Zambia nucleus Furon and others (1959) have shown a segment of ancient rock centred roughly on Lake Bangweulu in Zambia and sandwiched between the Kibara Belt on the northwest, the Muva–Irumide Belt on the southeast, and bounded by the Ubendian–Ruzizi Belt on the northeast and part of the Lufilian arc on the southwest. Although there are no supporting geochronological data, this region has been outlined as one of ancient stability (Ackermann, 1960; Vail, 1965) (see figure 13).

It may be that ages of detrital monazite from the region to the south of the nucleus have a bearing on this problem. Samples of this mineral from stream sands, and from Muva System quartzites in the Irumi Hills region of the Muva–Irumide Belt of Zambia, have given discordant ages of around 2300–2600 m.y. (Holmes and Cahen, 1955; Snelling and others, 1964b); the best estimate of their age, determined by graphical methods is 2720 m.y. Since this is a detrital mineral the source of the monazite 'presumably lies in the older Basement Complex' (Snelling and others, 1964b).

Dodoma–Nyanza nucleus The basement rocks of central Tanzania include: the Dodoman System, consisting of ferruginous quartzites, hornblende, quartz and sericite schists, amphibolites and granitic gneisses, and extensive migmatites, acid gneisses, granites and granodiorites (Haughton, 1963; Quennell and Haldemann, 1960).

These ancient rocks are bounded by the Mozambique Belt on the east, the Ubendian Belt on the south and the Kibaran Belt (with the younger Bukoban System) on the west (see figure 13). To the north, the rocks of the Nyanzian and Kavirondian Systems skirt the southern and eastern sides of Lake Victoria, and their relationship to the Dodoman System has been the subject of much discussion.

The Nyanzian System consists of banded ironstone, acid and basic volcanic rocks, phyllites, schists and quartzites; Haughton (1963) notes that it is 'possible that the Nyanzian and Dodoman may belong to the same cycle of deposition but represent different facies'. In western Kenya, the Nyanzian forms part of the Nyanza Shield and is believed to extend into southeastern Uganda as the Bulugwe Series.

Overlying the Nyanzian System, the Kavirondian System consists of a rudaceous, arenaceous and argillaceous sedimentary sequence believed to be equivalent to the Samia Series of southeastern Uganda. The Nyanzian and Kavirondian Systems are separated by an angular unconformity, representing a phase of orogenesis and granite emplacement (Cahen, 1961, p. 545).

These ancient systems with their intrusive granitic rocks thus occupy much of central Tanzania, the extreme western part of Kenya and the southeastern corner of Uganda (figure 13). Radiometric ages from the Tanzania section of this nucleus include: K–Ar ages of 2294 ± 90, 2246 ± 90 and 2555 ± 70 m.y. for biotites from granites from the Dodoma region, Isanga Hill and the Singida district respectively; a Rb–Sr age of 2313 ± 60 m.y. yielded by potassium feldspar from porphyritic granite in the Nzega district; and a Rb–Sr age of ca. 2300 m.y. obtained for lepidolite from pegmatite in the Dodoman System at Hombolo (Snelling, 1962, 1965; Holmes and Cahen, 1957; Davis and others, 1956). These ages give a minimum age limit of 2200–2500 m.y. for the rocks of the Dodoman System.

In Kenya, a number of ages have now been obtained for the rocks of the Nyanza Shield (the northward extension of the ancient rocks of Tanzania). These include ages of 2510 ± 120 and 2595 ± 250 m.y.

yielded by biotite and hornblende respectively from the Maragoli granite which intrudes and metamorphoses the rocks of the Kavirondian System (Snelling, 1963). In support, a whole rock Rb–Sr age of 2540 ± 280 m.y. was obtained for the post-Kavirondian Kitosh granite (Snelling, 1963). However, a rather lower whole rock Rb–Sr age of 2230 ± 200 m.y. was yielded by Nyanzian rhyolites (Snelling, 1963); the effects which produced this anomalously low age are still uncertain.

These ancient rocks from western Kenya are considered to extend into the southeastern corner of Uganda. In this region, the Lunyo granite postdates the supposed correlatives of the Nyanzian System and has given K–Ar ages as follows: zinnwaldite, 2316 ± 100 m.y.; and biotite 2290 ± 30 m.y. (Snelling, 1962, 1965). Even more detailed are the radiometric data for the Buteba granite (see table 7), which

Table 7 Age determinations of mineral and whole rock samples from the Buteba granite of Uganda. (After Snelling, 1962, 1964)

Dated sample	Age (m.y.)	
	Rb–Sr	K–Ar
K Feldspar	2210 ± 100	—
Muscovite	2600 ± 20	2175 ± 200
Biotite	—	1879 ± 80
Whole rock	2515 ± 250	—

intrudes the rocks of both the Bulugwe (≡ Nyanzian) and Samia (≡ Kavirondian) Series. Although mineral age determinations for this granite give discordant results ranging from 1800 to 2600 m.y., there is agreement between whole rock and muscovite ages, of 2500–2600 m.y. and this date has been accepted as the age of emplacement of this granite; the other discordant ages have been attributed to the effects of the post-Buganda–Toro metamorphism which is widely expressed by ca. 1800 m.y. ages to the north and west of Lake Victoria (see figure 10).

The Precambrian of the Dodoma–Nyanza nucleus thus shows a consistent concentration of ages greater than 2100 m.y.; whether the apparent subdivision into ages in the 2200–2300 m.y. range and ages in the 2500–2600 m.y. range has any structural significance is not yet known. That *very* old rocks may be involved in this nucleus is at least suggested by an isolated K–Ar age of 3150 ± 80 m.y. obtained for biotite from granite of the Nyanza Shield close to the western margin of the Mozambique Belt in Kenya; this is one of the oldest ages so far reported from East Africa (Snelling, 1963; Sanders, 1965).

Kasai nucleus In Kasai (see figure 12), in the southern Congo, a threefold division of the Precambrian has been recognized and is (from north to south): the Dibaya-type basement; the intermediate region; and the Luiza-type basement. These zones have a roughly ENE trend (Ledent and others, 1962).

The Dibaya-type basement consists largely of gneissic–granitic rocks with an average granodioritic composition, and two events in their evolution have been recognized: a granitization or migmatization; and a subsequent episode of deformation and recrystallization.

The Luiza-type basement is principally a charnockitic complex and consists of: (i) noritic gabbros intruded into an old paragneissic substratum; and (ii) charnockites (enderbites, charnockites, granulites and leptynites) considered to be the product of granitization and charnockitization of that old substratum. Subsequent migmatization has affected this complex.

In the intermediate region between these two basement types, the rocks of the Lulua Group and the Luiza Series of metasediments occur. The metamorphism of the latter around 2000–2100 m.y. has already been discussed (see p. 358), and our principal concern is the events recorded in the Luiza and Dibaya basement types. The pegmatite associated with gneisses of pre-charnockitization age in the Luiza-type basement has given a microcline Rb–Sr age of ca. 3310 m.y.; this is taken as a minimum age for the ancient substratum. Moreover, on the basis of microcline and muscovite ages of 2600 m.y. and 2700 m.y. respectively (see table 8), it has been considered that the granitization of the Dibaya-type basement took place around 2700 m.y. ago (Ledent and others, 1962). Ledent and others (1962) consider that the migmatization of the Luiza-type basement to the south probably corresponds to the time of granitization of the Dibaya-type basement and, furthermore, that 'charnockitisation et migmatitisation paraissent appartenir à un même cycle'; on this reasoning a provisional age around 2700 m.y. can be assigned to granitization (Dibaya type) and to migmatization and charnockitization (Luiza type). In this regard, it is of additional interest to note that Mendes (1964) has obtained ages of 2300–2480 m.y. for biotite from granite and quartz diorite in the probable extension of the Dibaya-type basement in northeastern Angola (see table 8).

Table 8 Rb–Sr age determinations from Kasai (Congo)[a] and northeastern Angola[b]

Rock	Region	Basement type	Dated sample	Age (m.y.)
Pegmatite	Kasai	Luiza	Microcline	3310
Granite	Kasai	Dibaya	Microcline	2380
Pegmatite	Kasai	Dibaya	Microcline	2600
Pegmatite	Kasai	Dibaya	Microcline	2280
			Muscovite	2710
Mylonitized granodioritic rock	Kasai	Dibaya	Biotite	2470
Granite	NE Angola	Dibaya (?)	Biotite	2479 ± 22
Quartzitic diorite	NE Angola	Dibaya (?)	Biotite	2479 ± 40
Granite	NE Angola	Dibaya (?)	Biotite	2297 ± 70

[a] After Ledent and others (1962).
[b] After Mendes (1964).

The extent of this nucleus of ancient rocks (>2500 m.y.) in this part of Africa is, at present, poorly defined by radiometric data. However, an impressive zone of charnockitic rocks of the Luiza-type basement has been shown to extend from the type locality in Kasai, WSW for some 750 miles to the coast of Angola (Delhal and Fieremans, 1964) (see figure 12).

Gabon–Cameroons nucleus In the Republic of the Congo, Gabon and the southern part of the Cameroons a heterogeneous granitic massif with migmatites forms an extensive part of the exposed Precambrian. Cahen (1963) notes that 'au moins certaines parties de ce massif remontent à 2500 m.a.' (see figure 12). The most consistent radiometric evidence for this ancient activity is from the Ebolowa granite and associated pegmatite in the southern Cameroons (Bessoles and Roques, 1959; Lasserre, 1964); the results are given in table 9. The Ebolowa granite lies within the Ntem metamorphic complex and its antiquity is supported by Rb–Sr ages of 2200 and 2360 m.y., and a zircon Pb/α age of 2490 m.y. Lasserre considers that it is possible that the 1980 m.y. biotite age from the Ebolowa Quarry may represent an event of rejuvenation which has not affected the coarser crystals of the pegmatite in the same quarry.

Table 9 Ancient ages from southern Cameroons

Rock	Locality	Dated sample	Age (m.y.) Pb/α[a]	Age (m.y.) Rb–Sr[b]
Granite	Ebolowa Quarry	Biotite	—	1980 ± 70
Pegmatite	Ebolowa Quarry	Biotite	—	2357 ± 50
Granite	Sangmélima Quarry	Biotite	—	2205 ± 85
Granite	2°55′N; 11°08′E	Zircon	2490	—

[a] After Bessoles and Roques (1959).
[b] After Lasserre (1964).

At present it is uncertain how extensive these ancient rocks are in this part of Africa; however, a Pb/α age of 2500 m.y. has been obtained for zircon from a Gabon granite (Bessoles and Roques, 1959; Cahen, 1961).

Mauritania nucleus The ancient Precambrian of Mauritania includes at least three rock groups: (i) the *granitic complex of Sattle Ogmane*, with leucocratic and migmatitic granites and pegmatites; (ii) the *Tasiast Group* comprising granites, pegmatites, migmatites and a metamorphosed sedimentary–volcanic sequence including amphibolites, gneisses, serpentines, etc.; and (iii) the *Amsaga Series*, a metamorphic complex with migmatites, and concordant and discordant granites.

Age data from the Sattle Ogmane granitic complex and the Tasiast Group show a concentration of Rb–Sr mineral and whole rock ages around 2500–2600 m.y. (see figure 12 and table 10) (Bonhomme, 1962; Vachette, 1964b). Bonhomme (1962) recognized, however, that although the Sattle Ogmane lepidolite gave an age of ca. 2600 m.y., the biotite from the leucocratic and migmatitic granites with which it is associated yielded Rb–Sr ages of ca. 1850 m.y. (see table 10) related to the emplacement of the Amsaga (= Eburnian) granites. These divergent results are reconciled by the view that there exists, in the Amsaga region, a massif whose age is ca. 2600 m.y. but on which a subsequent metamorphism has imposed ages of 1850 m.y. It is possible that the muscovite and spodumene ages (2000–2200 m.y.) recorded in the Tasiast Group may also have resulted from the effects of Eburnian granite emplacement; and that the somewhat younger ages of ca. 1600 m.y. may reflect an even more youthful metamorphism (see table 10) (Bonhomme, 1962; Vachette, 1964b).

Table 10 Rb–Sr age determinations from Mauritania

Rock	Dated sample	Age (m.y.)
Sattle Ogmane granitic complex[a]		
Migmatitic granite	Biotite	1812 ± 20
Leucocratic granite	Biotite	1862 ± 26
Pegmatite	Lepidolite	2580 ± 90
Pegmatite	Whole rock	2400 ± 80
Tasiast Group[b]		
Migmatite	Muscovite	2629 ± 54
	Biotite	1610 ± 34
	Whole rock	2591 ± 156
Granite	Biotite	1668 ± 51
Pegmatite	Muscovite	2637 ± 44
Pegmatite	Whole rock	2565 ± 72
Pegmatite	Muscovite	2553 ± 26
	Biotite	593 ± 18
	Whole rock	2407 ± 176
Pegmatite	Muscovite	2571 ± 74
Pegmatite	Whole rock	2810 ± 225
Pegmatite	Ferrophengite	2220 ± 43
Pegmatite	Microcline	2539 ± 66
Pegmatite	Spodumene	2079 ± 163

[a] After Bonhomme (1962).
[b] After Vachette (1964b).

The bearing of these age data on the definition of the Amsaga Series is still uncertain. From the dating of Amsaga (≡ Eburnian) granites, the series is clearly older than 2000 m.y. If, as has been suggested on geological grounds, the Amsaga Series is part of the gneissic complex, it is older than 2600 m.y. If, however, it overlies the basement complex, the Amsaga Series falls within the 2000–2600 m.y. time interval. The question is still open.

Ancient ages elsewhere in West Africa   A number of ancient Pb/α ages have been obtained for zircons from the Ivory Coast; the host rocks, and ages, are as follows: amphibole granite, 2670 m.y.; Man granite, 2680 m.y.; Grabo pegmatite, 2310 m.y.; and embrechite, 2420 m.y. (Roques, 1959; Bonhomme, 1962). The zircon age of the Man granite is in marked contrast to the biotite age of 2028 m.y. obtained by the Rb–Sr method (Bonhomme, 1962); this granite, which is associated with the charnockites of Man, has in fact been grouped with the Baoulé granites believed to have been emplaced during the post-Birrimian Eburnian episode some 1830–2030 m.y. ago (Bonhomme, 1962). Similarly, the Grabo pegmatite is considered to be post-Birrimian (Holmes and Cahen, 1957). On the basis of these data it appears that these ancient dates may be relict ages (2300–2700 m.y.) in younger (Eburnian) rocks, and that they may indicate the existence of important pre-Birrimian phenomena.

A similar age pattern has emerged from studies in the Kerdous massif of southern Morocco where the Oued Amarhous granite has yielded a whole rock Rb–Sr age of 2595 ± 130 m.y. and a muscovite Rb–Sr age of 1806 ± 55 m.y. The whole rock age has been taken as the real age of the granite; the muscovite age indicating subsequent rejuvenation.

Finally, it should be noted that alluvial monazite from central Sierra Leone has given a $^{207}Pb/^{206}Pb$ age of 2940 m.y. (Holmes and Cahen, 1955, 1957). The monazite is believed to be from pegmatites associated with concordant granites in the Sula Mountains schists; the regional stratigraphic and structural significance of this age is, as yet, unknown.

C   *Ancient ages preserved in zones of late Precambrian–early Palaeozoic orogenesis*

Malagasy Republic   A number of very old ages have been recorded from parts of the crystalline complex of the Malagasy Republic (see figure 12). A decade ago, a very discordant age of 2420 m.y. was obtained for alluvial monazite probably derived from the granitized rocks of the Graphite System (see p. 322); Besairie (1959) utilized this age as a minimum limit for that system. More recently, Delbos (1964) has recorded a number of old ages from the Androyan and Vohibory

Systems; on a basis of these data, listed in table 11, he considers that a major orogeny (dated around 2500 m.y.) affected both systems, but that it is now largely masked by the widespread younger ages around 500 m.y.

Table 11   Ancient ages from Malagasy Republic. (After Delbos, 1964)

Rock	Stratigraphic position	Dated sample	Age (m.y.)[a]
Granite	Androyan System	Whole rock	2160 ± 23
		Biotite	490 ± 20
Leptynite	Androyan System	Whole rock	2057 ± 20
		Biotite	500 ± 6
Granite	Androyan System	Whole rock	3020 ± 50
Gneiss	Vohibory System	Zircon	2749
		Biotite	456 ± 21

[a] All mica and whole rock ages were determined by the Rb–Sr method.

Lower Congo   Despite the complex subsequent history in the Lower Congo region (see p. 315 and p. 362) particularly during the Mayumbian and West Congolian orogenies, microcline from a pegmatite has 'retained' an age of ca. 2760 ± 500 m.y. This isolated age has been interpreted as representing an older orogenic cycle to which the name Pre-Mayumbian orogeny has been given (Cahen and others, 1963).

Rhodesia   In the northeastern part of Rhodesia, one of the schist belts largely composed of rocks of the Bulawayan System, is intruded by the Mtoko Batholith belonging to the gregarious batholith swarm of Rhodesia (Macgregor, 1951); associated with this granitic mass are important beryllium, lithium and tantalum pegmatites. Age data from this region, given in table 12, show a significant variety of dates (Snelling and others, 1964a). The lepidolite, and granite whole rock, Rb–Sr ages suggest that these rocks were emplaced over 2300 m.y. ago; this view is supported by the mineralogy of the pegmatites which closely resembles that of pegmatites characterized by ages of 2650 m.y. in the Rhodesia nucleus. This schist belt lies along the southern

Table 12 K–Ar and Rb–Sr ages from the Mtoko region of Rhodesia. (After Snelling and others, 1964a)

Rock	Dated sample	Age (m.y.)	
		K–Ar	Rb–Sr
Pegmatite	Muscovite	450 ± 20	1090 ± 50
	Lepidolite	965 ± 40	2370 ± 55
Granite	Whole rock	—	2825 ± 150
	Biotite	595 ± 25	—

margin of the Zambesi Belt, characterized by ages in the ca. 500 m.y. range reflecting the Damaran orogeny (Johnson and Vail, 1965). It has been suggested (Snelling and others, 1964a) that the low Rb–Sr and K–Ar ages for muscovite and K–Ar ages for lepidolite and biotite are due to complete or partial loss of radiogenic daughter products during that event.

A similar sequence of mixed ages has also been obtained from the Miami granite intruding the Piriwiri System to the west; whole rock and separated-mineral fractions from a single granite sample gave ages as follows (Clifford, Rex and Snelling, 1967).

	K–Ar age (m.y.)	Rb–Sr age (m.y.)
Biotite	480 ± 20	500 ± 30
Microcline	—	2894 ± 110
Whole rock	—	2800 ± 110

The significance of the microcline age is, as yet, unknown.

West Africa In West Africa, the Adrar des Iforas appears to be structurally continuous with the main portion of the Hoggar to the east (see Furon, 1963, p. 91); however, no trace has been found of the 500–600 m.y. episode that affected the major part of the Hoggar (see p. 326). Indeed, Ferrara and Gravelle (1966) report radiometric data from the northern extension of the Adrar des Iforas as follows: a Rb–Sr whole rock isochron age of 2680 m.y. for granite, syenite and gneiss samples; and Rb–Sr and K–Ar ages of 1700–1840 m.y. for biotites from pyroxenite, granite and gneiss samples. In the light of these data, Ferrara and Gravelle (1966) consider that the northern extension of the Adrar des Iforas belongs to the eastern part of the West African Craton (see p. 336). It is, however, equally possible that it lies within the western margin of the late Precambrian–early Palaeozoic orogenic belt, but that the thermal imprint, which accompanied orogeny at that time in that area, was too weak to effect a significant 'rejuvenation' of the micas.

D *Summary*

Ancient ages (>2100 m.y.) are widely represented in southern Africa with some representatives preserved in West Africa. In general, three age groupings seem to be emerging, separated by intervals of 200–300 m.y. as shown in the following table.

Whether these three age ranges represent separate orogenies is not

Group	Age patterns	Examples
I	2200–2300 m.y.	Dodoma–Nyanza nucleus
II	2500–2700 m.y.	Dodoma–Nyanza nucleus Rhodesia nucleus Transvaal nucleus Kasai nucleus (Hoggar) S. Morocco and Mauritania Great Dyke (Rhodesia)
III	3000 m.y.	Transvaal nucleus Sierra Leone (?)

known, although the weight of present evidence suggests that II and III reflect important orogenic events. However, irrespective of their interpretation, these ages constitute a number of bench marks which are useful in delimiting ancient stratigraphic sequences. They provide, for example, *minimum* ages of: 2200–2300 m.y. (and probably 2500 m.y.) for the Dodoman System (Tanzania); 2500 m.y. for the Nyanzian–Kavirondian Systems (East Africa) and possibly the Androyan–Vohibory Systems (Malagasy Republic); 2600–2700 m.y. for the Sebakwian–Bulawayan–Shamvaian Systems (Rhodesia), the Tasiast

Group (Mauritania), and the Luiza- and Dibaya-type basements (Congo); and 2900–3000 m.y. for the Moodies–Swaziland Systems (Transvaal). In all of these cases the ancient deformation was accompanied by the extensive emplacement of granitic rocks. Although many of these sequences are unmetamorphosed, there are important examples of amphibolite grade (e.g. Tasiast Group of Mauritania, Swaziland System of the Transvaal and Dodoman System of Tanzania), and even granulite–charnockite grades (e.g. Luiza-type basement of the Congo), of metamorphism associated with the granite emplacement.

From a consideration of stratigraphy these dated ancient events provide bench marks for certain overlying sequences. For example, radiometric determinations provide *maximum* ages of 2600 m.y. for the Piriwiri–Lomagundi–Deweras Systems (the Piriwiri is known to be older than ca. 1100 m.y.) of Rhodesia, and the Luiza Series (known to be older than ca. 2100 m.y.) of the Congo; and 2800–3000 m.y. for the Transvaal, Ventersdorp, Witwatersrand and Dominion Reef Systems (known to be older than 1950 ± 200 m.y.) of the Transvaal. Of these sequences, the latter deserve especial mention since, together, they represent 50,000–60,000 ft of sediments and volcanic rocks as shown in table 13.

Table 13 The stratigraphic sequence of the Transvaal

System	Lithology	Thickness (ft)
TRANSVAAL	Dolomitic limestones, shales, quartzites, some volcanic rocks and a tillite	15,000–20,000
---------- unconformity ----------		
VENTERSDORP	Mainly andesitic volcanic rocks with subordinate sediments (conglomerates, sandstones, cherts and shales)	10,000–12,000
---------- unconformity ----------		
WITWATERSRAND	Quartzites, shales, conglomerates, and thin volcanic horizons	25,000
---------- unconformity ----------		
DOMINION REEF	Acid and basic lavas, quartzose sediments, arkoses and conglomerates	2000–3000
---------- unconformity ----------		

This succession is of particular interest in that it contains: (i) some 10,000 ft of volcanic rocks; and (ii) the important gold-bearing horizons of the Witwatersrand and Dominion Reef Systems which provide more than 90% of Africa's gold production. However, from a sedimentological point of view it is perhaps worthy of record that this succession also contains some of the oldest known extensively developed *differentiated* sediments (e.g. quartzites and carbonate rocks) in Africa; in this respect, it stands in marked contrast to the largely *undifferentiated* sedimentary sequences of greywacke and shale (with only meagre pure quartzite and carbonate horizons) which characterize the older Moodies–Swaziland Systems, and the Sebakwian–Bulawayan–Shamvaian Systems (see Macgregor, 1951, p. xxix–xxx).

In regard to the ages of these four systems, zircons from a quartz porphyry lava high up in the Ventersdorp System has given a U–Pb age of 2300 ± 100 m.y. (Van Niekerk and Burger, 1964). In addition to these data for the Ventersdorp lavas, there is evidence: that the Witwatersrand System is younger than 2700 m.y.; that the Dominion Reef contains detrital uranium-bearing minerals dated at ca. 2950 m.y.; and that felsites from the Dominion Reef System are around 3000 m.y. old.

VI Conclusions

Until the application of radiometric dating techniques, geological investigations dealing with general stratigraphic correlations and the fundamental structural pattern of the pre-Silurian of the major part of Africa, were based on qualitative criteria. In the past 20 years the use of these techniques has, in the absence of fossils, greatly advanced the study of the geology of the continent. However, the value of these modern methods would have been limited but for the intensive and extensive field and laboratory studies carried out by geologists particularly since the latter part of the 19th century; and serious geologists and geochronologists alike recognize the complementary nature of their respective approaches.

Radiometric dating has resulted in the recognition of a number of important bench marks defined by thermal imprints largely related to igneous activity and orogenic processes. In regard to the latter, a number of orogenic events have been dated as follows: (i) 450–680 m.y. (Damaran–Katangan or Pan-African orogeny); (ii) 1100 ± 200 m.y. (Kibaran orogeny); (iii) 1850 ± 250 m.y. (including the Eburnian orogeny, 1850–2050 m.y. ago); (iv) 2500–2700 m.y. (Shamvaian orogeny); and (v) ca. 3000 m.y.

Of these structural events, the most extensively preserved evidence is for the Damaran–Katangan (Pan-African) orogenic events which apparently affected about half of the continent. Discussing ages within this range, Cahen (1963) maintains that only dates older than ca. 600 m.y. reflect orogenesis and that the more youthful ages, particularly around ca. 500 m.y., represent posttectonic events. However, considering all available data from southern and West Africa, it has been argued that the 450–680 m.y. ages reflect two orogenic pulses as follows.

Southern Africa[a]	West Africa[b]
Damaran episode (450–570 m.y.)	Main phase (± 500 m.y.)
Katangan episode (580–680 m.y.)	Early phase (± 600 m.y.)

[a] Clifford (1963, 1967).
[b] Kennedy (1964, in preparation).

The late Precambrian–early Palaeozoic orogenesis was, on this reasoning, biepisodic with episodes at end-Precambrian and at Cambro–Ordovician times. In this regard, it is perhaps worthy of note that regional orogenic belts are, in general, biepisodic or polyepisodic; for example, the Taconic and Acadian orogenies (separated by 50 m.y.) in the northern Appalachians, and the Nevadan and Laramide orogenies in western U.S.A.

The critical question is whether, in fact, the older Precambrian orogenies of Africa are similarly divisible. Certainly, in the case of the Kibaran orogeny, reflected by 1100 ± 200 m.y. ages, Cahen and others (1966) consider that the orogeny is reflected by ages somewhat older than ca. 1100 m.y., while ages younger than that represent post-tectonic events. In contrast to this view, however, ages in the 900–1100 m.y. range in the Cape Province of South Africa are considered to reflect 950–1000 m.y. old tectonism (Nicolaysen, 1962; Nicolaysen and Burger, 1965). These data imply two separate orogenic pulses within the 1100 ± 200 m.y. event. Moreover, within the 1850 ± 250 m.y. range, Eburnian orogenesis is well established at 1850–2050 m.y. ago and represents the *main phase* of tectonism in Africa around that time; the significant concentration of ages of 1600–1700 m.y. (cf. Zenaga region of southern Morocco) may however represent a *late phase* of tectonism within the 1850 ± 250 m.y. range (Kennedy, in preparation).

Clearly, more radiometric determinations and field data are required before the significance of these age ranges can be fully understood, but the thermal imprints dated at 450–680 m.y., 1100 ± 200 m.y., 1850 ± 250 m.y., 2500–2700 m.y. and ca. 3000 m.y. clearly reflect the major periods of mountain building, though it is still possible that these important ranges are masking less extensive orogenic activity of other ages.

These orogenic events were accompanied by the widespread emplacement of acid plutonic rocks. In addition, the pre-Silurian history of the continent was marked by a variety of anorogenic intrusive phases; those for which radiometric dates are available are shown in table 14, together with some of the more extensive volcanic sequences. From these data it is evident that there is a marked absence of extensive extrusive vulcanicity in the 600–2000 m.y. time interval (except for more limited sequences such as the Nzilo volcanics of the Kibara Group); indeed, there is a paucity of such rocks from ca. 2000 m.y. ago until the onset of Karroo vulcanism in early Mesozoic times. However, perhaps more significant is the fact that certain of the principal *plutonic* events in stable regions were clearly synchronous with orogenic events taking place elsewhere in mobile zones, particularly around 1100 ± 200 m.y., 1850 ± 250 m.y. and ca. 2500–2700 m.y. times (see table 14). An example of this synchroneity is shown by the Bushveld complex and the Palabora carbonatite of the Transvaal, emplaced while Eburnian orogenic activity was affecting large parts of West, East and Central Africa.

Of all the pre-Silurian anorogenic igneous events, perhaps the most impressive features are the Great Dyke and the Bushveld complex. The radiometric studies of these intrusives have not only been of stratigraphic importance, but also contribute to the understanding of crustal processes at the time of intrusion. In this regard, dating of minerals from the Bushveld complex indicates that it was emplaced 1950 ± 200 m.y. ago, while similar ages from the same general region have also been yielded by: biotite from 3000 m.y. old basement granite; 2950 m.y. old detrital uranium-bearing minerals in the Witwatersrand and associated systems; and granite and basic granophyre associated with the Vredefort Ring to the south of the Bushveld complex. These ages suggest that the emplacement of the complex may have had a profound effect not only in the reheating of underlying rocks, but also in crustal tectonics of local significance but orogenic intensity, as in the case of the inversion around the Vredefort Ring. It is perhaps worthy of note, in this regard, that even in its

Table 14 Elements of Precambrian igneous activity in Africa

Volcanic events	Anorogenic intrusive events
	690–750 m.y. MBOZI ALKALINE COMPLEX (Tanzania) and NKUMBWA CARBONATITE (Zambia)
	900–1100 m.y. BASIC SILLS (Transvaal, Botswana)
Pilanesberg volcanics—phonolites and trachytes	1200–1400 m.y. ALKALINE COMPLEXES (Pilanesberg, Leeuwfontein) (Transvaal)
Bushveld volcanics—basic lavas, tuffs and felsites	1950 ± 200 m.y. BUSHVELD IGNEOUS COMPLEX and PALABORA CARBONATITE (S Africa).
Transvaal System volcanics—andesites Ventersdorp System—ca. 10,000 ft of andesites and pyroclastics ------?--------?-------?------------	
	2400 ± 400 m.y. GREAT DYKE (Rhodesia)
------?--------?-------?------------ Bulawayan volcanics (Rhodesia)— basic and intermediate lavas, older than 2600 m.y.; Nyanzian volcanics (E Africa)—acid and basic lavas, older than 2500 m.y.; Tasiast volcanics (Mauritania)—older than 2600 m.y.; and Onverwacht volcanics (Transvaal)—acid and basic lavas, older than 2900 m.y.	

present form the Bushveld complex represents about 100,000 cubic miles of magma evacuated from the crust or upper mantle; this volume is sufficient to represent a circular mass of rock 1–2 miles in thickness and with a radius of 140 miles—an area which could embrace all of the regions of Bushveld igneous activity, and extend as far south as the Vredefort Ring.

It has long been considered that the emplacement of the Bushveld complex was synchronous with the intrusion of the Great Dyke of Rhodesia to the north (Faure and others, 1963). However, ages from the former (largely for samples from the Bushveld granite) indicate an emplacement age of 1950 ± 200 m.y. whereas mineral and whole rock data from the Great Dyke (largely for picrite samples) indicate an intrusion age of 2400 ± 400 m.y. Although the basic and ultrabasic rocks of these intrusives are very similar, the two complexes differ in that the Bushveld contains an extensive granite phase which postdates the basic–ultrabasic plutonic suite (Du Toit, 1954). It is just possible that, even though these two complexes seem now to be of different ages, it is the emplacement of the granite phase which is reflected by ca. 2000 m.y. ages in the Bushveld complex and surrounding terrains, and that the emplacement of the basic–ultrabasic suite was earlier, and synchronous with the intrusion of the Great Dyke. This possibility remains to be tested. It has important implications, particularly in regard to the age of the Transvaal, Ventersdorp, Witwatersrand and Dominion Reef Systems; any consideration of the ages of these sequences must, however, take account of the 2300 ± 100 m.y. age obtained for Ventersdorp lavas.

The structural and igneous bench marks which have been erected on the basis of radiometric data place limits of various degrees of precision on a number of African sequences; the more precisely dated units are shown in table 15. The Katanga System was affected by tectonothermal events reflected by 450–680 m.y. ages but rests on the eroded roots of the Kibara mountain chain yielding ages of ca. 1100 ± 200 m.y.; the Outjo System in South-West Africa is defined by similar limits. The Kibara–Urundi–Karagwe Ankolean Groups are clearly older than 1100 ± 200 m.y. and rest unconformably on the eroded remnants of mountain chains yielding ca. 1800 m.y. ages in Uganda (Lwanda unconformity) and elsewhere in East and Central Africa (Ubendian–Ruzizi and the Lukoshian Belts). Moreover, in the Transvaal, the Waterberg–Loskop Systems rest unconformably on the Bushveld complex and are believed to predate granophyres yielding ca. 1700–1800 m.y. ages. In addition, the Bushveld complex

Table 15 Some dated events which define selected stratigraphic sequences in Africa

Orogenic	Anorogenic [c]
BAKEL SERIES (Senegal)[a]	
450–680 m.y.	
KATANGA SYSTEM (Central Africa) and the OUTJO SYSTEM [Otavi and Damara Facies] (South-West Africa)[b]	
1100 ± 200 m.y.	
	1420 ± 70 m.y.
KIBARA, URUNDI and KARAGWE ANKOLEAN GROUPS (East and Central Africa)	WATERBERG SYSTEM
	1790 ± 70 m.y.
	WATERBERG(?)–LOSKOP SYSTEMS
ca. 1800–2000 m.y.	1950 ± 200 m.y.
LUIZA SERIES (Congo)	TRANSVAAL SYSTEM VENTERSDORP SYSTEM WITWATERSRAND SYSTEM DOMINION REEF SYSTEM
ca. 2800–3000 m.y.	
MOODIES AND SWAZILAND SYSTEMS (South Africa)	

[a] The depositional interval of the Bakel Series in Senegal is defined apparently by maximum and minimum ages of 600 m.y. and 350 m.y. respectively.
[b] The Malmesbury Series (South Africa) is believed to have been deposited around 560–600 m.y. ago (Allsopp and Kolbe, 1965); the Bukoban System of Central and East Africa contains lavas dated at ca. 675 m.y.
[c] South Africa and Botswana.

postdates the Transvaal, Ventersdorp, Witwatersrand and Dominion Reef Systems which, in turn, overlie unconformably the ancient basement which has yielded ages of 2800–3000 m.y., reflecting the orogenic and granite emplacement events which affected the Moodies and Swaziland Systems.

It is clear that even these sequences can only be defined within broad age limits. A number of plutonic, and metamorphic events (largely orogenic) help in the definition of other sequences as follows:

(i) Large-scale granite and pegmatite emplacement dated at ca. 2500–2700 m.y. is known from a number of parts of Africa and places a *minimum* age on the following: (a) the Kavirondian System ($\equiv$ Samia Series) and the Nyanzian System ($\equiv$ Bulugwe Series) of Uganda, Kenya and Tanzania; (b) the Sebakwian–Bulawayan–Shamvaian Systems of Rhodesia; (c) the Luiza- and Dibaya-type basement of Kasai (Congo); (d) the Dodoman System of Tanzania; (e) the Ntem metamorphic complex of the Cameroons; (f) the Tasiast Group of Mauritania; and (g) the Androyan and Vohibory Systems of Malagasy Republic. Amongst these sequences, there is radiometric evidence that the Luiza-type basement includes ancient rocks older than 3300 m.y. which were subsequently granitized 2600 m.y. ago. In addition, on lithological grounds (largely the fact that they are similar volcanic–sedimentary sequences) the Nyanzian–Kavirondian and Sebakwian–Bulawayan–Shamvaian Systems have been broadly correlated with the Moodies–Swaziland Systems known to be older than ca. 2900 m.y. (see table 15); the Tasiast Group of Mauritania is also a volcanic–sedimentary sequence.

(ii) Widespread granite and pegmatite emplacement and metamorphism during the Eburnian episode is reflected by ages of ca. 1850–2050 m.y. and places *minimum* ages on: (a) the Lukoshi Formation of the Congo; (b) the Luiza Series of the Congo; (c) the Ubendian System (and the Ruzizi Group) of Zambia, Tanzania and the Congo; (d) the Kibali Group of the Congo; (e) the Buganda–Toro System of Uganda; (f) the Messina Formation of Botswana, Rhodesia and the Transvaal; (g) the Ogooué gneiss (Mayumbe System (?)) of Gabon and the Lower Congo; (h) the Birrimian System (Yatakala Series) of Ivory Coast, Upper Volta and Mali; (i) the Suggarian 'System' of the Hoggar; and (j) the Amsaga Series of Mauritania. Despite these minimum ages, the true age range of the majority of these sequences is not known. The Luiza Series of metasediments, which has yielded ages of 2100 m.y., rests unconformably on Dibaya-type basement and is, thus, probably younger than ca. 2600 m.y. Moreover, ancient ages of 2400–2900 m.y., considered to represent pre-Birrimian events in West Africa, suggest that the Birrimian System was deposited between 2000 and 2400–2500 m.y. ago; the Amsaga Series, the Ogooué gneiss (and the Mayumbe System), and the Buganda–Toro System (Kibali Group) *may* represent a similar time range (Bonhomme, 1962; Cahen, 1961). If, however, the Amsaga Series is to be correlated with the Tasiast Group (see Bonhomme, 1962) and the Buganda–Toro Group is correlative with the Bulugwe Series (Nyanzian System) and the Samia Series (Kavirondian System)] (see McConnell, 1959), then these sequences must be older than 2500–2600 m.y., and thus *significantly* older than the regional tectonothermal imprint, dated at 1800–2000 m.y., which affects them. Similar difficulties arise in placing a *maximum* age on the Suggarian 'System', the Ubendian System and the Messina Formation. The latter, at least, has many similarities with the Swaziland–Moodies Systems (see table 15); however, Nicolaysen (1962) has pointed out that certain of the metasediments of the Messina Formation may actually be correlative with the Transvaal System deposited in the more youthful part of the 2000 to 3000 m.y. time interval.

(iii) Pegmatites, and granitic and metamorphic rocks yielding ages of 1100 ± 200 m.y. place a *minimum* age on a number of sequences as follows: (a) the Kheis System of the Cape Province of South Africa; (b) the crystalline rocks of Natal; (c) the Piriwiri System (and the Lomagundi–Deweras sequence; of Rhodesia and (d) an important anorthosite–norite mass in southern Angola and northern South-West Africa. Of these, the rocks of the Kheis System have been considered to be much older than the ca. 1000 m.y. ages suggest, and have been equated with the *lower* part of the basement systems of the Transvaal nucleus, and the Sebakwian System of Rhodesia (Du Toit, 1954, p. 54). The Piriwiri–Lomagundi–Deweras Systems are younger than ca. 2600 m.y. since they rest unconformably on the basement of the Rhodesia nucleus. Equally, no direct measurement is available for the age of the important anorthosite body of Angola and South-West Africa (Simpson and Otto, 1960), but it is certainly older than pegmatites yielding a muscovite Rb–Sr age of 1185 ± 80 m.y. (Simpson and Otto, 1960);

it may be related to the Bushveld complex or to the Great Dyke (see figure 11).

(iv) Ages in the 450–680 m.y. range, reflecting widespread metamorphism, granite–pegmatite emplacement and mineralization, place *minimum* ages on the following sequences: (a) rocks (including the Usagaran System of Tanzania) of the Mozambique and Zambesi Belts of East and Central Africa; (b) the 'Lower and Middle' Precambrian of the Cameroons and the Central African Republic; (c) the Dahomeyan of Dahomey, Togo and Nigeria; (d) the Marampa schists and the Kasila gneiss of Sierra Leone; (e) the Pharusian System of the Hoggar; (f) the Boje Series of Ethiopia; (g) the Inda Ad Series and the Basement Complex of the Somali Republic; (h) the Hammamat Series (and underlying sequences) of Egypt; (i) the Western Congo System of the Lower Congo–western Angola region; and (j) the Muva System of Zambia. In the case of (a)–(d) inclusive, no maximum age sare known, though these structural units clearly include a variety of old rocks since isolated ages of 1950 m.y. and 1650 m.y. have been obtained from the crystalline rocks of the Central African Republic and the Dahomeyan respectively. Equally, no maximum ages can be given to (g) and (h) above, although the sequences have been confidently ascribed to the late Precambrian ($\equiv$Katanga System). In other cases, maximum ages are of some limited assistance in defining certain sequences: for example, the Pharusian System is younger than ca. 1800 m.y.; the Muva System is considered to be younger than ca. 1600 m.y. and, since it is overlain unconformably by the Katanga System, has been equated with the Kibara Group (see table 15); and the Western Congo System is younger than ca. 1480–1800 m.y.

It is clear, therefore, that although sequences (g)–(i) are widely referred to the late Precambrian, the quantitative limits presently available are too wide to prove such a classification, although qualitative lithological reasoning suggests that they are, in part at least, equivalent to the Katanga System.

Analysing these data in the light of the major pre-Silurian structural pattern of Africa, it is evident that older basement formed the floor of many of the orogenic zones. For example, in addition to Upper Precambrian geosynclinal sequences, the following ancient rock units have been involved in the late Precambrian–early Palaeozoic orogenic zones: Suggarian (Hoggar), pre-Mayumbian (Congo), Mayumbian (Congo), Huabian (South-West Africa), Bulawayan System and Shamvaian granites (Rhodesia), Muva System (Zambia), Kibara Group (Congo) and the Androyan and Vohibory Systems (Malagasy). These represent a variety of rocks ranging in age from ca. 1100–1600 m.y. to older than 2500 m.y.; it is, moreover, widely believed that rocks of a similar variety of ages may also be involved in the Mozambique Belt sector of this 450–680 m.y. orogen (see Cahen, 1961).

In addition, older sialic rocks forming the basement ($\equiv$1800 m.y.?) to the Kibara Group (and its correlatives) were remobilized during the Kibaran orogeny (1100 $\pm$ 200 m.y. ago) in Uganda. If, moreover, the present correlation of the Kheis System with the lower part of the Swaziland System is correct, then the Cape Province portion of the 1100 $\pm$ 200 m.y. orogenic zone was 'floored' by rocks of high antiquity (see Nicolaysen and Burger, 1965).

Furthermore, regions affected by the Eburnian (1850–2050 m.y.) episode not only reveal deformation of geosynclinal sequences deposited immediately prior to the orogeny, but also show the involvement of ancient rocks (older than ca. 2500 m.y.); for example, in Mauritania, Kasai and, perhaps, the Limpopo Belt.

Clearly, then, African orogenic zones not only show characteristic deformed geosynclinal sequences, but also the participation of the older sialic basements with their record of earlier orogenic histories. Indeed, in many instances, little of the geosynclinal sequence is present and only the original basement floor is preserved with the record of later tectonism; these regions are *vestigial facets* of orogeny (vestigeosynclines) and form a striking part of the pre-Silurian pattern of African orogenesis. Such is, for example, the case in the late Precambrian–early Palaeozoic orogenic zone in which the Mozambique Belt, the Cameroons sector, the Central African Republic sector, the Mayumbian Belt (Lower Congo) and the pre-Karroo of the Malagasy Republic all apparently represent vestigial facets of 450–680 m.y. orogenesis in which no *extensive* sequences of Upper Precambrian geosynclinal rocks have been recognized. Whether those sequences were ever deposited is not known but Lowdon and others (1963), in discussing the Grenville Province of Canada (a similar vestigial facet), maintain that the Grenville orogeny (ca. 1000 m.y. ago) 'may have taken place without the usual immediately preceding geosynclinal phase.' Of course, in the case of the vestigial facets of the late Precambrian–early Palaeozoic orogenic zones of Africa it is equally possible that the geosynclinal rocks have been removed since deposition because, in the southern half of the continent at least, an important depositional break exists between the end of the Damaran episode and the mid-Carboniferous. On the basis of estimates of *past* erosion rates of 1 metre/100,000 yr (Holmes, 1965), that 200 m.y. interval provides sufficient time for the removal of 6000 ft of rock; moreover, if *past* erosion rates were similar to those of the *present* day, something like 18,000 ft may have been removed during the same period (Holmes 1965). In view of the fact that the type Upper Precambrian—Katanga System—is 12,000 to 20,000 ft in thickness in Central Africa it is certainly feasible that the absence of these rocks in parts of the orogenic zone is due to subsequent erosion (see figure 4). This line of reasoning can be applied to similar vestigial facets of even older orogenic belts and is facilitated by the apparently even lengthier time gaps in earlier Precambrian sedimentation.

These vestigial facets of orogeny, and the more limited areas of basement floor appearing from beneath deformed geosynclinal sequences, indicate that pre-Silurian orogenies since, and including, the Eburnian tectonism involved the reactivation of older tectonic units of sialic basement, and *did not result in the marginal accretion of new crust*; this conclusion receives additional support from the pattern of some of the orogenic belts shown in generalized form in figure 15. Indeed, when the more recent (ca. 1500 m.y. to present) structural events of the African continent are considered as a progressive sequence (figure 16), it is obvious that the overall trend throughout geological time has been one of *extension of regions of remnant stability* with: ca. 35% of the continent stable since at least ca. 1500 m.y. ago; ca. 40% stable since ca. 900 m.y. ago; 95% stable since ca. 450 m.y. ago; and 98% stable since the end of the Middle Palaeozoic–early Mesozoic orogenesis in the Mauritanides, Anti-Atlas and Cape Fold Belt (see figure 17). *Clearly the progress of African structural development has been the advance of cratonism with the consolidation of zones of mobility* (*orogens*). Although the data for pre-1500 m.y. times are less complete, and the analysis more qualitative, it is known that remnants of the oldest nuclei (undeformed orogenically during the past ca. 2100 m.y.) make up but a small percentage of the continent.

In the light of the view that the major crustal patterns and processes reflect convection in the earth's mantle (see Runcorn, 1962), the sizes of cratons during various orogenies offer some comparative measure of the sizes of the convection cells. For example, the Congo Craton (figures 15 and 16b) was stable during the Damaran–Katangan

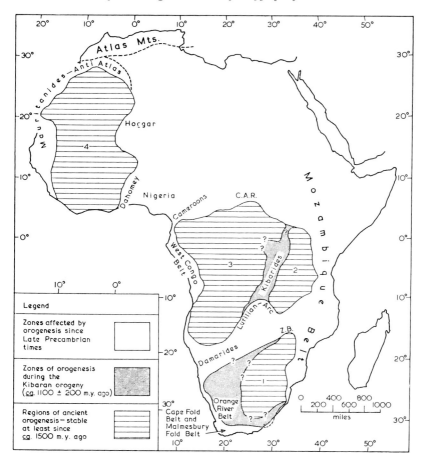

Figure 15 Generalized map relating the structural regions of Africa to the time of their last orogenic deformation; the specific regions affected by individual episodes of late Precambrian–Phanerozoic and pre-1500 m.y. orogenesis are not differentiated.

The four older cratons are: 1. the Rhodesia–Transvaal Craton; 2. the Tanzania Craton; 3. the Angola–Kasai Craton; 4. the West African Craton.

After ca. 1100 ± 200 m.y. orogenesis, the *consolidated* Kibaran Belt was added to the stable Angola–Kasai and Tanzania Cratons to form a single stable unit, the *Congo Craton*; the Orange River (Namaqualand) and Natal Belts were likewise added to the Rhodesia–Transvaal Craton to form the *Kalahari Craton*. These two, larger cratons represent units which were stable *after* Kibaran orogenesis and *during* late Precambrian–early Palaeozoic orogenesis.

C.A.R. Central African Republic; Z.B. Zambesi Belt.

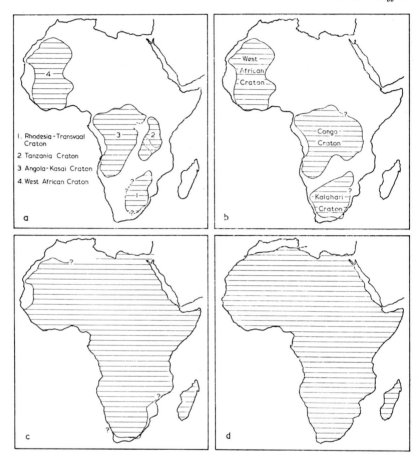

Figure 16 Stages in the structural consolidation of Africa.
a. Antique regions (ruled), stable since at least 1500 m.y. ago (the Ubendian Belt is shown with broken-ruleornament—see p. 343); b. regions (ruled) stable since the end of the Kibaran orogeny, ca. 800–900 m.y. ago; c. Regions (ruled) stable since the end of the Damaran–Katangan (Pan-African) orogenesis, ca. 400–500 m.y. ago; d. Regions (ruled) stable since the end of the Middle Palaeozoic–early Mesozoic orogenesis.

(Pan-African) orogeny, and apparently completely surrounded by a zone of orogenic activity at that time (450–680 m.y. ago). When that craton is compared with the size of the craton (virtually the whole of Africa) during the Hercynian and Alpine orogenies (figure 16c and d), it is clear that if convection cells are the fundamental cause of structural patterns, they have *increased in size* from late Precambrian–early Palaeozoic times to Upper Palaeozoic–Tertiary times. This conclusion conflicts apparently with the views of Runcorn (1962) who has suggested that, with the growth of the core of the earth, the size of convection cells has *decreased* (and their number has increased) since ca. 3000 m.y. ago (see figure 18); it is, moreover, difficult to

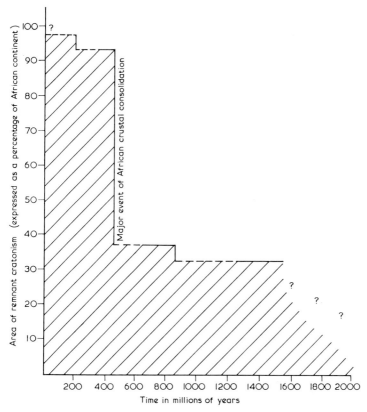

Figure 17 Changes in the percentage area of *remnant* stability in Africa with time.

reconcile the size of the Congo Craton, during late Precambrian–early Palaeozoic times, with the concept of *through-mantle* convection.

An *increase* in the size of the presumed convection cells from older Precambrian to the present time is *implied* by the increase in the sizes of remnant cratons shown in figure 16. However, it is not possible to prove a *progressive* increase on presently available data,

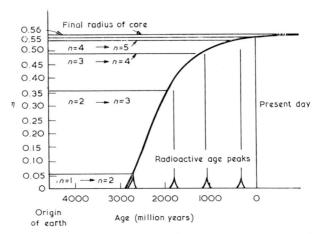

Figure 18 Runcorn's (1962) interpretation of the growth of the earth's core compared with the radioactive age determination peaks.

because the exact size of cratons during individual orogenies prior to ca. 800 m.y. ago cannot be measured. It may be, for example, that parts of regions which were *cratonic during the Kibaran orogeny* were subsequently involved in the Damaran–Katangan orogeny; under these circumstances the remnant regions of cratonism (figure 16a) represent *minimum* sizes for the stable anorogenic regions at the time of the Kibaran orogeny. The future application of radiometric dating techniques may help to define the possible further extensions of these cratons into zones of later tectonism. In the meantime, the available evidence at least *suggests* that *if mantle convection is the mechanism of major crustal events*, then the convection cells have increased in size from oldest Precambrian to the present time, and must have been very small in ca. 3000 m.y. and earlier times, since nowhere in the world are there extensive undeformed sequences of rocks older than 3000 m.y.; it may be imprudent to suggest that the *size* of cells at that time is reflected by that of the gregarious batholiths recognized by Macgregor (1951) in the Rhodesia nucleus (see figure 19).*

The application of radiometric dating techniques has, therefore, not only defined structural and stratigraphic bench marks in the pre-Silurian 'basement' of Africa, it has also opened up wide horizons

---

* The origin of domal features of this kind, and the possible influence of convection, is currently being studied by Talbot (1966).

in the study of crustal and mantle processes. In the field of economic geology, geochronology provides a basis for the recognition of major tectonometallogenic units characterized by distinctive suites of economic products (Clifford, 1966), and facilitates the understanding of the control exercised by Precambrian–Lower Palaeozoic structure in the localization of salt-bearing basins of Mesozoic and Tertiary

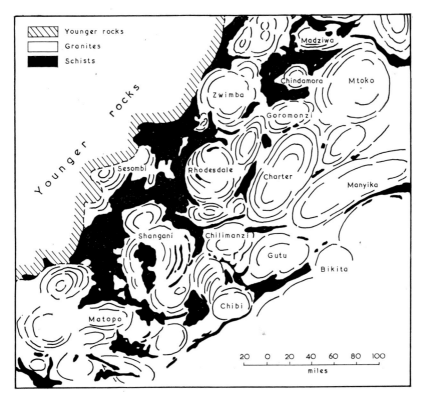

Figure 19 Rhodesian basement structure explained by gregarious batholiths. (After Macgregor, 1951)

rocks (Kennedy, 1965). In both cases, the posthumous control exercized by pre-Silurian structure is striking. Equally, the dating of orogenic belts is critical to the understanding of the major fault patterns of Africa since *all* of the major rift and related fault systems (Dixey, 1956) have been posthumously controlled by the *linear* orogenic belts of Damaran–Katangan (Pan-African), Kibaran and, to a lesser extent, Eburnian (?) (cf. Limpopo and Ubendian Belts) age (see figure 20). The crustal heterogeneity imposed during those orogenies is, clearly, of more than structural significance when it is realized that the major part of Africa's anorogenic plutonic, and *alkaline* volcanic, activity since Upper Palaeozoic times is concentrated within those earlier orogenic zones.

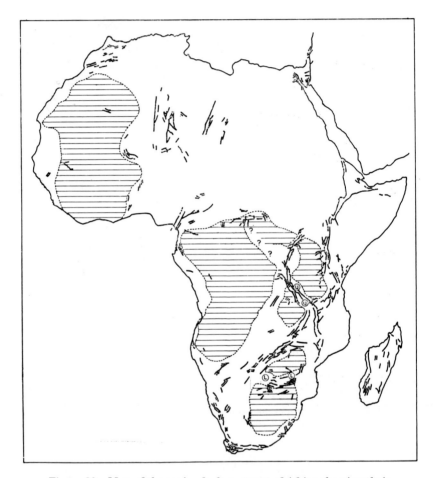

Figure 20 Map of the major fault systems of Africa showing their concentration within the Kibaran and younger orogenic belts, and in the Limpopo (L) and Ubendian (U) linear (ca. 1850 ± 250 m.y.) orogenic zones. Regions unaffected by orogenesis since the end of 1850 ± 250 m.y. tectonism (excluding the Ubendian and Limpopo Belts) are ornamented with horizontal ruling.

References

Ackermann, E. (1950). Ein neuer Faltengürtel in Nordrhodesien und seine tektonische Stellung im afrikanischen Grundgebirge. *Geol. Rundschau*, **38**, 24.
Ackermann, E. (1960). Strukturen im Untergrund eines interkratonischen Doppel-orogens (Irumiden, Nordrhodesien). *Geol. Rundschau*, **50**, 538.
Aldrich, L. T., Wetherill, G. W., Davis, G. L. and Tilton, G. R. (1958). Radioactive ages of micas from granitic rocks by Rb–Sr and K–A methods. *Trans. Am. Geophys. Union*, **39**, 1124.
Allsopp, H. L. (1961). Rb–Sr age measurements on total rock and separated mineral fractions from the Old Granite of the central Transvaal. *J. Geophys. Res.*, **66**, 1499.
Allsopp, H. L. (1965). Rb–Sr and K–Ar age measurements on the Great Dyke of Southern Rhodesia. *J. Geophys. Res.*, **70**, 977.
Allsopp, H. L. and Kolbe, P. (1965). Isotopic age determinations on the Cape granite and intruded Malmesbury sediments, Cape Peninsula, South Africa. *Geochim. Cosmochim. Acta*, **29**, 1115.
Allsopp, H. L., Roberts, H. R., Schreiner, G. D. L. and Hunter, D. R. (1962). Rb–Sr age measurements on various Swaziland granites. *J. Geophys. Res.*, **67**, 5307.
Bassot, J.-P., Bonhomme, M., Roques, M. and Vachette, M. (1963). Mesures d'âges absolus sur les séries précambriennes et paléozoiques du Sénégal oriental. *Bull. Soc. Geol. France*, **5**, Ser. 7, 401.
Besairie, H. (1959). La géochronologie à Madagascar en 1956. *Intern. Geol. Congr., 20th, Mexico, 1956, Asoc. Serv. Geol. Africanos*, 27.
Bessoles, B. and Roques, M. (1959). Ages apparents par la méthode plomb-alpha de zircons extraits de roches cristallines d'Afrique Equatoriale Française et du Cameroun. *Intern. Geol. Congr., 20th, Mexico, 1956, Asoc. Serv. Geol. Africanos*, 35.
Black, R. (1966). Sur l'existence d'une orogénie riphéene en Afrique occidentale. *Compt. Rend.* **262**, Ser. D, 1046.
Boissonnas, J., Duplan, L., Maisonneuve, J., Vachette, M. and Vialette, Y. (1964). Étude géologique et géochronologique de roches du compartiment suggarien du Hoggar central Algérie. *Ann. Fac. Sci. Univ. Clermont*, Num. 25, 73.
Bonhomme, M. (1962). Contribution à l'étude géochronologique de la plate-forme de l'ouest africain. *Ann. Fac. Sci. Univ. Clermont*, Num. 5, Part 5.
Boocock, C. and Van Straten, O. J. (1962). Notes on the geology and hydrogeology of the central Kalahari region, Bechuanaland Protectorate. *Trans. Geol. Soc. S. Africa*, **65**, 125.
Brock, B. B. (1956). Structural mosaics and related concepts. *Trans. Geol. Soc. S. Africa*, **59**, 149.
Burger, A. J. (1960). A new age determination on a Southern Rhodesian monazite from Mount Darwin. *Proc. Fed. Sci. Cong.*, 53.
Burger, A. J. (1962). *Natl. Phys. Res. Lab., Ann. Rept. 1961–62.* C.S.I.R., Pretoria, S. Africa. p. 19.
Burger, A. J. (1963). *Natl. Phys. Res. Lab., Ann. Rept. 1962–63.* C.S.I.R., Pretoria, S. Africa. p. 34.

Burger, A. J. (1964). *Natl. Phys. Res. Lab., Ann. Rept. 1963–64.* C.S.I.R., Pretoria, S. Africa. p. 51.
Burger, A. J., Von Knorring, O. and Clifford, T. N. (1965). Mineralogical and radiometric studies of monazite and sphene occurrences in the Namib Desert, South-West Africa. *Mineral. Mag.*, **35**, 519.
Cahen, L. (1954). *Géologie du Congo belge.* H. Vaillant-Carmanne, Liége.
Cahen, L. (1961). Review of geochronological knowledge in middle and northern Africa. *Ann. N.Y. Acad. Sci.*, **91**, Art. 2, 535.
Cahen, L. (1963). Grands traits de l'agencement des éléments du soubassement de l'Afrique centrale. Esquisse tectonique au 1 : 5,000,000. *Ann. Soc. Geol. Belg.*, **85**, B183.
Cahen, L., Choubert, G. and Ledent, D. (1964). Premiers résultats de géochronologie sur le Precambrien de l'Anti-Atlas (Sud marocain) par la méthode strontium–rubidium. *Compt. Rend.*, **258**, 635.
Cahen, L., Delhal, J., Ledent, D. and Reinharz, M. (1963). L'âge des migmatites de Boma et de l'orogenèse Ouest-Congolienne. Indications préliminaires sur l'âge des formations mayumbiennes et antérieures. *Ann. Soc. Geol. Belg.*, **86**, B229.
Cahen, L., Delhal, J. and Monteyne-Poulaert, G. (1966). Age determinations on granites, pegmatites and veins from the Kibaran belt of central and northern Katanga (Congo). *Nature*, **210**, 1347.
Cahen, L., Pasteels, P., Ledent, D., Bourguillot, R., Van Wambeke, L. and Eberhardt, P. (1961). Recherches sur l'âge absolu des minéralisations uranifères du Katanga et de Rhodésie du Nord. *Mus. Roy. Afrique Centrale Ann., Sc. Geol.*, **41**, 1.
Choubert, G. (1951). Note sur la géologie de l'Anti-Atlas. *Intern. Geol. Congr., 18th, Great Britain, 1948*, Part 14, 29.
Choubert, G. (1959). Les études de géochronologie au Maroc. *Intern. Geol. Congr., 20th, Mexico, 1956, Asoc. Serv. Geol. Africanos*, 39.
Clifford, T. N. (1963). The Damaran episode of tectono-thermal activity in South West Africa, and its regional significance in southern Africa. *Res. Inst. African Geol., Univ. Leeds, 7th Ann. Rept., 1961–62*, 37.
Clifford, T. N. (1965). Structural units and tectono-metallogenic provinces within the Congo and Kalahari cratons of southern Africa. *Res. Inst. African Geol., Univ. Leeds, 9th Ann. Rept., 1963–64*, 30.
Clifford, T. N. (1966). Tectonometallogenic units and metallogenic provinces of Africa. *Earth Planetary Sci. Letters*, **1**, 421.
Clifford, T. N. (1967). The Damaran episode in the Upper Proterozoic–Lower Paleozoic structural history of southern Africa. *Geol. Soc. Am., Spec. Papers*, **92**.
Clifford, T. N., Nicolaysen, L. O. and Burger, A. J. (1962). Petrology and age of the pre-Otavi basement granite at Franzfontein, northern South-West Africa. *J. Petrol.*, **3**, 244.
Clifford, T. N., Rex, D. and Snelling, N. J. (1967). Radiometric age data for the Urungwe and Miami granites of Rhodesia. *Earth Planetary Sci. Letters*, **2**, 5.
Cox, K. G., Johnson, R. L., Monkman, L. J., Stillman, C. J., Vail, J. R. and Wood, D. N. (1965). The geology of the Nuanetsi Igneous Province. *Phil. Trans. Roy. Soc. London, Ser. A*, **257**, 71.

Darnley, A. G., Horne, J. E. T., Smith, G. H., Chandler, T. R. D., Dance, D. F. and Preece, E. R. (1961). Ages of some uranium and thorium minerals from East and Central Africa. *Mineral. Mag.*, **32**, 716.

Dars, R. and Sougy, J. (1964). Sur les relations entre la série épimétamorphique de Bakel et les roches cristallines du Guidimaka dans la région de Mbout (Mauritanie méridionale). *Compt. Rend.*, **258**, 6195.

Davis, G. L., Aldrich, L. T., Tilton, G. R., Wetherill, G. W. and Jeffrey, P. M. (1956). The age of rocks and minerals. *Carnegie Inst. Wash. Yearbook*, No. **55**, 161.

Delbos, L. (1964). Mesures d'âges absolus sur les séries précambriennes de Madagascar. *Compt. Rend.*, **258**, 1853.

Delhal, J. and Fieremans, C. (1964). Extension d'un grand complexe charnockitique en Afrique centrale. *Compt. Rend.*, **259**, 2665.

De Swardt, A. M. J., Drysdall, A. R. and Garrard, P. (1964). Precambrian geology and structure in central Northern Rhodesia. *Northern Rhodesia Geol. Surv., Mem.*, **2**.

Dixey, F. (1956). The East African rift system. *Colonial Geol. Mineral Resources (Gt. Brit.), Suppl. Ser., Bull. Suppl.*, No. **1**.

Du Toit, A. L. (1954). *The Geology of South Africa*, 3rd ed. Oliver and Boyd, Edinburgh.

Eberhardt, P., Ferrara, G., Glangeaud, L., Gravelle, M. and Tongiorgi, E. (1963). Sur l'âge absolu des séries métamorphiques de l'Ahaggar occidental dans la région de Silet-Tibehaouine (Sahara central). *Compt. Rend.*, **256**, 1126.

Emberger, A. (1958). Les granites stratoïdes du Pays Betsileo (Madagascar). *Bull. Soc. Geol. France*, **8**, Ser. 6, 537.

Faure, G., Hurley, P. M., Fairbairn, H. W. and Pinson, W. H. (1963). Age of the Great Dyke of Southern Rhodesia. *Nature*, **200**, 769.

Ferrara, G. and Gravelle, M. (1966). Radiometric ages from western Ahaggar (Sahara) suggesting an eastern limit for the West African craton. *Earth Planetary Sci. Letters*, **1**, 319.

Furon, R. (1963). *Geology of Africa*, English ed. Oliver and Boyd, Edinburgh.

Furon, R. and others. (1959). *Esquisse structurale provisoire de l'Afrique au 1:10,000,000, avec notice explicative*. Intern. Geol. Congr., Assoc. Serv. Géol. Africains, Paris.

Gellatly, D. C. (1963). The geology of the Darkainle nepheline syenite complex, Borama district, Somali Rep. *Ph.D. Thesis*. Leeds Univ.

Gérard, G. (1956). *Carte géologique de l'Afrique Equatoriale Française au 1 : 2,000,000, avec notice explicative*. Gouvernement Général de l'A.E.F., Direction des Mines et de la Géologie.

Gheith, M. A. (1961). Age of basement rocks in eastern United Arab Republic and northern Sudan. *Ann. N.Y. Acad. Sci.*, **91**, Art. 2, 530.

Giraudon, R. and Sougy, J. (1963). Position anormale du socle granitisé des Hajar Dekhen sur la série d'Akjoujt et participation de ce socle à l'édification des Mauritanides hercyniennes (Mauritanie occidentale). *Compt. Rend.*, **257**, 937.

Greenwood, J. E. G. W. (1961). The Inda Ad Series of the former Somaliland Protectorate. *Overseas Geol. Mineral Resources (Gt. Brit)*, **8**, 288.

Guernsey, T. D. (1950). A summary of the provisional geological features of Northern Rhodesia. *Colon. Geol. Mineral Resources (Gt. Brit.)*, **1**, 121.

Hall, A. L. (1932). The Bushveld Igneous Complex of the central Transvaal. *S. Africa Geol. Surv. Mem.*, **28**.

Handley, J. R. F. (1956). Banded ironstones and associated rocks. *C.C.T.A., East-Central Reg. Comm., Dar-Es-Salaam*, 39.

Hanekom, H. J., Van Staden, C. M. v. H., Smit, P. J. and Pike, D. R. (1965). The geology of the Palabora Igneous Complex. *S. Africa Geol. Surv. Mem.*, **54**.

Hargraves, R. B. (1961). Shatter cones in the rocks of the Vredefort Ring. *Trans. Geol. Soc. S. Africa*, **64**, 147.

Haughton, S. H. (1963). *The Stratigraphic History of Africa South of the Sahara.* Oliver and Boyd, Edinburgh.

Higazy, R. A. and El-Ramly, M. F. (1960). Potassium–argon ages of some rocks from the eastern desert of Egypt. *U.A.R. Geol. Surv. Mineral Resources Dept.*, Paper No. 7.

Holmes, A. (1950). The age of uraninite from Gordonia, South Africa. *Am. J. Sci.*, **248**, 81.

Holmes, A. (1951). The sequence of Pre-Cambrian orogenic belts in south and central Africa. *Intern. Geol. Congr., 18th, Great Britain, 1948*, Part 14, 254.

Holmes, A. (1965). *Principles of Physical Geology*. Nelson, London.

Holmes, A. and Cahen, L. (1955). African geochronology. *Colonial Geol. Mineral Resources (Gt. Brit.)*, **5**, 3.

Holmes, A. and Cahen, L. (1957). Géochronologie africaine 1956. *Acad. Roy. Sci. Coloniales (Brussels), Classe Sci. Nat., New Ser., Mem. in 8°*, **5**, Part 1.

Jacobson, R. R. E., Snelling, N. J. and Truswell, J. F. (1964). Age determinations in the geology of Nigeria, with special reference to the older and younger granites. *Overseas Geol. Mineral Resources (Gt. Brit.)*, **9**, 168.

Johnson, R. L. and Vail, J. R. (1965). The junction between the Mozambique and Zambesi orogenic belts; north-east Southern Rhodesia. *Geol. Mag.*, **102**, 489.

Jones, D. L. and McElhinny, M. W. (1966). Paleomagnetic correlation of basic intrusions in the Precambrian of southern Africa. *J. Geophys. Res.*, **71**, 543.

Kennedy, W. Q. (1964). The structural differentiation of Africa in the Pan-African (± 500 m.y.) tectonic episode. *Res. Inst. African Geol., Univ. Leeds, 8th Ann. Rept., 1962–63*, 48.

Kennedy, W. Q. (1965). The influence of basement structure on the evolution of the coastal (Mesozoic and Tertiary) basins of Africa. In *Salt Basins Around Africa*. Inst. Petroleum, London. p. 7.

Kennedy, W. Q. (in preparation). The West African craton.

Krenkel, E. (1925). *Geologie Afrikas*, Vol. 1. Gebrüder Borntraeger, Berlin. (See also: Vol. 2, 1928; Vol. 3, 1934.)

Krenkel, E. (1957). *Geologie und Bodenschätze Afrikas*. Geest und Portig K.-G., Leipzig.

Lasserre, M. (1964). Mesures d'âges absolus sur les séries précambriennes et paléozoiques du Cameroun (Afrique équatoriale). *Compt. Rend.*, **258**, 998.

Lay, C. and Ledent, D. (1963). Mesures d'âge absolu de minéraux et de roches du Hoggar (Sahara central). *Compt. Rend.*, **257**, 3188.

Lay, C., Ledent, D. and Grögler, N. (1965). Mesure d'âges de zircons du Hoggar (Sahara central) par la méthode uranium/plomb. *Compt. Rend.*, **260**, 3113.

Ledent, D., Lay, C. and Delhal, J. (1962). Premières données sur l'âge absolu des formations anciennes du 'socle' du Kasai (Congo méridional). *Bull. Soc. Belge Geol., Paleontol., Hydrol.*, **71**, 223.

Ledent, D., Picciotto, E., Poulaert, G., Eberhardt, P., Geiss, J., Von Gunten, H. R., Houtermans, F. G. and Signer, P. (1956). Determination de l'âge de l'yttrocrasite de Mitwaba (Katanga) par la méthode au plomb. *Bull. Soc. Belge Geol., Paleontol., Hydrol.*, **65**, 233.

Lowdon, J. A., Stockwell, C. H., Tipper, H. W. and Wanless, R. K. (1963). Age determinations and geological studies. *Geol. Surv. Can. Paper*, **62**-17.

McConnell, R. B. (1959). The Buganda Group, Uganda, East Africa. *Intern. Geol. Congr., 20th, Mexico, 1956, Asoc. Serv. Geol. Africanos*, 163.

Macgregor, A. M. (1951a). Some milestones in the Precambrian of Southern Rhodesia. *Trans. Geol. Soc. S. Africa*, **54**, xxvii–lxxi.

Macgregor, A. M. (1951b). A comparison of the geology of Northern and Southern Rhodesia and adjoining territories. *Intern. Geol. Congr., 18th, Great Britain, 1948*, Part 14, 111.

Martin, H. (1961). The Damara System in South-West Africa. *C.C.T.A., South. Reg. Comm., Pretoria, Publ.*, No. **80**, 91.

Mendelsohn, F. (1961). *The Geology of the Northern Rhodesian Copperbelt.* MacDonald, London.

Mendes, F. (1961). Determinaçao pelo método do estrôncio da idade absoluta de algumas roches e minerais de Portugal continental e ultramarino. *Garcia Orta*, **9**, No. 4, 767.

Mendes, F. (1964). Ages absolus par la méthode au strontium de quelques roches d'Angola. *Compt. Rend.*, **258**, 4109.

Mohr, P. A. (1962). *The Geology of Ethiopia*. Poligrafico, Asmara.

Monteyne-Poulaert, G., Delwiche, R. and Cahen, L. (1962a). Ages des minéralisations pegmatitiques et filoniennes du Rwanda et du Burundi. *Bull. Soc. Belge Geol., Paleontol., Hydrol.*, **71**, 210.

Monteyne-Poulaert, G., Delwiche, R., Safiannikoff, A. and Cahen, L. (1962b). Ages des minéralisations pegmatitiques et filoniennes du Kivu méridional (Congo oriental). Indications préliminaires sur les âges de phases pegmatitiques successives. *Bull. Soc. Belge Geol., Paleontol., Hydrol.*, **71**, 272.

Nicolaysen, L. O. (1962). Stratigraphic interpretation of age measurements in southern Africa. *Petrologic Studies—a volume in honour of A. F. Buddington.* Geol. Soc. Am. pp. 569–598.

Nicolaysen, L. O., Allsopp, H. L., Ulrych, T. J. and Burger, A. J. (1965). Age measurements on rocks from Swaziland and the area around Barberton. Abstract. *Geol. Soc. S. Africa, 8th Ann. Congr.*

Nicolaysen, L. O. and Burger, A. J. (1965). Note on an extensive zone of 1000 million-year old metamorphic and igneous rocks in southern Africa. *Sci. Terre*, **10**, Nos. 3–4, 497.

Nicolaysen, L. O., Burger, A. J. and Liebenberg, W. R. (1962). Evidence for the extreme age of certain minerals from the Dominion Reef conglomerates and the underlying granite in the western Transvaal. *Geochim. Cosmochim. Acta*, **26**, 15.

Nicolaysen, L. O., Burger, A. J. and Van Niekerk, C. B. (1963). The origin of the Vredefort ring structure in the light of new isotopic data. Abstract. *Intern. Union Geod. Geophys.*, **9**, 40.

Nicolaysen, L. O., De Villiers, J. W. L., Burger, A. J. and Strelow, F. W. E. (1958). New measurements relating to the absolute age of the Transvaal System and of the Bushveld Igneous Complex. *Trans. Geol. Soc. S. Africa*, **61**, 137.

Oosthuyzen, E. J. and Burger, A. J. (1964). Radiometric dating of intrusives associated with the Waterberg System. *Rept. Ann. Geol. Surv. S. Africa*, **3**, 87.

Page, B. G. N. (1962). The stratigraphical and structural relationships of the Abercorn sandstones, the Plateau Series and Basement rocks of the Kawimbe area, Abercorn district, Northern Rhodesia. *Res. Inst. African Geol., Univ. Leeds., 6th Ann. Rept., 1961–62*, 29.

Phillips, K. A. (1959). Some interpretations arising from a remapping of the Katanga System southeast of Mumbwa, Northern Rhodesia. *Intern. Geol. Congr., 20th, Mexico, 1956, Asoc. Serv. Geol. Africanos*, 213.

Quennell, A. M. and Haldemann, E. G. (1960). On the subdivision of the Precambrian. *Intern. Geol. Congr., Rept. 21st Session, Norden, 1960*, Part 9, 170.

Reichelt, R. (1966). Métamorphisme et plissement dans Gourma et leurs âges (République du Mali). *Compt. Rend.*, **263**, 589.

Rocci, G. (1964). Ages absolus, histoire et structure de l'Ouest du Bouclier africain. *Compt. Rend.*, **258**, 2859.

Rocci, G. (1965). Essai d'interprétation des mesures géochronologiques. La structure de l'Ouest Africain. *Sci. Terre*, **10**, Nos. 3–4, 461.

Rogers, A. S., Miller, J. A. and Mohr, P. A. (1965). Age determinations on some Ethiopian Basement rocks. *Nature*, **206**, 1021.

Roques, M. (1959). Ages apparents de quelques zircons de l'Afrique Occidentale Française et du Togo. *Intern. Geol. Congr., 20th, Mexico, 1956, Asoc. Serv. Geol. Africanos*, 41.

Roubault, M., Delafosse, R., Leutwein, F. and Sonet, J. (1965). Premières données géochronologiques sur les formations granitiques et cristallophylliennes de la République Centre-Africaine. *Compt. Rend.*, **260**, 4787.

Runcorn, S. K. (1962). Palaeomagnetic evidence for continental drift and its geophysical cause. In Runcorn, S. K. (Ed.), *Continental Drift*. Academic Press, London.

Sanders, L. D. (1965). Geology of the contact between the Nyanza Shield and the Mozambique Belt in Western Kenya. *Kenya Geol. Surv. Bull.*, **7**.

Scholtz, D. L. (1946). On the younger Precambrian granite plutons of the Cape Province. *Proc. Geol. Soc. S. Africa*, **49**, xxxv–lxxxii.

Schreiner, G. D. L. (1958). Comparison of the ^{87}Rb–^{87}Sr ages of the red granite of the Bushveld complex from measurements on the total rock and separated mineral fractions. *Proc. Roy. Soc. (London), Ser. A*, **245**, 112.

Schreiner, G. D. L. and Van Niekerk, C. B. (1958). The age of a Pilanesberg dyke from the central Witwatersrand. *Trans. Geol. Soc. S. Africa*, **61**, 197.

Schürmann, H. M. E. (1961). The Riphean of the Red Sea area. *Geol. Foren. Stockholm Forh.*, **83**, 2, 109.

Schürmann, H. M. E. (1964). Rejuvenation of Pre-cambrian rocks under epirogenetical conditions during old Palaeozoic times in Africa. *Geol. Mijnbouw*, **43**, 196.

Simpson, E. S. W. and Otto, J. D. T. (1960). On the Pre-Cambrian anorthosite mass of southern Angola. *Intern. Geol. Congr., Rept. 21st Session, Norden, 1960*, Part 13, 216.

Simpson, J. G., Drysdall, A. R. and Lambert, H. H. J. (1963). The geology and groundwater resources of the Lusaka area. *Northern Rhodesia Geol. Surv., Rept.*, **16**.

Slater, D. (1965). Structural and stratigraphical relationships in the Umkondo System of eastern Southern Rhodesia. *Ph.D. Thesis*. Leeds Univ.

Snelling, N. J. (1962). In *Overseas Geol. Surv. Ann. Rept., 1960-61*. Her Majesty's Stationery Office, London. pp. 27–35.

Snelling, N. J. (1963). In *Overseas Geol. Surv. Ann. Rept., 1962*. Her Majesty's Stationery Office, London. pp. 30–39.

Snelling, N. J. (1964). In *Overseas Geol. Surv. Ann. Rept., 1963*. Her Majesty's Stationery Office, London. pp. 30–40.

Snelling, N. J. (1965). In *Overseas Geol. Surv. Ann. Rept., 1964*. Her Majesty's Stationery Office, London. pp. 28–38.

Snelling, N. J., Hamilton, E. I., Rex, D., Hornung, G., Johnson, R. L., Slater, D. and Vail, J. R. (1964a). Age determinations from the Mozambique and Zambesi orogenic belts, central Africa. *Nature*, **201**, 463.

Snelling, N. J., Hamilton, E. I., Drysdall, A. R. and Stillman, C. J. (1964b). A review of age determinations from Northern Rhodesia. *Econ. Geol.*, **59**, 961.

Söhnge, P. G., Le Roex, H. D. and Nel, H. J. (1948). The geology of the country around Messina: an explanation of Sheet No. 46 (Messina). *S. Africa Geol. Surv.*

Sougy, J. (1962). West African fold belt. *Geol. Soc. Am. Bull.*, **73**, 871.

Stille, H. (1955). Recent deformations in the earth's crust in the light of those of earlier epochs. *Geol. Soc. Am., Spec. Papers*, **62**, 171.

Talbot, C. J. (1966). Progress report on the geology near the eastern end of the Zambezi orogenic belt. *Res. Inst. African Geol., Univ. Leeds, 10th Ann. Rept., 1964-65*, 15.

Vachette, M. (1964a). Ages radiométriques des formations cristallines d'Afrique Equatoriale (Gabon, République Centrafricaine, Tchad, Moyen Congo). *Ann. Fac. Sci. Univ. Clermont*, Num. 25, 31.

Vachette, M. (1964b). Essai de synthèse des déterminations d'âges radiométriques de formations de l'Ouest Africain (Côte d'Ivoire, Mauritanie, Niger). *Ann. Fac. Sci. Univ. Clermont*, Num. 25, 7.

Vachette, M (1964c). Nouvelles mesures d'âges absolus de granites d'âge éburnéen de la Côte d'Ivoire. *Compt. Rend.*, **258**, 1569.

Vail, J. R. (1965). An outline of the geochronology of the late Precambrian formations of eastern Central Africa. *Proc. Roy. Soc. (London), Ser. A*, **284**, 354.

Van Breemen, O., Dodson, M. H. and Vail, J. R. (1966). Isotopic age measurements on the Limpopo Orogenic Belt, southern Africa. *Earth Planetary Sci. Letters*, **1**, 401.

Van Eeden, O. R., Visser, H. N., Van Zyl, J. S., Coertze, F. J. and Wessels, J. T. (1955). The geology of the eastern Soutpansberg and the Lowveld to the north: an explanation of Sheet 42 (Soutpansberg). *S. Africa Geol. Surv.*

Van Niekerk, C. B. and Burger, A. J. (1964). The age of the Ventersdorp System. *Rept. Ann. Geol. Surv. S. Africa*, **3**, 75.

Visser, D. J. L. and others (1956). The geology of the Barberton area. *S. Africa Geol. Surv. Spec. Publ.*, No. 15.

Wasserburg, G. J., Hayden, R. J. and Jensen, K. J. (1965). A^{40}–K^{40} dating of igneous rocks and sediments. *Geochim. Cosmochim. Acta*, **10**, 153.

Wiles, J. W. (1961). The geology of the Miami Mica Field. *Southern Rhodesia Geol. Surv. Bull.*, **51**.

Worst, B. G. (1960). The Great Dyke of Southern Rhodesia. *Southern Rhodesia Geol. Surv. Bull.*, **47**.

Charged particle tracks: tools for geochronology and meteorite studies

ROBERT L. FLEISCHER,* P. BUFORD PRICE* and ROBERT M. WALKER†

I *Introduction*, 417. II *Geochronology*, 418. III *Meteorites*, 422.
IV *Conclusion*, 434. *Acknowledgments*, 434. *References*, 435.

I Introduction

Tiny holes etched in crystals and other solids are keys which unlock such diverse information as ages of rocks, the mass spectrum of unusually heavy cosmic rays, and the rates at which primitive matter cooled during the birth of the solar system. We first describe these holes and then describe how these varied results are deduced from natural materials.

The holes identify the sites of narrow cylinders of radiation damage created by heavy, charged particles. In almost any insulating material, fragments from the fission of uranium will create damage regions—typically 30Å in diameter and 10^{-3} cm long. These regions, which are essentially continuous trails of displaced atoms, have a greatly increased chemical reactivity relative to that of the surrounding unaffected material. Immersing a material in a suitable chemical reagent results in the damaged material being preferentially dissolved, creating etch figures which (as shown in figure 1) can vary in geometry from long, narrow cylinders to shallow conical or pyramidal shapes, each one of which marks the site of the passage of a charged particle track (Fleischer, Price and Walker, 1965a,b).

* General Electric Research and Development Center, Schenectady, New York, U.S.A.
† Formerly at General Electric Research and Development Center, now at Washington University, St. Louis, Mo., U.S.A.

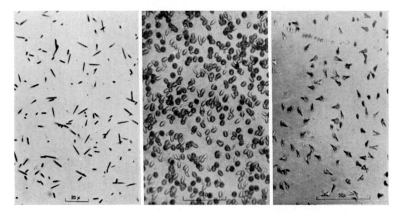

Figure 1 Examples of the varied geometry of fission-track etch pits in different materials. (a) Nearly cylindrical pits in a plastic —these are similar to tracks in mica and many other minerals; (b) Conical pits in a soda silicate glass; (c) Spike-shaped pits in orthoclase. Each pit marks the site of a fission track. (a) and (c) are typical of pits in minerals and (b) of pits in natural glasses.

Various natural sources of heavy charged particles can produce tracks in solids. In the next section we will discuss one such source. In section III, dealing with meteorite studies, we will describe other possibilities.

II Geochronology

The most common of the uranium isotopes, ^{238}U, undergoes spontaneous fission with a decay constant of 6.9×10^{-17} yr. As a result, if a mineral contains an atom fraction of one part per million of uranium, and if we consider any hypothetical surface within the material, 2000 fission fragments will cross each square centimeter of surface every million years. Provided these fragments produce tracks that are permanent, a simple count of the track density together with a measure of the uranium content allows the age of the sample to be measured. This assumes that no other sources leave tracks which might be mistaken for ^{238}U spontaneous fission tracks. Fortunately, calculation shows that for terrestrial samples, protected from cosmic radiation by the blanket of the Earth's atmosphere, this source is the only one which need be considered (Price and Walker, 1963). This assertion has an even more decisive basis. The dating technique—called *fission-track dating*—has been applied to a wide range of materials, and, as figure 2 shows, has yielded ages that are in agreement with 'known' ages over a wide time span, extending from man-made objects as little as 20 years old to natural crystals formed more than 1000 million years ago. Below $\sim 2 \times 10^8$ yr, concordant ages are the rule, while for greater ages, frequent discordancies are observed (Fleischer, Price and Walker, 1956a,b).

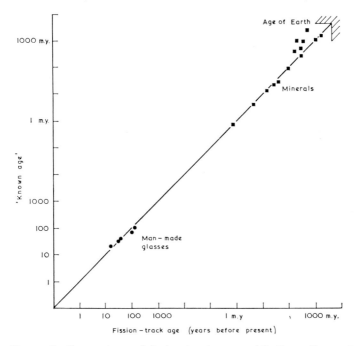

Figure 2 Comparison of fission-track ages with those 'known' by other methods, either (as in the case of man-made glasses) actually known dates or dates estimated from artistic style or, as in the case of natural minerals, dates measured by means of radiometric dating methods.

The experimental procedures used in fission-track dating are extremely simple. After determining the density ρ_s of fossil (ancient) tracks, the sample is sent to a nuclear reactor and exposed to a flux of thermal neutrons. These produce a density ρ_i of new tracks by inducing uranium fission. The age, T, of the sample is then determined from the ratio of new and old track densities by means of the following equation:

$$\frac{\rho_s}{\rho_i} = [\exp(\lambda_D T) - 1]\frac{\lambda_F}{\lambda_D f} \qquad (1)$$

where λ_D and λ_F are the total decay and spontaneous fission decay constants. The quantity f is the fraction of uranium atoms fissioned during the irradiation; it is conveniently determined by making track counts on a sample of calibrated, uranium-containing glass (Fleischer, Price and Walker, 1965c).

The materials chosen for dating are, of course, not selected at random; in order to be datable they must satisfy certain conditions of track permanence and track density. We now describe in more detail how these considerations apply.

Tracks will fade in any material that is heated to a sufficiently high temperature. Figure 3 shows that this fading temperature varies widely from substance to substance. In some, it is so low that tracks would fade in geological times even at ambient temperatures; one such mineral is calcite. In others, we conclude that tracks should last for much longer than the age of the solar system at 300°K and, hence, are effectively 'permanent'.

In any material, the age measured is the time since the sample was at a high enough temperature to remove previously formed

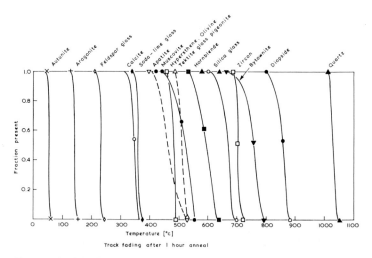

Figure 3 Track fading conditions. The fission-track density relative to that at room temperature is shown as a function of the annealing temperature for various minerals. In each case, the heating was for a one hour period. Materials in which tracks fade at temperatures below 400°C are undesirable candidates for dating. Those in which tracks do not fade until higher temperatures are reached are likely to store tracks 'permanently' i.e. over periods equal to the age of the solar system at 300°K.

tracks. In principle, dating a suite of minerals from the same site allows a thermal history to be constructed. Section III, in which we consider tracks in meteorites, includes an example of such an experiment.

There must be a sufficiently high track density to allow the compiling of adequate statistics. In large homogeneous samples, as few as 5 tracks/cm² can be adequate, while in small samples 10^{-2} cm on a side 100,000 tracks/cm² may be needed. It follows that it is not possible to set precise, general criteria on the required density. We, nevertheless, present figure 4 as an indication of the uranium content needed in a material of a given age to allow either an easy age determination or an awkward, but possible measurement. Figure 4 also includes some examples of specific materials that are potentially appropriate for measuring different ages.

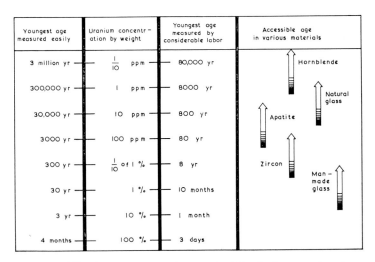

Figure 4 Uranium concentration required for dating as a function of the age of the sample. Values are given for two conditions depending on whether the observer is planning on a short, easy experiment or a more tedious one. For an 'easily' measured age, it is assumed that the observer will spend one hour at the microscope counting fission-track etch pits. For a determination 'by considerable labor' nearly 40 hours of such work are assumed. Glass that is free from inclusions and bubbles is considered for estimating the time necessary for counting tracks. On the right are indicated the lowest ages which should be measurable in various minerals on the basis of typical uranium contents, or, as in the man-made glasses, available uranium contents.

The fission-track technique has both advantages and difficulties relative to other techniques. Its major merits are its great simplicity, the wide time span to which it is applicable, and its suitability for dating extremely small samples. One major disadvantage is the necessity for manual counting of tracks, no adequate procedure for automating the counting of tracks having yet been devised.

An important and likely future application of the technique lies in dating Pleistocene events. In this period K–Ar dating is difficult and carbon dating is not applicable. There is also a wealth of information to be uncovered in this, the period of the evolution of man.

III Meteorites

Large densities of fossil particle tracks have been found in the silicate phases of a variety of meteorites (Maurette, Pellas and Walker, 1964; Fleischer, Price and Walker, 1965a,b,d; Maurette, 1965; Fleischer and others, 1965, 1967a,b) (for an example, see figure 5). Whereas spontaneous fission of ^{238}U is the only important source of fossil tracks in terrestrial samples, it is rarely, if ever, the dominant source of fossil tracks in meteorites. The differences arise for two reasons: the exposure of meteorites to the primary cosmic ray beam and the extreme antiquity of meteorites relative to terrestrial materials.

Cosmic rays will act in the following ways to give nuclear particles that are heavy enough to produce etchable tracks:

(a) Very heavy cosmic ray primaries will produce tracks as they slow down in the meteorite to the point where their rate of specific ionization exceeds the critical value necessary to form an etchable track.

(b) Interactions of cosmic ray primaries with heavy nuclei in the meteorite will give spallation recoils which will register, provided the recoils are sufficiently massive.

(c) High-energy cosmic ray interactions will induce fission in heavy element impurities such as gold and lead.

(d) Low-energy secondary neutrons will induce fission in uranium impurities.

(e) High-energy interactions will produce meson jets that could give tracks under certain conditions.

(f) Bizarre particles such as triply charged Dirac monopoles should also produce tracks—if they exist in the primary beam.

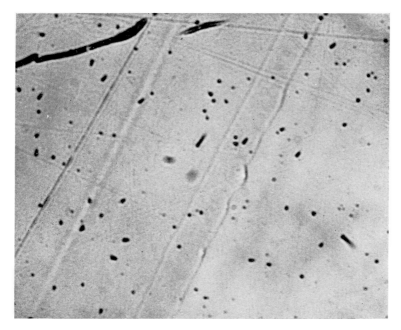

Figure 5 Fossil tracks in the Pallasite, Eagle Station. Polished surfaces of meteoritic olivine were alternately etched in hot KOH and dilute HF acid solutions to reveal the tracks. (After Maurette, 1965)

The extreme antiquity of the meteorites ($\sim 4.6 \times 10^9$ yr) raises the further possibility of finding spontaneous fission tracks from extinct transuranic elements, such as ^{244}Pu, in addition to those from ^{238}U.

The initial challenge in studying meteoritic tracks is to find which of the possible sources is responsible for the tracks in a given meteorite. The further challenge is to extract useful physical information from the tracks once their origin is known. Although it is beyond the scope of this article to treat these problems in complete detail, we hope to show to the reader the methods we have used to approach these problems, and to present a picture of the current state of understanding in what is still a rapidly developing field.

A variety of quantities can be measured, including the density, the length distribution and the angular distribution of tracks and, equally important, the variation of these quantities as a function of the position within the meteorite. Other important information is obtained by comparing these parameters for different types of minerals removed from adjacent locations. In addition, there are a number of physical tests that can be performed on the samples. For example, by irradiation with different particles their response to related irradiations in nature can be calibrated.

Combining the results of the possible measurements with theoretical predictions, it is possible to identify unambiguously the sources of tracks. Table 1 indicates the behavior expected of the various measurables in common meteoritic minerals for different assumed track sources. In what follows, we consider several specific cases in more detail.

Qualitatively, it is found that there exist two general classes of meteoritic tracks which we will call Type I and Type II. This distinction is clear cut in some cases and less so in others.

The prototype meteorite showing Type I behavior is the mesosiderite, Estherville. The corresponding prototype for Type II tracks is the iron meteorite, Toluca. The length and angular distributions are shown schematically in figures 6 and 7. Type I tracks are aniso-

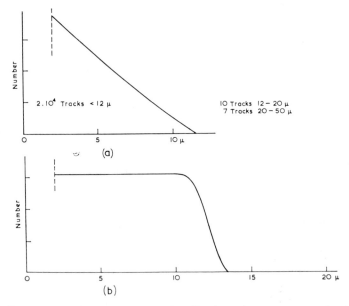

Figure 6 Schematic length distributions in two meteorites: (a) Nonuniform length distribution found in the mesosiderite Estherville (Type I); (b) Uniform distribution found in a silicate inclusion in the iron meteorite Toluca (Type II).

tropic and have a nonuniform length distribution. Type II tracks are isotropic and have a uniform length distribution.

Consider first the explanation of the tracks in Estherville. Their appearance alone serves to rule out spontaneous or induced fission. This conclusion is confirmed by neutron irradiations which show that the uranium concentrations are far too small to account for the observed densities. Extended proton irradiations in the Brookhaven cosmotron show further that neither spallation recoils nor induced fissions can account for the results. Dirac monopoles should give very long tracks if present. Meson jets can be eliminated on theoretical grounds. The only remaining possibility is heavy cosmic ray primaries. Tracks from this source should give equal track densities in different minerals removed from the same location. The track density should also decrease markedly with increasing distance into the meteorite (see figure 8) due to the large nuclear absorption cross-section of very heavy primaries. For test samples located away from the meteorite center, the strong absorption should also give anisotropically directed tracks. All these positive identifying features are observed (see figure 9) and there can be little doubt that the tracks in Estherville are predominantly due to slowed-down, very heavy cosmic ray primaries.

It can be shown theoretically that the maximum etchable track length of a slowing-down heavy ion increases rapidly with mass

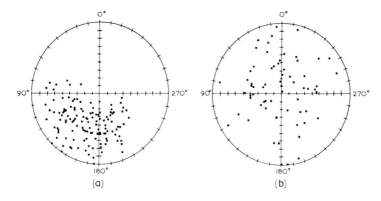

Figure 7 Angular distributions in two meteorites: (a) Distribution of track orientations projected onto an arbitrary plane in the mesosiderite Estherville (Type I). Each point represents one track; (b) Same for silicate inclusion in the iron meteorite Toluca (Type II).

Table 1 Identifying characteristics of fossil tracks

Source	Length distribution[a]	Angular distribution	Depth variation	Comments
Spont. fission of ^{238}U	$L_{max} \approx 13\ \mu$	Isotropic	None	Should vary directly as U concentration. Identifiable (statistically) by reactor irradiation
Spont. fission of ^{244}Pu	$L_{max} \approx 13\ \mu$	Isotropic	None	Should depend on U concentration and thermal retentivity of different silicate phases
High-energy fission	$L_{max} \approx 13\ \mu$	Slightly anisotropic[b]	Less than $2 \times$ in 60 cm	V-tracks in variable density depending on impurity content of samples. Identifiable (statistically) by proton irradiation
Thermal neutron fission	$L_{max} \approx 13\ \mu$	Isotropic	Slight	Should vary directly as U concentration
Spallation recoils	Peak at low end; $L_{max} \approx 5\ \mu$	Anisotropic[b]	Less than $2 \times$ in 60 cm	Identifiable (statistically) by proton irradiation. Should vary with impurity content of different samples
Heavy cosmic primaries	Occasional very long tracks	Very anisotropic[b]	Very marked	At a given location, densities should be constant for different silicate phases with equal sensitivities
Magnetic monopoles	Long tracks	Isotropic	Slight	Probably nonexistent

[a] Length distribution is uniform from 0 to L_{max} unless otherwise noted.
[b] Anisotropy may result from noncentral sample location in meteorite.

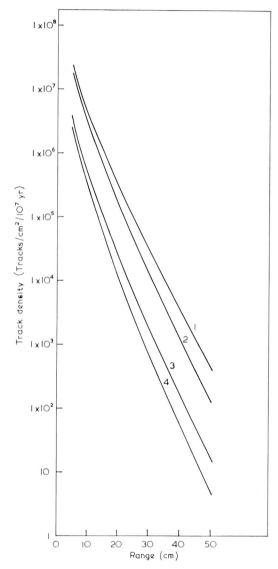

Figure 8 Density of visible iron group tracks as a function of depth in the meteorite: Curves 1 and 2 give values calculated for the center of meteorites of different radii. Curves 3 and 4 give values calculated for different depths below an infinite plane surface. Odd-numbered curves have a fragmentation correction to take into account the fact that some nuclear interactions give product nuclei that are still capable of registering tracks. All curves calculated for a meteorite that is 50% Fe and 50% silicate by weight.

beyond a threshold value which is somewhere in the vicinity of iron. Once it has been demonstrated that the tracks in a given meteorite are predominantly due to very heavy primaries, it is possible to measure the flux of ions that are much heavier than iron by looking for occasional very long tracks. Positive evidence for such extremely heavy ion tracks has recently been presented (Maurette, 1965; Fleischer, Price and Walker, 1965e; Fleischer and others, 1967b). At present it has not been possible to measure the flux of these particles by using conventional techniques, and a precise determination presents itself as one of the most promising objectives of future track studies in meteorites.

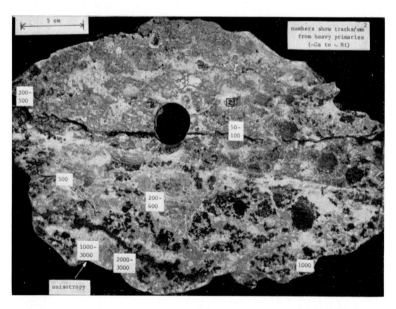

Figure 9 Measured track densities at different locations in Estherville. The direction of track anisotropy at one point in the meteorite is indicated by an arrow.

The Type II tracks in Toluca evidently originate in a different way. Neutron irradiation of test samples shows that uranium is definitely present and that the track density in a given location varies considerably from mineral to mineral; it is independent of the physical location of the sample within the meteorite. The track densities are, however, much too high to be accounted for by the uranium present, even assuming a track storage time of 4.6×10^9 yr. As we have shown previously (Fleischer, Price and Walker, 1965d) the best explanation of these excess tracks is the spontaneous fission of ^{244}Pu, which was present when the meteorite was first formed, but is now extinct.

Provided an assumption is made concerning the ratio of Pu : U at the end of nucleosynthesis, the number of Pu tracks ρ_{Pu} can be used to infer the interval of time ΔT between the end of nucleosynthesis and the formation of cooled down meteoritic material. Specifically, this time is given by the following relation:

$$\frac{\rho_{Pu}}{\rho_U} = \left(\frac{^{244}Pu}{^{238}U}\right)_0 \exp[-(\lambda_{DPu} - \lambda_{DU})\Delta T] \cdot \frac{\lambda_{FPu}\lambda_{DU}}{\lambda_{DPu}\lambda_{FU}} \frac{[1 - \exp(-\lambda_{DPu}T)]}{[1 - \exp(-\lambda_{DU}T)]} \quad (2)$$

where the λ's are spontaneous fission and total decay constants for ^{244}Pu and ^{238}U, $(^{244}Pu/^{238}U)_0$ is the initial Pu to U ratio at the end of nucleosynthesis and T is the total age of the meteorite taken as $\sim 4.5 \times 10^9$ yr. Assuming $(^{244}Pu/^{238}U)_0 = 1/45$, we previously inferred that $\Delta T \approx 200$ m.y. for Toluca (Fleischer, Price and Walker, 1965d).

More recent experiments, involving a comparison of excess track densities in different types of crystals from the same meteorite, lend strong support to the plutonium interpretation (Fleischer, Price and Walker, 1968). In minerals which retain tracks at high temperatures, permanent tracks are formed earlier in a cooling process than they are in less retentive minerals. If the cooling of the parent body of Toluca took place in a time comparable to the decay life of ^{244}Pu ($\tau_{1/2} \approx 80$ m.y.), we would expect to find differences in track densities correlated with the thermal retentivities of tracks in different minerals. Consistent with this idea, we find increasing track excesses, attributable to plutonium, in the sequence enstatite–albite–diopside, for which the extrapolated 100 m.y. track-annealing temperatures are respectively 535°K, 725°K, and 925°K. From the data summarized in table 2, we calculate a cooling rate of 1.1°C $(^{+.6}_{-.5})$/m.y. This value is independent of the ratio $(Pu/U)_0$ and in agreement with independent values obtained from electron probe studies of the metallic phases in this same meteorite (Wood, 1964).

All fossil tracks that have been found in meteorites can be ascribed either to slowed-down heavy primaries or to spontaneous fission of uranium and plutonium. Cases of the latter are quite rare as it is only in very large meteorites that the heavy ion tracks will not swamp out the contribution due to the plutonium. Uranium concentrations

Table 2 Summary of data in Toluca

	Mineral	ρ_{total} (cm^{-2})	ρ_U (cm^{-2})	ρ_{Pu}/ρ_U	$\Delta T_0(\frac{1}{45})$ (yr)[a]	$\Delta T_0(\frac{1}{2})$ (yr)[a]
Location A	Diopside	4.7×10^6	8.0×10^4	59	0.95×10^8	4.7×10^8
	Albite	7.0×10^4	1.0×10^4	6	3.7×10^8	7.4×10^8
	Enstatite	5.0×10^3	1.3×10^3	3	4.6×10^8	8.3×10^8
Location B	Diopside	5.6×10^6	1.1×10^5	51	1.1×10^8	4.8×10^8
	Albite	1.8×10^5	1.7×10^4	9	3.2×10^8	6.9×10^8
	Enstatite	1.0×10^4	2.0×10^3	4	4.2×10^8	7.9×10^8

[a] ΔT_0 values calculated from equation (2) of text. Assumed values of $(Pu/U)_0$ are given in parentheses.

are also very low in most meteorites. Table 3 gives a summary of the observations which have been made to date.

Table 3 Summary of fossil track observations in meteorites

Meteorite and mineral	ρ_{total} (cm^{-2})	Angle distribution[a]	Length distribution[a]	Uranium conc. (ppb)
Admire Olivine	$<10^3$	—	—	$<4 \times 10^{-3}$
Brahin Olivine	4×10^2	—	—	$<4 \times 10^{-3}$
Brenham Olivine	$<10^3$	—	—	$<4 \times 10^{-3}$
Eagle Station Olivine	2×10^6	A[b]	P[b]	$<4 \times 10^{-3}$
Krasnojarsk Olivine	$\sim 10^3$	—	—	$<4 \times 10^{-3}$
Magnesia Olivine	3×10^3	—	—	$<4 \times 10^{-3}$
Molong Olivine	$<10^3$	—	—	$<4 \times 10^{-3}$
Pavlodar Olivine	10^6	A	P	$<4 \times 10^{-3}$
Santa Rosalia Olivine	3×10^5	—	—	$<4 \times 10^{-3}$
Springwater Olivine	5×10^3	—	—	$<4 \times 10^{-3}$
Thiel Mt. Olivine	2×10^4	—	—	$<4 \times 10^{-3}$
Crab Orchard				
Bytownite	2×10^6	A	P	0.6
Olivine		A	—	
Estherville	6×10^5	A	P	10^{-2} to 0.3
Hypersthene				
Lowicz	5×10^4	A	P	$\lesssim 2 \times 10^{-2}$
Mincy	6×10^5	A	—	<0.2
Udei Station	1×10^7	A	P	3
Albite				
Patwar	7×10^6	I[b]	P	10^{-2}
Steinbach				
Bronzite	4×10^6	A	—	0.3
Tridymite	3×10^6	—	—	6
Lodran	6×10^6	I	—	4×10^{-2}
Bronzite				
Vaca Muerta				
Hypersthene	6×10^3	—	P	<0.06
Anorthite	2×10^3	—	P	<0.06

Table 3—*Continued*

Meteorite and mineral	ρ_{total} (cm^{-2})	Angle distribution[a]	Length distribution[a]	Uranium conc. (ppb)
Clovis				
Bronzite	4×10^5	A	P	10
Olivine	4×10^5	—	—	
Bishopville	$\sim 5 \times 10^5$	A	P	<0.8
Enstatite				
Johnstown	$\sim 3 \times 10^5$	—	P	<0.8
Hypersthene				
Moore County				
Bytownite	3×10^6	I	P	2–4
Pigeonite	2×10^6	I	P	~ 50
Lafayette	2×10^6	I	—	~ 22
Diopside				
Norton City	1.5×10^5	A	P	<0.3
Enstatite				
Khor Temiki	7×10^5	A	P	<0.1
Enstatite				
Utzensdorf	2.5×10^6	A	P	<0.1
Olivine				
Bondon	$\sim 1 \times 10^3$	—	—	0.8
Hypersthene				
Simondium	$<6 \times 10^3$	—	—	$<6 \times 10^{-3}$
Hypersthene				
Veramin	6×10^3	—	—	~ 0.1
Hypersthene				
Four corners				
Diopside	6×10^6	—	—	~ 15
Enstatite	3×10^6	—	—	$\lesssim 5$
Kodaikanal	2×10^5	—	—	4 to 20
Diopside				
Linwood	2×10^6	—	—	15
Diopside				
Weekeroo Station	$\lesssim 1 \times 10^5$	—	—	15
Diopside	2×10^5	—	—	
Nakhla	Low	—	—	$\lesssim 20$
Diopside				
Odessa	5×10^6	I	U[b]	20
Diopside				
Toluca				
Diopside	5×10^6	I	U	20–30
Albite	$(7–20) \times 10^4$	I	U	1–3
Enstatite	$(5–20) \times 10^3$	I	U	0.4–1.0
Quartz	$<10^3$	—	—	<0.1
Olivine	$<10^3$	—	—	Not measured

[a] Failure to give an entry under 'Angle distribution' or 'Length distribution' means that insufficient data have been analyzed.
[b] A Anisotropic; I Isotropic; P Peaked; U Uniform.

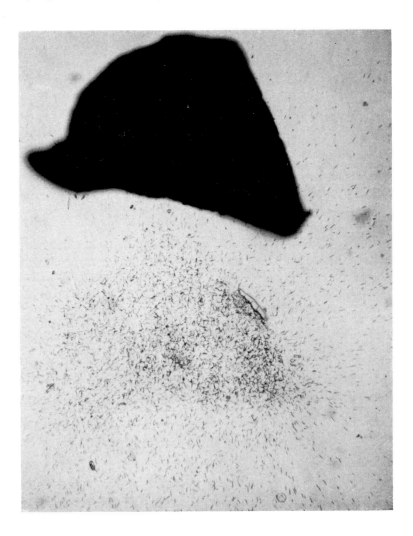

Figure 10 Photomicrograph illustrating a method for measuring the uranium concentration in meteoritic material. The dark grain is a tiny crystal of zircon that was sandwiched between two sheets of mica and irradiated with thermal neutrons. After etching for 12 minutes in hydrofluoric acid, the mica contains a pattern of tracks resulting from uranium fissions in the portion of the zircon that was next to the mica. The zircon was moved to one side of the track pattern when the picture (magnified × 90) was taken.

Solid-state track detectors are also useful for measuring the distribution of trace quantities of uranium and thorium in meteorites. The samples to be measured are placed next to detectors such as mica or silicate glasses and then exposed to large neutron doses in a nuclear reactor. In figure 10 we show results for a small crystal of zircon from the mesosiderite Vaca Muerte. Uranium concentrations in this meteorite were found to vary by a factor of $>10^6$ for different mineral phases (Fleischer and others, 1965). Although no experimental determinations have yet been reported, it is also possible to measure the Th/U ratio by properly choosing the energy spectra of bombarding neutrons. Of course the technique described above is equally useful for studying terrestrial materials; however, the extreme sensitivity of the method, which makes it possible in principle to measure a uranium concentration $<10^{-14}$ atom/atom, renders it particularly useful for meteorites.

IV Conclusion

Etched tracks in natural crystals give a promising new method of dating terrestrial samples and of measuring uranium and thorium distributions. In extraterrestrial samples, they are being used to study the relative abundance of extremely heavy cosmic rays as well as the early history of the solar system.

Acknowledgments

We are grateful to our colleagues M. Maurette of the Laboratory for Physical Chemistry, University of Paris, and G. Morgan of Brookhaven National Laboratory for permission to quote some of our joint results. The samples of Estherville were provided by P. Gast of the Lamont Geological Observatory and those of Toluca by U. Marvin of the Smithsonian Astrophysical Observatory. Most of the other samples were supplied by B. Mason and E. P. Henderson of the National Museum, Washington, and Carleton Moore of Arizona State University. Without their fine cooperation, this work would not have been possible. We are happy to thank H. Couch and G. Nichols of our laboratory for experimental assistance.

References

Fleischer, R. L., Naeser, C. W., Price, P. B., Walker, R. M. and Marvin, U. B. (1965). *Science*, **148**, 629.
Fleischer, R. L., Price, P. B. and Walker, R. M. (1965a). *Science*, **149**, 383.
Fleischer, R. L., Price, P. B. and Walker, R. M. (1965b). *Annual Reviews of Nuclear Science*, Vol. 15. Academic Press, New York.
Fleischer, R. L., Price, P. B. and Walker, R. M. (1965c). *J. Nucl. Sci. Engr.*, **22**, 153.
Fleischer, R. L., Price, P. B. and Walker, R. M. (1965d). *J. Geophys. Res.*, **70**, 2703.
Fleischer, R. L., Price, P. B. and Walker, R. M. (1965e). *Proc. Intern. Conf. Cosmic Rays, 9th, London, 1965*. The Physical Society, London.
Fleischer, R. L., Price, P. B. and Walker, R. M. (1968). *Geochim. Cosmochim. Actak*, **32**, 21.
Fleischer, R. L., Price, P. B., Walker, R. M. and Maurette, M. (1967a). *J. Geophys. Res.*, **72**, 331.
Fleischer, R. L., Price, P. B., Walker, R. M., Maurette, M. and Morgan, G. (1967b). *J. Geophys. Res.*, **72**, 355.
Maurette, M. (1965). *Thesis*. Laboratoire de Chimie-Physique, Univ. Paris.
Maurette, M., Pellas, P. and Walker, R. M. (1964). *Nature*, **204**, 821.
Price, P. B. and Walker, R. M. (1963). *J. Geophys. Res.*, **68**, 4847.
Wood, J. A. (1964). *Icarus*, **3**, 429.

The isotopic composition of strontium applied to problems of the origin of the alkaline rocks

E. I. HAMILTON*

I *Introduction*, 437. II *The significance of the isotopic composition of strontium in rocks and minerals*, 440. III *The isotopic composition of strontium in basalts and their differentiates*, 444. IV *The isotopic composition of strontium in continental plutonic alkaline rocks*, 451. V *Strontium isotopes applied to the origin of carbonatites*, 452. VI *The isotopic composition of strontium in sea water*, 457. VII *Conclusions*, 459. *References*, 460.

I Introduction

The natural radioactive decay of ^{87}Rb, with a half-life of between 4.7 and 5.0×10^{10} years, to the stable daughter nuclide ^{87}Sr results in an increase in the abundance of ^{87}Sr relative to the other isotopes of strontium in rocks and minerals. In a particular rock or mineral the degree of enrichment in radiogenic strontium will depend upon the amount of rubidium that has been associated with strontium throughout the duration of the rock's existence. In the first instance, the accumulation of ^{87}Sr (radiogenic) from the time a mineral is first formed to the present day is a direct measure of geological age. In order to measure the concentration of radiogenic strontium it is first necessary to subtract the contribution from ^{87}Sr present as a component of the common strontium. Variations in the isotopic composition of strontium as a result of the addition of radiogenic ^{87}Sr

---

* Late of Dept. of Geology and Mineralogy, Oxford. Present address: Radiological Protection Service, Sutton, Surrey, England.

are expressed in terms of the invariant abundance of ^{86}Sr by means of the ratio $^{87}Sr/^{86}Sr$. It has been shown that the initial $^{87}Sr/^{86}Sr$ ratio for terrestrial rocks having a low Rb/Sr ratio is about 0.700 ± 0.002 which is similar to that obtained from meteorites (Gast, 1960, 1965; Wasserburg and others, 1964) and suggests that at $\sim 4.5 \times 10^9$ years ago the initial $^{87}Sr/^{86}Sr$ ratio of terrestrial and extraterrestrial materials were similar.

In terms of dating the time in the past when a rock was metamorphosed or remobilized, it is the $^{87}Sr/^{86}Sr$ ratio at the time the event occurred that is required. If prior to the event the rock had a high Rb/Sr ratio, the subsequent complete or partial homogenization of the strontium isotopes during the event will often result in a new initial $^{87}Sr/^{86}Sr$ ratio for the common strontium that is distinctly greater. In dating the time when a geological event occurred (e.g. crystallization of a magma, a metamorphism) it is essential to use methods which are capable of defining the initial $^{87}Sr/^{86}Sr$ ratio.

If the Rb/Sr ratio of a sample is very high, the initial $^{87}Sr/^{86}Sr$ ratio is not too important, but restricting Rb–Sr dating to such material limits the usefulness of the method. In order to apply the method to samples having low Rb/Sr ratios, the isotopic composition of the common strontium is of importance.

Apart from dating geological events, the isotopic composition of strontium has become a very useful tool in studies involving petrogenetic relations between rock series. Consider a well-mixed magma that after emplacement gives rise through differentiation to three rock types, gabbro–granodiorite–granite, forming one intrusion. It is possible to consider that either all three rock types crystallized from the same magma or that the original magma could only give rise to gabbros, and the other rock types represent refusion and mixing of country rock with gabbro magma. If the $^{87}Sr/^{86}Sr$ ratio at the time of intrusion of all three rocks is identical, then this would support fractionation from a single magma, while if the granodiorite and granite have higher initial $^{87}Sr/^{86}Sr$ ratios, it would suggest contamination by rubidium-enriched country rock. The techniques used for obtaining initial $^{87}Sr/^{86}Sr$ ratios are discussed in section II.

In fundamental problems of rock genesis by differentiation from a source region in the mantle, the Rb/Sr ratio and the isotopic composition of strontium in basalts and other related basic rocks are of obvious importance. The ensuing discussion considers the isotopic composition of strontium in such rocks and describes some variations resulting from fractional crystallization in which some rocks become enriched in rubidium and depleted in strontium. Attention is focused upon the alkaline rocks whose origin is currently believed in many instances to be derived from subsialic basic magmas.

The alkaline rocks comprise only a very small volume of all the igneous rocks, but because of their rather unusual mineral assemblies, and the presence of characteristic accessory minerals enriched in elements such as Sr, Ba, Rare Earths (R.E.), Nb, Zr, U and Th, they have attracted the attention of geologists and have been described in great detail. Often the unusual mineralogy of such rocks has given rise to a bewildering array of rock types; while some may refer to rather unique suites of minerals, others are named solely on a geographical locality. In order to obviate the complex nomenclature of the alkaline rocks, they are considered in terms of their petrochemistry related to problems of origin and the possibility of a genetic relation to primitive mantle magmas. A feature common to many of the highly undersaturated alkaline rock provinces is the occurrence of significant amounts of quartz syenite and alkaline granites. Such rocks pose problems of origin in terms of chemical differentiation of a parental basic magma along two lines, one towards a high degree of undersaturation, the other of saturation with respect to silica. As an alternative, these rocks may be viewed in terms of a secondary origin by the assimilation at depth of acid basement rocks, combined with the removal by 'selective leaching' of some elements of acid rocks by invading basic magma.

Finally, there is the problem of the origin of the carbonatites. These rocks are found in many alkaline provinces, particularly in association with ijolites (pyroxene and nepheline) and pyroxenites. The major minerals consist of calcium, magnesium and iron carbonates together with accessory minerals enriched in elements characteristic of the alkaline rocks. Most carbonatites are restricted to zones of deep fracture of the Earth's crust, particularly cutting Precambrian platforms. The origin of carbonatites has been extremely controversial, although current opinion supports a magmatic origin, in the past the predominance of carbonates has been taken to indicate an origin by the remobilization of sedimentary carbonates. A measurement of the isotopic composition of strontium in various types of alkaline rocks would appear to be capable of answering problems pertaining to a crustal or mantle origin. However, in many instances there is ample evidence to show that the replacement of country rock by an invading alkaline magma has taken place. While this may be restricted to roof areas and the margins of magma chambers, in some cases local conditions may have allowed extensive mixing between the alkaline magma and the surrounding country rock, particularly at depth.

Studies by Yoder and Tilley (1962) indicate that a large number of rocks, such as basalt, gabbro, pyroxenite, amphibolite, pyroxene, hornblendite and eclogite can, when melted completely, give rise to magmas of basaltic composition. The primary source 'rock' in terms of chemical composition may be homogeneous or relatively non-homogeneous depending upon the depth at which melting occurred. An alternative view by Kuno (1959) proposes that the major variations in magma types are due to variations in the bulk composition of the source rock at various depths. Of the two major basic volcanic rock types, tholeiite and alkaline basalt, that have been derived from below the sial, the alkaline basalt liquids are regarded by some as being derived from areas of higher pressure and greater depth than those of the tholeiitic liquids.

Apart from an origin of basaltic rocks derived from a liquid mantle magma, it is also possible to view the mantle as a mass that can be physically considered as a 'solid', but one that contains components of solid, liquid and gaseous constituents (Leet and Leet, 1965). Under conditions of shear movement in the mantle, volatiles and low-temperature assemblages are removed to give rise to basaltic liquids.

II The Significance of the Isotopic Composition of Strontium in Rocks and Minerals

Of the two naturally occurring rubidium isotopes, ^{87}Rb and ^{85}Rb, with relative abundances of 27.85 and 72.15% respectively, the isotope ^{87}Rb is radioactive and decays by beta-particle emission to the stable isotope, strontium-87. As a consequence of this, ^{87}Sr will accumulate with the passage of time at the sites of rubidium atoms in a mineral. Common strontium consists of four isotopes at mass 88–87–86–84, with natural relative abundances of 82.56, 7.02, 9.87 and 0.55% respectively. In any strontium-rich rubidium-poor mineral, the ratio ^{87}Sr/^{86}Sr $= 0.710 \pm 0.007$, but if rubidium is present, the ratio ^{87}Sr/^{86}Sr will increase as an almost linear function of time and at a rate proportional to the decay constant of ^{87}Rb. The ease with which the enrichment in ^{87}Sr can be measured in the laboratory will depend upon the Rb/Sr ratio of the particular mineral. Only small variations can be expected if the Rb/Sr ratio is low, but in the case of potassium-rich minerals, the simultaneous enrichment in rubidium and the tendency towards the depletion of strontium as a result of chemical fractionation can result in very high ^{87}Sr/^{86}Sr ratios. Some figures given in table 1 illustrate some ^{87}Sr/^{86}Sr ratios for various minerals from an ancient granite. In the case of a recent granite derived from a single homogeneous magma, the ^{87}Sr/^{86}Sr would typically be ~ 0.71 irrespective of the Rb/Sr ratio.

Table 1 The relation between measured ^{87}Sr/^{86}Sr ratios and Rb, Sr content in different minerals of an ancient granite

Mineral	(^{87}Sr/^{86}Sr)$_{today}$	Rb(ppm)	Sr(ppm)
Biotite	10.36	1867	8
Muscovite	0.91	129	26
Feldspar	0.86	256	54
Apatite	0.71	5	208

Variations have not been observed in the ratio ^{87}Sr/^{86}Sr other than those resulting from the addition of radiogenic strontium by the radioactive decay of rubidium-87. From theoretical considerations variations as a result of natural isotopic fractionation would not be expected to be observed for a heavy involatile element like strontium. Natural fractionation remains a slight possibility for strontium isotopes that have passed through several sedimentary stages particularly if biological cycles are involved. In determining the isotopic composition of strontium by mass spectrometry, strontium separated from a sample is heated on a metal filament in order to produce strontium ions. During the evaporation process from the filament, and the subsequent passage of the positive ions through slit systems, fractionation of the strontium isotopes does occur. In order to correct for this effect, the observed ^{87}Sr/^{86}Sr ratio is normalized to an assumed invariant value of 0.1194 for the ratio ^{86}Sr/^{88}Sr (Faure and Hurley, 1963).

The relation between the radioactive decay of rubidium-87 to its stable daughter product strontium-87 in terms of geological age and the isotopic composition of common strontium (Compston, Jeffery and Riley, 1960; Gast, 1960; Nicolaysen, 1961; Faure and Hurley, 1963) can be described by the following equation:

$$\left(\frac{^{87}Sr}{^{86}Sr}\right)_{today} - \left(\frac{^{87}Sr}{^{86}Sr}\right)_{initial} = \left(\frac{^{87}Rb}{^{86}Sr}\right)_{today}(e^{\lambda t} - 1) \qquad (1)$$

Equation (1) can be rearranged in terms of the age, t, of the sample

$$t = \frac{1}{\lambda} \ln\left[1 + \left(\frac{^{87}Sr}{^{86}Sr}\right)_{today} - \left(\frac{^{87}Sr}{^{86}Sr}\right)_{initial} \bigg/ \left(\frac{^{87}Rb}{^{86}Sr}\right)_{today}\right] \quad (2)$$

where t = age of the sample in years.

λ = decay constant for ^{87}Rb.

(a) $\left(\dfrac{^{87}Sr}{^{86}Sr}\right)_{today}$ = atomic ratio of $^{87}Sr : {}^{86}Sr$ today, measured by mass spectrometry.

(b) $\left(\dfrac{^{87}Sr}{^{86}Sr}\right)_{initial}$ = atomic ratio of $^{87}Sr : {}^{86}Sr$ at the time the rock cooled and became a closed system with respect to Sr and Rb, i.e. at time $t = 0$.

(c) $\left(\dfrac{^{87}Rb}{^{86}Sr}\right)_{today}$ = atomic ratio of $^{87}Rb : {}^{86}Sr$ today, measured by isotope-dilution analysis, or if the concentrations are high (>10 ppm) by x-ray fluorescence analysis or atomic absorption spectrophotometry.

If values can be determined for (a), (b) and (c) together with the decay constant, an age can be readily calculated. In most cases the ratio $(^{87}Sr/^{86}Sr)_{initial}$ cannot be measured directly and, therefore, together with the age, t, equation (2) contains two unknowns. A solution to this equation can be obtained if two samples from the same system, having the same age but different Rb/Sr ratios, can be obtained; thus there are two equations with two unknowns. In using this approach it is customary to use the isochron method (Nicolaysen, 1961), by rearrangement of equation (1) to the form

$$(^{87}Sr/^{86}Sr)_{today} = (^{87}Sr/^{86}Sr)_{initial} + \frac{^{87}Rb(e^{\lambda t} - 1)}{^{86}Sr}$$

where,
$$y = (^{87}Sr/^{86}Sr)_{today}$$
$$m(\text{slope}) = e^{\lambda t} - 1 \quad (3)$$
$$x = (^{87}Rb/^{86}Sr)_{today}$$

c = intercept on y axis = $(^{87}Sr/^{86}Sr)_{initial}$

Equation (3) is now one of a straight line of the form $y = mx + c$ as illustrated in figure 1. In geological samples, it is quite common

The isotopic composition of strontium

to obtain a wide range of rocks of different chemical compositions arising by chemical differentiation from a single magma. Variations in the ratio Rb/Sr will result in variations in the measured ratio $(^{87}Sr/^{86}Sr)_{today}$. When both these parameters are plotted as illustrated in figure 1 all the intercepts will lie along a straight line, an isochron, i.e. the slope corresponds to the age; as all the samples have the same age the points will all be distributed along a single straight line. To

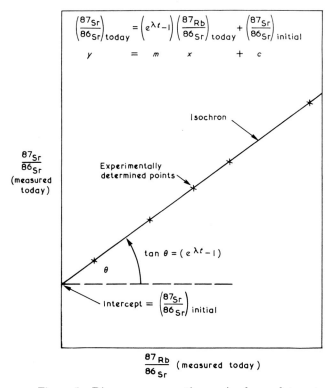

Figure 1 Diagram representing an isochron plot.

obtain an isochron a minimum of two points is required although it is far better to have as many points as possible so that the best statistical line can be fitted through the points. In order for an isochron plot to be valid, the following requirements are essential.

(1) Since the rock was formed Rb and Sr have remained, in terms of the hand specimen, within closed chemical systems.

(2) Any variations in the ratio $^{87}Sr/^{86}Sr$ are solely the result of the radioactive decay of ^{87}Rb.

(3) If the age of intrusion is required, it is essential that the cooling period should not be prolonged, otherwise the measured age may only be that at which the diffusion of rubidium and strontium ceased.

Measurements of present-day $^{87}Sr/^{86}Sr$ ratios for basic volcanic rocks indicate that, while oceanic basalts have, in general, a lower $^{87}Sr/^{86}Sr$ ratio than their continental counterparts, both lie within the range of ~ 0.702–0.706. Hurley and others (1962) have calculated the initial $^{87}Sr/^{86}Sr$ ratio for 82 igneous granites and obtained values of between 0.707–0.710 which showed no preferred variation with time. As more initial $^{87}Sr/^{86}Sr$ ratios for granites become available, the results suggest that the spread is somewhat greater; exceptions are found in the case of remobilized sial, metamorphic rocks and sediments. The average Rb/Sr ratio for the sialic layer is about 0.25; ancient sial with an initial $^{87}Sr/^{86}Sr$ ratio of ~ 0.707 would increase to ~ 0.720–0.730 in 1–2 billion years. If a magmatic rock has, in fact, been formed by the reworking of ancient sial, the initial $^{87}Sr/^{86}Sr$ ratio would clearly be greater than the calculated values. This does not rule out the possibility that some sialic igneous rocks have been formed by the reworking of sial, but necessitates that any assimilated matter must be of recent age in relation to the time of formation of the magmatic phase. This does not support the idea of reworked ancient sial to form crustal rocks. The $^{87}Sr/^{86}Sr$ ratios for clastic sediments formed by the erosion of newly formed mountain chains lie between 0.706–0.710, which are similar to the measured values for modern sea water (0.709–0.710) (Faure, Hurley and Powell, 1965).

From the foregoing discussion, it is, therefore, possible to recognize the comagmatic nature of a suite of igneous rocks having different Rb/Sr ratios, providing there has been no exchange of either Rb or Sr from other environments. This may not always be valid for some facies of an alkaline rock series particularly those that can be classed as fenites.

III The Isotopic Composition of Strontium in Basalts and their Differentiates

The isotopic composition of strontium in oceanic and continental basalts would appear to be quite uniform, with a value of 0.704 ± 0.002, as illustrated in figure 2 (Faure and Hurley, 1963; Hedge and Walthall, 1963; Hamilton, 1963; McDougall and Compston, 1965;

Gast, Tilton and Hedge, 1964; Lessing and Catanzaro, 1964; Heier, Compston and McDougall, 1965; Moorbath and Walker, 1965; Hamilton, 1965a,b,c).

The isotopic composition of strontium in acid granophyres that occur in basalt provinces is of interest as, if they are true differentiates, they should have initial $^{87}Sr/^{86}Sr$ ratios that are identical to those found in the comagmatic basic rocks. The acid granophyre from the Palolo Quarry, Hawaii, (Hamilton, 1965a), and some granophyres from Iceland (Moorbath and Walker, 1965) have $^{87}Sr/^{86}Sr$ ratios that can be interpreted in terms of differentiates from a basalt magma.

Slight, but possibly significant differences from the average value have been observed in the case of the Tasmanian dolerites (Heier, Compston and McDougall, 1965), acid granophyres of the Skaergaard intrusion, East Greenland (Hamilton, 1963) and some alkaline rocks of the Mid-Atlantic Ridge (Gast, Tilton and Hedge, 1964).

The $^{87}Sr/^{86}Sr$ ratios for seven Tasmanian dolerites are given in table 2 representing various differentiation products which have an

Table 2 The strontium $^{87}Sr/^{86}Sr$ and Rb/Sr ratios in some Tasmanian dolerites. (After Heier, Compston and McDougall, 1965)

Rock type	Rb (ppm)	Sr (ppm)	Initial $^{87}Sr/^{86}Sr$
Red Hill			
Chilled contact	33	130	0.7115
Central zone quartz dolerite	37	180	0.7112
Fayalite granophyre	90	138	0.7122
Granophyre	103	98	0.7107
	138	103	0.7122
Great Lake			
Dolerite Lower Zone	18	88	0.7123
Fayalite granophyre	86	113	0.7106

initial $^{87}Sr/^{86}Sr$ ratio of 0.7115 ± 0.0007. An acid granophyre from Red Hill gave a value of 0.7107 and 0.7122, clearly illustrating that this particular acid differentiate has not been formed by the assimilation of any appreciable amounts of crustal material, because the large volume of rock involved, and the uniformity of the $^{87}Sr/^{86}Sr$ ratio would not be in keeping with local contamination. The authors regard the high $^{87}Sr/^{86}Sr$ ratios for the basic dolerites as a reflection of the selective diffusion of certain elements into the dolerite magma. Further, consideration of some geochemical ratios, U/Th, U/K, Th/K

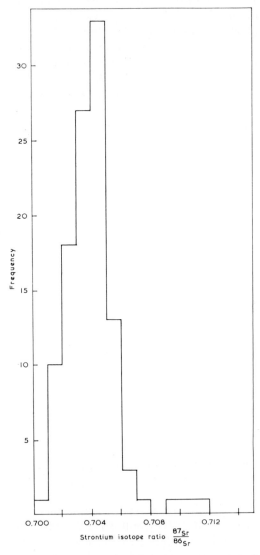

Figure 2 The $^{87}Sr/^{86}Sr$ ratio for basalts and genetically related rocks.

and K/Rb, are not in keeping with a hypothesis requiring the assimilation of sialic rocks in order to account for the acid granophyres.

Compared with the Tasmanian dolerites, the rocks of the Skaergaard intrusion represent a very small volume of basic magma trapped in a high crustal level magma chamber. The intrusion is Tertiary in age and is emplaced into Precambrian gneisses. The average $^{87}Sr/^{86}Sr$ ratio for the basic rocks is 0.7035 ± 0.002* which may be compared with a value of 0.7028 obtained for a contemporaneous basalt lava. Unlike the Tasmanian dolerites, the acid granophyres from Skaergaard have $^{87}Sr/^{86}Sr$ ratios of between 0.707–0.727. Compared with the ratios found for the basic rocks, and correcting for the small quantity of ^{87}Sr produced by the decay of ^{87}Rb during the last 50 m.y., the acid granophyres cannot have been formed by direct differentiation from the parent basic magma and it is necessary to invoke the assimilation of country rock. Further evidence that the granophyres from Skaergaard have been formed by the remobilization of country rock is obtained from a study of the isotopic composition of lead (Hamilton, 1966). Compared with the Skaergaard intrusion, the neighbouring Tertiary alkaline intrusion of Kangeralussuaq has an initial $^{87}Sr/^{86}Sr$ ratio of 0.704 which suggests an origin from a source region that has a Rb/Sr ratio similar to that present in the source regions of basalts.

Of all the alkaline rocks that are presumed to have been derived by differentiation of basalt magma those from the Hawaiian Islands are so far removed from possible sources of continental material that contamination by the assimilation of crustal granites would appear to be most improbable. The isotopic compositions of strontium for a selected series of Hawaiian rocks described by Powell, Faure and Hurley (1965) and Hamilton (1965b) are given in table 3. Powell, Faure and Hurley (1965) have correlated the slightly lower $^{87}Sr/^{86}Sr$ ratios with rocks of low silica content and suggest that these variations may reflect an origin from different source regions of the upper mantle. This would be in accordance with the views of Kuno (1960) and Yoder and Tilley (1962), who suggest from petrological evidence two or more independent parent magmas for the Hawaiian rocks.

Unlike the Hawaiian islands, Réunion, some 375 miles to the east of Malagasy (formerly Madagascar) in the Indian Ocean, is in the

---

* Strontium isotope ratios given in Hamilton (1963) have been reduced, as a result of comparison to a standard strontium solution.

Table 3 Strontium $^{87}Sr/^{86}Sr$ and Rb/Sr ratios in some Hawaiian basalts. (After Hamilton, 1965b; Powell, Faure and Hurley, 1965)

Locality	Rock type	Rb (ppm)	Sr (ppm)	Rb/Sr	$^{87}Sr/^{86}Sr^a$	$^{87}Sr/^{86}Sr^b$
Palolo Quarry, Oahu	Quartz dolerite	8	391	0.02	0.7047	—
Palolo Quarry, Oahu	Vein in dolerite	11	335	0.03	0.7041	—
Keauhov Beach, Hawaii	Alkali olivine basalt	16	413	0.04	0.7036	0.7040
Popo Gulch, Hawaii	Ankaramite	10	515	0.01	0.7044	—
S. Popo Gulch, Hawaii	Hawaiite	21	1436	0.02	0.7040	0.7041
1881 Flow Mauna Loa, Hawaii	Tholeiite	5	297	0.02	0.7048	0.7041
Honolulu, Volcanics Oahu	Nepheline basalt	23	1249	0.02	0.7023	0.7031
Molliili, Oahu	Melilite nepheline basalt	5	1763	0.02	0.7030	0.7030
Puu, Hawaii	Trachyte	124	64	1.94	0.7048	0.7043

[a] E. I. Hamilton (1965b).
[b] J. L. Powell (1965).

proximity of continental rocks. Apart from basic intrusives and extrusives, alkaline basalts and thick sills of quartz syenite are also present: the isotopic compositions of strontium for rocks from Réunion (Hamilton, 1965a; McDougall and Compston, 1965), given in table 4, are very similar to those obtained for the Hawaiian Islands.

Table 4 The isotope ratio $^{87}Sr/^{86}Sr$ for strontium in some volcanic rocks from Réunion

Rock type	Sample type	$^{87}Sr/^{86}Sr$	Reference[a]
Olivine eucrite	Whole rock	0.7037	1
Oceanite	Whole rock	0.7038	1
	Olivine	0.7037	1
Mugearite	Whole rock	0.7035	1
	Plagioclase	0.7029	1
Mugearite	Whole rock	0.7031	1
Feldsparphyric basalt	Whole rock	0.7035	1
Quartz syenite	Whole rock	0.7050	1
	Feldspar	0.7040	1
	Mean	0.7037	
Olivine basalt	Whole rock	0.7043	2
Alkaline basalt	Whole rock	0.7040	2
Syenite	Whole rock	0.7046	2
Alkali andesite	Whole rock	0.7044	2
Picrite basalt	Whole rock	0.7046	2
	Mean	0.7044	

[a] 1. Hamilton (1965a); 2. McDougall and Compston (1965).

It is noteworthy that the quartz syenite which contains ~5% K_2O has a $^{87}Sr/^{86}Sr$ ratio which is very similar to that found in the more basic rocks. From these measurements it would appear that the Réunion rocks have been derived from a source region having a Rb/Sr ratio similar to that of the basalts, and the measurements do not support the idea of the assimilation of rubidium-rich sialic material.

In the deep ocean areas the floor consists of tholeiitic basalt, while small islands capping ridge structures are predominantly of alkali basalts and trachytes. The peridotites and serpentinites that comprise St. Pauls Rocks (Atlantic Ocean) are an exception and may represent a cumulative mineral phase formed by differentiation of the common tholeiitic basalt magma or, alternatively, are thrust masses of a more primitive deeper rock type. The isotopic composition of strontium for some rocks from islands of the Mid-Atlantic Ridge are given in table 5. Gast, Tilton and Hedge (1964) have shown that

Table 5 The isotopic composition of strontium for basalts, alkaline rocks and Ascension xenoliths from the Mid-Atlantic Ridge

Locality	Rock type	$^{87}Sr/^{86}Sr$	Reference[a]
Basic rocks			
Ascension Is.	Basalt	0.7043	1
Iceland	Basalt	0.7024	2
Iceland	Basalt	0.7071	1
Tristan da Cunha	Basalt	0.7035	3
Azores	Basalt	0.7058	1
Gough Is.	Olivine-poor basalt	0.7045	4
Ascension Is.	Olivine-poor basalt	0.7025	4
		0.7028	4
		0.7028	4
East Greenland	Tholeiite	0.7028	3
St. Pauls Rock	Dunite–mylonite	0.7046	5
Alkaline rocks			
Gough Is.	Trachyte	0.7094	4
	Trachyte	0.7069	4
Ascension Is.	Trachyte	0.7073	4
	Trachyte	0.7045	4
	Trachyte	0.7050	4
Bouvet	Trachyandesite	0.7050	6

Plutonic xenoliths[b] from Ascension Is.	$^{87}Sr/^{86}Sr$	Rb(ppm)	Sr(ppm)	Reference[a]
Vlasovite granite	0.7080	88	~3	6
Alkaline granite	0.7092	162	5	6
Alkaline granite	0.7057	75	3	6
Alkaline granite	0.7090	157	3	6
Hornblende granite	0.7048	52	290	6
Fayalite syenite	0.7050	<10	14	6

[a] 1. Faure and Hurley (1963); 2. Moorbath and Walker (1965); 3. Hamilton (1963); 4. Gast, Tilton and Hedge (1964); 5. Hurley, Fairbairn and Pinson (1964); 6. Hamilton (unpublished data).
[b] Samples kindly provided by J. Cann and D. Bell.

the $^{87}Sr/^{86}Sr$ for alkaline rocks from Ascension and Gough Islands are more radiogenic than those of the basic rocks. The alkaline rocks of Ascension consist predominantly of dome-like masses of trachyte, which in some cases pass laterally into thick flows; but, perhaps of more interest is the occurrence of coarse- and fine-grained ejected blocks of igneous rock (Darwin, 1876; Atkins and others, 1964). The blocks consist of layered basic rock, diorites, syenites and alkaline granites, which occur in a redistributed agglomeritic deposit. In the hand specimens the blocks retain textures identical to those found in coarse-grained granites from continental areas. Apart from water rounding, the lack of a metamorphic fabric and marginal remelting shows that they have not been immersed in mantle magma, and are probably derived from a shallow source region. The isotopic composition of strontium and the Rb, Sr content from a few of the acid xenoliths are given in table 5. The low strontium content of the alkali granites is typical for such rocks, but the rubidium contents are low when compared with the values of 300–500 ppm so characteristic for other alkaline granites (Hamilton, 1964). The origin of these unique xenoliths still remains to be solved. Gast, Tilton and Hedge (1964) have shown that interisland variations in the isotopic composition of strontium and lead indicate that Rb/Sr, U/Pb ratios for the source regions of these rocks exhibit regional variations. The magnitude and extent of any such variations are of utmost importance; they may be the result of mantle diffusion processes, a reflection of volcanic processes or related to major mantle convection flow patterns. Variations in the isotopic composition of strontium could be explained as a result of these dynamic processes or, alternatively, a rubidium-rich strontium-poor phase may be formed by crystallization differentiation at depth, and at a later date remobilized during a subsequent volcanic phase with little mixing with the parent basic magma.

IV The Isotopic Composition of Strontium in Continental Plutonic Alkaline Rocks

The initial $^{87}Sr/^{86}Sr$ ratio for continental alkaline rock series have been but little studied. Fairbairn and others (1963) have measured the initial ratio for the well-known alkaline rock series of the Monteregian Hills. These small plutonic intrusions consist of ten small masses emplaced along a NNW–ESE line into Palaeozoic sediments of the St. Lawrence Lowlands, Canada. The field relations and petrology of the various rocks described by Dresser and Denis (1944) are in keeping with a comagmatic origin. The initial $^{87}Sr/^{86}Sr$ ratio of 0.7040 ± 0.0002 is similar to that found for the alkaline basalts, although the spread of results between $0.7032–0.7049 \pm 0.0005$ may reflect small variations in the initial magma. Radiometric ages for these rocks are given in table 6; the younger ages obtained for the biotites may be the result

Table 6 Comparison between Rb–Sr and K–Ar ages on biotites, Monteregian Hills, Quebec. (After Fairbairn and others, 1963)

Locality	Rock type	Rb–Sr age (m.y.)	K–Ar age (m.y.)
Oka	Carbonatite	114 ± 7	95 ± 5
Johnson	Essexite	111 ± 6	110 ± 10
Megantic	Granite	100 ± 1	115 ± 9
St. Hilaire	Essexite	99.5 ± 4	—
Brome	Essexite	88.3 ± 3	—
	Nordmarkite	84.4 ± 2	122 ± 10

of the diffusion of strontium as has been suggested by Kulp and Engels (1963) for metamorphic rocks, or, alternatively, of an enrichment in rubidium. If a metamorphic event has been responsible for the lower biotite ages under such conditions, it would be expected that the liquid inclusions so common to alkaline rocks would not be retained within the minerals and this could be used to support a metamorphic period.

A brief study of the initial $^{87}Sr/^{86}Sr$ ratios for two rocks from the Oslo Alkaline Province by Czamanske (1965) gave values of 0.706 and 0.703 (Moorbath, see Czamanske, 1965) for an akerite and granodiorite respectively. These results would not support the views of Barth (1954) for the mobilization of a Precambrian basement prism, but rather those of Oftedahl (1959) who has suggested an origin by gravitational differentiation. However, the results are far too few to answer this outstanding geological problem.

V Strontium Isotopes Applied to the Origin of Carbonatites

Carbonatites are found in association with alkaline rocks and generally have the form of plug-like bodies either cutting the country rock or as central masses in ring complexes. Apart from vertical pipe-like bodies, others typically include inward or outwardly dipping cone sheets or dyke-like masses. The major minerals consist predominantly of calcium, magnesium and iron carbonate, together with varying amounts of alkaline silicates and a characteristic suite of accessory minerals enriched in elements such as Ba, Sr, Nb, P, R.E., Zr, U and Th. In some intrusions local concentrations of these elements reach ore grade and are profitably mined.

A problem inherent to the origin of carbonatites by the remobilization of sedimentary carbonates has been the very high temperatures that are required to melt anhydrous carbonates. Paterson (1958) and Wyllie and Tuttle (1959) showed that calcite under conditions of high pressures and low temperatures could be maintained as a liquid, in the presence of water. The recognition of low-temperature melts in the system $CaO-CO_2-H_2O$ is regarded by Wyllie and Tuttle (1960a,b) as evidence to support a magmatic origin for the carbonatites, provided the field relationships are consistent with such an interpretation.

In the field, the intrusive sequences of the various carbonatites are in accordance with the expected crystallization differentiation sequence of a mixed carbonate magma, namely, calcite and dolomite, followed by ankerite and, finally, siderite; this is well illustrated by the alkaline intrusives of Chilwa Island (Garson, Smith and Campbell, (1958) and Alnö (Eckermann, 1948) and such a differentiation trend would be further enhanced by the explosive method of intrusion and the movement of volatiles.

The reactive nature of a carbonatite in the source region, and during the process of crystallization and cooling, would inevitably result in chemical reaction with the enclosing rocks. Groos and Wyllie (1963) and Wyllie (1965) have shown that in the presence of alkali and basic oxides, liquid immiscibility exists between a carbonate and a silicate melt. The extent to which reaction with basement rocks occurs will depend upon the particular intrusive form of the carbonatite. During the explosive intrusive phase, little reaction is likely to occur except for a certain amount of carbonatization; reaction with the host rocks is likely to be most pronounced during the final stage of cooling. The high concentration of the characteristic trace elements could be derived by complex transfer during the differentiation of the carbonatite from a mantle source region. Alternatively, as these elements are so characteristic of the sial, it is feasible to consider that they have been concentrated into carbonatites by extraction from deeply seated sialic material. From consideration of the isotopic composition of strontium in carbonatites any such source region would have to have a low Rb/Sr ratio.

Powell, Hurley and Fairbairn (1962) determined the isotopic composition of strontium in a variety of carbonatites, and the measured $^{87}Sr/^{86}Sr$ ratios of between 0.702 and 0.705* with an average of 0.7035 suggest that they have had no prior relation with crustal rocks enriched in radiogenic strontium. These results were soon confirmed by Hamilton and Deans (1963) for a series of carbonatites of different geological age from Africa. The results given in table 7

Table 7 Initial $^{87}Sr/^{86}Sr$ ratios for some rocks of the Monteregian Hills, Quebec. (After Fairbairn and others, 1963)

Locality	Rock type	Initial $^{87}Sr/^{86}Sr$
Oka	Carbonatite	0.7032
Mt. Royal	Tinguaite dyke	0.7043
St. Bruno	Gabbro	0.7040
St. Hilaire	Essexite	0.7036
Rougement	Peridotite	0.7047
Mt Johnson	Essexite	0.7036
	Pulaskite	
Yamaska	Yamaskite	
	Syenite	0.7049
	Essexite	
	Gabbro	
Brome	Essexite	0.7040
	Mean	0.7040 ± 0.002

illustrate that the initial $^{87}Sr/^{86}Sr$ ratios for carbonatites are at least similar to those found in basalts and are not compatible with an origin by the remobilization of crustal rocks.

The isotopic composition of strontium in sedimentary limestones is of obvious importance when considering the origin of carbonatites. Insufficient measurements have been made to generalize, although current measurements show that most limestones have $^{87}Sr/^{86}Sr$ ratios greater than 0.707. However, the isotopic composition of strontium in the early Precambrian Bulawayan limestone of South Africa has a significantly lower $^{87}Sr/^{86}Sr$ ratio than that observed for other limestones. Gast (1960) and Hamilton (1963) obtained a value of 0.704 while Hedge and Walthall (1963) quote a value of 0.7004. Limestones that were formed early in the Earth's history would be expected to have low initial $^{87}Sr/^{86}Sr$ ratios, which would be retained

---

* Note that the original analyses have been reduced in order to correct for instrumental error (Hurley, 1963).

until present times in view of the very low Rb/Sr ratio of such rocks. The Bulawayan limestone is greater than 2000 m.y. old and is associated with vast thicknesses of ironstones and basic lavas, all of which are compatible with an origin by the decomposition of basic rocks, together with the contemporaneous outpouring of basic lavas. The South African limestones and dolomites include some horizons which contain detrital minerals, some of which have significant quantities of rubidium. The pure carbonate facies contains trace quantities of Ba, Sr, R.E., Nb and Zr and, although an origin of carbonatites by the remobilization of deep-seated limestone is not in favour at the moment, it should be borne in mind that some of the characteristic trace elements found in carbonatites are present in sedimentary carbonates. During a process of remobilization, at depth, these elements would become concentrated in a mobile carbonate phase as a result of chemical fractionation. The predominance of strontium over rubidium in sedimentary limestones of different geological ages is illustrated in table 8.

Although most carbonatites have low $^{87}Sr/^{86}Sr$ ratios, some do not conform to the general pattern. The Keshya limestone in Zambia (formerly Northern Rhodesia) described by Bailey (1961) is the only one that has been studied in sufficient detail to date. The $^{87}Sr/^{86}Sr$

Table 8 Strontium and rubidium content of some limestones and dolomites

Geological age	Locality	Sr (ppm)[a]	Rb (ppm)[a]
Eocene	London Clay, Berkshire, Eng.	230	<2
Cretaceous	L. Chalk, Berkshire, Eng.	344	<5
	M. Chalk, Berkshire, Eng.	242	<5
Jurassic	Taynton Stone Gt. Oolite, Oxford, Eng.	220	<8
Permian	Magnesium Lst., Durham, Eng.[c]	46	<2
Carboniferous	S$_2$ Zone, Glamorgan, Eng.	98	<10
Devonian	U. Dev., Iowa, U.S.A.	114	<2
Silurian	Wenlock, Dudley, Worc., Eng.	710	<2
Ordovician	Trondjheim, Norway	434	<5
Cambrian	Schodack, N.Y., U.S.A.	110	<5
Precambrian	Arabia	216	<10
	Dolomite, Transvaal, S.A. (av.17 samples) (Mean)	156–490 (420)	<2–80[b] (<10)

[a] Analyses by x-ray fluorescence E. I. Hamilton.
[b] Significant concentrations of rubidium have been found in impure limestone horizons.
[c] Possibly a reworked deposit.

ratios measured by Hamilton and Deans (1963) and later by Powell (1965) are significantly greater than those found for carbonatites, but similar to those of sedimentary limestones. In order to account for the $^{87}Sr/^{86}Sr$ ratios it is suggested that the intrusive Keshya limestones have been formed by the remobilization of sedimentary limestones that have assimilated granitic material during the intrusive phase. The remobilized limestones contain between 110–230 ppm of strontium (Deans, 1964) compared with between 1000–10,000 ppm found in many carbonatites. The validity of such criteria to prove or disprove whether or not a carbonate rock is in fact a carbonatite is open to criticism.

A further problematic limestone from the Songwe Scarp, which occurs in the Mbeya mountain range of southwest Tanzania (formerly Tanganyika), Africa, described by Brown (1964), consists of an ankeritic carbonate rock having a fragmental and brecciated texture. The fragmented portion consists of feldspathized and carbonated country rock which occurs along an elongated belt parallel to the regional schistosity. The Songwe Scarp carbonatite is regarded as a high-level facies of the crystallization of a parent carbonatite magma at depth. The $^{87}Sr/^{86}Sr$ ratios (Hamilton, see Brown, 1964) for two specimens of 0.709 and 0.712 are greater than those found in most carbonatites. The carbonatite is intruded into potassium- and rubidium-rich feldspathized mica schists which could possibly provide a source of radiogenic strontium released during the intrusive carbonatite phase. Although large quantities of radiogenic strontium would be available, any high initial $^{87}Sr/^{86}Sr$ ratios would be lowered by the high content of common strontium in the carbonatite. The carbonate rock exhibits an anomaly in that although characteristic trace elements such as Sr, Ba, La, Nb, U and Th are present, there is a noticeable absence of these elements in accessory minerals. The total chemistry of the rock may not have been suitable for the formation of these minerals or, alternatively, these elements may have been extracted preferentially from the surrounding mica schists during the extrusive phase, together with radiogenic strontium-87.

From the available evidence, it would appear that the $^{87}Sr/^{86}Sr$ ratio will prove to be a valuable criteria for distinguishing between sedimentary limestones and carbonatites, although the extent to which undisturbed ancient deep-seated limestones could conceivably be the source material for carbonatites remains, if only a faint possibility. Recently, Hayatsu and others (1965) have shown that for the $^{87}Sr/^{86}Sr$ ratio to be an effective criterion for the recognition of carbonatites two requirements must be fulfilled.

(1) There must be a negligible overlap of the strontium isotope ratios of carbonatites and limestones.

(2) The $^{87}Sr/^{86}Sr$ ratio should not be altered by a metamorphic event. The authors have noted that some limestones of the Grenville province in Ontario and Quebec, Canada, show a significant decrease in the $^{87}Sr/^{86}Sr$ ratio with increase in the degree of metamorphism.

Carbonatites are restricted to stable ancient blocks and are not found in orogenic belts where any sedimentary carbonate would be assimilated during any orogenic processes. If the mantle is the source region of the carbonatites, carbon dioxide at least would be available as a product of mantle outgassing apart from those quantities existing as a gas phase in liquid inclusion present in most common, rock-forming minerals.

VI The Isotopic Composition of Strontium in Sea Water

Finally, no discussion would be complete without some reference to the isotopic composition of sea water from which limestones are precipitated. The distribution of the ratio $^{87}Sr/^{86}Sr$ in carbonatites and sea water is given in figure 3. Measurements of the $^{87}Sr/^{86}Sr$ ratio in modern sea water by Ewald, Garbe and Ney (1956), Herzog, Pinson and Cormier (1958), Gast (1961), Compston and Pidgeon (1962), Hedge and Walthall (1963), Faure, Hurley and Powell (1965) and Hamilton (1967) vary between 0.708 and 0.710, with a mean value of 0.709. No regional differences have been observed in the main oceans, which suggests that the water masses are well mixed, which is to be expected considering the constancy in the composition of salts dissolved in the oceans (Turekian, 1964; Sverdrup, Johnson and Fleming, 1942). However, variations in the concentration of the rare earths with depth have been observed (Balashov and Khitrov, 1961; Goldberg and others, 1963).

Faure, Hurley and Powell (1965) have estimated that a minimum of 30% of the strontium in the oceans is derived by the weathering of marine carbonates present on the continents. Significantly higher $^{87}Sr/^{86}Sr$ ratios than those found in most water masses have been observed by Compston and Pidgeon (1962) and Faure, Hurley and Powell (1965), but in both cases a direct and local relation to radiogenic ^{87}Sr enriched continental run-off water has been established. Faure, Hurley and Powell (1965) have reported a $^{87}Sr/^{86}Sr$ ratio of 0.7308 for the Pleistocene shells of *Sphaerium sulcatum*, which presumably reflects the isotopic composition of strontium released into solution during the chemical weathering of ancient silicate rocks.

If the strontium content of the ancient limestones is taken to reflect the strontium content of the ancient seas it is apparent from table 8 that very large amounts of radiogenic strontium would be required to significantly alter the isotopic composition of strontium in sea water. Prior to any dilution stage in the marine environment, any

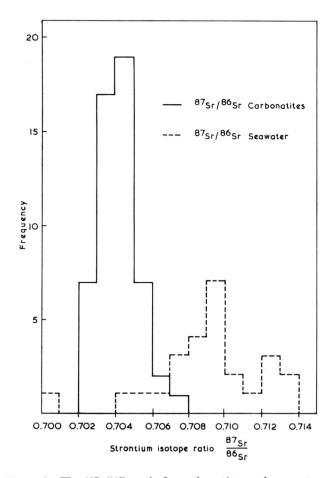

Figure 3 The $^{87}Sr/^{86}Sr$ ratio for carbonatites and sea water.

radiogenic strontium derived from the decay of ^{87}Rb would first of all be diluted in the common strontium pool released by the chemical breakdown of continental rocks.

VII Conclusions

The similarity between the initial $^{87}Sr/^{86}Sr$ ratios in both basic and alkaline rocks suggests that these rocks have been derived from an environment having very similar Rb/Sr ratios. Whether or not this can be taken as direct evidence that the alkaline rocks have been derived from basaltic mantle magma is still an open question. If it is accepted that the characteristic trace elements common to the alkaline rocks can be concentrated by chemical fractionation of basic magma, the case for alkaline rocks as differentiation products of basic magma is strengthened. However, it is possible to view the crust as an area in which there is a depletion of rubidium and a decrease in the ratio Rb/Sr with depth, as has been suggested by Hedge and Walthall (1963), Hamilton (1965a) and Heier (1964, 1965), or that primary differences were established early in the history of the Earth. Provided that deep areas of the crust have not been remobilized, it would be reasonable to expect that such rocks, perhaps typified by granulite facies, would be depleted in volatiles and elements such as rubidium and uranium.

In oceanic areas 'mobile' elements would be free to migrate upwards and become concentrated in the residual liquids. In continental areas, however, any differentiates would become trapped below the sial unless they were carried along together with the release of basic magma. Apart from alkaline differentiates of basic magma at depth, it is necessary to consider the junction between lower sialic granulite facies and basic upper mantle. In the absence of pressure release by crustal block fracture, it is suggested that reaction must occur at the sial/mantle interface at depth. Under these conditions, mantle outgassing products and local melting would give rise to alkaline liquids in which complex ions of elements such as Be, R.E., Nb, Ta, Zr and U would become concentrated. In any potential alkaline rock liquid, any original differentiation is disturbed as a result of explosive emplacement into the upper sial. Although mixing between an alkaline and basaltic liquid fraction would occur, it may be physical rather than chemical such that the alkaline fraction retains its identity and, after cooling, gives rise to various plutonic and volcanic alkaline rocks. The explosive emplacement of most alkaline rocks suggests an origin from great depths and the low $^{87}Sr/^{86}Sr$ implies practically no reaction with ancient sialic rocks. In this context, the low $^{87}Sr/^{86}Sr$ ratio for carbonatites may be interpreted as a result of the source rocks having a Rb/Sr ratio similar to basalts, not as a result of original magma type but rather as a result of depletion in rubidium. The high calcium and magnesium content of carbonatites places these rocks in a unique position and an origin by the remobilization of ancient limestones present as a facies of the lower sial should still be considered. This is conceivable if only limited miscibility exists between carbonatite liquids and silicate melts in which case, through complex ion transfer from the lower sial, the carbonate phase would act as a sink for many elements characteristic of the carbonatites.

References

Atkins, F. B., Baker, P. E., Bell, J. D. and Smith, D. G. W. (1964). Oxford Expedition to Ascension Island, 1964. *Nature*, **204**, 722.

Bailey, D. K. (1961). Intrusive limestones in the Keshya and Mkwisi Valleys, Northern Rhodesia. *Quart. J. Geol. Soc. London*, **117**, 419.

Balashov, Y. A. and Khitrov, L. M. (1961). Distribution of the rare earths in the waters of the Indian Ocean. *Geochemistry (USSR) (English Transl.)*, No. **9**, 877.

Barth, T. F. W. (1954). The igneous rock complex of the Oslo region. XIV. Provenance of the Oslo magmas. *Skrifter Norske Videnskaps-Akad. Oslo, I: Mat.-Naturv. Kl.*, No. **4**, 20.

Brown, P. E. (1964). The Songwe scarp carbonatite and associated feldspathization in the Mbeya Range, Tanganyika. *Quart. J. Geol. Soc. London*, **120**, 223.

Compston, W., Jeffery, P. W. and Riley, G. H. (1960). Age of emplacement of granites. *Nature*, **186**, 702.

Compston, W. and Pidgeon, R. T. (1962). Rubidium–strontium dating of shales by the total-rock method. *J. Geophys. Res.*, **67**, 3493.

Czamanske, G. K. (1965). Petrologic aspects of the Finnmarka igneous complex, Oslo area, Norway. *J. Geol.*, **73**, No. 2, 293.

Darwin, C. (1876). *Geological Observations on Volcanic Islands*, 2nd ed. Smith Elder, London.

Deans, T. (1964). Reply to paper. Isotopic composition of strontium in carbonatite. *Nature*, **201**, 599.

Dresser, J. A. and Denis, T. C. (1944). Geology of Quebec, descriptive geology. *Geol. Rept. Dept. Mines, Prov. Quebec*, **2**.

Eckermann, H., von (1948). The alkaline district of Alnö Island. *Sveriges Geol. Undersokn., Ser. Ca*, No. **36**.

Ewald, H., Garbe, S. and Ney, P. (1956). Die Isotopen-Zusammensetzung von Strontium aus Meerwasser und aus Rubidium-armen Gesteinen. *Z. Naturforsch.*, **11A**, 521.

Fairbairn, H. W., Faure, G., Pinson, W. H., Hurley, P. M. and Powell, J. L. (1963). Initial ratio of strontium-87 to strontium-86, whole rock age, and discordant biotite in the Monteregian Igneous Province, Quebec. *J. Geophys. Res.*, **68**, 24, 6515.

Faure, G. and Hurley, P. M. (1963). The isotopic composition of strontium in oceanic and continental basalts: application to the origin of igneous rocks. *J. Petrol.*, **4**, No. 1, 31.

Faure, G., Hurley, P. M. and Powell, J. L. (1965). Isotopic composition of strontium in surface water from the North Atlantic Ocean. *Geochim. Cosmochim. Acta*, **29**, 209.

Garson, M. S., Smith, W. and Campbell, (1958). Chilwa Island. *Mem. Geol. Surv. Nyasaland*, **1**.

Gast, P. W. (1960). Limitations on the composition of the upper mantle. *J. Geophys. Res.*, **65**, No. 4, 1287.

Gast, P. W. (1961). The rubidium–strontium method. In Kulp, J. L. (Ed.), Geochronology of Rock Systems. *Ann. N.Y. Acad. Sci.*, **91**, 181.

Gast, P. W. (1965). Terrestrial ratio of potassium to rubidium and the composition of the Earth's mantle. *Science*, **147**, 858.

Gast, P. W., Tilton, G. R. and Hedge, C. (1964). Isotopic compositions of lead and strontium from Ascension and Gough Islands. *Science*, **145**, 1181.

Goldberg, E. D., Koide, M., Schmitt, R. A. and Smith, R. H. (1963). Rare earth distributions in the marine environment. *J. Geophys. Res.*, **63**, No. 14, 4209.

Groos, A. F. K., van and Wyllie, P. J. (1963). Experimental data bearing on the role of liquid immiscibility in the genesis of carbonatites. *Nature*, **199**, 801.

Hamilton, E. I. (1963). The isotopic composition of strontium in the Skaergaard intrusion. *J. Petrol.*, **4**, 383.

Hamilton, E. I. (1964). The geochemistry of the northern part of the Ilímaussaq intrusion, S. W. Greenland. *Medd. Grønland*, **162**, No. 10, 1.

Hamilton, E. I. (1965a). Isotopic composition of strontium in a variety of rocks from Reunion Island. *Nature*, **207**, 1188.

Hamilton, E. I. (1965b). Distribution of some trace elements and the isotopic composition of strontium in Hawaiian lavas. *Nature*, **206**, 251.

Hamilton, E. I. (1965c). *Applied Geochronology*. Academic Press, New York. p. 267.

Hamilton, E. I. (1966). The isotopic composition of lead in igneous rocks. *Earth Planetary Sci. Letters*, **1**, No. 1, 30.

Hamilton, E. I. (1967). The isotopic composition of strontium in Atlantic Ocean water. *Earth Planetary Sci. Letters*, **1**, No. 6, 435.

Hamilton, E. I. and Deans, T. (1963). Isotopic composition of strontium in some African carbonatites and limestones and in strontium minerals. *Nature*, **198**, 776.

Hayatsu, A., York, D., Farquhar, R. M. and Gittens, J. (1965). Significance of strontium isotope ratios in theories of carbonate genesis. *Nature*, **207**, 625.

Hedge, C. E. and Walthall, F. G. (1963). Radiogenic strontium-87 as an index of geologic processes. *Science*, **140**, 1214.

Heier, K. S. (1964). Rubidium/Strontium and Strontium-87/Strontium-86 ratios in deep crustal material. *Nature*, **202**, 477.

Heier, K. S. (1965). Radioactive elements in the continental crust. *Nature*, **208**, 479.

Heier, K. S., Compston, W. and McDougall, I. (1965). Thorium and uranium concentrations, and the isotopic composition of strontium in the differentiated Tasmanian dolerites. *Geochim. Cosmochim. Acta*, **29**, 643.

Herzog, L. F., Pinson, H. W. and Cormier, R. F. (1958). Sediment age by Rb/Sr analysis of glauconite. *Bull. Am. Assoc. Petrol. Geologists*, **42**, 717.

Hurley, P. M. (Supervisor) (1963). *M.I.T. Annual Rept.* U.S.A.-A.E.C. CON. AT (30-1) 1381. p. 161.

Hurley, P. M., Fairbairn, H. W. and Pinson, W. H., Jr. (1964). Rb–Sr relationships in serpentinite from Mayaguez, Puerto Rico and dunite from St. Paul's Rocks. *A Progress Report in a Study of Serpentinite*, Publ. 1188. Natl. Acad. Sci.–Natl. Res. Council. p. 149.

Hurley, P. M., Hughes, H., Faure, G., Fairbairn, H. W. and Pinson, W. H., Jr. (1962). Radiogenic strontium-87 model of continent formation. *J. Geophys. Res.*, **67**, 5315.

Kulp, J. L. and Engels, J. (1963). Discordances in K–Ar and Rb–Sr isotopic ages. In *Radioactive Dating*. Intern. Atomic Energy Agency, Vienna. p. 219.

Kuno, H. (1959). Origin of Cenozoic petrographic provinces of Japan and surrounding areas. *Bull. Volcanol., Ser. 2*, **20**, 37.

Kuno, H. (1960). High-alumina basalt. *J. Petrol.*, **1**, 121.

Leet, L. D. and Leet, F. J. (1965). The Earth's mantle. *Bull. Seismol. Soc. Am.*, **55**, No. 3, 619.

Lessing, P. and Catanzaro, E. J. (1964). $^{87}Sr/^{86}Sr$ Ratios in Hawaiian lavas. *J. Geophys. Res.*, **69**, No. 8, 1599.

McDougall, I. and Compston, W. (1965). Strontium isotope composition and potassium–rubidium ratios in some rocks from Réunion and Rodriguez, Indian Ocean. *Nature*, **207**, 252.

Moorbath, S. and Walker, G. P. L. (1965). Strontium isotope investigations of igneous rocks from Iceland. *Nature*, **207**, 837.

Nicolaysen, L. O. (1961). Graphic interpretation of discordant age measurements on metamorphic rocks. In Kulp, J. L. (Ed.), Geochronology of Rock Systems. *Ann. N.Y. Acad. Sci.*, **91**, 198.

Oftedahl, C. (1959). Volcanic sequence and magma formation in the Oslo region. *Geol. Rundschau*, **48**, 18.

Paterson, M. S. (1958). The melting of calcite in the presence of water and carbon dioxide. *Am. Mineralogist*, **43**, 603.

Powell, J. L. (1965). Isotopic composition of strontium in carbonate rocks from Keshya and Mkwisi, Zambia. *Nature*, **206**, 288.

Powell, J. L., Faure, G. and Hurley, P. M. (1965). Strontium-87 abundance in a suite of Hawaiian volcanic rocks of varying silica content. *J. Geophys. Res.*, **70**, 1509.

Powell, J. L., Hurley, P. M. and Fairbairn, H. W. (1962). Isotopic composition of strontium in carbonatites. *Nature*, **196**, 1085.

Sverdrup, H. U., Johnson, M. W. and Fleming, R. H. (1942). *The Oceans, their Physics, Chemistry and General Biology*. Prentice-Hall, New York.

Turekian, K. K. (1964). The marine geochemistry of strontium. *Geochim. Cosmochim. Acta*, **28**, 1479.

Wasserburg, G. J., MacDonald, G. J. F., Hoyle, F. and Fowler, W. (1964). Relative contributions of uranium, thorium and potassium to heat production in the earth. *Science*, **143**, 465.

Wyllie, P. J. (1965). Melting relationships in the system $CaO–MgO–CO_2–H_2O$, with petrological applications. *J. Petrol.*, **6**, No. 1, 101.

Wyllie, P. J. and Tuttle, O. F. (1959). Melting of calcite in the presence of water. *Am. Mineralogist*, **44**, 453.

Wyllie, P. J. and Tuttle, O. F. (1960a). The system $CaO-CO_2-H_2O$ and the origin of carbonatites. *J. Petrol.*, **1**, Part 1, 1.

Wyllie, P. J. and Tuttle, O. F. (1960b). Experimental verification for the magmatic origin of carbonatites. *Intern. Geol. Congr., Rept. 21st Session, Norden, 1960*, Part 13, 310.

Yoder, H. S., Jr. and Tilley, C. E. (1962). Origin of basalt magmas: an experimental study of natural and synthetic rock systems. *J. Petrol.*, **3**, No. 3, 342.

Author Index

Abelson, P. H. *222*
Ackermann, E. 342, 380, *409*
Adams, J. A. S. 204, *223*
Adams, N. E. 151, *219*
Ahrens, L. H. 230, 232, 233, *255*, *257*
Albee, A. L. 74, *110*
Aldrich, L. T. 7, 8, 25, 40, *64*, *71*, 74, 80, 84, 89, 94, *109*, *110*, 114, 115, 117, 119, 120, *144*, 175, *216*, 229, 252, 254, *255*, *257*, 270, *295*, 315, 339, 343, 374, 381, *409*, *411*
Allan, D. W. *220*
Allen, V. T. 197, *216*
Allsopp, H. L. 315, 334, 367, 374, 375, 379, 380, 397, *409*, *413*
Alpher, R. A. 165, 166, *216*
Amirkhanoff, K. I. 19, 22, 25, 26, *64*, 90, *109*
Anderson, C. A. 35, 36, *64*
Anderson, J. G. C. 260, 262, 268, 276, *295*
Armstrong, R. L. 52, 56, *65*
Atkins, F. B. 451, *460*
Austin, C. F. 196, 197, 204, *221*

Baadsgaard, H. 7, 16, 25, 58, *65*, *67*, 90, *109*, 119, 121, 124, 125, 143, *144*, 189, *216*, *218*, 245, *256*
Bailey, D. K. 455, *460*
Baker, P. E. 451, *460*
Balashov, Y. A. 457, *460*
Barnes, I. L. 17, 69
Barth, T. F. W. 452, *460*
Bartnitsky, E. N. 19, 22, 25, *64*, 90, *109*
Bass, M. N. 175, *216*, 249, *257*
Bassett, W. A. 19, *65*, 115, 117, *144*
Bassett, W. H. 19, *68*
Bassot, J.-P. 329, 330, 345, 352, *409*
Bate, G. L. 185, 196, 204, *216*
Behre, C. H. 197, *216*
Bell, J. D. 270, *297*, 451, *460*

Bell, K. 263, 264, 270, 285, 287, *295*
Bennett, R. 16, 26, 32, 37, 38, 41, 46, 51, *69*
Besairie, H. 332, 387, *409*
Bessoles, B. 360, 384, 385, *409*
Bikerman, M. 46, 47, 48, 50, *65*, *66*
Black, R. 325, 333, *409*
Blomquist, N. G. 159, *223*
Boelrijk, A. I. M. 58, *69*
Boissonnas, J. 327, 328, 344, *409*
Boltwood, B. B. 148, *216*
Bonhomme, M. 325, 329, 330, 345, 350, 351, 352, 353, 364, 369, 385, 386, 387, 399, *409*
Boocock, C. 375, *409*
Bourguillot, R. 311, *410*
Brancazio, P. J. *71*
Brandt, S. B. 19, 22, 25, *64*, 90, *109*
Brock, B. B. 304, *409*
Broecker, W. S. 25, *65*
Brown, H. 182, *222*, 226, *257*
Brown, P. E. 264, 265, 276, 287, *295*, *296*, 456, *460*
Browne, W. R. 175, *216*
Burger, A. J. 306, 308, 314, 315, 318, 340, 341, 342, 344, 345, 347, 362, 367, 370, 391, *409*, *410*, *413*, *414*, 416
Burke, W. H. 52, *65*
Burwash, R. A. 119, 121, 143, *144*, 189, *216*
Butler, B. S. 35, *65*
Butler, J. R. 137, 138, *144*
Butler, R. 244
Butterfield, J. A. 227, *255*

Cahen, L. 301, 302, 306, 311, 314, 315, 316, 317, 318, 322, 323, 333, 339, 342, 343, 344, 346, 452, 354, 355, 357, 359, 360, 361, 362, 363, 375, 380, 381, 385, 387, 388, 393, 399, 401, *410*, *412*, *413*
Cameron, A. G. W. *71*

Campbell, C. J.   260, 262, 268, 275, 297, 453, *461*
Carr, D. R.   25, *65*
Carroll, D.   228, *255*
Carruthers, D. S.   175, *216*
Catanzaro, E. J.   36, 50, *65*, 115, 119, 120, 122, 134, 135, 138, 139, 143, 144, *145*, 182, 189, 190, *216*, 226, 234, 237, 238, 239, 240, 241, 242, 244, 245, 246, 247, 248, 249, 253, *255*, 445, *462*
Chadwick, J.   27, *69*
Chandler, T. R. D.   311, 318, *411*
Cherdyntsev, V. V.   17, *65*
Chew, R. T.   51, *65*
Choubert, G.   353, 354, *410*
Chow, T. J.   162, 177, 178, 179, 182, 184, 204, *216*
Clifford, T. N.   307, 308, 310, 314, 315, 316, 332, 334, 336, 342, 344, 362, 382, 393, 407, *410*
Cobban, W. A.   44, *65*
Coertze, F. J.   359, *415*
Collins, C. B.   171, 175, 192, *216*
Compston, W.   175, *216*, 263, 265, 268, 274, 275, 282, 283, *295*, 441, 444, 445, 449, 457, *460*, *461*, *462*
Cooper, J. A.   9, *65*, 175, *220*
Cormier, R. F.   457, *462*
Cox, A.   12, 16, 31, *65*, *66*
Cox, K. G.   359, *410*
Craig, H.   *218*, *220*
Creasey, S. C.   52, *65*
Cree, A.   76, 77, *109*
Cruse, M. A. J. B.   267, *295*
Cumming, G. L.   185, 189, 213, *216*, *221*
Cunningham, B. B.   151, *217*
Curtis, G. H.   11, 12, 14, 15, 16, 19, 20, 23, 24, 31, 53, *66*, *67*, 89, 99, 101, 102, *109*
Czamanske, G. K.   452, *460*

Dalrymple, G. B.   12, 15, 16, 31, *65*, *66*
Damon, P. E.   10, 12, 14, 15, 16, 17, 18, 26, 31, 32, 36, 37, 38, 39, 41, 42, 46, 50, 51, 52, 53, 54, 55, 56, 58, 59, 60, 62, 63, *69*, *249*, *256*

Dance, D. F.   311, 318, *411*
Darnley, A. G.   311, 318, *411*
Dars, R.   330, *411*
Darwin, C.   451, *460*
David, Sir T. W. E.   175, *216*
Davis, G. L.   25, 40, 64, 71, 74, 80, 84, 89, 94, *109*, *110*, 114, 115, 117, 119, 120, *144*, 175, 182, 197, *216*, 222, 229, 249, 250, 252, 254, *255*, *256*, *257*, 270, *295*, 315, 339, 343, 381, *409*, *411*
Deans, T.   454, 456, *460*, *461*
Denis, T. C.   451
Delafosse, R.   321, 322, 340, 344, 363, *414*
Delbos, L.   322, 323, 361, 387, *411*
Delhal, J.   306, 315, 316, 339, 346, 358, 359, 362, 363, 383, 384, 388, 393, *410*, *411*, *413*
Delwiche, R.   339, *413*
De Swardt, A. M. J.   313, 342, *411*
Deutsch, S.   37, *70*, 234, 247, 248, 249, 253, *257*
De Villiers, J. W. L.   366, *414*
Dewey, J. F.   261 262, 266, 267, 268, 274, 275, 276, *295*
Dimitriyev, L. V.   204, *216*
Dixey, F.   407, *411*
Dodd, R. T.   14, *68*
Dodson, M. H.   276, *295*, 359, *415*
Doe, B. R.   74, 92, 93, 94, *109*, 156, 176, 184, *217*
Doell, R. R.   12, 16, 31, *65*, *66*
Dresser, J. A.   451, *460*
Drysdall, A. R.   311, 313, 318, 342, 361, 380, *411*, *415*
Du Bois, R. L.   249, *256*
Duplan, L.   327, 328, 344, *409*
Du Toit, A. L.   299, 346, 366, 369, 373, 375, 377, 379, 396, 399, *411*

Eardley, A. J.   43, *66*
Eaton, J. P.   171, *217*
Eberhardt, P.   150, *217*, *218*, 311, 339, 363, *410*, *411*, *413*
Eckelmann, W. R.   114, 122, 135, 136, *145*, 227, 229, 243, 249, *256*, *257*
Eckermann, H. von   453, *460*

Author Index

Edmund, R. W. 141, *145*
Edmunds, W. M. 267, *296*
Edwards, G. 162, *217*
Ellis, C. D. 27, *69*
El-Ramly, M. F. 319, *411*
Emberger, A. 307, 323, *411*
Endt, P. M. 7, *66*
Engels, J. 17, 19, *68*, 280, *297*, 452, *462*
Erickson, R. C. 38, 39, 41, 42, 51, *66*
Evans, B. W. 266, *296*
Evans, R. D. 204, *217*
Evernden, J. F. 11, 12, 14, 15, 16, 19, 20, 23, 24, 31, 53, *66*, *67*, 89, 99, 101, 102, *109*, 124, 125, 175, *217*, 265, 271, *296*
Ewald, H. 457, *460*

Fairbairn, H. W. 19, 32, *68*, 166, 175, 189, 197, 200, 202, *216*, *217*, *218*, 228, *256*, 263, 274, *296*, 379, 380, 396, *411*, 444, 450, 451, 452, 453, 454, *460*, *462*
Farquhar, R. M. 151, 155, 158, 159, 171, 172, 175, 176, 185, 186, 187, 189, 190, 192, 200, 201, 204, 212, 213, *216*, *217*, *218*, 220, 221, 456, *461*
Farrar, E. 264, 265, 271, 287, *295*, *296*
Faul, H. *70*
Faure, G. 166, 175, 189, 200, *217*, *218*, 270, 282, *296*, 379, 380, 396, *411*, 441, 444, 447, 448, 450, 451, 452, 454, 457, 460, *461*, *462*
Fechtig, H. 19, 25, *67*
Ferrara, G. 150, *218*, 363, 389, *411*
Fick 20
Fieremans, C. 384, *411*
Fitch, J. F. 264, 265, 276, *296*
Fleischer, R. L. 417, 419, 420, 422, 428, 429, 434, *435*
Fleming, G. H. 151, *217*
Fleming, R. H. 457, *462*
Flynn, K. F. 162, *217*, 270, *296*
Folinsbee, R. E. 25, *67*, 124, 125, *144*
Fowler, W. 438, *462*
Fraser, W. E. 264, 287, *295*

Friedlander, G. 27, *67*
Frondel, C. 228, *256*
Funkhouser, J. G. 17, *69*
Furon, R. 299, 302, 318, 322, 325, 329, 338, 341, 342, 346, 360, 363, 380, 389, *411*
Fyfe, W. S. 107, 108, *109*

Gast, P. W. 43, *67*, 74, 94, 106, *109*, *110*, 114, 115, 117, 119, 120, 131, 134, 140, *145*, 162, 175, 180, 182, 185, 196, 204, *216*, *217*, *222*, 243, *256*, 438, 441, 445, 450, 451, 454, 457, *461*
Gastil, G. 189, *217*
Geijer, P. 159, *223*
Garbe, S. 457, *460*
Garlick, 342
Garrard, P. 313, 342, *411*
Garson, M. S. 453, *461*
Geiss, J. 170, 185, *217*, 220, 221, *222*, 339, *413*
Gellatly, D. C. 319, *411*
Gentner, W. 14, 18, 19, 25, *67*, *68*
Gérard, G. 340, *411*
Gerling, E. K. 3, 9, 17, 18, 19, 25, *67*, 103, *109*, 153, 175, 212, *216*, *220*
Gheith, M. A. 319, *411*
Ghiorso, A. 151, *217*
Griffin, C. E. 84, *109*
Giletti, B. J. 36, 38, 41, 46, *66*, *69*, 74, *109*, 115, 117, 119, 120, 124, 125, 127, 129, 131, 132, 140, *144*, *145*, 249, *256*, 260, 263, 265, 270, 273, 282, 295, *296*
Gill, J. E. 189, *218*
Gilluly, J. 43, 45, 55, 59, 61, *67*
Giraudon, R. 331, *411*
Gittens, J. 456, *461*
Glangeaud, L. 363, *411*
Glendenin, L. E. 7, *67*, 162, *217*, 270, *296*
Goddard, E. N. 75, 76, *110*
Goldberg, E. D. 220, 221, 222, 457, *461*
Goldich, S. S. 7, 16, 25, *67*, 90, *109*, 123, *145*, 189, *218*, 245, 247, *256*, *257*

Goldsmith, J. R.   81, *109*
Golubchina, M. N.   182, *218*
Gonyer, F. A.   227, *256*
Goodman, C.   204, *217*
Gottfried, D.   228, *256*
Grasty, R. L.   264, 287, *295*, 314, 341
Gravelle, M.   363, 389, *411*
Green, D. H.   14, *69*
Green, W. D.   17, *66*
Greenhalgh, D.   175, 192, *218*
Greenwood, J. E. G. W.   318, *411*
Grimaldi, F. S.   269, *296*
Grögler, N.   327, 328, 363, *412*
Groos, A. F. K. van   453, *461*
Guernsey, T. D.   342, *411*
Gulbrandsen, R. A.   123, *145*
Grunning, H. C.   175, *221*

Haldemann, E. G.   343, 357, 381, *414*
Hall, A. L.   366, *412*
Halva, C. J.   54, *67*
Hamilton, E. I.   3, 9, *67*, 311, 313, 317, 318, 344, 361, 380, 388, 389, *415*, 444, 445, 447, 448, 449, 450, 451, 454, 455, 456, 457, 459, *461*
Handley, J. R. F.   304, 305, *412*
Hanekom, H. J.   366, *412*
Hansen, E.   52, *65*
Hanson, G. N.   43, *67*, 106, *109*
Hargraves, R. B.   367, *412*
Harker, A.   227, *256*
Harland, W. B.   265, 270, 286, *296*
Harper, C. T.   264, 276, 287, *296*
Harris, A. L.   264, *296*
Harris, P. M.   265, 271, *296*
Harris, R. L.   114, *145*
Hart, S. R.   14, 16, 19, 23, 26, 43, *68*, *69*, 74, 82, 86, 87, 88, 91, 93, 94, *109*, *110*, 126, *145*, 184, *217*, 274, *296*
Haughton, S. H.   299, 320, 321, 322, 324, 326, 337, 341, 350, 352, 353, 354, 359, 381, *412*
Hayatsu, A.   456, *461*
Hayden, R. J.   25, *71*, 89, *110*, 117, 119, *145*, 182, *222*, 226, *257*, 379, *416*

Hebeda, E. H.   58, *69*
Hedge, C. E.   51, 52, *68*, 119, 121, 143, *146*, 180, 182, *217*, 245, *256*, 270, *296*, 444, 445, 450, 451, 454, 457, 459, *461*
Heier, K. S.   81, *110*, 445, 459, *461*
Heinrich, E. W.   114, 127, *145*
Herman, R. C.   165, 166, *216*
Herzog, L. F.   457, *462*
Hess, D. C.   162, 182, *219*, *222*, 226, *257*
Hess, D. C.   162, *217*, *218*
Hess, H. H.   113, *145*
Heyl, A. V.   197, *216*
Higazy, R. A.   319, *412*
Hills, A.   115, 120, 134, *145*
Hinds, N. E. A.   36, *68*
Hintenberger, H.   18, *70*
Hoffman, J. H.   7, 16, 25, *67*, 189, *218*, 245, *256*
Holland, H. D.   228, *256*
Hollander, J. M.   6, *70*
Holmes, A.   153, 164, *218*, 265, 270, 286, *296*, 300, 301, 307, 313, 314, 316, 317, 318, 322, 323, 333, 339, 340, 343, 345, 346, 352, 355, 357, 359, 361, 366, 375, 380, 381, 387, *412*
Hopson, C. A.   74, 94, *109*, *110*, 156, 184, *217*
Horne, J. E. T.   311, 318, *411*
Hornung, G.   317, 344, 388, 389, *415*
Houston, R. S.   114, 115, 120, 134, *145*
Houtermans, F.   150, 155, 161, 170, 185, 210, *217*, 218, 339, *413*
Howland, A.   113, 137, *145*, *146*
Hoyle, F.   438, *462*
Hughes, H.   19, 32, *68*, 166, 189, *218*, 274, *296*, 444, *462*
Hunt, G. H.   175, *218*
Hunter, D. R.   374, *409*
Hurley, P. M.   19, 32, *68*, 166, 175, 189, 197, 200, 202, *216*, *217*, *218*, 228, 256, 263, 270, 274, 282, *296*, *298*, 379, 380, 396, *411*, 441, 444, 447, 448, 450, 451, 452, 453, 454, 457, *460*, *461*, *462*

Author Index

Inghram, H.   151, 182, *218, 222,* 220, *257*
Ingold, L. M.   266, *296*
Irving, J.   227, *256*

Jacobson, R. R. E.   307, 324, 326, 335, *412*
Jaeger, J. C.   126, *145*
Jager, E.   74, *110*
James, G. T.   12, 16, 31, 53, *67*
James, H. L.   74, *109*
Jeffery, P. M.   175, 192, *218*, 263, *295*, 381, *411*, 441, *460*
Jensen, K. J.   25, *71*, 89, *110*, 379, *416*
Johnson, M. R. W.   264, *296*
Johnson, M. W.   457, *462*
Johnson, R. L.   317, 344, 345, 359, 388, 389, *410, 412, 415*
Johnstone, M. S.   177, *216*
Jones, D. L.   369, *412*
Jones, W. R.   113, *146*
Jost, W.   20, *68*

Kalbitzer, S.   19, 25, *67*
Kanasewich, E. R.   151, 156, 157, 158, 159, 160, 165, 166, 168, 176, 185, 186, 187, 190, 191, 193, 194, 200, 201, 202, 204, 215, *218, 221*
Karakeda, Y.   227, *256*
Kay, M.   43, *68*
Keevil, N. B.   204, *218*
Kennedy, J. W.   27, *67*
Kennedy, W. Q.   307, 308, 309, 310, 323, 324, 328, 331, 334, 369, 393, 407, *412*
Kenney, G. S.   52, *65*
Kerr, P. F.   122, *146*
Khitrov, L. M.   457, *460*
Kilburn, C.   266, *296*
Kilroe, J. R.   *297*
King, H. F.   175, 193, *218*
King, P. B.   198, *219*
Kistler, R. W.   12, 19, 20, 23, 24, 52, *65, 66*, 89, 99, 101, 102, *109*
Kluyver, J. C.   7, *66*
Kley, W.   18, *67*
Knill, J. L.   261, *297*
Knopf, A.   124, 125, *146*

Koide, M.   457
Kolbe, P.   315, 334, 397, *409*
Kollar, F.   157, 192, 193, 204, 215, *219*
Koltsova, T. V.   103, *109*
Koschmann, A. H.   197, *219*
Kouvo, O.   *110*, 160, 175, 191, *219, 222*, 251, 252, *256*
Kovarik, A. F.   151, *219*
Krenkel, E.   299, *412*
Krenkle, E.   *223*
Kreuger, H. W.   7, 16, 25, *67*, 189, *218*, 245, *256*
Kulp, J. L.   10, 12, 14, 17, 18, 19, 25, 50, *65, 66, 68*, 74, 84, *109, 110*, 114, 115, 117, 119, 122, 123, 134, 135, 138, 139, 143, 144, *145, 146*, 160, 175, 183, 190, 191, 194, 195, 204, *216, 219, 222*, 226, 229, 234, 238, 240, 241, 242, 243, 244, 245, 247, 248, 249, 253, *255, 256*, 265, 270, 280, 286, 452, *462*
Kuno, H.   440, 447, *462*
Kurbatov, V. V.   18, 19, *67*

Ladd, H. S.   61, *69*
Lagaay, R. A.   58, *69*
Lambert, H. H. J.   313, *415*
Lambert, R. St. J.   74, *109*, 129, *145*, 260, 263, 265, 270, 273, 274, 282, 283, 285, 295, *296, 297*
Lance, J. F.   51, *68*
Lanphere, M. A.   37, 39, 40, *71*, 74, *110*
Larsen, D. S.   227, *256*
Larsen, E. S.   182, *222*, 226, 227, *256, 257*
Larsen, L. H.   227, *256*
Laserre, M.   320, 321, 360, 384, 385, *412*
Laughlin, A. W.   14, 16, 26, 32, 37, 38, 41, 46, 51, *68, 69*
Laves, F.   81, *109, 110*
Lawson, A. C.   207, *219*
Lay, C.   327, 328, 358, 359, 363, 383, 384, *412, 413*
Leake, B. E.   260, 265, 266, 268, 274, 275, 282, 283, *297*

Ledent, D.   306, 311, 315, 316, 327, 328, 339, 344, 358, 359, 362, 363, 383, 384, 388, *410, 412, 413*
Leet, F. J.   440, *462*
Leet, L. D.   440, *462*
Leggo, P.   265, 268, 274, 275, 282, *297*
Leonova, L. L.   166, 204, *216, 219*
Le Roex, H. D.   359, *415*
Lessing, P.   445, *462*
Leutwein, F.   321, 322, 340, 344, 363, *414*
Liebenberg, W. R.   367, *413*
Liebhafsky, 20, 25
Lindgren, W.   61, *68*
Lippolt, H. J.   14, *68*
Lipson, J.   25, 58, *65, 67*, 124, 125, *144*
Livingston, D. E.   16, 26, 32, 37, 38, 39, 41, 42, 46, 51, 52, *66, 68, 69*
Long, L. E.   74, *110*, 114, 115, 117, 123, 134, *145*, 243, *256*, 283, 285, *297*
Loughlin, G. F.   197, *219*
Lovering, J. F.   14, 16, 17, 30, *69*, 164, *219, 220*
Lovering, T. S.   75, 76, *110*
Lovtzyus, A. V.   162, *221*
Lovtzyus, G. P.   162, *221*
Lowdon, J. A.   175, *219*, 401, *413*

McConnell, R. B.   355, 356, 399, *413*
MacDonald, G. J. F.   438, *462*
McDougall, I.   14, *69*, 444, 445, 449, *461, 462*
McDowell, F. W.   53, *69*
McElhinny, M. W.   369, *412*
Macgregor, A. M.   311, 377, 379, 388, 391, 406, 407, *413*
MacIntyre, R. M.   14, *71*, 264, 265, 271, 287, *295, 296*
McKee, E. D.   43, *69*
MacKenzie, W. S.   81, *110*
McKerrow, W. S.   260, 262, 268, 275, *297*
McKnight, E. T.   197, *216*
McManus, J.   266, *295*

Maisonneuve, J.   327, 328, 344, *409*
Martin, H.   *413*
Martin, I. D.   9, *65*
Martin, W. C.   175, *219*
Marshall, R. R.   162, 182, *218, 219*
Marvin, U. B.   434, *435*
Masuda, A.   179, 198, *219*
Mauger, R. L.   16, 26, 32, 37, 38, 41, 46, 51, 52, 53, 54, 55, 56, 58, 59, 60, 63, *66, 69*
Maurette, M.   422, 423, 428, *435*
Maurice, O. D.   228, *256*
Mayo, E. B.   57, 58, *69*
Menard, H. W.   61, 63, *69*
Mendelsohn, F.   342, 361, *413*
Mendes, F.   315, 317, 344, 358, 359, 360, 383, 384, *413*
Menzie, J. C.   131, 140, *146*
Mercy, E. L. P.   283, *297*
Mielke, J. E.   54, *69*
Miller, D. S.   185, 196, 204, *216*
Miller, J. A.   264, 265, 271, 276, 287, *295, 296, 297*, 318, 319, *414*
Miller, J. M.   27, *67*
Miller, S. L.   *218*, 220
Miller, T. G.   286, *297*
Mohler, F. L.   159, *219*
Mohr, P. A.   318, 319, *413, 414*
Monkman, L. J.   359, *410*
Monteyne-Poulaert, G.   339, 346, 393, *410*
Moorbath, S.   74, 109, 129, *145*, 159, 166, 199, *219*, 260, 263, 265, 270, 273, 282, 283, 295, *296*, 445, 450, 452, *462*
Morgan, G.   422, 428, *435*
Morgan, J. W.   164, *219, 220*
Morozova, I. M.   18, 19, 25, *67*
Mumpton, F. A.   228, 254, *256*
Murthy, V. R.   151, 161, 162, 163, 177, 182, 193, *220*

Naeser, C. W.   434, *435*
Naughton, J. J.   17, *69*
Nel, H. J.   359, *415*
Newell, M. F.   247, *257*
Ney, P.   457, *460*
Nicolaysen, L. O.   232, *256*, 263, *297*, 302, 308, 314, 315, 340, 341,

Author Index

342, 344, 346, 347, 359, 362, 365, 366, 367, 374, 375, 377, 378, 379, 393, 399, 401, *410*, *413*, *414*, 441, 442, *462*
Nier, A. O.   6, 7, 16, 25, *67*, *69*, 90, *109*, 189, *218*, 245, *256*

Obradovich, J.   19, 20, 23, 24, *66*, 89, 99, 101, 102, *109*
Oftedahl, C.   452, *462*
Olmsted, F. H.   53, *69*
Oosthuyzen, E. J.   345, 370, *414*
Oouvo, O.   175, *222*
Osmond, J. C.   43, 55, *69*
Ostic, R. G.   172, 183, 190, 201, 204, 215, *220*, *221*
Otto, J. B.   52, *65*
Otto, J. D. T.   344, 399, *415*
Ozard, J. M.   166, *221*

Page, B. G. N.   357, *414*
Pantoja Alor, J.   38, 41, *66*
Parker, R. B.   114, *145*
Parwel, A.   159, *223*
Pasteels, P.   37, 38, 42, *69*, 311, *410*
Paterson, M. S.   453, *462*
Patterson, C. C.   151, 160, 161, 162, 163, 177, 178, 179, 182, 183, 184, 193, 204, *216*, *220*, *221*, *222*, 226, *257*
Pauly, H.   159, *219*
Pellas, P.   422, *435*
Peoples, J. W.   113, *146*
Peterman, Z. E.   119, 121, 143, *144*, *146*, 189, *216*
Petrov, B. V.   103, *109*
Pettijohn, F. J.   228, *257*
Phillips, K. A.   313, *414*
Phillips, W. E. A.   262, 266, 267, 268, 274, 275, 276, *295*
Philpott, T. H.   196, *220*
Picciotto, E.   151, *220*
Pidgeon, R. T.   175, 194, *220*, 282, 283, *295*, 457, *460*
Pike, D. R.   366, *412*
Pinson, W. H.   19, 32, *68*, 166, 175, 189, 197, 200, 202, *216*, *217*, *218*, 263, 274, *296*, 379, 380, 396, *411*, 444, 450, 451, 452, 454, 457, *460*, *462*

Pitcher, W. S.   266, 283, *296*, *297*
Poldervaart, A.   114, 135, 136, *145*, 226, 227, 243, *256*, *257*
Polkanov, A. A.   175, *217*, *220*
Poulaert, G.   339, *413*
Powell, J. L.   444, 447, 448, 451, 452, 453, 454, 456, 457, *460*, *461*, *462*
Pratten, R. D.   175, *216*
Preece, E. R.   311, 318, *411*
Price, P. B.   417, 418, 419, 420, 422, 428, 429, 434, *435*
Priem, H. N. A.   58, *69*

Quennell, A. M.   343, 357, 381, *414*

Rabbit, J. C.   114, 127, *145*
Rabinovich, A. V.   182, *218*
Ragland, P. C.   195, 204, 213, *220*
Rama, S. N. I.   14, *69*
Rankama, K.   4, *69*, 227, *257*
Reed, G. W.   183, *222*
Reed, J. C.   140, *146*
Reichelt, R.   326, *414*
Reid, R. R.   114, 128, 129, 134, *146*
Reinharz, M.   306, 315, 316, 362, 363, 388, *410*
Rex, D.   314, 317, 342, 344, 362, 382, 388, 389, *410*, *415*
Reynolds, J. H.   19, 25, *67*, *69*
Rice, H. M. A.   175, *220*
Richards, J. R.   14, 16, 17, 30, *69*, 150, 175, 194, *217*, *220*, 265, 271, *296*
Rickard, M. J.   281, *297*
Riley, G. H.   263, *295*, 441, *460*
Roberts, H. R.   374, *409*
Rocci, G.   335, *414*
Roedder, E.   14, *69*
Rogers, A. S.   318, 319, *414*
Rogers, J. J. W.   195, 204, 213, *220*, *223*
Roques, M.   329, 330, 345, 352, 360, 384, 385, 387, *409*, *414*
Ross, C. P.   121, 122, *146*
Rothstein, A. T. V.   266, *298*
Roubault, M.   321, 322, 340, 344, 363, *414*
Roy, R.   228, 254, *256*

Runcorn, S. K.   402, 405, 406, *414*
Russell, R. D.   155, 157, 158, 159, 166, 171, 172, 175, 176, 185, 189, 192, 193, 200, 201, 202, 204, 212, 213, 215, *216*, *217*, *219*, *220*, *221*, *222*, 230, 232, 233, *257*
Rutherford, E.   27, *69*

Safiannikoff, A.   339, *413*
Sahama, T. G.   227, *257*
Sanders, L. D.   383, *414*
Sardarov, S. S.   89, *110*
Savage, D. E.   12, 16, 31, 53, *67*
Schilling, J. H.   53, *70*
Schmitt, H.   61, *70*
Schmitt, R. A.   457, *461*
Scholtz, D. L.   315, *414*
Schreiner, G. D. L.   6, *70*, 345, 366, 374, *409*, *414*
Schreyer, W.   13, *70*
Schuchert, C.   *70*
Schürmann, H. M. E.   319, 320, *414*
Schütze, W.   157, *221*
Schwartzman, D. W.   138, *146*
Seaborg, G. T.   6, *70*
Shackleton, R. M.   264, 266, *296*
Shatz, M. M.   162, *221*
Shaw, D. M.   166, *221*, 283, *298*
Shields, R. M.   162, *221*
Shillibeer, H. A.   18, *70*
Signer, P.   339, *413*
Signer, P.   106, *109*
Silver, L. T.   37, 38, 39, 42, *69*, *70*, 234, 247, 248, 249, 253, *257*
Silverman, A. J.   123, *146*
Simpson, E. S. W.   344, 399, *415*
Simpson, J. G.   313, *415*
Sims, P. K.   58, *70*
Sinclair, A. J.   165, 200, 215, *221*
Sinha, R. C.   283, *297*
Skinner, R.   175, *221*
Skirrow, G.   266, *297*
Slater, D.   317, 344, 388, 389 *415*,
Slawson, W. F.   156, 159, 190, 191, 196, 197, 200, 204, *218*, *221*
Smales, A. A.   9, *70*
Smit, P. J.   366, *412*
Smith, A. G.   7, 8, *70*, 265, 270, 286, *296*

Smith, C. H.   175, *221*
Smith, D. G. W.   451, *460*
Smith, G. H.   311, 318, *411*
Smith, J. V.   13, *70*
Smith, R. H.   457, *461*
Smith, W.   453, *461*
Smithson, F.   227, *257*
Snelling, N. J.   307, 311, 313, 314, 317, 318, 319, 320, 324, 326, 331, 332, 335, 339, 340, 342, 343, 344, 346, 355, 356, 357, 358, 361, 362, 368, 375, 380, 381, 382, 383, 388, 389, *410*, *412*, *414*, *415*
Sobotovich, A. V.   162, *221*
Soddy, F.   148, *221*
Söhnge   359, *415*
Sonet, J.   321, 322, 340, 344, 363, *414*
Soper, N. J.   264, 287, *295*
Sougy, J.   329, 330, 331, *411*, *415*
Spencer, E. W.   114, *146*
Stanton, R. L.   172, *221*, 261, 268, *298*
Staryk, E. E.   162, *221*
Steiger, R.   74, 82, 94, *110*
Stern, T. W.   106, *110*, 247, 254, *257*
Stevens, J. R.   18, *70*
Stille, H.   315
Stillman, C. J.   311, 313, 318, 359, 361, 380, *410*, *415*
Stockwell, C. H.   29, *70*, 175, 189, *219*, *221*, 401, *413*
Strelow, F. W. E.   366, *414*
Strock, L. W.   228, *257*
Strominger, D.   6, *70*
Sutton, J.   129, *146*, *298*
Sverdrup, H. U.   457, *462*
Swanson, G. O.   175, *221*

Talbot, C. J.   406, *415*
Tarasov, L. S.   157, 160, 182, 204, *222*
Tatsumoto, M.   178, 179, 180, 198, *221*
Taubeneck, W. H.   227, *257*
Taylor, O. J.   54, 55, *70*
Thomas, H. H.   123, *145*
Thomas, P. R.   267, *296*

Thompson, B. P. 175, 193, *218*, *222*
Thomson, R. 175, *222*
Thurber, D. L. 180, *222*
Tilley, C. E. 440, 447, *463*
Tilton, G. R. 8, *71*, 74, 80, 84, 94, 105, 109, 110, 114, 115, 117, 119, 120, 144, 156, 175, 180, 182, 183, 184, 197, *216*, *217*, *222*, 226, 229, 232, 234, 245, 249, 250, 252, 253, 254, *255*, *256*, *257*, 270, *295*, 315, 339, 343, 381, *409*, *411*, 445, 450, 451, *461*
Tipper, H. W. 175, *219*, 401, *413*
Tomisaka, T. 81, *110*
Tongiorgi, E. 363, *411*
Truswell, J. F. 307, 324, 326, 335, *412*
Tupper, W. M. 175, *222*
Turekian, K. K. 10, *70*, 457, *462*
Tuttle, O. F. 453, *462*, *463*
Tweto, O. 58, *70*

Ubisch, H. V. 159, *223*
Ulrych, T. J. 157, 165, 192, 193, 200, 201, 202, 204, 215, *219*, 375, *413*
Umbgrove, J. H. F. 58, 59, *70*

Vaasjoki, O. 175, *222*
Vachette, M. 321, 322, 326, 327, 328, 329, 330, 331, 344, 345, 350, 351, 352, 353, 360, 362, 385, 386, *409*
Vail, J. R. 317, 344, 345, 359, 380, 388, 389, *410*, *412*, *415*
Van Breemen, O. 359, *415*
Van Eeden, O. R. 359, *415*
Van Niekerk, C. B. 345, 367, 391, *413*, *414*, *416*
Van Staden, C. M. v. H. 366, *412*
Van Straten, O. J. 375, *409*
Van Wambeke, L. 311, *410*
Van Zyl, J. S. 359, *415*
Verbeek, A. A. 6, *70*
Vernon, M. J. 9, *65*
Verschure, R. H. 58, *69*
Vialette, Y. 327, 328, 344, *409*
Vinogradov, A. P. 157, 160, 182, 204, *222*

Visser, H. N. 359, 373, *415*, *416*
Vitanage, P. W. 227, *257*
Von Gunten, H. R. 339, *413*
Von Knorring, O. 314, *410*
Voshage, H. 18, *70*

Wager, L. R. 9, *70*, 266, *298*
Walker, G. P. L. 445, 450, *462*
Walker, R. D. 52, *65*
Walker, R. M. 417, 418, 419, 420, 422, 428, 429, 434, *435*
Walthall, F. G. 270, *296*, 444, 454, 457, 459, *461*
Wampler, J. M. 183, *222*
Wanless, R. K. 175, *219*, 401, *413*
Wasserburg, G. J. 8, 10, 22, 25, 33, 34, 37, 39, 40, 70, *71*, 74, 89, 105, *110*, *218*, *220*, 232, 234, 236, 240, 253, 257, *258*, 379, *416*, 438, *462*
Watson, J. 129, *146*
Webb, A. W. 175, *220*
Weber, 53
Wehrenberg, J. P. 117, 119, *145*
Welin, E. 159, *223*
Wessels, J. T. 359, *415*
Wetherill, G. W. 7, 8, 25, 40, 64, 71, 74, 80, 84, 94, *109*, *110*, 114, 115, 117, 119, 120, *144*, 175, 197, *216*, *222*, 229, 230, 234, 249, 250, 252, 254, *255*, *256*, *257*, *258*, 270, *295*, 315, 339, 343, 381, 409, 411
Whitfield, J. M. 204, *222*
Whitney, P. R. 282, *298*
Whittles, A. B. L. 157, 160, 215, *223*
Wickman, F. E. 159, *223*
Wilcock, B. 265, 270, 286, *296*
Wiles, J. W. 314, *416*
Wilgain, S. 151, *220*
Wilson, E. D. 2, 35, 36, 39, 44, 45, *65*, *71*
Wilson, J. T. 144, 185, 189, 207, 213, *216*, *221*, *223*
Wilson, M. E. 189, *223*
Wood, D. N. 359, *410*
Wood, J. A. 429, *435*
Worst, B. G. 379, *416*
Wyllie, P. J. 453, *461*, *462*, *463*

Yoder, H. S. 440, 447, *463* 287,
York, D. 14, *71*, 264, 265, 271, *295*, *296*
Zähringer, J. 19, *67*

Zartman, R. E. 176, *223*
Zulfikarova, Z. K. 103, *109*
Zykov, S. I. 157, 160, 182, 204, *222*

Subject Index

Acadian orogeny 393
  correlation 331
Addition–dilution method, for experimental determination of K 269
Africa,
  advance of cratonism 402
  ages, 800–1300 m.y. 337–347
    1300–2100 m.y. 347–370
    > 2100 m.y. 370–392
    of Precambrian igneous activity 394
    preserved 360–364
  ancient nuclei 372, 373–387
  Central, geosynclinal facet 310
  change in area of remnant stability 405
  cratons 309
  definition of events 397
  distribution of ages, 450–680 m.y. 309
    800–1300 m.y. 338, 346
    1300–2100 m.y. 348
    > 2100 m.y. 371
  early Palaeozoic history 299
  Eburnian effects 368
  geology 299
  K–Ar ages 306
  late Precambrian–early Palaeozoic orogenesis 392
  major fault systems 408
  major pre-Silurian structural pattern 400, 401
  northeastern, orogenic effects 318–320
  orogenic effects 310–331
  Phanerozoic history 299
  Precambrian 303, 304
    –early Palaeozoic orogenic zones 305, 306
    history 299
    orgenic belts 301
    –Palaeozoic cycles 302

  pre-Silurian, basement 300
    bench marks 304
    geology 299–416
  radiometric ages 302
    400–700 m.y. 306–337
  radiometric dating 299–416
    of orogenic events 391, 392
  Rb–Sr ages 306, 308
    in 1610–1770 m.y. range 352
  relation between structure and time of orogenic deformation 403
  southern, age patterns 389
    Damaran episode 308
    distribution of Upper Precambrian age rocks 310–311
    elements of stratigraphy 365
    Katangan episode 307
    Upper Precambrian 312
  stage in structural consolidation 404
  subdivision of Precambrian 304
  U–Th–Pb ages 306, 308
  western, orogeny 307
African orogenesis, vestigial facets 401
Age measurements,
  analytical procedure 79–80
  precision 113
  sample collection 79–80
Age of minerals,
  relation to distance from contact 107
  relation to metamorphic facies 106
  variation 87
Age stability, relation to metamorphic facies 106–108
Age variations as diffusion phenomena 97–106
Air, $^{40}Ar/^{36}Ar$ ratio 10
Air argon 9, 10
Alberta, Laramide plutons 58

475

16*

Alkaline rock provinces 439
Alkaline rocks,
  accessory minerals 439
  liquid inclusions 452
  origin 438, 439
    from great depths 459
      Sr isotopic composition 437–460
    $^{87}Sr/\ ^{86}Sr$ ratio 459
Alpine Fault Zone (New Zealand),
  K–Ar dating 32
  metamorphism 32
  mineralogy 32
  Paringa River granite 32
Amphiboles, retrograde metamorphism 129
Amphibolite facies metamorphism 134
Amphibolites 80–81
  Ar content 139
  melting 440
Amsaga Series,
  age 353
  correlation with Dahomeyan 353
  Rb–Sr biotite age 352, 353
Androyan System,
  age 387, 388
  Rb–Sr age 361
Angola,
  age 337
  anorthosite–norite mass, age 399
  Kibaran ages 344
  Rb–Sr age 358, 360, 384
Anomalous lead—*see* Pb
Anomalous lead line 211
Ansongo Series 324
Anti-Atlas Mountains 349
  West African Craton 336
Antilles, Laramide magmatism 58
Antler Orogenic Belt 43
Apache Group, dating 42
Apatite, excess radiogenic Sr 91
Aplite, Rb–Sr age 361
Appalachian geosyncline 130
Appalachian Mountains,
  ages 129
  orogeny 393

Appalachian Province 249
  zircon data 250
Appalachian System 36
Appalachian zone 304
^{36}Ar, as indicator of atmospheric contamination 9
^{38}Ar,
  in isotope-dilution method 269
  as isotopic diluent 10
Ar 4
  in abyssal igneous rocks 17
  age effect,
    beryl 17
    cordierite 17
  air 9
    isotopic composition 10
    in minerals 10
  air correction 10,12
    in minerals 10
  atmospheric 10, 29
  atmospheric correction 10
  built in age 30
  concentration gradient 30
  continuous leakage model 35
  determination, mass spectrometer 9
  diffusion coefficient 99
  diffusive losses 20, 21, 22
  effect of metamorphism 32
  episodic loss model 35
  excess 6, 13, 14, 29, 31
    in abyssal environments 63
    in amphiboles 139
    anomalous ages due to 32
    in deep-seated environments 17
    in hydrothermal veins 15
    in intrusive igneous rocks 15
    in margarite 13, 14
    in minerals, 12, 14, 63
    in Precambrian minerals 29
    problem 12
  experimental determination 269–270
  external pressure 29
  extraneous 12–18
    in feldspar 11
  ^{4}He as indicator of excess 17
  inherited 13, 30
    in amphiboles 139

Subject Index

for interpretation of cooling history 63
isotopic abundance 4, 5
isotopic composition, in atmosphere 10
lithospheric 10
measurement, effect of contamination 12
in mica lattice 277
occluded 13
open system 29
partial pressure under abyssal conditions 25
in Phanerozoic rocks 30
in phlogopite 30
in Precambrian feldspars 12
production, by radioactive decay 13, 17, 22
radioactive decay 6
rate of diffusion 20
relative abundance 3
secular equilibrium 28, 30
thermal diffusion 19
transient equilibrium 27, 30
in the 'unsteady zone' 25
in xenoliths 30
Ar activation, effect of crystal chemical transformations on 22
Archaean 115
anomalous K–Ar ages 17, 18
sediments 135
Arctic Ocean 44
Ar degassing,
in Pleistocene rocks 31
rate 10
Ar diffusion 25
effect of defect structures 25
effect of desorption 25
in feldspars 25
Arenig, isotopic age of the base 265
^{40}Ar error, in minerals 13, 14
Argillite zone, K–Ar dating 32, 33
Arizona,
Bagdad region 37
Basin and Range—see Basin
Catalina gneiss, 247
zircon data 248
copper porphyry plutons 58
Cretaceous 44

Dragoon quadrangle 37
Glance conglomerate 44
Grand Wash cliffs 37
Johnny Lyon granodiorite 247
zircon data 248
Mazatzal Mountains 37
Nevadan orogeny 44
Oracle 48, 50
Palaeozoic 39
Phanerozoic magmatism 45
Pinal Mountains 37
Precambrian 35–37
Rincon Mountains 50
Roskruge Range 48
K–Ar dating 48
Rb–Sr dating 37
Safford Dacite Neck 47
Safford Peak 47
Santa Catalina Mountains—see Santa
Tortillita Mountains 50
Tucson Mountains—see Tucson
Arizonan Revolution 37
Arkansas, Ozark Dome 196
^{40}Ar/^{40}K ratio 8
chemical alteration 18
determination 9–12
effect of base exchange 19
effect of high-grade metamorphism 31
effect of induced nuclear reactions 18
effect of mechanical abrasion 18
effect of thermal diffusion 18
in potash feldspars 11
precision of measurement 11
time of refrigeration 27
Ar leakage 23
Ar loss,
in biotite 23, 153
from crystal damage in mica 18, 19
in fault zones 19
as function of temperature 24
in hornblende 23
Ar retentivity 23
under high-temperature conditions 26
under low-temperature conditions 26

after magma extrusion 15
  in mica 43
  in xenoliths 12
Aruba, Laramide plutons 58
Ascension Island,
  alkali granites,
    Rb content 451
    Sr content 451
  alkaline rocks 450
  igneous rocks 451
  Pb from 180
  Sr isotopic composition 450
  trachyte 450
  xenoliths 450, 451
Atacorian Series 324
Atlantic Ocean, isotopic composition of Pb 180, 182
Atlas Mountains 349
Australia, Broken Hill—see Broken Hill
Azguermerzi granite, Rb–Sr ages 354
Azores, Sr isotopic composition 450

Bakel Series, Rb–Sr ages 329
Bald Mountain Batholith (Oregon) 227
Baltic Shield 157
  as an example of an isochron 157
  ordinary Pb 191
Baoulé granite 350, 387
  effect of Eburnian orogenic episode 350
  Rb–Sr age,
    biotite 352
    associated pegmatite 350
Barberton Mountainland,
  ages 373
  schist belt sequence 373
Basalt,
  alkaline 440, 449
  melting 440
  oceanic 438
  Pb 181
  Pb isotopes 178
  Sr isotopic composition 444–450
  tholeiitic 449
Basaltic lava, postorogenic extrusion 55

Base exchange in mica 280
Basement Complex (Nigeria), K–Ar age 326
Basin and Range area 46
Basin and Range orogeny 45
Basin and Range Province 1, 2
  Antler orogenic belt 43
  copper mineralization 53, 54, 55
  Cretaceous 44
  date of basalts 54
  as part of East Pacific Rise 63
  Eocene 55
  faulting 47
  K–Ar data 46, 56
    histograms 52
  Laramide 54, 55
  magma 54
  magmatic quiescence 55
  magmatism 54, 55
  Manhattan geanticline 43
  Mesozoic 44
  Miocene 55
  in Oregon 52
  orogeny 55
  Palaeozoic 43
  Phanerozoic history 43–45
  Pliocene 55
  plutons 52
  rift faulting 52
  tectonism 55
  Tertiary 45, 63
  topographic setting 46
  Turonian–Hemphillian 53
  volcanic rocks 52
Basin Range 58
  magmatism 2
  ore deposits 2
Bathurst (New Brunswick) 172
  galena 183
Beartooth Range 114, 115, 134–138, 241, 247, 253
  concordia plot 134
  episodic Pb loss interpretation 135
  K–Ar data 134
  Minnesota compared with 247
  Precambrian 120
  sediments 243, 245

Subject Index

Stillwater complex 137
 U–Pb data 134
 uraninite 247
 zircon data 243
Belt sedimentary rocks, time of mineralization 122
Belt sediments, deposition 122
Belt Series 123
 Algonkian 121
 in Coeur d'Alene 122
 deposition 143
 glauconite 123
 La Hood Formation 122
Beryl,
 age effects 17
 excess ^{40}Ar 14
 K–Ar dating 13
Betafite, U–Th–Pb age 318
Bighorn Range (Wyoming) 143
Biotite 81
 age 85–87
 in Rhodesia 362
 Ar loss 23, 153
 colour change 81
 contact, diffusion in 98–99
 diffusion coefficient,
  of Ar 99
  of Sr 99
 diffusion data 100
  Ar 102
 diffusion loss of Sr 86
 diffusion results, laboratory measurement 99–102
 generation 153
 grain-size effect 85
  experimental demonstration 86
 K–Ar age 43, 84, 153, 237
  Nigeria 325, 326
 Rb–Sr age 84
  Baoulé granite 352
  Bondoukou granite 352
  Pharusian granite 327, 328
  variation 88
 secular equilibrium 32
 thermal diffusion of Ar 63
Birrimian System 350, 398
 biotite Rb–Sr age 350
 minimum age 398
 U–Th–Pb age 350

Black Hills (South Dakota) 114
 Bob Ingersoll Mine, mineral ages 114
Blacktail Range 143
Bluebell Mine—see British Columbia
Bondoukou granite 350
 discordant ages 352
 mineral–whole rock age 352
Bonne Terre leads 196
Botswana,
 age determination on Gaberones-type granite 376
 anorogenic effects 337, 345
 Gaberones granite, age 367
 K–Ar age of dolerite 345
 Precambrian stratigraphy 364
 Shushong sill 332
 stratigraphy 376
Boulder Batholith (Montana) 123, 124, 129
 emplacement 143
 mineralization near 124
Bourré gneiss (Mali) 324
Bouvet, Sr isotopic composition 450
Brahma Schist 42
British Columbia,
 Bluebell Mine 157, 158
  Comfort zone 165
  Kootenany Chief zone 165
 Laramide magmatism 58
British Isles, Precambrian sedimentary sequences 285
Broken Hill (Australia) 192–195
 age determinations 194
 galena 192
 Pb standard 173, 176, 186
Buem Formation 324
Buganda Group, K–Ar measurement 356
Buganda–Toro Belt, orogenic effects 349
Buganda–Toro folding 354
Buganda–Toro Systems 355
 Karagwe–Ankolean relation with 356
Bukoban System (Kenya) 332
Bulawayan System,
 K–Ar age 377
 strata 377

Bulugwe Series 355
Burundi, Rb–Sr ages 339
Bushveld complex 364, 369, 379
   age 366, 394, 395, 396, 398
   age of emplacement 366, 368, 394, 395
   effects of emplacement 370, 372
Buteba granite 382
   K–Ar age 357
   Rb–Sr–age 357

^{40}Ca, relative abundance 3
Ca,
   in igneous rocks 3
   isotopic abundance 4
Calcite, fission tracks 420
Caledonides, Scottish,
   cooling hypothesis 287
   K–Ar age 287
   overprinting hypothesis 287
Cambrian, Flathead quartzite 126
Cameroons 360
   ages 349
   ancient ages 385
   Congo Craton zone 336
   Ntem metamorphic complex 398
   orogenic activity 323
   orogenic effects 320–322
   Pb/α age 384, 385
   Precambrian 400
   Rb–Sr age 321, 335, 360, 384, 385
   vestigial facet 310
Campanian, magmatic quiescence 54
Canada, Cobalt–Noranda area 187–191
Canadian Shield 143, 144
   classification 189
   continental growth 144
   Superior Province, age 143
Cape Fold Belt, rejuvenated facet 310
Cape granite,
   Rb–Sr isochron age 315
   Rb–Sr whole rock isochron 334
   zircon age 315
Cape Province,
   age of pegmatitic uraninite 340
   geosynclinal facet 310
   Kheis System 341

   minimum age 399
   Malmesbury System 315
Carbonate melts 453
Carbonates, CaO–C$_2$O –H$_2$O system 453
Carbonatites,
   accessory minerals 453
   African 454
   complex transfer during differentiation 453
   ijolite association with 439
   intrusive sequence 453
   major minerals 439, 452
   origin 439
      by remobilization of sedimentary carbonates 439, 453
      Sr isotopes applied to 452–457
   pyroxenite association with 439
   ^{87}Sr/^{86}Sr ratio 454
Carn Chinneag intrusion 283
Cashel (Ireland), major folding 267
Cenozoic,
   early 54
   isotopic composition of air Ar 10
   K–Ar dating 8
Cenozoic magmatism 45
Cenozoic minerals,
   excess ^{40}Ar 17
   excess ^{4}He 17
Cenozoic orogeny 61
Cenozoic sediments, in Mexico 63
Central Africa, orogenic effects 310–311
Central African Republic 363
   ages 337
   in 800–1300 m.y. range 344
   Congo Craton zone 336
   Lower Precambrian amphibolite age 340
   orogenic effects 320–322
   Ouango massif 339
   Precambrian, minimum age 400
   Rb–Sr age 321, 335, 344
   vestigial facet 310
Cherry Creek gneiss, initial ^{87}Sr/^{86}Sr ratio 132
Chilwa Island, alkaline intrusives 453
Chlorite, excess ^{40}Ar 16

Subject Index

Chlorite schist zone, K–Ar dating 32, 33
Clocks, radioactive—*see* Radioactive
Closed systems 6, 149, 155
Cobalt–Noranda area (Canada) 187–191
  anomalous Pb 204
  geological features 188
Coeur d'Alene, Pb isotope composition 123
Colorado,
  Bryan Mountain 83
  Duluth gabbro 106
  Front Range—*see* Front Range
  Ute Mountain 82, 83
Colorado mineral belt, Laramide plutons 58
Colorado Plateau 1, 2, 55
  as part of East Pacific Rise 63
  Triassic Chinle Formation 45
Columbite–tantalite pegmatites, in Nigeria 324
Concordia diagram 229, 230, 231, 232, 233, 236, 241, 247, 252
  for Beartooth Range 134
  diffusion curves 235
  for zircon 136
Congo,
  Dibaya-type basement 391
  Kibali Group 354
  Luiza Series 391
  Luiza-type basement 391, 398
  Lukoshi Formation 398
  maximum age of Bukoban System 347
  minimum age for systems 347
  Ruzizi Belt 343
Congo Craton 309, 336, 337
Connemara (western Ireland),
  geochronological studies 259–295
  geochronology 287
  geology 260
  granite intrusions 268
  K–Ar age, 272–273
    pre-Arenig overprinting 276
  mineral ages 271–274
  Newer Granites 282
  retrogressive metamorphism 275
  $^{87}Sr/^{86}Sr$ initial ratio 281
Connemara antiform 268, 269
Connemara metasediments,
  age of deposition 282–286
  Rb–Sr isotopic analysis 282
  Rb–Sr whole rock age 283, 284
Connemara Migmatites,
  description 294
  K–Ar hornblende age 263
Connemara Schists 259, 261, 262
  age of metamorphism 270, 285
  K–Ar ages 261, 265, 270–277
  mineral age, interpretation 281, 286
  Rb–Sr ages 261, 277–282
  Scottish Dalradian correlation 286
Connemara Series 285
  age of deposition 284
  description 290–295
  K–Ar ages 274
    comparison with Rb–Sr ages 280
Contact metamorphic zone,
  age results 84
  mineral ages 86
Contact metamorphism,
  isotopic mineral age variations induced by 73–108
  petrologic changes induced by 73–108
  time 85
Continental growth, by lateral accretion 144
Continuous diffusion Pb loss 234, 235, 236, 253—*see also* Pb loss
Copper mineralization, relation to Laramide magmatism 55
Cordierite,
  age effects 17
  excess ^{40}Ar 14
  K–Ar dating 13
Cordillera, Laramide magmatism 58
Correction common Pb, 122
Corvock granite 286
  biotite–whole rock age 282
  K–Ar age 276–277
Cosmic ray bombardment 18
Cosmic ray primaries 425
Cratonism, advance in Africa 402
Cretaceous, Maestrichtian 54

Cretaceous magmatism 43
Crustal rigidity, propagation of faults due to 55
Cyrenaica, K–Ar ages 320
Cyrtolite ages of Kivu 339

Dahomey,
  orogenic activity 323
  Rb–Sr age 325, 364
Dahomeyan,
  age 325
    of Kusheriki granite 325, 326
    Amsaga Series correlation with 353
    discordant U–Th–Pb age of monazite 326
    K–Ar hornblende age 326
    minimum age 400
    in Nigeria 324
    Rb–Sr age, lepidolite 326
      muscovite 326
Dakota, Black Hills—see Black Hills
Dalradian,
  metamorphosed—see Connemara Schists
  Rb–Sr ages 263
  Scottish, K–Ar age, 264
    Rb–Sr age 264
Dalradian metamorphism,
  K–Ar age of biotite 264
  main 264
  pre-Arenig age 264
Dalradian schists 262
Dalradian sediments, $^{87}Sr/^{86}Sr$ ratio 264
Dalradian sequence, 263
Dalradian Series,
  age 259
  age of deposition 284
  Connemara region 259
  folding 259
  Galway granite 261
  granite intrusion 261
  Lower Dalradian 261
  major metamorphism 265
  metamorphism 259
  Middle Dalradian 261
  Oughterard granite 261
  Scottish 285, 287
  total thickness 259
  Upper Dalradian 261
Damara Facies 314, 315, 336
  Rb–Sr ages 314
  U–Th–Pb age 314
Damaran episode 316, 334, 335, 336
  age data 318
Damaran–Katangan orogeny,
  date 392
  zones affected 335, 336
Damaran orogenic episode 323, 402
Damara System 346
Darwin Rise 61, 62
Dating of ancient sedimentary rocks 121
Daughter–parent ratio 8
Davidite,
  $^{207}Pb/^{206}Pb$ age 192
  U–Th–Pb age 318
Decay, radioactive 21
Decay constant 5, 21
  determination 6–9
  for ^{40}K 7
Decay system,
  ^{238}U–^{206}Pb 148
  ^{235}U–^{207}Pb 148
  ^{232}Th–^{208}Pb 148
Deterministic model,
  for K–Ar interpretation 26–33
Devonian,
  Mauritania 329, 330
  Spanish Sahara 329, 330
Diamonds, in ultrabasic igneous intrusives 196
Dibaya-type basement, age 398
Diceratherium 51
Diffusion,
  Fick's laws 20
  slab model 23, 25
Diffusion models 235
Dirac monopoles 422
Discordant minerals,
  age relations 230
  ages 73
Discordant zircons,
  concordia plot 245, 249
  studies 236–252

Subject Index

Dodoma–Nyanza nucleus 370, 381–383
  Precambrian 382
Dodoma System, Rb–Sr lepidolite age 381
Dolomite,
  detrital minerals in South African 455
Dominion Reef sedimentation, maximum age 375
Dominion Reef System 373, 375, 376
  ages 397
    auriferous conglomerate 375
    detrital U-bearing minerals 391
  minimum age 369
  Rb–Sr whole rock age 375
Donegal, Dalradian sequence 263

Earth, age 161, 192
Earth's core, growth 406
East Africa,
  Kavirondian System age 398
  Nyanzian–Kavirondian Systems ages 390
  Nyanzian System age 398
  vestigial facet 310
East Pacific Rise 61–63
  Basin and Range Province 63
  Colorado Plateau 63
  topography 62
  Turonian 63
Eburnian effects in Africa 368
Eburnian orogenic episode 349, 353, 387, 393, 394
  date 392
  effects 401
Eburnian tectonothermal event, age data 350
Eclogite,
  melting 440
  trace Pb 182
Egypt,
  Dokhan Series 319
  Fundamental Complex 319
  Gattarian granite series 319
  geosynclinal facet 310
  Hammamat Series 400
  Mitiq Series 319
  Old Paraschist Series 319
  pre-Carboniferous rocks 319
  Shadli Series 319
  Shaitian granite series 319
Eldora (Colorado) 74, 82, 107
  Tertiary intrusives 77
Eldora contact zone, zircon ages 97
Eldora Stock 74, 76–77, 79, 126
  composition 77
  configuration 77
  contact metamorphic effects 126
  effects,
    on Ar systems 126
    on Sr systems 126
  feldspar types 83
Eocene, magmatic quiescence 54
Epeirogenic versus orogenic cycles 59–61
Epicontinental seas, extent 60
Epidote, excess radiogenic Sr 91
Epimetamorphism 319
Episodic lead loss 236, 239, 253— see also Pb loss
Epomeneo tuff, analysis of feldspar 11
Essexite, initial $^{87}Sr/^{86}Sr$ ratio 454
Estherville meteorite 424
  tracks 425
  Type I tracks 424
Ethiopia,
  Boje Series age 400
  geosynclinal facet 310
  K–Ar age 319
  orogenic effects 318–320
  Precambrian 318

Feldspar 81–84
  activation energy 104
  ages 89–94
  analysis 12
  areal distribution 82
  Ar loss 252
  Ar retentivity 25
  contact 103–105
  diffusion of Ar 25
  diffusion coefficient for Pb 104

diffusion model 103
effect of base exchange 25
exchangeable K 25
K–Ar age 12, 89
model Pb age 93, 103
nature of Ar 90
Pb isotope composition 184
pegmatitic 91
  amount of Pb 91
  Ar loss 18
perthitic, release of Ar 18
  Ar error 13, 14
plagioclase, K–Ar age 15
potash, K–Ar age 89
  model lead age 89
  as Palaeozoic geochronometers 15, 16
  Pb content 184
  Pb ratio 92
  as Precambrian geochronometers 15, 16
  Rb–Sr age 89
pre-Mesozoic, Ar content 16
Rb–Sr age 84, 90
Rb–Sr isochron 91
Sr loss 252
trace Pb 182
xenolithic, ^{40}Ar retention 12
Feldspar lead system, effect of metamorphism on 91, 92
Fick's laws of diffusion 20
Finland,
  Bodom granite 252
  discordant zircon results 252
  Onas granite 252
  Pb isotope ratios 191
  Pb ores 172
  Rapakivi granite 252
  zircon data 251
Fission-track ages, comparison with known ages 419
Fission-track dating,
  choice of materials 420
  experimental procedure 419
Fission-track etch pits, geometry 418
Fission-track method, application to meteorites—*see* Meteorites
Fission-track technique 422

Flame photometer, determination of K 269
Fluorite, ^{40}Ar error 13, 14
Fold belt, idealized section 305
Frequently mixed lead model—*see* Pb model
Front Range (Colorado) 74, 75, 85
  geologic setting 74–76
  Idaho Springs Formation 74, 77
  —*see also* Idaho

Gabbro,
  initial $^{87}Sr/^{86}Sr$ ratio 454
  melting 440
Gaberones-type granite,
  age determination 376
  Rb–Sr whole rock age 375
Gabon 360
  ages 349
  Pb/α age of zircon 344
  Rb–Sr age 360
Gabon–Cameroons nucleus 370, 384–385
  Ebolowa granite 385
Galena—*see also* Pb
  ages 322, 323
  Broken Hill 192
  common Pb correction 122
  isotopic analysis 201
  isotopic composition 172
  model Pb age 379
  Pb isotope ratio 165, 183, 186
Gallatin Canyon (Montana) 130
  K–Ar biotite age 142
  K–Ar hornblende age 142
Galway granite 286
  age 268
Geochronology, use of charged particle tracks 417–435
Geochronometer 5
  ideal 18
  perturbation 18
Geological Survey, U.S. 36
Geologic model 26, 27
Geosynclinal facet, strata involved 333
Ghana,
  orogenic activity 324
  Tarkwaian sequence 352

Subject Index 485

Glauconite,
  Belt Series 123
  diffusion of Ar 22
Gough Island,
  Pb 180
  Sr isotopic composition 450
Grand Canyon disturbance 39, 43
Granite,
  alkaline 439, 451
  dating 373
  initial $^{87}Sr/^{86}Sr$ ratio 282
  Old, Rb–Sr age 375
  trace Pb 182
Gravelly Range (Montana) 130
  detailed geological study 114
Great Dyke (Rhodesia),
  K–Ar age 379
  Rb–Sr age 379, 380
  whole rock isochron age 379
Greenland,
  East, Sr isotopic composition 450
  Ivigtut 157, 158, 159
  Mesozoic 60
Grenville orogeny 201, 207, 410
Grenville Province (Canada),
  $^{87}Sr/^{86}Sr$ ratio 457
  as vestigial facet 401
Guinea 329

Hawaii,
  Pb 198
  Pb isotope ratios 178, 179, 181
  $^{87}Sr/^{86}Sr$ ratio 445
Hawaiian Islands, Sr isotopic composition 447, 448
He,
  age effect, in beryl 17
    in cordierite 17
  excess 17
  radioactive decay to 17
Heat flow calculations 98
Heat flow model 77
  geometry 79
Heat flow theory, contact temperatures 77–79
Hercynian cycle 302
Hercynian orogeny 331

Hispaniola, Laramide plutons 58
Hoggar,
  ages 337
  interpretation of age data 328
  Nigritian 324
  orogenic activity 323
  $^{207}Pb/^{206}Pb$ zircon age 326
  Pharusian System 324, 400
  Precambrian 363
  Rb–Sr ages 326
    Kibaran 344
  Suggarian 324, 333
    minimum age 398
  tectonic 'horsts' 325
  vestigial facet 310
  West African Craton 336
Holmes–Houtermans common lead development curve 115
Hook Batholith, ages 313
Hornblende,
  activation energy 103
  ages 87
  Ar loss 23
  Ar retention 89
  contact, diffusion 102–103
  diffusion coefficient of Ar 104
  heating experiments 88, 89
  K–Ar age 38, 84, 87, 237, 263
  use in K–Ar dating 16
  variation in age of 87
Hornblendite, melting 440
Houtermans' isochron 150, 158
Huabian episode 362

Iceland,
  Sr isotopic composition 450
  $^{87}Sr/^{86}Sr$ ratio in granophyre 445
Idaho Springs Formation 85, 97
  ages 84
  composition 77
Igneous rocks, use in K–Ar dating 16, 17
Ijolites, association of carbonatites with 439
Inda Ad Series (Somali Rep.) 319
Inert gases,
  in mantle 17
  mobilization 17

Iowa 44
Ireland,
  Dalradian Series 259
  western, dolerite sill intrusion 266
    geochronological studies 259–295
      isoclinal folding 266
      late metamorphism 268
      Maam Valley faults 269
      metamorphism 266
      Ordovician 268, 269
      Oughterard granite—see Oughterard
      pegmatite development 268
      Rb–Sr age of Dalradian 263
      sequence of events 266–269, 288–289
      Silurian 268, 269
Islay, Dalradian sequence 263
Isochron 155, 161, 170
  meteoritic 161, 171
  $^{206}Pb/^{204}Pb$ 155
  $^{207}Pb/^{204}Pb$ 155
  Rb–Sr 270
  single-stage 157, 158
    experimental results 206
  Sr 200
  whole rock Rb–Sr 282
  zero 161
Isochron equation 158
Isochron method, Rb–Sr ratios 442
Isochron plot 443
Isotope-dilution method 79, 183, 269
  determination of Rb 269
  determination of Sr 269
  use of ^{38}Ar 269
Ivigtut (Greenland) 157, 158, 159
  age of the crust 212
  Pb analysis 161
Ivory Coast,
  Eburnian orogenic episode 349
  Pb/α zircon ages 387
  Precambrian 350
  Rb–Sr age 350, 351, 352
  West African Craton 336
Iwo Jima, Pb 180

Jamaica, Laramide plutons 58
Johannesburg, Transvaal System 367
Joplin leads 196

^{40}K 3, 4
  β-emission 3
  decay constant 7, 113, 269
  determination, accuracy 9
    addition–dilution method 269
    chemical 5
    flame photometry 9, 269
    isotope-dilution technique 5
    neutron-activation technique 5
  electron capture 3
  half-life 3
  isotopic abundance 4, 269
    determination 6
  isotopic composition 3
  physical constants 7
  radioactive decay 6
  specific β-emission 9
  specific γ-emission 9
  transmutation to ^{40}Ar 3
  transmutation to ^{40}Ca 3
Kalahari Craton 309, 336, 337
Kampala–Mubende region, Rb–Sr age 355
Kangeralussuaq, $^{87}Sr/^{86}Sr$ ratio 447
K–Ar age 194, 195
  African 306, 307
    > 200 m.y. 303
    400–800 m.y. 307, 308
  amphibolite 319
  anomalous Archaean 17, 18
  Basement Complex 326
  biotite, cordierite–biotite gneiss 355
  Gaberones granite 367, 368
  hornblende—see Hornblende
  Kibara Mountains 339
  Masaba granite 356
  Mubende granite 356
  Nigerian charnockite 326
  posttectonic gabbro 264
  in Rhodesia 362
  Stuhlmann Pass Series 355
Buganda Group 356
Bulawayan System 377

Subject Index

Buteba granite 357
chlorite from gneiss 319
Connemara 271
Connemara Schists 261, 265, 270–272
Connemara Series 274
Corvock granite 276–277
Cyrenaica 320
Dahomeyan biotite 325
Damaran sequence 334
discordancy 19
feldspar 16, 317, 381
Fundamental Complex 319, 320
Gattarian granite 320
hornblende—*see also* Hornblende
    Gaberones granite 367, 368
    gneiss 319
        Nigerian charnockite 326
        syenite 326
Karagwe Ankolean 339
Kate granite 357
Kisii basalts 331
Kubuta pegmatite 374
Mbozi syenite–gabbro complex 332
metamorphic crystalline rocks 319
mica 245
    augen gneiss 319
    effect of base exchange 19
    muscovite gneiss 319
    post-Shamvaian pegmatites 379
    quartz diorite 319
Mitiq–Atalla sequence 320
Monteregian Hills 451, 452
Mozambique Belt 317
muscovite 249
    pre-Katanga System 361
Nagongera granite 357
Nigerian muscovite 326
Nkumbwa Hill carbonatite 332
pegmatitic mica 317
pegmatitic minerals 137
pegmatitic muscovite 354
perthitic feldspars 25
phyllite 319
plagioclase feldspar 15
pre-Nama System 341
pyroxenite whole rock 319

quartz–muscovite metasediments 339
Rb–Sr ages compared with 280
refrigeration 29
Scottish Dalradian 264
Shaitian granite 320
Sierra Leone 330
slate 275–276
slate mudstone 339
Somali Rep. 319
Suggarian 363
Ubendian–Ruzizi Belt 343
Uganda 332
volcanic K feldspar 16
weathering effect 19
Zambesi Belt 317
zircon 247, 255
Karagwe Ankolean cycle 302
Karagwe Ankolean Systems 302, 346
    K–Ar age 356
    Rb–Sr whole rock age 356
    sediments 340
    ages 339
K–Ar analysis, Black Hills 114
K–Ar clock,
    in biotite 63
    departure from ideal behaviour 15
    in muscovite 63
    radioactive 39
    resetting during thermal event 43
K–Ar data 143
    Basin and Range Province 55
    Cretaceous–Cenozoic 53
    Laramide plutons 57
K–Ar date 8
    Alpine Fault 32
    Arizona, effect of Cretaceous–Cenozoic events 39
    biotite 43
    built in age 28
    Cenozoic 8
    effect of diagenesis 23
    effect of errors 8
    effect of metamorphism 23
    hornblende 38
    mica 26
    Oracle granite 50

Pleistocene 8
Precambrian mica 40
  relation to geologic event 6
  Sierra Ancha Mountains 40
  Tertiary 8
K–Ar dating 153, 189
  accuracy 9, 10
  application to Basin Range 1–64
  beryl 13
  cordierite 13
  excess ^{40}Ar effect 14
  hornblende 16
  igneous rocks 1–64
  late Mesozoic–Cenozoic events 45–58
  Mesozoic 59
  metamorphic rocks 1–64
  meteorites 63
  mica 16
  physical constants 9
  Pleistocene 12
  Precambrian 14
    of Arizona 39–43
  Precambrian micas 39
  volcanic glass 19
  Volcanic rocks 19
K–Ar measurements,
  age of slaty cleavage 275
  Karagwe Ankolean 356
K–Ar method 3–35
  applications 3
  limitations 2, 63
K/Ar ratio,
  determination of errors 269
  feldspar 245
  refrigeration 29
K–Ar results, interpretation,
  deterministic models 26–33
  probabilistic models 33–35
  two-stage geologic model 26
Karroo 341, 342
  vulcanism 394
Kasai (Congo), Rb–Sr age 384
Kasai nucleus 370, 383–384
  Dibaya-type basement 383
  Luiza-type basement 383
Kasai Province 358–359
Katametamorphism 319
Kataga, uraninite ages 335

Katangan cycle 302
Katangan orogenic episode 306, 316, 334
  age data 335
Katanga Province 333
  Kibara–Karagwe Ankolean Belt 338
  orogenic effects 311
Katanga System 311, 338, 400
  maximum age 347
  mineralization 311
Kate granite 357
Kavirondian System 355
Kenya,
  Kisii basalt K–Ar age 331
  Mozambique Belt 317, 333
  Pb–U monazite age 344
  Zambesi Belt 317
Kerdous massif,
  age studies 353, 354
  Rb–Sr age 387
Keshya limestone (Zambia) 455
K feldspar—*see* Feldspar
Kheis System 341, 346
Kibalian folding 354
Kibalian orogenic episode 354
Kibali–Buganda Toro Belt 354–357
Kibali Group, minimum age 398
Kibara Belt, Rb–Sr age 339
Kibara Group 401
Kibara–Karagwe Ankolean Belt 337–340
  age 346
Kibaran ages 337
  in younger orogenic zones 343–345
  pegmatite 332
Kibaran deformation 311
Kibaran orogenic episode 347, 393, 401
  date 392
  effects 337
Kibara System 346
Kibara–Urundi–Karagwe Ankolean Groups 396
Kilembe Mine (Uganda), K–Ar age 340
Kinkéné granite 352
K isotopes,

fractionation 6
in sedimentary country rock 6
in xenoliths 6
Kivu, ages 339
K/Rb ratio 445, 447

Labbazenga Series (Mali) 324
Laramide 74, 75, 237
 Basin and Range Province 54
 felsic porphyry dikes 243
 Fort Union Formation 54
 Laramie Formation 54
 as a pulse of the Earth 58, 59
 Wasatch Formation 54
Laramide batholiths 45, 112
Laramide dikes 45
Laramide granite 47
Laramide intrusives 111, 113, 123–127
 age 124
Laramide magmatic pulse, Gaussian distribution of data 55
Laramide magmatic rocks 55
Laramide magmatism 55, 58
Laramide orogeny 45, 393
 retreat of epicontinental seas 44
Laramide plugs 45
Laramide plutons,
 in Alberta 58
 Aruba 58
 Colorado mineral belt 58
 Hispaniola 58
 Jamaica 58
 K–Ar data 57
 Puerto Rico 58
 Yukon 58
Laramide regional metamorphism 85
Laramide Revolution 243
Laramide sequence 47
Laramide volcanic rocks 45
Laxfordian metamorphism 129
Lead—see Pb
Lepidolite, Rb–Sr age 317
Limestone,
 ancient, deep-seated 456
 Songwe Scarp 456
 South African 455

Sr content 455
Sr isotopic composition 454
Limpopo Belt 359, 376
 K–Ar age 359
 orogenic effects 349
 U–Th–Pb age 359
Limpopo cycle 302
Little Belt Mountains (Montana) 122, 138–140, 143, 237, 239, 240, 243, 247, 253
 episodic Pb loss interpretation 139
 K–Ar ages 120, 121, 139
 K content of amphibole 139
 metamorphic event 139
 Rb–Sr data 121
 U–Pb data 121
 zircon data 238
Loskop System 376
Lough Wheelaun intrusion 292
Lower Cambrian cycle 302
Lower Congo 315, 333
 ages 337, 390
 geosynclinal facet 310
 Mayumbian Belt 362
 orogenic effects 310, 311
 Pb/α zircon age 344
 Rb–Sr age 344
Luiza Series,
 minimum age 398
 orogenic effects 349
 sequence 358
Luiza Series of metasediments 383
Lukoshian Belt,
 age 396
 orogenic effects 349
Lukoshi Formation 358
Lukoshi Region 358–359
Lunya district, Rb–Sr age 355
Lunyo granite (Kenya), K–Ar age 382
Lusaka granite, age 313
Lwanda unconformity 356

Madi Quartzites 355
Madison Range (Montana) 130, 143
Maestrichtian 44, 54
Magma,
 alkaline differentiates 459
 degassing 15

Magmatic pulses, analysis 61
Magmatism, correlation between epicontinental seas and 59
Malagasy Republic 361
  ancient ages 387, 388
    Androyan System 322
      minimum age 398
    Androyan–Vohibory Systems 390
    Cipolin Series 322
    discordant ages of alluvial monazite 287–388
    geology 322–323
    Graphite System 322
    orogenic effects 307, 323
    Quartzite Series 322
    Rb–Sr age 323
    Sahatany System 322
    thorianite ages of pre-Karroo 323
    vestigial facet 310
    Vohibory System 322
    zircon ages 322, 323
Malawi 317
Mali,
  Ansongo Series 324
  Bourré gneiss 324
  correlation of strata 326
  Labbazenga Series 324
  orogenic activity 323, 349
Malmesbury Beds 336
Malmesbury orogenesis 315
Malmesbury sediments 315
  Rb–Sr isochron 334
Malmesbury System 315
  Rb–Sr isochron 315
Man charnockites 387
Manganese nodules, isotopic Pb ratios 177
Man granite,
  Pb/α zircon age 350
  Rb–Sr biotite age 387
  zircon age 387
Manhattan geanticline 43
Manitouwadge, anomalous Pb 172
Mantle 185
  concentration of complex ions 459
  outgassing products 459
  ratio of inert gases to active volatiles 17

Mantle convection 402, 406
Masaba granite,
  K–Ar biotite age 356, 357
  Rb–Sr biotite age 356
Mass spectrometer,
  gas-source 203
  A.E.I. MS–2 269
  Reynolds-type 269
  Sr determination 441
  solid-source 156
Matsap System 373
Mauritania,
  Amsaga Series 352, 385, 386
  Bakel–Akjoujt Series 352
  fossiliferous Devonian 329, 330
  orogenic effects 349
  Rb–Sr ages 330, 386
  Sattle Ogmane, Rb–Sr biotite age 385, 386
    Rb–Sr lepidolite age 386
  Tasiast Group 385
    age 385, 386
    amphibolite grade metamorphism 391
    minimum age 391, 398
  West African Craton 336
Mauritania nucleus 370
  Precambrian 385
Mauritanide zone 329
  correlations with 331
  Falemian rocks 329
Mayumbe Belt, Rb–Sr data 335
Mayumbian orogeny 362
Mazatzal Revolution 36, 37
Medicine Bow Range 143
  ages 120
  geology 114
  Rb–Sr isotope composition 115
  Rb–Sr whole rock age 134
Mesozoic,
  of the Basin Range 2
  K–Ar dating of events 59
Mesozoic–Cenozoic,
  epicontinental seas 62
  K–Ar dating 45–58
Mesozoic orogeny 45, 61
Messina Formation, minimum age 398

Subject Index 491

Metamictization 245, 252, 254
  as a factor in zircon Pb loss 234
  effect in age determination 228
  in zircon 228
Metamorphic history of an area 73
Metamorphism,
  Ar loss 252
  contact—see Contact
  Pb loss 252
    episodic 253
  Sr loss 252
Metasediments, Precambrian 74
Metavolcanic rocks, Precambrian 74
Meteorites 160–164
  age 162
  charged particle tracks 417–435
  as closed systems 18
  cosmic ray primaries 422
  Estherville—see Estherville
  fission tracks 421
  fossil particles 422
  fossil track observations 423, 431, 432
  high-energy interactions 422
  isotopic composition 161
  K–Ar dating 63, 161
  low-energy secondary neutrons 422
  Pb isotope data 163, 164
  $^{206}Pb/^{204}Pb$ ratio 162
  $^{208}Pb/^{204}Pb$ ratio 162
  primordial $^{87}Sr/^{86}Sr$ ratio 162
  Rb–Sr age 162
  silicate phases 422
  spallation recoils 422
  $^{87}Sr/^{86}Sr$ variation 162
  stone, isotopic composition 161
  Th content 434
  Th/U ratio 162
  Toluca—see Toluca
  Type I tracks 424
  Type II tracks 424
  U content 425, 429, 431, 434
  U–He dating 161
Meteoritic isochron 171
Mexican geosyncline 44
Mexico,
  Cenozoic sediments 63
  Gulf 44

Mica,
  Ar loss 252
  Ar retention 41
  deterministic two-stage model 26
  effect of base exchange 280
    on K–Ar ages 19
  excess ^{40}Ar 16
  expulsion of Ar 277
  K–Ar dates 13, 26
  Precambrian K–Ar date 40
  Sr loss 252
  use in K–Ar dating 16
Mica clock,
  date 29
  K–Ar age 27
Microcline,
  retentivity 22
  transition to orthoclase 81, 83
Microlite, U–Th–Pb age 379
Mid-Atlantic Ridge,
  Pb concentration 180
  Sr isotope composition 450
  $^{87}Sr/^{86}Sr$ ratio 445
Mineral age variations 84–97
Mineralogy, effect of contact metamorphism on 80–84
Minnesota 245
  ages of events 247
  southwestern 245
    zircon data 246
  true age of metamorphic events 247
Miocene, Basin and Range Province 55
Mississippi valley 62
Missouri,
  mineral ages 197
  Ozark Dome 196
Model lead 153
  feldspar age 93
Model lead analysis 189
Models, intrusive 78
Mohorovicic boundary 55
Moine Series (Scotland) 285
  Carn Chuinneag intrusion 283
  isochron 285
Monazite,
  concordant 237
  detrital origin 375

discordant U–Pb age   230
  Dominion Reef System age   375
  $^{207}Pb/^{206}Pb$ age   387
  U–Th–Pb age   318, 329, 344, 366
  in Zambia nucleus   380
Montana,
  ages   116, 118
  Boulder Batholith   123
  chronology   142
  continental growth   144
  Gallatin Canyon   130
  isotopic geochronology   111–144
  K–Ar ages   121, 142
  Little Belt—see Little Belt
  Madison Range   130
  Precambrian events   247
  Rb–Sr ages   121, 142
  southwestern, Gravelly Range   127
    K–Ar mica ages   132
    Precambrian   127–134
    Ruby Range   127
    Stillwater complex   135
  Tendoy Range   130
  Tobacco Root Batholith   123, 124, 129, 143
  Tobacco Root Range—see Tobacco
  U–Pb zircon ages   142
Monteregian Hills,
  ages of rocks   452
  comparison of K–Ar and Rb–Sr ages   451, 452
  initial $^{87}Sr/^{86}Sr$ ratio   454
  $^{87}Sr/^{86}Sr$ ratio   451
Moodies System   373, 376
  age   397
  minimum age   374
  undifferentiated sediments   391
Morocco   371, 373
  Azguermerzi granite   353
  Kerdous massif   387
  orogenic effects   349
  Ouarzazate Series   353
  Tazanakt granite   353
  Zenaga Series   353
Mozambique, Zambesi Belt   317
Mozambique Belt   316–318, 320, 336, 376
  ages   337
  K–Ar ages   317

minimum age   400
Rb–Sr age   317
U–Th–Pb age   317
vestigial facet   310
Mozambiquian orogeny   333
Mubende granite,
  K–Ar biotite age   356
  Rb–Sr whole rock age   356
Multi-stage equation   164
Multi-stage model   208
Murchison Range   373
Murrisk(western Ireland),
  geochronological studies   259–295
  lower Palaeozoic sediments   275–276
  Ordovician sediments   261
  Silurian sediments   262
Muscovite,
  built in age   32
  K–Ar age   326
  Na–K ratio   51, 52
  Rb–Sr age   38, 84
  thermal diffusion of Ar   63
Musozi, K–Ar biotite age   355
Muva–Irumide Belt   337, 342–343, 380
  comparison with Kibara–Karagwe Ankolean   342
  effect of Kibaran orogeny   343
  K–Ar age   342
  Rb–Sr age   342
  stratigraphic significance   342
  structural significance   342
Muva–Irumide segment   346
Muva System   342, 380
  maximum age   361

Na–K ratio, in muscovite   51, 52
Namaqualand–Natal Belt   337, 340–342, 346
  age 346
  U–Th–Pb ages   341
Natal, age of crystalline rocks   399
Neutron-activation technique   164, 183
Nevada, Rb–Sr whole rock age   37
Nevadan orogeny   44, 45, 393
  Jurassic plutonism   45
Nevadan Revolution   45

Subject Index

New Brunswick 172
Newer Granites (Connemara) 286
   Galway granite 282
New Mexico, anomalous Pb mineralization 196, 197–198, 204
Nigeria,
   Basement Complex, K–Ar age 326
   columbite–tantalite pegmatites 324
   Dahomeyan ages 325
   Dahomeyan basement 324
   orogenic activity 307, 323,
   vestigial facet 310
Nigritian, sequence 325
North America,
   geological provinces 190
   geosynclinal troughs 36
   Precambrian geology 191
   rising of western 62
North Dakota,
   K–Ar data 121
   Rb–Sr data 121
Nuclear reactions,
   induced, in iron meteorites 18
   in the lithosphere 18
Nyanza Shield (Kenya) 381, 382, 383
Nyanzian System 355, 382

Oceanic basins, Pb composition 177–182
Oceanic lead 171, 178
Oceans,
   rare earth concentration 457
   Sr content 457
Ogooué gneiss 398
Ogooué System (Gabon) 360
Oklahoma, Ozark Dome 196
Olivine, trace Pb 182
Oracle granite,
   K–Ar date 50
   Rb–Sr date 50
Orange River region, Rb–Sr age 340, 341
Ordinary lead—*see* Pb
Ordovician,
   Lower, *Didymograptus extensus* Zone 262

Didymograptus hirundo Zone 262
   Upper, Llandeilo–Caradoc age 262
   western Ireland 262
Didymograptus extensus 268
Ordovician sediments of Murrisk 261
Oregon, Bald Mountain Batholith 227
Orogenesis, facets 306
Orogenic versus epeirogenic cycles 59–61
Orthoclase isograd 82
Orthoclase–microcline transition 83
Oslo, Permian igneous rocks 81
Oslo Alkaline Province, $^{87}Sr/^{86}Sr$ ratio 452
Otavi Facies (South-West Africa) 314, 315, 362
   K–Ar ages 314
   Rb–Sr ages 315, 343, 344
Ouango massif, maximum age 347
Oughterard granite,
   age 268
   Pb age 263
   Rb–Sr whole rock isochron 263
Ozark Dome area 196–197

Pacific Ocean,
   depth 60
   sinking of the floor 62
Pacific Ocean basin 63
Pacific Rise, East—*see* East Pacific Rise
Palabora complex, U–Pb age 366
Palabora intrusion 364
Palaeoatmosphere of the Earth 10
Palaeocene 54
Palaeogene–Neogene boundary 54
Palaeozoic,
   Arizona 39
   Basin and Range Province 2, 43
   Lower, Murrisk 275–276
   Sonora 39
Pallasite, Eagle Station 423
Pan-African orogeny,
   date 392
   zones affected 335, 336

Pan-African thermotectonic episode   334
^{204}Pb 148
  error line   157
  errors in measurement   156, 160, 171, 187
  isotopic abundance   155
^{206}Pb 148
^{207}Pb 148
  loss   252
^{208}Pb 148
Pb,
  anomalous   164, 180, 197, 204
    β-type   170
    Broken Hill   204
    Central Finnish Zone   204
    classification   185–187
    Cobalt–Noranda area   204
    Goldfields   204
    higher order multi-stage   185, 203
    interpretation of results   167, 185–207
    J-type   170
    Manitouwadge   172
    mixture of two ordinary Pb   185
    Ozark Dome   204
    $^{206}Pb/^{207}Pb$–$^{207}Pb/^{204}Pb$ relation   185
    Lead Roseberry   172, 204
    Russell–Farquhar model   195
    short period   185, 198–199
    simple three-stage   185, 200
    Sudbury   204
    Thunder Bay   204
    Th/U ratio   204–205
    two-stage model   185, 191–198
    West-Central New Mexico   204
    White Sea Region   204
  Ascension Island   180
    in basalts   181
    in basic volcanic rocks   181
    Bonne Terre   196
    closed system   149
  common   148, 152, 160
    Holmes–Houtermans curve   115
    in Vohibory System   322
  concentration, Gough Island   180
    Mid-Atlantic Ridge   180
    oceanic   179

phytoplankton   179
continental   171
continuous diffusion   232, 240, 244, 245
deep mantle source   180
derivation   180
episodic loss   136, 236, 239, 253
growth curve   150
Hawaiian   198
Ivigtut   200
Iwo Jima   180
Joplin   196
in K feldspar   115
Kootenay Arc   200
measurement   183
multi-stage   151–153, 207
oceanic   171
  isotopic evolution   178
ordinary   153–160, 210
  isotopic composition   176
  mixture   187–191
  $^{206}Pb/^{204}Pb$ ratio   176
  $^{208}Pb/^{204}Pb$ ratio   176
  the search for   171–185
primeval   148, 161
  isochron through   170
primeval ratio   150, 192
in pyrites   183
in pyroclastic rocks   198
radiogenic   148
  apparent excess   171
in sedimentary pyrite   183
simple three-stage anomalous   199, 200
single-stage   176, 188, 210
  growth curve   174
Sudbury   200
table for plotting single-stage growth curve   214
Thackaringa type   192, 194
  $^{206}Pb/^{204}Pb$ ratio   193
  $^{207}Pb/^{204}Pb$ ratio   193
  Th/U ratio   195
theory of multi-stage models   208–213
three-stage model   213
Th/U ratio   204
trace,
  in meteorites   162, 164

Subject Index

in minerals 182
in ores 182
in rocks 182
studies 182–184
two-stage interpretation 210
two-stage model 211
two-stage system 154
in ultramafic rocks 183
in volcanic rocks 198
zero isochron 162, 177, 180
Pb/α zircon age,
Gabon 344
Ivory Coast 387
Lower Congo 344, 362
Man granite 350
Ogooué region 360
pegmatite 361
Pb determination, discrepancy 162
Pb isochron 150
Pb isotope age, granophyre 370
Pb isotope composition,
Coeur d'Alene 123
feldspar 184
manganese nodules 177, 178, 182
New Mexico 197
pegmatitic feldspar 120
pelagic sediments 178, 182
south Atlantic 180
sulphide deposits 179
Pb isotope data, to define single-stage growth curve 215
Pb isotope ratio 165
Baltic Shield 160
Bluebell Mine 158
Broken Hill standard 173, 176, 193, 194
galena 145, 183
Hawaii 181, 182
volcanic rocks 178, 179
Ivigtut 159
Kootenay Arc 165
Lake District 199
meteoritic 162
Ozark Dome 196
single-stage model 155, 156
Sudbury 201
theoretical model 168, 169, 170
West Karelian zone 191

Pb isotopes,
anomalous 164–166
correlation with geologic structure 197
development in closed system 153
geological significance 147–215
growth curve 152
interpretation 147–215
Pb loss 230, 231, 240
chemical process 230
continuous diffusion 249
correlation with U content 253
episodic 136, 233, 239
during metamorphism 253
by ground water solutions 234
hypothesis 229–236
by leaching of metamict zircons 254
low-temperature 240, 253
zircon 137, 229, 252, 253
Pb model,
frequently mixed 166–171
single-stage 170
two-stage 167
Pb model age 263
Pb ores 171–177
$^{206}Pb/^{204}Pb$–$^{208}Pb/^{204}Pb$ diagram 92
$^{206}Pb/^{204}Pb$ ratio 156, 166, 171, 177, 195
two-stage anomalous 191
$^{206}Pb/^{238}U$ age of granites 327
$^{207}Pb/^{206}Pb$ age,
davidite 192
Hoggar 326, 363
monazite 387
Pharusian granite 327
Suggarian granite 327
$^{207}Pb/^{204}Pb$–$^{206}Pb/^{204}Pb$ diagram 92, 173, 186
$^{207}Pb/^{204}Pb$ ratio 150, 156, 166, 171, 209
$^{207}Pb/^{206}Pb$ ratio, errors 237
$^{207}Pb/^{235}U$ age of granites 327
$^{208}Pb/^{204}Pb$–$^{206}Pb/^{204}Pb$ relation 173, 191, 192
in galena 187
$^{208}Pb/^{204}Pb$ ratio 150, 156, 189, 195
Pb–U age, monazite 344
Pb–U analysis 189

Pegmatite,
  dating 42
  discordant 77
  Rb–Sr age 339
    lepidolite 355
  secretion 77
  sweat 77
Pegmatitic minerals, K–Ar ages 137
Peridotite, St. Pauls Rocks 449
Perthshire, Dalradian sequence 263
Phanerozoic,
  Basin and Range Province 44
  effect of external Ar pressure on 30
  precise boundaries 299
Phanerozoic history,
  Africa 299
  Basin Ranges 43–45
Phanerozoic magmatism, in U.S.A. 45
Pharusian,
  Rb–Sr age of granites 327, 328
  sequence 325
Phlogopite,
  diffusion of Ar 23
  retentivity 22
  slab model 20
  two-stage model 28
Phyllosilicates, use in K–Ar dating 16
Phytoplankton, Pb concentration 179
Piedmont, ages 130
Piriwiri System,
  ages 389
  Rb–Sr age 362
Pleistocene,
  effect of Ar degassing 30
  isotopic composition of air Ar 10
  K–Ar dating 8, 12
Pliocene, Basin and Range Province 55
Plutonium, in enstatite–albite–diopside 429
Polonnarnenke charnockites 227
Polymetamorphic history of an area 73
Potassium—see K
Precambrian 74

Africa 303, 304
Arizona 35–37
  Alder Group 36
  Apache Group 35
  Brady Butte granodiorite 36
  geosynclinal troughs 36
  Grand Canyon Series 35
  K–Ar dates 38
  K–Ar interpretation 36
  Mazatzal quartzite 35
  Pinal Series 35
  Vishnu Series 35
  Yavapai Series 35, 36
Basin Range 2
Beartooth Range 113
Belt Series—see Belt
  correlation in Sonora 41
Dodoma–Nyanza nucleus 382
effect of excess ^{40}Ar in minerals 29
Hoggar 363
K–Ar dating 14, 15
Montana 247
Oracle granite 50
regional metamorphism 81
Sonora 35–37
  Bamori district 37
  Granito Aibó 39
  K–Ar dates 38
Stillwater igneous complex 113
subdivision of Africa 304
U–Pb zircon dating 50
Upper, in southern Africa 312
Wyoming 247
Precambrian activity in Africa, ages 394, 395
Precambrian–early Palaeozoic in Africa,
  ages 302, 303
  orogenic zones 301, 305, 306
Precambrian geology of North America 191
Precambrian–Palaeozoic cycles in Africa 302
Precambrian–Phanerozoic orogenic belts in West Africa 331
Precambrian sedimentary sequences of the British Isles 285
Pre-Cherry Creek,

Subject Index

K–Ar age   133
  whole rock gneiss age   133, 142
Probabilistic models   33–35
^{244}Pu,
  spontaneous fission   429
  spontaneous fission tracks   423
Puerto Rico, Laramide plutons   58
Pulaskite, initial $^{87}Sr/^{86}Sr$ ratio   454
Pu/U ratio   429
Pyroxenes,
  ^{40}Ar error   13, 14
  retrograde metamorphism   129
Pyroxenites,
  association with carbonatites   439
  melting   440

Quad Creek   243
Quartz, ^{40}Ar error   13, 14
Quartz syenite, in undersaturated rock provinces   439

Radiation damage   105
  continuous diffusion model   253
  correlation with Pb loss   253
Radiation damage model   234, 235, 236, 240, 244, 245
Radioactive clocks   5–6
  setting   33
Radioactive decay   21, 149
Radioactivity, geophysical application   148
Radiogenic daughter loss   247
Radiometric dating techniques, definition of bench marks   406
Rare-gas loss   232
^{85}Rb   440
^{87}Rb   440
  decay constant   113, 114, 270, 437, 441, 442
    choice of value   280
  decay to ^{87}Sr   441
  half-life   437
Rb
  closed system   443
  decay constant, uncertainties   85
  determination   209
  in dolomite   455
  in limestone   455
Rb–Sr ages   194, 195, 249, 270

in Africa   306, 308
  > 200 m.y.   303
  400–800 m.y.   307, 308
Amsaga Series biotite   353
Androyan System   361
Angola   358, 360
Bakel Series   330
Baoulé granite   350
Burundi   339
Cameroons   321, 360
Central African Rep.   321, 340, 344
comparison with K–Ar age   280
Congo pegmatite   339
Connemara   271
Connemara Schists   261, 277–282
Dahomey   364
Dahomeyan lepidolite   326
Dahomeyan muscovite   326
Damara Facies   314
Damaran sequence   334
Dodoma System lepidolite   381
Ebolowa granite   385
errors   270
feldspar   245
Gabon   360
Ghana   352
Great Dyke picrite   379
Hoggar   326, 363
Ivory Coast granite   350, 351
Kaap Valley granite   374, 375
Kampala–Mubende region   355
Kate granite   357
Kerdous massif muscovite   387
Kibara Belt pretectonic granite   339
Kitosh granite   382
Kivu   339
Koulountou Series   330
Kubuta pegmatite   374
Lower Congo   363
Luiza-type basement   383
Lunya district lepidolite   355
Malagasy Rep.   323
Masaba granite biotite   356
Mauritania   330
mica   245
Monteregian Hills   451, 452
Mozambique lepidolite   317

Mozambique Belt 317
Mtuga aplite 361
Namaqualand 340, 341
Orange River region mica 340, 341
pegmatitic muscovite 375
Pharusian granites 327, 328
Pharusian metamorphic rocks 328
post-Moodies migmatitic granite 374, 375
post-Shamvaian pegmatite 379
pre-Katanga 361
Rhodesian biotite 362
Rwanda 339
Sattle Ogmané biotite 385, 386
Sattle Ogmané lepidolite 386
Scottish Dalradian metasediments 264
Senegal 330, 345, 352
single mineral 263
Suggarian 327, 363
Swaziland granites 374
Tazenakht granite 354
Tchad 322
Transvaal alkaline complexes 345
Transvaal Old Granite 367
Ubendian–Ruzizi Belt 343
Uganda 332
Upper Volta granite 350, 351
whole rock,
    Connemara metasediments 283, 284
    Dominion Reef System 375
    Gaberones granite 367, 368
    Gaberones-type granite 375
    Karagwe Ankolean 356
    Kerdous massif 387
    Mubende granite 356
    Nevada 37
    Nyanzian rhyolite 382
    Suggarian granite 327
Zambesi Belt 317
Zenaga region 354
zircon results 247, 255
Rb–Sr analysis,
  Black Hills 114
  Connemara metasediments 282
  Mayumbe Belt 335
Rb–Sr data 39

Rb–Sr date,
  muscovite 38
  Oracle granite 50
Rb–Sr dating 189
  in Arizona 37
Rb–Sr determinations from Mauritania 386
Rb–Sr geochronometry 263
Rb–Sr isochron 37, 270, 284
  Cape granite 315, 334
  Eaton gneissic granodiorite 37
  errors 270
  Fig Tree Shales 374
  Madera quartz diorite 37, 38
  Malmesbury sediments 334
  mineral 41
  Ruin granite 37
  schist 279
  whole rock 41, 263, 282, 283
Rb/Sr ratio 438, 441
  decrease with depth 459
  Hawaiian basalts 447, 448
  isochron method 442
  measurement 442
  minerals 264
  sialic layer 444
  terrestrial rocks 438
Read Roseberry, anomalous Pb 172
Regression analysis, meteoritic 161
Rejuvenated facet, strata involved 333
Retentivity of minerals 32
Réunion,
  $^{87}Sr/^{86}Sr$ ratio 449
  volcanic rocks 447
Reynolds-type mass spectrometer 269
Rhodesia 362
  anorogenic effects 337
  Great Dyke 394, 395
    age 379, 380
  gregarious batholiths 407
  K–Ar lepidolite age 344
  K–Ar whole rock age 345
  maximum age of Sijirira Beds 347
  minimum age of Umkondo System 347
  model Pb ages 378
  Mtoko Batholith 388

Mtoko region age 390
orogenic effects 313, 314
Piriwiri–Lomagundi–Deweras Systems 391, 400
Rb–Sr lepidolite age, 344
Sebakwian–Bulawayan–Shamvaian Systems 390, 398
Sebakwian System 399
tectonic sketch map 378
Wankie district, common Pb age 342
Rhodesia nucleus 370, 376–380
Rincon Mountains 52
K–Ar dating 50
Mineta Formation 51
Riphean orogeny 333
Rocky Mountain geosyncline 44
Rochy Mountain ranges 112
Rocky Mountains 115
Little 121
Roundstone granite 286
Ruby–Gravelly Range,
Cherry Creek gneiss 131
geology 127
mineral isochron 132
Dillon granite gneiss 127, 128
Pony gneiss 127, 128
Pre-Cherry Creek gneiss 127
Ruby Range (Montana) 127, 130, 143
detailed geological study 114
Russell–Farquhar model for anomalous Pb 195
Ruwenzori Mountain region 355
Ruwenzori Mountains (Uganda), K–Ar mica age 340
Rwanda,
minimum age of systems 347
Rb–Sr muscovite age 339

Sahara, Spanish 329, 330
St. Magloire (Quebec) 157
St. Pauls Rocks (Atlantic) 449
Samarskite, U–Th–Pb age 318
Samia–Bulugwe Series 357
Samia Hills Series 355
Santa Catalina Mountains 50, 52
K–Ar dating 50

Santa Catalina–Rincon block, K–Ar dating 51
Santa Lucian orogeny 45
Sattle Ogmane (Mauritania),
granitic complex 385
Rb–Sr lepidolite age 386
Scotland,
Dalradian Series 259
K–Ar isotopic chronology 264
Rb–Sr ages 263
Rb–Sr isotopic chronology 264
Scourian metamorphism, comparison with southwestern Montana 129
Sea water, Sr isotopic composition 263, 457
Sebakwian–Bulawayan–Shamvaian Systems,
sediments 391
strata 377
Sedimentary cycle, zircon 255
Senegal,
ages 337
Bakel–Akjoujt Series 329
Bakel Series 330
equivalents of Birrimian System 352
Koulountou Series, Rb–Sr age 330
orogenic effects 349
Rb–Sr age 330, 352
whole rock 345
Serpentinites, St. Pauls Rocks 449
Shamvaian cycle 302
Shamvaian orogeny, date 392
Shamvaian System, strata 377
Sialic rocks, differentiation 166
Sierra Ancha Mountains (Arizona) 40
Sierra Leone,
K–Ar age 330
Kasila gneiss 328
minimum age 400
Marampa schists 328
minimum age 400
monazite $^{207}Pb/^{206}Pb$ age 387
Rokel River Series 328
Sierra Nevada 44
Silurian,

sediments of Murrisk 262
  Upper Llandovery age 262
  Wenlock age 262
  western Ireland 268, 269
    post-Wenlock metamorphism 262
Single-stage growth curve 203—*see also* Pb
Single-stage model—*see* Pb
Single-stage system 158
Skaergaard, acid granophyre 447
Skaergaard intrusion 447
Slab model for phlogopite 20
Slate, whole rock K–Ar age 275–276
Solar system,
  history 434
  origin 161
Somali Republic 333
  geosynclinal facet 310
  K–Ar age 319
  minimum age 400
  orogenic effects 318–320
  Precambrian 318
Songwe Scarp (Tanzania),
  ankeritic carbonate 456
  anomalous limestone 456
Sonora,
  Basin and Range Province—*see* Basin
  correlation of Precambrian strata 41
  Cretaceous 44
  Palaeozoic 39
  Precambrian 35–37
Sonoran geosyncline 44
South Africa,
  correlation of Rhodesian schist belts 377
  Gordonia region 340
  orogenic effects 310, 311
  Pecambrian stratigraphy 364
  Sr isotopic composition of Bulawayan limestone 454
South Dakota, ages 118
South-West Africa 341, 362
  age pattern 315
  ages 337
  anorthosite mass, age 399
  Damaran orogenic episode 315

geosynclinal facet 310
Kibaran ages 343
maximum ages of systems 347
Mayumbian orogeny, age 363
orogenic effects 310, 311
Outjo System 396
Upper Precambrian 314
Specific activity, determination 7
Sphaerium sulcatum, $^{87}Sr/^{86}Sr$ ratio 457
^{87}Sr,
  abundance 263
  inherited 282
  decay to 437
Sr,
  closed system 443
  common 437
    isotopic composition 440
    relative abundance 440
  determination 269
  in dolomite 455
  isotopic composition 437–460
  oceanic 457
  in sea water 263
  in sediments 263
Sr isochron, age of post-Karagwe Ankolean metamorphism 339
Sr isotope homogenization, effect of mild thermal event 277
Sr isotopes,
  applied to origin of the carbonatites 452–457
  determination 441
  fractionation 441
Sr isotopic composition,
  acid granophyres 445
  basalts 444–450
  continental basalts 444
  continental plutonic rocks 451, 452
  Hawaiian Islands 447, 448
  limestone 455
  Mid-Atlantic Ridge 450
  oceanic basalts 444
  sea water 457–459
  sedimentary limestone 454
  significance 440–444
$^{86}Sr/^{88}Sr$ ratio 270, 441

Subject Index

$^{84}Sr + {}^{86}Sr$ tracer, in isotope-dilution method 269
$^{87}Sr/^{86}Sr$ ratio 166, 283, 441
  African carbonatites 454
  basalts 446
  basic rocks 447
  basic volcanic rocks 444
  carbonatites 454, 456, 458
  clastic sediments 444
  common Sr 438
  continental alkaline rocks 451
  Dalradian sediments 264
  extraterrestrial material 438
  granophyre 445
  Grenville Province 457
  Hawaiian basalts 447, 448
  igneous granites 444
  initial 132, 263, 438
    alkaline basalts 451
    assumed 284
    Monteregian Hills 454
    true 284
  Kangeralussuaq 447
  limestone 455
  metamorphic rocks 444
  Mid-Atlantic Ridge 445
  normalization 270
  Oslo Alkaline Province 452
  Red Hill acid granophyre 445
  relation of Rb content 441
  relation of Sr content 441
  remobilized sialic rocks 444
  Réunion quartz syenite 449
  Réunion volcanic rocks 449
  sea water 458
  similarity in alkaline and basic rocks 457
  Skaergaard intrusion 447
  Sphaerium sulcatum 457
  Tasmanian dolerites 445
  terrestrial rocks with low Rb/Sr ratio 438
  use in determining petrogenesis 438
Stability of mineral ages 106
Stillwater complex (Montana) 135
  age 138
  effect of regional metamorphism 138
  hornfelsed zone 137
  K–Ar age 138
  metamorphism adjacent to 137
  pegmatitic biotite age 137
  ultramafic zone 138
Stratigraphic bench marks, use of radiometric techniques to define 406
Structural bench marks, use of radiometric techniques to define 406
Stuhlmann Pass Series, K–Ar age 355
Subsial, $^{238}U/^{204}Pb$ ratio 176
Sudbury region 202
  galena 213
  Pb 200–203
Suggarian,
  In Ouzzal Facies 324, 325
  K–Ar ages 363
  Rb–Sr ages 327, 363
Sulphide deposits, Pb isotope composition 179
Sulphide mineralization 153
Swaziland, dating of granites 373
Swaziland granites, Rb–Sr ages 374
Swaziland System 373, 375, 376, age 397
Syenite,
  Ascension Island 451
  initial $^{87}Sr/^{86}Sr$ ratio 454

Taconic orogeny 393
Tanzania,
  basement rocks 381
  Dodoma System 381
    minimum age 390, 398
  K–Ar hornblende age 358
  K–Ar muscovite age 357, 358
  Kavirondian System 381
  Mbozi syenite gabbro complex, age 332
  Mozambique Belt 317
  Nyanzian System 381
  Rb–Sr biotite age 381
  Ubendian Belt 343
  Zambesi Belt 317
Tasmanian dolerites 447
Tazenakht granite, Rb–Sr age 354

Tchad, Rb–Sr age 322
Tendoy Range (Montana) 130
Tertiary,
  Arizona 51
  Basin and Range Province 54, 63
  Kangeralussuaq alkaline intrusion 447
  K–Ar dating 8
  Skaergaard intrusion 447
Tertiary intrusives 75
Teton fault 141
Teton Range (Wyoming) 115, 140–142, 143
  epidotization 141
  geological study 114
  geothermal gradient 141
  K–Ar ages 140
  mineral ages 131
  Precambrian 141
  quartz monzonite intrusion 140
  Rb–Sr age 140
  sequence of events 140
^{230}Th 148
^{232}Th 148
Th,
  closed system 149
  in volcanic rocks 198
Thackaringa, Pb 192, 193
Th isotopes, decay 149
Th/K ratio 445
Tholeiite 440
Thorianite ages, pre-Karroo of Malagasy Rep. 323
^{232}Th/^{204}Pb ratio 152, 155, 171, 177, 185, 195, 203, 209
  discordancy problem 229–254
Th/U ratio 163, 189, 198, 203, 204, 205
  acid rocks 204, 206
  anomalous Pb 204–205
  intermediate rocks 206
  meteoritic 164
  variation 165
Tinguaite dyke, initial ^{87}Sr/^{86}Sr 454
Tobacco Root Batholith 123, 124, 129
  emplacement 143
Tobacco Root Range (Montana) 122, 129, 130
  detailed geological study 114
  orogenic activity 323
  sequence of metamorphic events 128
Togo Series 324
Toluca 424, 429, 430
  Type II tracks 425, 428
Toro quartzites, age 340
Toro System, Igara Schist Series 355
Torridonian Series of Scotland 285
Tourmaline, excess ^{40}Ar 14
Trachytes 449
Tracks,
  charged particles 417–435
  density 427
  in Estherville 428
Track fading conditions 420
  identifying characteristics 423, 426
  storage time 428
Transvaal,
  anorogenic effects 337, 345
  basic activity 366
  Bushveld complex 364–366, 394, 395
  Dominion Reef System 391
  granite emplacement 366
  Palabora carbonatite 394, 395
  Palabora complex 366
  plutonic activity 364
  Rb–Sr age 345
  stratigraphic sequence 376, 392
  Swaziland System 391
  Transvaal System 391
  ultrabasic activity 366
  Ventersdorp System 391
  Waterberg–Loskop Systems 396
  Witwatersrand System 391
Transvaal nucleus 370, 372, 373–375, 379, 399
  Precambrian sequence 373
Transvaal System 336, 373, 376
  ages 397, 398
  minimum 368
  Old granite 367
Tristan da Cunha, Sr isotopic composition 450

Subject Index

Tri-State region 196
Troilite 161
Tucson Mountains (Arizona) 48
  Basin and Range faulting 48
  Cat Mountain rhyolite 49
  Cretaceous sedimentary rocks 47
  Cretaceous volcanic rocks 46
  K–Ar dating of Cenozoic 46
  Mesozoic–Cenozoic magmatism 48
  mid-Tertiary magmatism 48
  plutons 49
  volcanic rocks 47
Turonian 44
  East Pacific Rise 63
  magmatic quiescence 54
Two-stage lead model—*see* Pb

^{235}U 148
  decay constant 113
  half-life 226
^{238}U 148
  half-life 226
  intermediate daughter loss 229
  spontaneous fission 418
    as a source of fossil tracks 422
U 418
  closed system 149
  fission 417
U concentration,
  as a function of age 421
  effect of oxidizing conditions 164
  measurement 418
  in meteorites 434
    measurement 433
  oceanic 180
  in pyroclastic rocks 198
  in volcanic rocks 198
  in zircon 241
U isotopes, decay 149
U minerals 230
Ubendian Belt, K–Ar ages 357
Ubendian–Ruzizi Belt 337, 343, 357
  age 396
  K–Ar age 343
  orogenic effects 349
  Rb–Sr age 343
  thermal imprint 343
Ubendian System, minimum age 398

Uganda,
  Buganda–Toro Systems age 398
  Buteba granite age 382
  K–Ar age 332
  minimum age of systems 347
  Mozambique Belt 317
  pre-Karagwe Ankolean rocks 355
  Rb–Sr age 332
  Zambesi Belt 317
U/K ratio 445
Umbgrove, pulse of the Earth theory 58, 59
United States, western,
  geology 43
  topography 62
U–Pb ages,
  discordancy problem 229–254
  discordant, in Hoggar 363
  gneissoid rhyolite 38
  Little Belt Range 138
  Palabora complex 366
  Precambrian zircon 50
  Sierra Ancha Mountains 37
  uraninite 122
  zircon age,
    Ebolowa granite 385
    Suggarian granites 327
U–Pb method, zircon 41
U–Pb mineral isochron 37, 38
^{235}U/^{204}Pb ratio 155
^{235}U/^{204}Pb ratio 150, 152, 155, 172, 185, 198, 209
  binomial distribution 167
  subsialic 176
U/Pb ratio 189, 203
  variation in a crustal source 165
U–Pb system, in zircon 245
Upper Volta,
  Eburnian orogenic episode 349
  Precambrian 350
  Rb–Sr age of granite 350, 351
Uraninite,
  discordant U–Pb ages 230
  U–Pb age 122
  U–Pb analysis 114
  U–Th–Pb age 318, 350
Uraninite ages,
  Dominion Reef System 375
  Katanga 335

Kivu 339
Urundi System 346
Utah,
  ages 118
  Wasatch Range 120
U–Th–Pb ages,
  Africa 306, 308
    > 200 m.y. 303
    400–800 m.y. 307, 308
  betafite 318
  Damara Facies 314
  Damaran sequence 334
  davidite 318
  discordant 326
  Limpopo Belt 359
  microlite 379
  monazite 318, 379
  Mozambique Belt 317
  samarskite 318
  uraninite 318, 350
  Zambesi Belt 317
  zircon 318
U/Th ratio 445, 447

Vaca Meurte, zircon 434
Venezuela, Mesozoic 60
Ventersdorp System 373, 375, 376
  age 369, 397, 398
Vestigial facet, strata involved 333
Vohibory System (Malagasy Rep.),
  age 369, 387, 388
  common Pb ages 322
  minimum age 398
  Pb/α zircon age 361
Volatiles, in the mantle 17
Volcanic rocks,
  K–Ar dating 19
  Pb content 181
Vredefort Ring 364, 366–367, 369, 394
  shatter cones 366

Wasatch Range (Utah), Rb–Sr age 120
Waterberg–Loskop sequence 336
  minimum age 364
  maximum age 370
Waterberg System 332, 346, 376
  minimum age 347

Weathering, effect on K–Ar ages 19
West Africa 349–354
  1850 m.y. ages 368
  ancient ages 387
  late Precambrian–early Palaeozoic activity 323–331
  Precambrian basement 350
  Precambrian–Phanerozoic orogenic belt 331
  western region 328
West African Craton 309, 323, 336, 337
West Congolian orogeny 306
  date 316
Western Congo System 333
  common Pb galena ages 316
  minimum age 400
Western Ireland—see Ireland
Western Rift Valley 355, 356
West Karelian zone 191
Westport Grit Series, K–Ar age 276
Whole rock age 37—see also Rb–Sr age
Wind River Range (Wyoming) 115, 143
  ages 142
Witwatersrand System (Transvaal) 373, 375, 376, 394, 395
  ages 397, 398
  minimum age 369
Wyoming 241
  ages 116, 118
  Beartooth Range 115
  Bighorn Range 143
  chronology 142
  continental growth 144
  isotopic geochronology 111–144
  Medicine Bow Range 115, 134
  Precambrian events 247
  Teton Range 131, 140—see also Teton
  Wind River Range 115, 142, 143

Xenoliths,
  ^{40}Ar content 12
  K isotopic composition 6

Yamaskite, initial $^{87}Sr/^{86}Sr$ ratio 454

Subject Index

Yttrocrasite, age of Mitwaba 339
Yukon, Laramide plutons 58

Zambesi Belt 316, 318, 376
  K–Ar age 317
  minimum age 400
  Rb–Sr age 317
  U–Th–Pb age 317
  vestigial facet 310
Zambia 361
  K–Ar age 313
    biotite 358
    Nkumbwa Hill carbonatite 332
  Kate granite 357
  Mozambique Belt 317
  Muva System 361
    minimum age 400
  orogenic effects 311
  Zambesi Belt 317
Zambian copperbelt,
  relict ages 361
  uraninite age 311
Zambia nucleus 380–381
Zenaga Series, minimum age 368
Zero isochron—*see* Pb
Zinnwaldite,
  Lunyo granite 382
  Rb–Sr age of Pharusian granite 327
Zircon 136, 226–229, 237, 240, 245
  accessory 227
  activation energy 105
  age determination 226
  age discordance 245
  analytical aspects 226
  in arenaceous sediments 228
  in basaltic rocks 226
  change in radiogenic elements 97
  characteristics 227
  concordia curve 96, 136
  contact 105
  continuous diffusion—*see* Continuous diffusion
  decomposition 226
  detrital 136, 137, 227, 243, 244, 247, 249
  discordant 187, 229, 236–252, 253
    ages 189, 226, 240
    leaching studies 229
  discordant Pb–U ages 137
  in granitic rocks 226
  hydrothermal environment 229
  in intermediate rocks 226
  low-temperature overgrowths 227
  low-temperature susceptibility 225
  metamict, Pb loss 234
  metamictization 227, 228, 239
  metasedimentary 136, 227
  morphology 96
  overgrowths 136, 227, 243, 244
    importance 228
  Pb/α age,
    Gabon 344
    Ivory Coast 387
    Lower Congo 362
    Ogooué region 360
  Pb loss 225, 229, 236, 252, 253
  Pb–Pb ages 84
  $^{207}Pb/^{206}Pb$ age of Pharusian granites 328
  Pb–U age 153
  $^{206}Pb/^{238}U$ age 328
  $^{207}Pb/^{235}U$ age 328
  $^{206}Pb/^{238}U$ ratio 153
  $^{207}Pb/^{205}U$ ratio 153
  pegmatitic 227
  propensity for Pb loss 254
  radiation damage curve 228
  rejuvenated 225, 237
  rejuvenation 249
  rounded 226
  self-annealing rate 228
  shape 227
    in metamorphic rocks 226
  silica deficiency 228
  stability 228
  Th content 227
  theoretical composition 227
  $^{232}Th-^{208}Pb$ age 226
  U content 153, 227, 239, 241
  U–Pb age 120
    South-West Africa 362
    Suggarian granites 327
  use in determining spatial relations 227
  use in establishing generic relations 227
  U–Th–Pb age 225, 318

Zircon ages 94–97
  analytical errors 226
  Bushveld igneous complex 366
  Cape granite 315
  correlation with geologic history 254
  Finland 251
  geologic meaning 254–255
  Hoggar 363
  interpretation 225–255
  Man granite 387
  pre-Karroo of Malagasy Rep. 322, 323
  resistance to metamorphism 94
Zircon analysis, as an indication of metamorphism 142
Zircon data, episodic Pb loss 245
Zircon generation 153
Zircon suites,
  mixtures 255
  reduced major axis 227
Zircon–thorite group, hydrothermal stability 228

DATE DUE			
APR 22 1996			
MAY 15 1996			

GAYLORD No. 2333   PRINTED IN U.S.A.